IFIP Advances in Information and Communication Technology 522

Editor-in-Chief

Kai Rannenberg, Goethe University Frankfurt, Germany

Editorial Board

IFIP – The International Federation for Information Processing

IFIP was founded in 1960 under the auspices of UNESCO, following the first World Computer Congress held in Paris the previous year. A federation for societies working in information processing, IFIP's aim is two-fold: to support information processing in the countries of its members and to encourage technology transfer to developing nations. As its mission statement clearly states:

> IFIP is the global non-profit federation of societies of ICT professionals that aims at achieving a worldwide professional and socially responsible development and application of information and communication technologies.

IFIP is a non-profit-making organization, run almost solely by 2500 volunteers. It operates through a number of technical committees and working groups, which organize events and publications. IFIP's events range from large international open conferences to working conferences and local seminars.

The flagship event is the IFIP World Computer Congress, at which both invited and contributed papers are presented. Contributed papers are rigorously refereed and the rejection rate is high.

As with the Congress, participation in the open conferences is open to all and papers may be invited or submitted. Again, submitted papers are stringently refereed.

The working conferences are structured differently. They are usually run by a working group and attendance is generally smaller and occasionally by invitation only. Their purpose is to create an atmosphere conducive to innovation and development. Refereeing is also rigorous and papers are subjected to extensive group discussion.

Publications arising from IFIP events vary. The papers presented at the IFIP World Computer Congress and at open conferences are published as conference proceedings, while the results of the working conferences are often published as collections of selected and edited papers.

IFIP distinguishes three types of institutional membership: Country Representative Members, Members at Large, and Associate Members. The type of organization that can apply for membership is a wide variety and includes national or international societies of individual computer scientists/ICT professionals, associations or federations of such societies, government institutions/government related organizations, national or international research institutes or consortia, universities, academies of sciences, companies, national or international associations or federations of companies.

More information about this series at http://www.springer.com/series/6102

Abdelmalek Amine · Malek Mouhoub
Otmane Ait Mohamed · Bachir Djebbar (Eds.)

Computational Intelligence and Its Applications

6th IFIP TC 5 International Conference, CIIA 2018
Oran, Algeria, May 8–10, 2018
Proceedings

 Springer

Editors
Abdelmalek Amine
University of Saida
Saida
Algeria

Malek Mouhoub (iD)
University of Regina
Regina, SK
Canada

Otmane Ait Mohamed
Concordia University
Montreal, QC
Canada

Bachir Djebbar
University of Oran
Oran
Algeria

ISSN 1868-4238   ISSN 1868-422X (electronic)
IFIP Advances in Information and Communication Technology
ISBN 978-3-030-07844-7   ISBN 978-3-319-89743-1   (eBook)
https://doi.org/10.1007/978-3-319-89743-1

Printed on acid-free paper

This Springer imprint is published by the registered company Springer International Publishing AG
part of Springer Nature
The registered company address is: Gewerbestrasse 11, 6330 Cham, Switzerland

Preface

This volume contains research papers presented at the 6th IFIP International Conference on Computational Intelligence and Its Applications (CIIA 2018), held during May 8–10, 2018, in Oran, Algeria. CIIA 2018 continued the series of conferences whose main objective is to provide a forum for the dissemination of research accomplishments and to promote the interaction and collaboration between various research communities related to computational intelligence and its applications. These conferences have been initiated by researchers from Algeria and extended to cover worldwide researchers focusing on promoting research, creating scientific networks, developing projects, as well as facilitating faculty and student exchange of ideas, especially in Africa.

CIIA 2018 attracted 202 submissions from all over the world. Each submission was carefully reviewed by two to three members of the Program Committee. For the final conference program and for inclusion in this volume, 56 papers, with allocation of 12 pages each, were selected, which represents an acceptance rate of 27.7%.

Additionally, the conference hosted three keynote presentations, and this volume includes the abstracts of the respective keynote talks. In this regard, we would like to express our warmest thanks to the keynote speakers, namely, Professor Reda Alhajj (University of Calgary, Canada), Professor Mounir Boukkadoum (UQAM, Montreal, Canada), and Professor Ferhat Khendek (Concordia University, Montreal, Canada).

We would like to thank the Program Committee members for their time, effort, and dedication in providing valuable reviews in a timely manner. We sincerely thank all the authors for their respective submission to this conference. We were in particular pleased by the number and the quality of the submitted papers. We congratulate the authors of accepted papers, and we hope that the comments of the reviewers were constructive and encouraging for the other authors.

We would like to extend our gratitude to the International Federation for Information Processing (IFIP) for given us the opportunity to publish CIIA 2018 accepted papers, in the *IFIP Advances in Information and Communication Technology* (IFIP-AICT) series by Springer.

We would like to thank the USTO-MB of Oran for hosting the conference, Tahar Moulay University of Saida, and the GeCoDe Laboratory for providing all the needed support together with USTO-MB and Concordia University for hosting the conference website.

Last but not least, we thank our sponsors for their generous contribution, and the EasyChair team for making their conference management system available to CIIA 2018.

March 2018

Abdelmalek Amine
Malek Mouhoub
Otmane Ait Mohamed
Bachir Djebbar

Organization

The 6th IFIP International Conference on Computational Intelligence and Its Applications (IFIP CIIA 2018) will be held on May 8–10, 2018, at the University of Science and Technology of Oran "Mohamed Boudiaf" (USTO-MB) in Oran, Algeria. USTO-MB is the main organizer together with the support of the GeCoDe Laboratory, the University Moulay Tahar of Saida, Concordia University and the University of Regina in cooperation with the International Federation for Information Processing (IFIP).

Conference Committees

Honorary General Chairs

Nassira Benharrats   Rector of the University of Science and Technology of Oran-MB, Algeria

Fethallah Tebboune   Rector of the Taher Moulay University of Saida, Algeria

General Co-chairs

Abdelmalek Amine   University of Saida, Algeria

Bachir Djebbar   USTO-MB, Algeria

Program Committee Co-chairs

Otmane Ait Mohamed   Concordia University, Canada

Malek Mouhoub   University of Regina, Canada

Organizing Committee Chair

Rabea Azzemou   USTO-MB, Algeria

Tutorial Chair

Otmane Ait Mohamed   Concordia University, Canada

Workshop Chair

Eisa Alanazi   Umm Al-Qura University, Saudi Arabia

Web Co-chairs

Marwan Ammar   Concordia University, Canada

Shubhashis Kumar Shil   University of Regina, Canada

Organizing Committee

Mohamed Addou	USTO-MB, Algeria
Khaled Belkadi	USTO-MB, Algeria
Zakaria Bendaoud	University of Saida, Algeria
Abderrahim Belmadani	USTO-MB, Algeria
Hasna Bouazza	USTO-MB, Algeria
Hadj Ahmed Bouarara	University of Saida, Algeria
Mohamed Amine Boudia	University of Saida, Algeria
Latifa Dekhici	USTO-MB, Algeria
Mohamed Guendouz	University of Saida, Algeria
Fatiha Guerroudji	USTO-MB, Algeria
Ali Kies	USTO-MB, Algeria
Karima Kies	USTO-MB, Algeria
Ahmed Chaouki Lokbani	University of Saida, Algeria
Hanane Menad	University of Saida, Algeria
Nabil Neggaz	USTO-MB, Algeria
Myriam Noureddine	USTO-MB, Algeria
Sidi Ahmed Rahal	USTO-MB, Algeria
Amine Rahmani	University of Algiers 1, Algeria
Mohamed Elhadi Rahmani	University of Saida, Algeria
Hicham Reguieg	USTO-MB, Algeria
Mounir Tlemçani	USTO-MB, Algeria
Mebarka Yahlali	University of Saida, Algeria

Advisory Board

Abdelmalek Amine	University of Saida, Algeria
Otmane Ait Mohamed	Concordia University, Canada
Ladjel Bellatreche	ISAE-ENSMA, France
Bachir Djebbar	USTO-MB University, Algeria
Mahieddine Djoudi	SIC/XLIM, France
Malek Mouhoub	University of Regina, Canada
Carlos Ordonez	University of Houston, USA

Program Committee

Chouarfia Abdallah	University USTO-MB, Algeria
Moussaoui Abdelouahab	University of Sétif 1, Algeria
Sultan Ahmed	University of Regina, Canada
Esma Aimeur	University of Montreal, Canada
Yamine Ait Ameur	IRIT/INPT-ENSEEIHT, France
Otmane Ait Mohamed	Concordia University, Canada
Munira Al-Ageili	University of Regina, Canada
Eisa Ayed Awadh Alanazi	University of Regina, Canada
Mohand Said Allili	Université du Québec en Outaouais, Canada
Safa Alsafari	University of Regina, Canada

Abdelmalek Amine	University Tahar Moulay of Saida, Algeria
Abdelkrim Amirat	University of Souk Ahras, Algeria
Yacine Atif	Skövde University, Sweden
Bilami Azeddine	University of Batna2, Algeria
Atmani Baghdad	University of Oran, Algeria
Amar Balla	Ecole Supérieure d'Informatique, Algeria
Ghaith Bany Hamad	Concordia University, Canada
Abdelhakim Baouya	Saad Dahlab University, Algeria
Fatiha Barigou	Université d'Oran, Algeria
Ghalem Belalem	Université d'Oran, Algeria
Hafida Belbachir	University USTO-MB, Algeria
Salem Benferhat	Université d'Artois, France
Djamal Bennouar	Saad Dahlab University, Algeria
Sidi Mohamed Benslimane	University of Sidi Bel Abbes, Algeria
Virendra Bhavsar	University of New Brunswick, Canada
Mahdi Bidar	University of Regina, Canada
Ismaïl Biskri	Université du Québec à Trois-Rivières, Canada
Frédéric Boniol	ONERA, France
Leszek Borzemski	Wroclaw University of Technology, Poland
Thouraya Bouabana Tebibel	Ecole Superieure d'Informatique, Algeria
Karim Bouamrane	University of Oran 1, Ahmed Benbella, Algeria
Kamel Boukhalfa	University of Science and Technology Houari Boumediene, Algeria
Mourad Bouneffa	LISIC ULCO, France
Andres Bustillo	University of Burgos, Spain
Humberto Bustince	UPNA
Allaoua Chaoui	Constantine 1 University, Algeria
Salim Chikhi	Constantine 2 University, Algeria
Samira Chouraqui	LAMOSI
Rozita Dara	University of Guelph, Canada
Mourad Debbabi	Concordia University, Canada
Abdelkader Dekdouk	Dhofar University, Oman
Mahieddine Djoudi	University of Poitiers, France
Richard Dosselmann	University of Regina, Canada
Gerard Dreyfus	ESPCI ParisTech, France
Mohamed El-Darieby	University of Regina, Canada
Zakaria Elberrichi	University of Sidi-Bel-Abbes, Algeria
Maher Elshakankiri	University of Regina, Canada
Larbi Esmahi	Athabasca University, Canada
Jocelyne Faddoul	Saint Mary's University, Canada
Bendella Fatima	.
Hadria Fizazi	University of Science and Technology of Oran, Algeria
Enrico Francesconi	ITTIG-CNR, France
Fred Freitas	Universidade Federal de Pernambuco (UFPE), Brazil

Bandar Ghalib	University of Regina
Nacira Ghoualmi-Zine	Badji Mokhtar University, Algeria
Zahia Guessoum	LIP6, Université de Paris 6 and CReSTIC, Université de Reims Champagne Ardenne, France
Adlane Habed	ICube, University of Strasbourg, France
Haffaf Hafid	Université d'Oran, Algeria
Djamila Hamdadou	Université d'Oran, Algeria
Abdelwahab Hamou-Lhadj	Concordia Unversity, Canada
Khaza Anuarul Hoque	University of Missouri, USA
Ilya Ioshikhes	University of Ottawa, Canada
Aminul Islam	University of Louisiana at Lafayette, USA
Mohamed Ismail	University of Regina, Canada
Souhila Kaci	LIRMM, France
Faraoun Kamel Mohamed	Université de Sidi Belabbes, Algeria
Okba Kazar	Biskra University, Algeria
Christel Kemke	University of Manitoba, Canada
Ferhat Khendek	Concordia University, Canada
Lars Kotthoff	University of Wyoming, USA
Guy Lapalme	University of Montreal, Canada
Sekhri Larbi	University of Oran, Algeria
Fuhua Lin	Athabasca University, Canada
Pawan Lingras	Saint Mary's University, Canada
Samir Loudni	Université de Caen Basse-Normandie, France
Rene V. Mayorga	University of Regina, Canada
Mohamed El Bachir Menai	King Saud University, Saudi Arabia
Marie-Jean Meurs	Université du Québec à Montréal (UQAM), Canada
Abidalrahman Moh'D	Dalhousie University, Canada
Benyettou Mohamed	University USTO-MB, Algeria
Senouci Mohamed	University of Oran 1, Ahmed Ben Bella, Algeria
Benmohammed Mohammed	University of Constantine, Algeria
Malek Mouhoub	University of Regina, Canada
Abdi Mustapha Kamel	Université d'Oran 1, Ahmed Ben Bella, Algeria
Erich Neuhold	University of Vienna, Austria
Samir Ouchani	University of Luxembourg, France
Gerald Penn	University of Toronto, Canada
Dilip Pratihar	Indian Institute of Technology Kharagpur, India
Hamou Reda Mohamed	University Tahar Moulay of Saida, Algeria
Robert Reynolds	Wayne State University, USA
Kaushik Roy	NC A&T State University, USA
Samira Sadaoui	University of Regina, Canada
Fatiha Sadat	UQAM, Canada
Hamid Seridi	University of Guelma, Algeria
Shiven Sharma	University of Ottawa, Canada
Weiming Shen	National Research Council Canada
Shubhashis Kumar Shil	University of Regina, Canada
Yahya Slimani	INSAT Tunis, Tunisia

Joao Sousa	TU Lisbon, Portugal
Karim Tabia	Artois University, France
Noria Taghezout	LIO, Oran University, Algeria
Ahmed Tawfik	Microsoft
Abd-Ed-Daïm Tenachi	Université Laarbi Ben M'hidi, Algeria
Trevor Tomesh	University of Regina, Canada
Robert Wrembel	Poznan University of Technology, Poland
Dan Wu	University of Windsor, Canada
Ayman Yafoz	University of Regina, Canada
Qian Yu	University of Regina, Canada
A. N. K. Zaman	University of Guelph, Canada
Yan Zhang	University of Regina, Canada
Bing Zhou	Sam Houston State University, USA
Boufaida Zizette	University of Constantine 2, Algeria

Additional Reviewers

Alattas, Khalid	Dang, Anh	Mokhtari, Rabah
Almeida, Hayda	Djamel, Benmerzoug	Ouali, Abdelkader
Amina, Chikhaoui	Djellal, Asma	Rehab, Seidali
Atallah, Ayman A.	Hamami, Dalila	Souad, Bouaicha
Bachtarzi, Faycal	Haque, Md. Enamul	Tahar, Ziouel
Bendaoud, Zakaria	Henni, Fouad	Younsi, Fatima-Zohra
Boussalia, Serial Rayene	Khadidja, Yachba	Zakia, Challal
Brahim, Farou	Mohamed, Sayah	

Abstracts of Invited Talks

Plagiarizing Nature for Engineering Analysis and Design

(Extended Abstract)

Mounir Boukadoum

COFAMIC, Department of Computer Science,
University of Quebec at Montreal, QC, Canada
boukadoum.mounir@uqam.ca

Systems composed of multiple integrated processors and sensors, with the ability to autonomously satisfy their energy needs and intelligently adapt to their environment are becoming a reality. This example of system complexity brought forth by the current technological advances has created analysis and design problems that were either marginal or ignored in the past. Given the increasingly shorter design cycle and lifespan of the end products, the current formal tools for system design and analysis are often overwhelmed by these new problems, when not out of scope. This is not the case in Nature where complex systems are common and have been dealt with successfully by an approach based on pragmatism and self-preservation, as opposed to logic and/or analytical models. This very slow process at the human scale has produced remarkable results in Nature, although often offering suboptimal solutions and limited explanatory capacities in comparison to formal approaches. After discussing the limitations of current formal thinking with respect to complexity, the talk with outline some of today's computational intelligence paradigms, followed by illustrative applications using neural networks, fuzzy logic and evolutionary algorithms.

Model Based Software Management

(Extended Abstract)

Ferhat Khendek[1] and Maria Toeroe[2]

[1] Concordia University, Montreal, Canada
ferhat.khendek@concordia.ca
[2] Ericsson, Montreal, Canada
Maria.Toeroe@ericsson.com

Model Driven Engineering (MDE) [1] is now a widely accepted paradigm for software development. During the last two decades, the research on modeling, design, and validation of software systems, using the Unified Modeling Language (UML) [2], its profiles or other Domain Specific Modeling Languages (DSML), has gained significant interest from both academia and practitioners in the industry. There are many ongoing research projects and significant progress has been made for software development using the MDE paradigm. However, very little has been achieved in software management, like software configuration, dynamic reconfiguration and upgrade. This is the focus of our research for the past ten years.

We have been investigating MDE based techniques for the configuration of complex systems, in a cluster or cloud environment, techniques for dynamic reconfiguration as well as for live upgrade of such systems under the constraint of high availability, i.e. the service provided by the system should not experience more than five minutes 26 seconds of downtime per year including outage due to upgrade [3].

In this talk we will introduce the problem, discuss the challenges, and the proposed solutions. In particular, we will discuss some MDE based techniques for system configuration [5–8] and their benefits. We will see for instance how model weaving, introduced in [4], has been extended and used for system configuration design from configuration fragments [7]. We will also discuss how these techniques are applied in the domain of network function virtualization (NFV) [9] for Virtual Network Functions (VNF) configuration [8]. We will show how to use models at runtime for configuration validation [10] after an adjustment has been made in response to a change in the environment or for system fine-tuning. System's configurations can be very large and at runtime it is important to avoid the validation of the complete configuration and focus only on the portion affected by the changes. We will present how we achieve this. Moreover, we will present a technique that can fix automatically the configuration if found inconsistent by the validation technique after such changes [11]. The consistency of the configuration can be re-established with complementary changes [11]. On the other hand, software systems undergo several upgrades during their life time, e.g. because of new software versions or for performance fine-tuning. In this talk we will also discuss model based techniques for live system upgrade in a cluster [12, 13] and cloud environment [14, 15]. We will end the presentation with the benefits of MDE in the domain of software management and lessons learned throughout the past 10 years.

Acknowledgment. This work has been partially supported by Natural Sciences and Engineering Research Council of Canada (NSERC) and Ericsson. Many students and collaborators have been involved in this research as shown in the references [5–8, 10–15], we would like thank them all and acknowledge their contributions.

References

1. Brambilla, M., Cabot, J., Wimmer, M.: Model-Driven Software Engineering (MDE). Morgan & Claypool Publishers (2012)
2. OMG: OMG Unified Modeling Language^{TM} (OMG UML 2.5), March 2015
3. Toeroe, M., Tam, F.: Service Availability: Principles and Practice. Wiley and Sons, Chichester (2012)
4. Jossic, A., Del Fabro, M.D., Lerat, J.-P., Bezivin, J., Jouault, F.: Model integration with model weaving: a case study in system architecture. In: Proceedings of the International Conference on Systems Engineering and Modeling (ICSEM), pp. 79–84. IEEE CS Press (2007)
5. Kanso, A., Khendek, F., Toeroe, M., Hamou-Lhadj, A.: Automatic configuration generation for service high availability with load balancing. In: Concurrency and Computation: Practice and Experience. Wiley (2013)
6. Pourali, P., Toeroe, M., Khendek, F.: Pattern based configuration generation for highly available cots components based systems. In: Information and Software Technology (IST), vol. 74, pp. 143–159. Elsevier (2016)
7. Jahanbanifar, A., Khendek, F., Toeroe, M.: Semantic weaving of configuration fragments into a consistent system configuration. In: Information Systems Frontiers, vol. 18, issue 5, pp. 891–908. Springer (2016)
8. Rangarajan, P., Khendek, F., Toeroe, M.: Managing the availability of VNFs with the availability management framework. In: Proceedings of the 4th International Workshop on Management of SDN and NFV Systems, CNSM 2017, pp. 26–30, Tokyo, Japan, November 2017
9. ETSI: Network Functions Virtualization (NFV) Release 2; Management and Or-chestration; Report on NFV Inf. Model: ETSI GR NFV-IFA 015 V2.1.1, January 2017
10. Jahanbanifar, A., Khendek, F., Toeroe, M.: Partial validation of configurations at runtime. In: Proceedings of IEEE ISORC'2015, Auckland, New Zealand, pp. 13–17, April 2015
11. Jahanbanifar, A., Khendek, F., Toeroe, M.: Runtime adjustment of configuration models for consistency preservation. In: Proceedings of 17th IEEE International Symposium on High Assurance Systems Engineering (HASE), pp. 102–109, January 2016
12. Jebbar, O., Sackmann, M., Khendek, F., Toeroe, M.: Model driven upgrade campaign generation for highly available systems. In: Grabowski, J., Herbold, S. (eds.) SAM 2016. LNCS, vol. 9959, pp. 148–163. Springer, Cham (2016)
13. Jebbar, O., Khendek, F., Toeroe, M.: Upgrade campaign simulation and evaluation for highly available systems. In: Proceedings of DEVS/TMS 2017, SpringSim 2017. ACM Proceedings, Virginia Beach, USA, April 2017
14. Nabi, M., Toeroe, M., Khendek, F.: Rolling upgrade with dynamic batch size for IaaS cloud. In: Proceedings of IEEE CLOUD 2016, San-Francisco, June 2016
15. Nabi, M., Khendek, F., Toeroe, M.: Upgrade of the IaaS cloud: issues and potential solutions in the context of high-availability. In: Proceedings of IEEE ISSRE, October 2015

Contents

Machine Learning

Optimization

Planning and Scheduling

Wireless Communications and Mobile Computing

Internet of Things and Decision Support Systems

Pattern Recognition and Image Processing

Semantic Web Services

Advanced Technology and Social Media Influence on Research, Industry and Community

Reda Alhajj$^{(\boxtimes)}$ (iD)

Department of Computer Science, University of Calgary, Calgary, AB, Canada
alhajj@ucalgary.ca

Abstract. The rapid development in technology and social media has gradually shifted the focus in research, industry and community from traditional into dynamic environments where creativity and innovation dominate various aspects of the daily life. This facilitated the automated collection and storage of huge amount of data which is necessary for effective decision making. Indeed, the value of data is increasingly realized and there is a tremendous need for effective techniques to maintain and handle the collected data starting from storage to processing and analysis leading to knowledge discovery. This chapter will cite our accomplished works which focus on techniques and structures which could maximize the benefit from data beyond what is traditionally supported. In the listed published work, we emphasized data intensive domains which require developing and utilizing advance computational techniques for informative discoveries. We described some of our accomplishments, ongoing research and future research plans. The notion of big data has been addressed to show how it is possible to process incrementally available big data using limited computing resources. The benefit of various data mining and network modeling mechanisms for data analysis and prediction has been addressed with emphasize on some practical applications ranging from forums and reviews to social media as effective means for communication, sharing and discussion leading to collaborative decision making and shaping of future plans.

Keywords: Social media · Social networks · Data analysis · Big data
Frequent pattern mining · Clustering · Bioinformatics

1 Introduction

Data is a major resource for decision making. Its value and importance has never been ignored since the existence of mankind on earth. It has been collected, stored and maintained using a wide variety of affordable means ranging from primitive to advanced. Indeed, collecting, storing and maintaining data was a cumbersome task in the past, mainly prior to the development of various technologies that gradually helped humans in handling data. However, the recent development in technology rapidly influenced data collection, storage and maintenance. For instance, sensors are becoming

© IFIP International Federation for Information Processing 2018
Published by Springer International Publishing AG 2018. All Rights Reserved
A. Amine et al. (Eds.): CIIA 2018, IFIP AICT 522, pp. 1–9, 2018.
https://doi.org/10.1007/978-3-319-89743-1_1

popular in all aspects of the daily life; they have been installed in almost every indoor and outdoor equipment. They are widely available and equipped with wireless communication skills which allow them to feed huge amounts of data that should be captured, stored, cleaned, and processed for knowledge discovery as main ingredient of effective decision making.

In the past, humans used computing devices in a limited way. Database management systems were developed to facilitate flexibility in storing and retrieving data. Making sense of data was left to domain experts who are expected to retrieve and study data related to a specific problem in a way to draw some conclusions which may guide the decision-making process. Automating the knowledge discovery process was better realized towards the end of the 20th century when various machine learning and data mining techniques were developed and put in practice to serve a variety of application domains including business, health, security, etc.

To cope with the new era, researchers, developers and practitioners realized the need to develop new techniques and technologies capable handling growing volumes of data captured incrementally from heterogeneous sources. In other words, growth in volume and types of data expected to be processed suddenly witnessed a boom. Social networks and social media platforms are gaining increased popularity and are generating tremendous amounts of data. Surveillance devices are available almost everywhere. Even traditional archives are digitized. Consequently, storage media and techniques which were previously accepted as sufficient are no more capable of handling new needs. For instance, hard drives of personal computers were only couple megabytes in capacity when they were initially manufactured with less than one megabyte of main memory. People were happily competing to get the honor of owning and using such devices. It may be impossible to image using same computing platform in the current era where gigabytes of storage are no more sufficient. In fact, computing resources have improved rapidly to partially meet human needs but will never be satisfactory. Therefore, researchers and developers are always seeking new technologies and techniques, and hence conducting and advancing research will continue to attract more attention and investment.

Explicitly speaking, data volume, characteristics and associated expectations may be described as a moving target. This necessitates the availability of enough room for storage and sophisticated techniques for processing. People will continue to collect more data as time passes, but they will never afford to increase their computing power to handle their data effectively. Thus, the need for algorithms and techniques that can depend only on limited computing resources to deal with various aspects of data from dynamicity to volume, among others. Along this direction, we contributed various techniques and algorithms that could successfully satisfy a variety of applications which require handling large volumes of dynamic and stream data. These techniques are described in our published papers listed in the references at the end of this paper. Scalability is the main aspect considered by our techniques, including frequent pattern mining, clustering, network analysis, finding repeating patterns in long sequences, etc.

2 Partial Mentioning of Our Achievements

Our completed and ongoing research addresses various aspects of data from definition to construction to manipulation and analysis leading to knowledge discovery for decision making. Our initial contributions focused on traditional aspects related to handling and manipulation of data which were popular during the last two decades of the 20th century. We then gradually moved onto more advanced techniques which we realized as necessities since 1990s. These techniques, include, network analysis, data mining and machine learning techniques which have tremendously and visibly served various applications. We also realized scalability as a serious need especially in the current era of big data. We developed advanced techniques and adapt them to various domains, including:

- Bioinformatics and Health informatics
- Data partitioning and allocation
- Homeland security and terror/criminal network analysis
- Financial data analysis: from stock market to FOREX to fraud detection
- Web/network data analysis: from structure to content to usage
- Social media analysis and opinion mining including spam detection
- Recommendation and customer behavior analysis
- Network representation is a powerful mechanism for modeling many-to-many relationships.

A network consists of a set of nodes corresponding to the entities in the application domain and a set of links representing certain types of relationships between the entities. On the other hand, data mining includes a set of powerful techniques for studying the relationships/connections between various objects. Further, data mining may be used in the network construction phase. To construct a network data may be first analyzed using techniques like frequent pattern mining or clustering. Once a network is constructed, it can be analyzed for knowledge discovery.

Most of the traditional approaches for frequent pattern mining assume unlimited main memory which is not realistic. Therefore, scalability is a major concern when it comes to practical applications where data streams dynamically and available in large volume. To tackle these problems, we developed a novel approach which satisfies the following:

- The ability to mine in a bounded amount of memory space that may vary based on task priority. Thus, it is possible to mine using common PC.
- Improve external data access and make the mining process more I/O conscious.
- Introduce a specialized mining task aware memory manager for both RAM and the external memory.

We build a tree structure namely, Frequent-Pattern (FP) tree, which summarizes the given data and allows for effective discovery of frequent patterns. Each branch represents at least one transaction. We build the tree from left to right and from top to bottom as shown in Fig. 1.

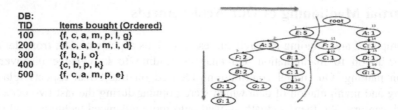

DB:	
TID	Items bought (Ordered)
100	{f, c, a, m, p, l, g}
200	{f, c, a, b, m, i, d}
300	{f, b, j, o}
400	{c, b, p, k}
500	{f, c, a, m, p, e}

Fig. 1. Construction of FP-tree top-down and left-to-right

This way, we can store on the disk left side of the tree as it grows to the right. Therefore, our upper bound is the size of free disk space rather than the available memory.

To facilitate effective data investigation and analysis, we build our own tool, namely NetDriller which is capable of analyzing raw data to derive a network. Then various network analysis techniques could be applied on the network to identify actors which may reveal some important aspects related to the analyzed network, like most knowledgeable employee, most dangerous criminal, least performing student, best team to undertake next project, etc. The basics of NetDriller are summarized in Fig. 2.

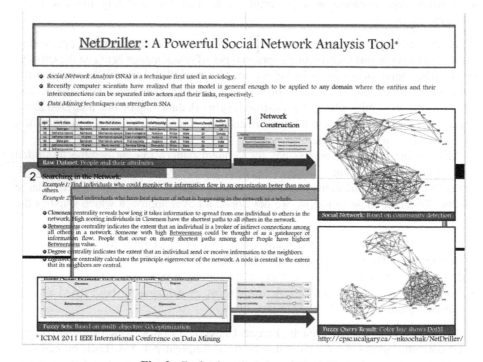

Fig. 2. Basic characteristics of NetDriller.

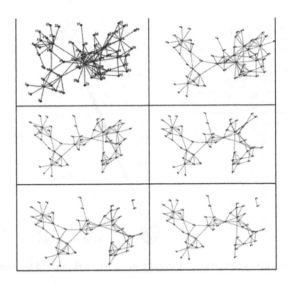

Fig. 3. Sep. 11 network changes after excluding each level nodes for eigenvector centrality measure.

We utilized NetDriller to analyze September 11 terror network (Fig. 3). It is surprising to realize that those who planned for the attacks considered all difficulties they could face. In other words, the network continues to be connected after removing terrorists who were identified as leaders down to level 6. The same is not true when the network of Madrid attacks was analyzed. The latter network became disperse only after removing second level leaders as shown in Fig. 4.

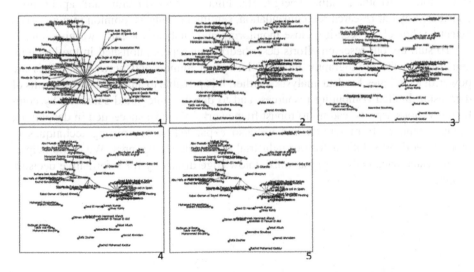

Fig. 4. Madrid network changes after excluding each level nodes for eigenvector centrality measure

Genes	Keywords
PCAP	regulate
HPC5	interact
MAD1L1	activate
HPC4	suppress
HIP1	prognostic
MSR1	biomarker
KLF6	network
PTEN	prognosis
MXI1	elucidate mechanism
CD82	prostate pathway
BRCA2	
CDH1	
ZFHX3	
ELAC2	
HPCQTL19	
HPC3	
CHEK2	
HPC6	
AR	

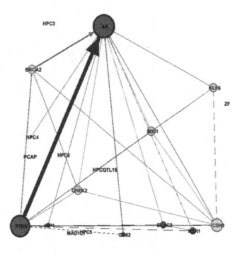

Fig. 5. A list of genes and part of the gene-gene network related to prostate cancer.

NetDriller was also employed using gene expression data related to prostate cancer to identify proteins attributed to the disease. The main result reported by NetDriller is shown in Fig. 5.

Finally, next are some of our ongoing and planned research activities based on the promising results reported in our already published papers. First, in bioinformatics we are tracking disease evolution: spatial and temporal aspects, drug repositioning, etc. Second, in health Informatics we are working on patient monitoring, referral optimization and prediction, etc. Third, we demonstrated the applicability and effectiveness of sequence analysis and prediction for various domains, including financial (e.g., stock, forex), weather, traffic, energy, etc. Finally, other domains and applications considered by our research recommendation, sentiment analysis, opinion mining, spam detection, homeland security, close monitoring and analysis for early warning, etc.

To sum up, our research efforts described in our papers published in the literature and listed in the bibliography illustrate how data mining and network analysis are powerful techniques for data analysis. Further, it is possible to analyze huge amounts of data using limited computing resources, and to develop integrated solutions by combining various aspects leading to robust framework. We have succeeded in developing some techniques from scratch and we also expanded some existing techniques to produce working solutions for our industrial and academic partners. We could help in sophisticated data analysis to maximize knowledge discovery for informative decision making.

References

1. Manber, U., Myers, G.: Suffix arrays: a new method for on-line string searches. In Proceedings of the First Annual ACM-SIAM Symposium on Discrete Algorithms, pp. 319–327 (1990)
2. Cormode, G., Hadjieleftheriou, M.: Methods for finding frequent items in data streams. VLDB J. (2009). https://doi.org/10.10007/s00778-009-0172-z. Unpaginated
3. Boyer, S., Moore, J.: A fast majority vote algorithm. Technical report ICSCA-CMP-32, Institute for Computer Science, University of Texas (1981)
4. Demaine, E.D., López-Ortiz, A., Munro, J.I.: Frequency estimation of internet packet streams with limited space. In: Möhring, R., Raman, R. (eds.) ESA 2002. LNCS, vol. 2461, pp. 348–360. Springer, Heidelberg (2002). https://doi.org/10.1007/3-540-45749-6_33
5. Karp, R., Papadimitriou, C., Shenker, S.: A simple algorithm for finding frequent elements in sets and bags. ACM Trans. Database Syst. **28**, 51–55 (2003)
6. Manku, G., Motwani, R.: Approximate frequency counts over data streams. In: International Conference on Very Large Data Bases, pp. 346–357 (2002)
7. Metwally, A., Agrawal, D., Abbadi, A.E.: Efficient computation of frequent and top-k elements in data streams. In: International Conference on Database Theory (2005)
8. Greenwald, M., Khanna, S.: Space-efficient online computation of quantile summaries. In: ACM SIGMOD International Conference on Management of Data (2001)
9. Bandi, N., Metwally, A., Agrawal, D., Abbadi, A.E.: Fast data stream algorithms using associative memories. In: ACMSIGMOD International Conference on Management of Data (2007)
10. Alon, N., Matias, Y., Szegedy, M.: The space complexity of approximating the frequency moments. In: ACM Symposium on Theory of Computing, pp. 20–29 (1996). Journal version in J. Comput. Syst. Sci. **58**, 137–147 (1999)
11. Cormode, G., Muthukrishnan, S.: An improved data stream summary: the count-min sketch and its applications. J. Algorithms **55**(1), 58–75 (2005)
12. Xylogiannopoulos, K.F., Karampelas, P., Alhajj, R.: Real time early warning ddos attack detection. Int. J. Cyber Warf. Terror. **7**(3), 44–54 (2017)
13. Üçer, S., Koçak, Y., Ozyer, T., Alhajj, R.: Social network analysis-based classifier (SNAc): a case study on time course gene expression data. Comput. Methods Program. Biomed. **150**(3), 73–84 (2017)
14. Aksac, A., Ozyer, T., Alhajj, R.: Complex networks driven salient region detection based on superpixel segmentation. Pattern Recogn. **66**, 268–279 (2017)
15. Jurca, G., Addam, O., Aksac, A., Gao, S., Ozyer, T., Demetrick, D., Alhajj, R.: Integrating text mining, data mining, and network analysis for identifying genetic breast cancer trends. BMC Res. Notes **9**(1), 236 (2016)
16. Xylogiannopoulos, K.F., Karampelas, P., Alhajj, R.: Repeated patterns detection on big data using classification and parallelism on LERP reduced suffix arrays. Appl. Intell. **45**(3), 567–597 (2016)
17. Ozsoy, M.G., Polat, F., Alhajj, R.: Making recommendations by integrating information from multiple social networks. Appl. Intell. (2016). https://doi.org/10.1007/s10489-016-0803-1
18. Addam, O., Chan, A., Hoang, W., Alhajj, R., Rokne, J.: Foreign exchange data crawling and analysis for knowledge discovery leading to informative decision making. Knowl. Based Syst. **102**, 1–19 (2016)

19. Chen, A., Elhajj, A., Gao, S., Afra, S., Sarhan, A., Kassem, A., Alhajj, R.: Approximating the maximum common subgraph isomorphism problem with a weighted graph. Knowl. Based Syst. **85**, 265–276 (2015)

20. Shafiq, O., Alhajj, R., Rokne, J.G.: On personalizing web search using social network analysis. Inf. Sci. **314**, 55–76 (2015)

21. Xylogiannopoulos, K.F., Karampelas, P., Alhajj, R.: Analyzing very large time series using suffix arrays. Appl. Intell. **41**(3), 941–955 (2014)

22. Rahmani, A., Chen, A., Sarhan, A., Jida, J., Rifaie, M., Alhajj, R.: Social media analysis and summarization for opinion mining: a business case study. Soc. Netw. Anal. Min. **4**, 171 (2014)

23. Xylogiannopoulos, K.F., Karampelas, P., Alhajj, R.: Experimental analysis on the normality of pi, e, phi and square root of 2 using advanced data mining techniques. Exp. Math. **23**(2), 105–128 (2014)

24. Rahmani, A., Afra, S., Zarour, O., Addam, O., Aljomai, R., Koochakzadeh, N., Kianmehr, K., Alhajj, R.: Graph-based approach for outlier detection in sequential data and its application on stock market and weather data. Knowl. Based Syst. **61**, 89–97 (2014)

25. Almansoori, W., Addam, O., Zarour, O., Sarhan, A., Elzohbi, M., Kaya, M., Rokne, J., Alhajj, R.: The power of social network construction and analysis for knowledge discovery in the medical referral process. J. Organ. Comput. Electron. Commer. **24**(2–3), 186–214 (2014)

26. Qabaja, A., Alshalalfa, M., Alanazi, E., Alhajj, R.: Prediction of novel drug indications using a network driven biological data prioritization and integration. J. Cheminform. **6**(1), 1 (2014)

27. Peng, P., Addam, O., Elzohbi, M., Özyer, S., Elhajj, A., Gao, S., Liu, Y., Özyer, T., Kaya, M., Ridley, M., Rokne, J., Alhajj, R.: Analyzing alternative clustering solutions by employing multi-objective genetic algorithm and conducting experiments on cancer data. Knowl. Based Syst. **56**, 108–122 (2014)

28. Kaya, M., Alhajj, R.: Development of multidimensional academic information networks with a novel data cube based modeling method. Inf. Sci. **265**, 211–224 (2014)

29. Rasheed, F., Alhajj, R.: A framework for periodic outlier pattern detection in time series. IEEE Trans. Syst. Man Cybern. **44**(5), 569–582 (2014)

30. Polash Paul, P., Gavrilova, M., Alhajj, R.: Decision fusion for multimodal biometrics using social network analysis. IEEE Trans. Syst. Man Cybern. **44**(11), 1522–1533 (2014)

31. Szeto, J., Lycett, A., Yi, X., Afra, S., Sarhan, A., Xylogiannopoulos, K.F., Karampelas, P., Alhajj, R.: Integrating data mining techniques into a user-friendly framework for visualization of health indicators. Health Inform. **3**, 63 (2014)

32. Alshalalfa, M., Alhajj, R.: Integrating protein networks for identifying cooperative miRNA activity in disease gene signatures. BMC Bioinform. **14**(Suppl. 12), S1 (2013). https://doi.org/10.1186/1471-2015-14-S12-S1

33. Öztürk, O., Aksaç, A., Elsheikh, A.M., Özyer, T., Alhajj, R.: A consistency-based feature selection method allied with linear SVMs for HIV-1 protease cleavage site prediction. PLoS ONE **8**(8), e63145 (2013)

34. Guerbas, A., Addam, O., Zarour, O., Nagi, M., Elhajj, A., Ridley, M., Alhajj, R.: Effective web log mining and online navigational pattern prediction. Knowl. Based Syst. **49**, 50–62 (2013)

35. Almansoori, W., Gao, S., Jarada, T.N., Elsheikh, A.M., Murshed, A.N., Jida, J., Alhajj, R., Rokne, J.: Link prediction and classification in social networks and its application in healthcare and systems biology. Netw. Model. Anal. Health Inform. Bioinform. **1**(1–2), 27–36 (2012)

36. Nagi, M., Elhajj, A., Addam, O., Qabaja, A., Zarour, O., Jarada, T., Gao, S., Jida, J., Murshed, A., Suleiman, I., Özyer, T., Ridley, M., Alhajj, R.: Robust framework for recommending restructuring of websites by analyzing web usage and web structure data. J. Bus. Intell. Data Mining **7**(1/2), 4–20 (2012)
37. Adnan, M., Nagi, M., Kianmehr, K., Ridley, M., Alhajj, R., Rokne, J.: Promoting where, when and what?: an analysis of web logs by integrating data mining and social network techniques to guide eCommerce business promotions. Soc. Netw. Anal. Min. **1**, 173–185 (2012)
38. Rasheed, F., Adnan, M., Alhajj, R.: Out-of-core detection of periodicity from sequence databases. Knowl. Inf. Syst. **36**(1), 277–301 (2013)
39. Khabbaz, M., Kianmehr, K., Alhajj, R.: Employing structural and textual feature extraction for semi-structured document classification. IEEE Trans. Syst. Man Cybern. C **42**(6), 1566–1578 (2012)
40. Rasheed, F., Alhajj, R.: Periodic pattern analysis of non-uniformly sampled stock market data. Intell. Data Anal. **16**(6), 993–1011 (2012)
41. Gao, S., Zeng, J., ElSheikh, A.M., Naji, G., Alhajj, R., Rokne, J., Demetrick, D.: A Closer look at "social" boundary genes reveals knowledge to gene expression profiles. Curr. Protein Pept. Sci. **12**(7), 602–613 (2011)
42. Adnan, M., Alhajj, R.: A bounded and adaptive memory-based approach to mine frequent patterns from very large databases. IEEE Trans. Syst. Man Cybern. B **41**(1), 154–172 (2011)
43. Rasheed, F., Alshalalfa, M., Alhajj, R.: Efficient periodicity mining in time series databases using suffix trees. IEEE Trans. Knowl. Data Eng. **23**(1), 79–94 (2011)
44. Alshalalfa, M., Özyer, T., Alhajj, R., Rokne, J.: Discovering cancer biomarkers: from DNA to communities of genes. Int. J. Netw. Virtual Organ. **8**(1/2), 158–172 (2011)
45. Rasheed, F., Alhajj, R.: STNR: a suffix tree based noise resilient algorithm for periodicity detection in time series databases. Appl. Intell. **32**(3), 267–275 (2010)
46. Rasheed, F., Alshalalfa, M., Alhajj, R.: Adaptive machine learning technique for periodicity detection in biological sequence. J. Neural Syst. **19**(1), 11–24 (2009)
47. Adnan, M., Alhajj, R.: DRFP-tree: disk-resident frequent pattern tree. Appl. Intell. **30**(2), 84–97 (2009)
48. Kaya, M., Alhajj, R.: Multi-objective genetic algorithms based automated clustering for fuzzy association rules mining. J. Intell. Inf. Syst. **31**(3), 243–264 (2008)
49. Kaya, M., Alhajj, R.: Online mining of fuzzy multidimensional weighted association rules. Appl. Intell. **29**(1), 13–34 (2008)

Data Mining and Information Retrieval

Basketball Analytics. Data Mining for Acquiring Performances

Leila Hamdad[(✉)], Karima Benatchba, Fella Belkham, and Nesrine Cherairi

Ecole nationale Supérieure en Informatique ESI,
BP 68M, 16309 Oued-Smar, Alger, Algeria
l_hamdad@esi.dz
http://www.esi.dz

Abstract. Choices of decision makers in a basketball team are not limited to the strategies to be adopted during games. The most important
ones are outside the field and concern team composition and talented
and productive players to acquire on which the team can rely to raise
its game level. In this paper, we propose to use data mining tasks to
help decision makers to make appropriate decisions that will lead to the
improvement of the performance of their players and their team. Tasks
such as clustering, classification and regression are used to detect weaknesses of a team; best players that can help overcome these weaknesses;
predict performance and salaries of players. These will be done on the
NBA dataset.

Keywords: Data mining · Basket ball analytics · Clustering
Classification · Performances

1 Introduction

Data mining (DM) is used in many fields and the sports industry is no exception. In fact, sport is an ideal area for the implementation of data mining tasks
and techniques. This is due to the amount of statistics collected on players,
teams, games and seasons. Team managers wish to understand these data in
order to extract information that will help them improve performances of their
players and their teams by predicting future performances and affecting players
in appropriate training group to make better profits. The Basket ball is one of
those sports. A basketball team is composed of five players. A player can have
a great impact on the efficiency of the team due to his talent. Indeed, in many
cases, a basketball team with new acquisitions has seen its game significantly
improved. Sometimes, the reverse phenomena can take place and observers might
say that the added talent could not adapt his game to the rest of the team. And
in other cases, one has seen teams with modest talents exceeding expectations.
The difficulty of finding talents that are coherent with the rest of the team and
the uncertainty of the stability of the acquired players' performances amplify

A. Amine et al. (Eds.): CIIA 2018, IFIP AICT 522, pp. 13–24, 2018.
https://doi.org/10.1007/978-3-319-89743-1_2

the risks taken by the team managers in their choices; especially if the league to which they belong is as competitive and exigent as NBA.

NBA is one of the three major leagues in the United States. The results and statistics that it provides make it an ideal subject of study. Indeed, the amount of existing data is accessible and concerns over 1200 games played in a single season (Sports Reference, 2000). Moreover, the measured performances of the players, relative to their different positions, are available. To extract significant knowledge from this amount of data, DM tasks such as clustering, classification, regression can be used. In this work, we have two objectives: 1. Our first goal is to study acquisition of new players. Before acquiring new players, it is necessary to determine team's needs. For this purpose, a clustering of the players is done according to their performances. It will give managers a description of players' performances' levels and hence allow detecting weaknesses that will help in deciding which talent to acquire. 2. The second goal is to insure the stability or performance improvement of the chosen player. This can be done by predicting his future performances and cost according to his efficiency. All the proposed tasks have been tested on the data of 10 NBA seasons (from 2005–2006 to 2014–15).

The rest of the paper is organized as follows: Sect. 2 presents related work in the literature. In Sect. 3, the data used to test different tasks is described. Then in Sect. 4, we present some DM tasks and techniques to be used for Basket ball and run them on NBA dataset in Sect. 5.

2 Related Works

The NBA is one of the major leagues of the US. It represents an extremely competitive environment for players and teams. Moreover, it is the subject of intensive research and perfect for the application of DM due to the amount of data generated by multiple meetings and sport events. In Basketball, it was Dean Oliver who introduced statistical basketball analysis in his book "Basketball on Paper" [4]. Nowadays, it is regarded as a revolution for the Basketball statistical analysis. The literature concerning Basketball analysis is quite abundant. Among these works, one can find those of [5]. They were interested in identifying the variables which had a potential effect on players' performance from a game to another, using linear regression. They, first, used a linear mixed model (LMM), to model Points (P) according to variables with both fix and random effects. Then, they used a generalized linear mixed model (GLMM) to model Win Score (WS). They concluded that minutes played, percent utilization, and team quality difference variables were among the ones that had an impact on P and WS.

[1] studied the impact of a group of two or three players (called *Big 2's* or *Big 3's*) on successive wins of a team. To this end, [1] grouped players, according to their level, in appropriate groups of two or three, using k-means algorithm. This clustering is done according to some features as points per game, offensive rebounds per game, defensive rebounds per game, assists per game, etc. A regression model was then used to measure the impact of the composition of

"*Big 2s*" and "*Big 3s*" on winning teams. He showed that the composition of a team's top 2 and top 3 players is a factor with a high statistical significant in the success of a team, and showed which combinations yielded over-performance, and which combinations yielded underperformance, relative to the team's talent and coaching quality.

The successive victories of teams have also been studied in [8], where their income and victories were evaluated according to the performance of their players. A regression method was used to study the relationship that may exist between *PER* (Player Efficiency Rating) and variables that determine team's wins and earnings. He concluded that the defensive abilities contributed to win games and thus, ensured greater income, knowing that every win brings 3% more revenue. Among the existing works, some focused on prediction. Among them, [6] studied which teams would make the NBA playoffs. They collected and analysed team data using Principal Components Analysis to reduce the dimensionality of the data set and then a Discriminant Analysis to predict the classifications of teams into playoffs or non-playoffs. In their paper, NBA Oracle [2], supervised and unsupervised learning methods are applied for predicting game outcomes and providing guidance and advice for common decisions in the field of professional basketball. For game outcome prediction, four different binary classification techniques are compared according to their accuracy prediction: Linear Regression, SVM (Support Vector Machines), Logistic Regression, and ANN (Artificial Neural Networks). In the same paper, k-Means is applied to infer optimal player positions and Outlier Detection to identify outstanding players. Neural networks have also been used for predictive end in [9]. Some others works focused on the prediction of points scored and therefore the results of games using regression models [11].

The clustering was mainly used in the construction of teams [1]. However, in our work, studying similarities between performances of players is only a first step in the acquisition process. Indeed, the resulting groups aim to determine the weaknesses of a team and help choose the right player(s) that will reinforce the team. This will be done using clustering and classification. Thereafter, our goal is to predict the performance of a player (assume selected) and his salary to determine whether he is a good option for the team. Regression and time series are used for this purpose.

3 Data

We have chosen to use the data source "Basketball-reference.com", a web site with a rich data base. One can find there, statistics on over 60 NBA seasons. They are reliable and contain few missing values [3]. The elements that point out the performance of a player during the game are divided into four categories:

- Defensive performance: these are indicators that ensure the defensive play of a player. They are represented by: blocks, defensive rebounds and steals.
- Offensive performance: indicators that show the offensive play. They are represented by: Offensive personal fouls and rebounds.

– Scoring: all indicators that have a relationship with the scoring and points: the attempts of field goals, attempts goals successful on the field, attempts goals from three points on the ground, the attempts of goals successful three-point field, attempts goals from two points on the ground, successful goals attempts goals from two points on the ground, free throws, the successful free throws.

– Play-making: It gives information on the participation of a player in building the game on the field. It is represented by: the assists, turnovers, number of games played and minutes spent on the field. The performance of a given team is represented by the sum of performance indicator values of all its players.

We worked on two types of indicators to best meet our needs in terms of players' performance description. The first type consists of basic statistics (rebound, interception points...) as they give a summary of the performance during a particular season. The second type allows a more detailed overview of players' performances (a zoom) according to a particular event (match or possession).

4   Data Mining Tasks for Basket Ball Analysis

In this section we will present the different DM tasks that we used for basket ball analysis.

4.1   Clustering

Clustering is a descriptive task of DM. It consists on partitioning the data into homogenous clusters using similarity measures. The objective of applying clustering on a set of NBA players is to form groups of players statistically homogenous as they differ in their playing style. In this study, the data used for clustering represents the 10 NBA seasons from 2005–06 until 2013–14.

The clustering allows to have an overview on Player's Skills and to compare their efficiency according to the partition they belong to. As each player plays in a particular position, clustering is also used to compare players of the same position according to their performance. It also provides decision makers with elements for future analysis that enable them to identify potential undervalued players or even to propose appropriate salaries to the players.

Moreover, the use of clustering could be extended for comparative purposes. This is done by applying clustering after the prediction in order to compare the salaries of the players based on their future statistics. This is useful when multiple players meet the need of the team. In this case, the one whose value increases will have an advantage over others, whether for performance or salary. We used K-means algorithm for its simplicity and efficiency on large dataset.

4.2   Classification

Classification is a predictive task of DM. It consists on affecting new objects to existing groups. In our case, we apply classification on the results of clustering.

It will determine to which cluster, a new player can belong according to his features. Hence, this classification allows to have an overview on the players competences depending on the cluster to which he belongs and the players therein. This classification is interesting when acquiring a new player specifically when one wants to substitute a player against a new one, having the same characteristics or to fill a gap in the team. Indeed, once the clusters have been defined using a sample of players according to their skills, the classification will determine to which cluster a new player will belong according to its descriptive features.

We have chosen to use Naïve Bayes algorithm (see [10] for more details). It is a supervised classification algorithm based on the Bayes theorem with strong features' independence hypothesis. Its principle is as follows: suppose, we have K clusters, and we went to know cluster of a new entries X. X is affected to a cluster C_k if:

$$P(C_k/X) = P(C_k)P(X/C_k) = max_{j=1,...,K}P(C_j)P(X/C_j). \qquad (1)$$

Where, $P(C_j)$ is the apriori probability of the cluster C_j and $P(X/C_j)$ the probability of X in cluster C_j. The advantage of this algorithm is that it requires relatively little training data.

4.3 Prediction

Basketball is a very competitive sport and a great business. As a result, prediction becomes important. Indeed, NBA statistics have particular relevance for team managers and coaches. When acquiring a new player or signing and renewing a contract, it is possible to project the player history in order to predict future trends of his performance and salary, and make the right decisions. This is done by studying and analysing the existing relationships in past occurrences of that player's performances. In this perspective, the multiple linear regression and exponential smoothing on a set of historical data retrieved from the database according to specific needs are used. We can use the results of predictive analysis to evaluate a given player and predict future values of his performance and salary. Moreover, we can predict team's costs.

Performance Prediction. Several metrics and measures are used to evaluate the performance of a player in the NBA. Some represents performance detailed on several axes (defensive, offensive, shooting rate), some reflect the player's contribution to the victory of his team and others are more general and global. We are interested, in this study by predicting the PER (Player Efficiency Rating). It is a performance measure that summarizes the performance of a player during a season in a single number. This metric reflects individual performance of a player in several sports. It was proposed by [7]. In Basketball, Hollinger's formula takes into account many variables that represent the positive achievements (the points scored on the ground (FG), defensive rebounds (DRB) ...), the negative effects (personal faults (FP), team faults (FT) ...) and adjust them according to games' time and games' rhythm. The value of the average PER

of the league is set to 15.00; it allows to compare the performances of players over the seasons. This task allows to predicts the future value of *PER* based on a set of performance indicators selected. To do so, we use linear regression model to predict this performance to evaluate players' performance indicators and selecting those that have a significant effect on the *PER* and predict their values.

Salaries Prediction. Players' salaries through the seasons are considered as time series. Each observation is associated to a particular season. The income of player at time t is a function of previous incomes. The future value of the player's salary can be predicted in short term, using exponential smoothing. This method is used when the series (T observations) are not seasonal.

5   Tests and Results

In this section we will show through different scenarios, how the different tasks of data mining, cited above, can help managers make decisions on acquiring efficient players for basket ball teams. Indeed, we have as objective to group players according to their performance, evaluate and predict player performance and predicting the salaries of players and franchises costs.

5.1   Clustering

We present two scenarios of clustering. In the first one, the players are grouped according to their performances and in the second one, we focus on the move of players from cluster to another across the seasons.

Scenario 1. We will present in what follows the result of a clustering which purpose is to have a comprehensive view of the talents distribution, either in the league or in a team. It also provides an overview of the skills of the players according to the cluster they belong to compared to their opponent. We applied K-means, with $k = 6$, on features of 337 players of season 2012–2013. The used feature were: Minutes played (*MP*), Field Goals (*FG*), Field Goals Attempt (FGA), basket of threes Points (*3P*), basket of threes Points Attempt (*3PA*), basket of two Points (*2P*), basket of two points attempt (2PA), Free Throw (*FT*), Free Throw Attempt (*FTA*), Offensive ReBound (*ORB*), Defensive ReBound (*DRB*), ASsisTs (*AST*), Interception (*STL*), Block (*BLK*), Turnover (*TOV*), Points (*PTS*), Player Efficiency Rating (*PER*). Table 1 represents this clustering by giving the average of each feature for all clusters We distinguish some clusters from others. Indeed cluster1 contains imposing statistics. It is a superstar cluster as the best NBA players belongs there. It contains 25 players from a total of 337. It includes players who have proven their talent by playing exceptional season and winning awards such as: Most Valuable Player (best player of the regular season) and NBA All-Defensive Second Team (best defensive player of the regular season). They participated in the NBA, All Star and won the NBA championship with their team. Among them, we have: Lebron James, Stephen Curry, Kobe Bryant, Kevin Durant ... According to the results of the Table 1, one can see that cluster 4 includes excellent rebounders who scored many two points baskets. However, cluster 2 is the one with lower statistics particularly in defence (STL, BLK...). Cluster 3 also displays statistics that are far from impressive, particularly in defence. It includes players like Louis Amundson and Joel Anthony who are considered by the site Bleacherreport.com as "50 Most Worthless Players in the NBA for the 2012–13 Season". We can also find Rodrigue Beaubois, a player from whom one expected a lot but he did not have great performances. The distribution of the players is given by the following Fig. 1. Almost half of the sample players are in clusters 2 and 3 with low to medium skills.

Table 1. Clustering results of 337 players on season 2012–2013.

Clus	MP	FG	FGA	3P	3PA	2P	2PA	FT	FTA	ORB	DRB	AST	STL	BLK	TOV	PTS	PER
Clus0	1580	219	419	42	117	177	374	89	119	70	192	131	48	30	83	570	13.17
Clus1	2766	528	1167	119	326	408	840	300	368	64	296	428	112	31	213	1477	14.99
Clus2	389	46	112	12	36	34	76	17	25	16	44	27	11	6	19	124	8.75
Clus3	840	106	241	21	61	58	179	37	54	47	111	53	25	20	40	271	12.35
Clus4	2192	382	751	3	13	378	738	173	248	190	423	126	61	90	126	940	16.28
Clus5	1999	312	723	104	277	207	445	140	172	47	188	209	67	21	115	870	14.74

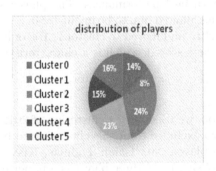

Fig. 1. Distribution of players.

Scenario 2. For this clustering, the goal was to observe a variation in players' performances from a season to another according to their productivity and substitution of players and talented players in the team. For this purpose, we have chosen to work on Phoenix Sun players for the two seasons 2011–12 and 2012–2013. We tested the clustering only on players that were part of the team during the two seasons. K-means with $K = 3$ was used. The results are given

Table 2. Results of K-means (K = 3).

Clusters	FG	FGA	3P	3PA	2P	2PA	FT	FTA	Pts
Cluster 0	420	863	30	98	389	765	169	269	1066
Cluster 1	78	178	7	26	71	152	38	51	202
Cluster 2	257	591	56	163	201	427	87	116	660

Table 3. Repartition of the players in the clusters according to the two seasons 2011–2012, 2012–2013.

Cluster 0	Cluster 1	Cluster 2
G. Dragic	J. Childress	M. Beasley
M. Gortat (2011–2012)	D. Garrett Robin	S. Brown [x2]
L. Scola	L. Kendall	J. Dudley [x2]
	M. Ronnie Price	C. Frye 11.12
		M. Gortat [2012–13]
		W. Johnson
		M. Morris [x2]

in Tables 2 and 3. We note that for the two selected seasons, the players Markieff Morris, Shannon Brown and Jared Dudley remain in the same cluster (2), which groups efficient players with the highest number of 3 points baskets scored. Cluster 0 is the one with best performances. The player Marcin Gortat whose 2011–12 season statistics allowed him to be in cluster 0, has shown a performance decrease during 2012–2013 and moved to cluster 2. His only improvement is seen in the three points baskets scored, which is a characteristic of cluster 2.

5.2 Classification

To test the classification, we used, first a clustering of 2012–13 season statistics. Some players have been excluded from this clustering to be used as test sample in this section. These players are: Isaiah Thomas, Klay Thompson and John Wall. The raison of choosing these three player is a paper in Bleacherreport.com published in June 2012, intituled "*15 NBA players who will be the stars of 2015*" and in which the three players are included. We wanted to check if with our classification, we would obtain the same prediction. For this purpose a Naive Bayes algorithm is used on the statistics of the three players in 2014–2015 season, the results confirmed Bleacherreport.com ranking. Indeed, Klay Thompson and John Wall were classified in the ≪cluster 1≫ containing NBA superstars (See Table 4). On the other hand, Isaiah Thomas, is assigned to cluster 5 for the same season: a cluster grouping players with a great number of marked points and very good defensive statistics. However, it should be noted that during the 2013–14 season, his excellent statistics affected him to cluster 1. The site

Bleacherreport.com also quoted some players as being the worst in the NBA, among them, Brian Cook. He averaged 5.7 points and 2.7 rebounds in less than 14 min per game and spent a lot of time on the bench during his career. The last season for Cook is 2011–12; so we took the statistics of this last season to predict to which cluster he will be assigned. The algorithm classified him in cluster2. A very natural result because Cook spent only 276 min on the playground which generated very weak statistics in defence as in offense (Table 6). The average salary of Cook is 1, 955, 569; an average salary that matches perfectly to the cluster to which he belongs, as the average salary of cluster 2 is 1, 774, 548 (Table 5).

Table 4. Thompson and Wall statistics in cluster1.

Clusters	MP	FG	FGA	3P	3PA	2P	2PA	FT	FTA	ORB	DRB	AST	STL	BLK	TOV	PTS	PER
K. Thompson	2455	602	1299	239	545	363	754	225	256	27	220	87	60	149	122	1668	?
J. Wall	2837	519	1166	65	217	454	949	284	362	36	330	792	138	45	304	1387	?

Table 5. Isaiah Thomas statistics in cluster 5.

Cluster	MP	FG	FGA	3P	3PA	2P	2PA	FT	FTA	ORB	DRB	AST	STL	BLK	TOV	PTS	PER
I.Thomas13-14	1726	335	797	129	346	206	451	302	348	33	120	284	57	5	143	1101	22.3
I.Thomas14-15	2497	496	1096	127	364	369	732	346	406	47	163	545	93	8	213	1465	20.5

Table 6. Isaiah Thomas statistics in cluster 2.

Cluster	MP	FG	FGA	3P	3PA	2P	2PA	FT	FTA	ORB	DRB	AST	STL	BLK	TOV	PTS	PER
Brian Cook	276	31	98	10	50	21	48	9	10	10	53	10	6	5	11	81	10.4

5.3 Performance Prediction

We have selected players who have played several seasons and apply predictions on their statistics.

1- To predict Performance of the players, we used the following multiple regression model:

$$PER = 21.436 - 0.0426G - 0.0068MP + 0.0419a_3FG - 0.0118FGA$$
$$+ 0.0215FT - 0.00467FTA + 0.0183ORB + 0.00433DRB$$
$$+ 0.0135AST + 0.0222STL + 0.0171BLK - 0.0196TOV - 0.0113PF.$$

Multiple linear regression tests conducted on R software showed that all variables are statistically significant with p-value under $\alpha = 5\%$.

Moreover, we also used previous performance in different seasons to predict players' performance in a given season by Holt Winters algorithm (exponential smoothing algorithm).

22   L. Hamdad et al.

2- In the following example, we chose to predict the performance of Darius Miller who played eight NBA seasons. We used statistics of the first four seasons as learning data and the rest to test a prediction to compare the predict values to the real ones. The results of prediction of PER are displayed in the Table 7 and represented in Fig. 2. Figure 2 shows that the prediction obtained across the season by the exponential smoothing are closer than the regression model to the true values. Note that, prediction value by regression are obtained by firstly predict by Holt Winters method the values of indicators that affect PER, than compute this latter.

Table 7. Results of performances prediction.

Seasons	2006–07	2007–08	2008–09	2009–10	2010–11	2011–12	2012–13	2013–14
	Learning values				Predicted values			
True values	14.2	11.8	10.8	14.4	12.4	15.3	16	14.1
Pred values: reg	14.2	11.8	10.8	14.4	16.8	18.49	20.79	23.09
Pred values: exp smooth	14.2	11.8	10.8	14.4	15.57	14.69	16.85	12.99

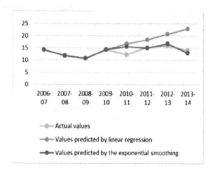

Fig. 2. Comparison between linear regression and exponential smoothing in performance prediction.

3- In other example, we tested performances of the NBA player, Michael Jordan. We focused on prediction of its interception and the number of points scored using simple exponential smoothing. We recall that Michael Jordan is the first player in NBA history who has scored 200 interceptions in a season. The results are displayed in Figs. 3 and 4.

From this two figures we see clearly that the obtained predicted value matches the true values of interceptions or Fields goal.

4- The prediction of players' salaries was also tested using multiple regression model and significant explanatory variables are selected. These variables are also been predicted using exponential smoothing since they occurred each seasons.

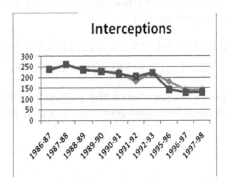

Fig. 3. Predicted and true values comparison of Michael Jordan interceptions across the seasons.

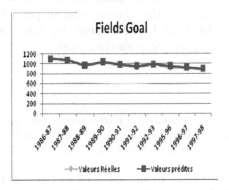

Fig. 4. Predicted Fields goal across the seasons.

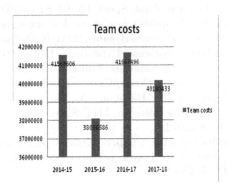

Fig. 5. Histogram of predicted cost of Phoenix Suns team.

Hence the salaries of the players of Phoenix Suns team are predicted using the following estimated regression model:

$$Sal = 15122431 - 7099MP + 38047FG - 11575FGA + 18615FT - 14558FTA$$
$$+ 22853DRB + 9222AST - 21840STL - 28384PF.$$

According of the predicted salaries of the players of Phoenix Suns team, we obtained the predicted team franchise costs as shown in Fig. 5.

6   Conclusions

In this work, we aim to assist NBA sports decision makers in the process of acquisition of players by exploiting and applying the data mining techniques as k_means, Naive Bayes algorithm, linear regression and time series on large amounts of data. These data are the statistics of ten seasons in the NBA (from 2005–06 to 2014-15). The results we have reached, allow decision makers to identify their needs, determine the players who respond to this need and to select the one or ones that work best for them depending on their current and future performance, taking into account their cost.

References

1. Ayer, R.: Big 2s and big 3s: analyzing how a team's best players complement each other. In: MIT Sloan Sports Analytics Conference, Boston, MA, USA (2012)
2. Beckler, M., Hongfei, W., Papamichael, M.: NBA oracle. Zuletzt besucht am 17, 2008–2009 (2013)
3. Cao, C.: Sports data mining technology used in basketball outcome prediction. Dissertation, Dublin Institute of Technology (2012)
4. Oliver, D.: Basket Ball on a Paper. Rules and Tools for Performance Analysis. Potomac Books Inc., Lincoln (2004). 392 pages
5. Casals, M., Martinez, J.A.: Modelling player performance in basketball through mixed models. Int. J. Perform. Anal. Sport **13**, 64–82 (2013)
6. Hoffman, L., Joseph, M.: A multivariate statistical analysis of the NBA (2003). http://www.units.miamioh.edu/sumsri/sumj/2003/NBAstats.pdf
7. Hollinger, J.: Pro Basketball Forecast: Paperback, 1900 (2005)
8. Li, H.: True value in the NBA: an analysis of on-court performance and its effects on revenues. Undergraduate Honor Thesis, University of California, Berkeley (2011)
9. Maheswaran, R., Chang, Y.-H., Henehan, A., Danesis, S.: Deconstructing the rebound with optical tracking data. In: MIT Sloan Sports Analytics Conference 2012, Boston, MA, USA (2012)
10. Mitchell, T.M.: Machine Learning, Chap. 6. McGraw-Hill Science, New York (1997)
11. Wheeler, K.: Predicting NBA player performance (2012). cs229.stanford.edu/proj2012/Wheeler-PredictingNBA

Similarity Measures for Spatial Clustering

Leila Hamdad$^{(\boxtimes)}$ ⓘ, Karima Benatchba, Soraya Ifrez, and Yasmine Mohguen

Ecole nationale Supérieure en Informatique ESI,
BP 68M, 16309 Oued-Smar, Alger, Algeria
l_hamdad@esi.dz
http://www.esi.dz

Abstract. The spatial data mining (SDM) is a process that extracts knowledge from large volumes of spatial data. It takes into account the spatial relationships between the data. To integrate these relations in the mining process, SDM uses two main approaches: Static approach that integrates spatial relationships in a preprocessing phase, and dynamic approach that takes into consideration the spatial relationship during the process. In this work, we are interested in this last approach. Our proposition consists on taking into consideration the spatial component in the similarity measure. We propose two similarity measures; d_{Dyn1}, d_{Dyn2}. We will use those distances on the main task of SDM, spatial clustering, particularly on K-means algorithm. Moreover, a comparaison between these two approaches and other methods of clustering will be given. The tests are conducted on Boston dataset with 590 objects.

Keywords: Spatial data mining · Dynamic approach
Similarity · Preprocessing approach · Clustering · K-means · DBSCAN

1 Introduction

These last decades have seen an explosion in the volume of spatial data. This is due to the various technological advances and the development of automatic data acquisition tools (GPS, satellite images,...). The wide use of these data has given rise to spatial data mining (SDM). It is a process that allows extracting knowledge and useful patterns from large volumes of spatial datasets [11,15]. Spatial data consist of two types of components; a descriptive component containing data of usual type: integer, real, boolean,... describing some features of the data and geometric component which describes the spatial localization of the data. The SDM is more difficult than classical data mining as it handles geo-spatial data characterized by autocorrelation. In SDM, spatial component can be taken into account according to three different approaches. The first one is basic and considers all attributes in the same way (spatial and non-spatial) and uses the techniques of classical data mining. As a result, the auto-correlation between objects is not taken into consideration. In fact, several studies have shown that

A. Amine et al. (Eds.): CIIA 2018, IFIP AICT 522, pp. 25–36, 2018.
https://doi.org/10.1007/978-3-319-89743-1_3

this process produces inconsistent results, [14,19]. The second approach (static), likewise, uses classical DM techniques to process geo-spatial data. However, a preliminary data preprocessing is required. It consists of extracting and representing the geo-spatial relationship between entities, in an explicit manner [4,17]. This approach has many drawbacks. First of all, it is time consuming as the preprocessing of the data takes time. Indeed, extraction may require a lot of computation time when many relationships must be considered. Second, the preprocessed and the original data have to be stored leading to a redundancy in storage. Moreover, if the original data is modified, the pre-processing has to be executed again. To overcome these drawbacks, the dynamic approach emerged. It consists on taking into consideration the geo-spatial component during the data-mining process. Our proposal lies in this last approach, as, in the literature, spatial dynamic processing is not explicitly defined (see [12]). Our approach consists on taking into consideration the spatial component in the similarity measure. We will be interested, in this work, in the main task of SDM, clustering. Formally, it consists on partitioning a set S of N objects, $S = \{O_1, O_2, ..., O_N\}$, based on a similarity metric, into a number of clusters $(C_1, C_2, ..., C_K)$, such as: $C_i \neq \emptyset$ for $i = 1, ..., K$, $C_i \cap C_j = \emptyset$ for $1, ..., k$ and $j = 1, ..., K$,$i \neq j$, $S = \cup_{i=1}^{k} C_i$. Objects of a cluster must be as similar as possible, while objects of different clusters must be as dissimilar as possible. It is an important task as it enables to show interesting objects grouping without a priori knowledge.

The objective of this paper is to propose two similarity distances that take into consideration the spatial component during the clustering process. To show the effectiveness of this proposal, we used these distances in K-means, a simple partitioning method yet efficient and widely used. We compared the obtained results to the first approache which we called static using K-means. Then we compare the three versions of K-means to density based clustering method DBSCAN [8] and CAH an ascending hierarchical clustering method.

This paper is organized as follows: In Sect. 2, we will introduce the static approach. Then in Sect. 3, we will present the similarity distances to be used for dynamic clustering. In Sect. 4, we present some clustering algorithms. Finally in Sect. 5, tests and results on the different approaches will be presented.

2   Static Approach

This approach allows taking advantage of traditional data mining methods to process spatial data. There are different types of spatial relationships: metric, topological and directional. In this paper, we deal with metric relations as the data considered are represented by points. The data preprocessing begins by extracting the metric relationships between the data. These relationships are then used to modify the attributes of data. There exist in the literature several methods to extract it. Moran [16] and Geary [9] were the first to propose a measure for spatial interaction. Since then, several measures have been proposed [3]. Most of them require a threshold as parameter (obtained from the objects' distance matrix) [13]. We have chosen to use the K nearest neighbors (KNN) as it is a simple non parametric method (does not need a threshold); moreover, it is the most used method [13]. KNN processes as follows: For a given object O_i of the dataset, KNN searches among all objects of the data set, the K-nearest neighbors of O_i according to a distance such as the geometric distance. The neighbor relations obtained by applying KNN on the original dataset are used to obtain neighborhood matrix. The last step of the static approach consists on building the new dataset (smoothing matrix) from the original one. In fact, each attribute value of an object is replaced by the mean of its neighbors attribute. The algorithm is given below:

Smoothing Matrix algorithm For each object O_i of S, $i = 1, \ldots, N$:

(1) Compute K-nearest- neighbors of O_i with KNN algorithm.
(2) For each attribute of O_i, compute the mean of the attributes of O_i and its corresponding neighbors' attribute.
(3) Create a new object O'_i which attributes values are the one computed in (2).
(4) Insert O'_i in S' where S' is the new dataset.

3   Similarity Distances for a Dynamic Approach

This approach, unlike the previous one, proposes dynamic spatial data processing. Its goal is to take into account the spatial component in the process of DM and not through a preprocessing. However, in the literature very little work is available on this approach [12]. This might be due to the fact that this research area is relatively young [17]. One of the goals of this paper is to take into consideration the spatial component, in a similarity distance. We propose two distances d_{Dyn1} and d_{Dyn2}. Both compute the similarity between spatial objects by considering the non-spatial attributes and spatial attributes simultaneously. They are based on two distances: Euclidean distance that measures the similarity between the non-spatial attributes and geographic distance that measures the metric relationship. The first one, d_{Dyn1} is formally defined as the product of the Euclidean distance between the non-spatial attributes and geographic distance between the spatial attributes:

$$d_{Dyn1}(O_i, O_k) = d_{euclidian}(O_i, O_k) * d_{geo}(O_i, O_k).$$

Where

$$d_{euclidian}(O_i, O_k) = \sqrt{\sum_{j=1}^{m}(O_i - O_k)^2}$$

and

$$d_{geo}(O_i, O_k) = 6371 * Acos[Cos(lat1) * Cos(lat2) * Cos(long2 - long1) + Sin(lat1) * Sin(lat.2)].$$

Where $(lat1, long1)$ and $(lat2, long2)$ are the spatial coordinates of respectively points O_i and O_k. In the second proposed distance d_{Dyn2}, weights are assigned to the two types of attributes (spatial and non-spatial). The formula of d_{Dyn2} is given below:

$$d_{Dyn2}(O_i, O_k) = \alpha d_{euclidian}(O_i, O_k) + \beta d_{geo}(O_i, O_k).$$

where α and β are respectively weights of Euclidian and geographic distance and verify that $\alpha + \beta = 1$. When $\alpha = 1(\beta = 0)$, clustering is done on descriptive attributes only. The spatial attributes are ignored. On the contrary, when $\alpha = 0(\beta = 1)$, only spatial attributes are considered and we obtain a regionalization.

4   Clustering Algorithms

Many methods dedicated to the clustering exist in the literature. They fall into two main classes: partitioning methods and hierarchical methods. They differ in the way they build clusters. While the second gradually build those clusters, the first discover them by moving objects between clusters [11]. In addition to these two fundamental approaches, other methods exist such as density and grid based methods. They use different mechanisms for data organization and processing, and for building the clusters [1].

4.1   Partitioning Methods

They seek to find the best k partitions for a set of n objects (data), while optimizing an objective function. This function aims to maximize the similarity between objects of the same cluster and to minimize the similarity between objects of different clusters. These methods improve iteratively clusters by moving objects between clusters. There exist several partitioning methods among which K-means, the first clustering method. K-means is by far the most popular clustering algorithm and widely used in scientific and industrial applications [5]. Its popularity is due to the fact that it is simple to implement and converges rapidly. However, it has some drawbacks. It is influenced by outliers and the obtained clustering depends on the initial one.

4.2   Hierarchical Methods

These clustering methods build hierarchical clusters gradually. They can be of two types ascending hierarchical clustering methods and descending ones. Hierarchical methods have many advantages such as flexibility regarding the level of granularity one wants to have. However they present some drawbacks such as the difficulty of setting a stopping criterion and their high execution time [6]. In what follows, we present, CAH, an ascending hierarchical clustering method. CAH builds a hierarchy of clusters assuming that initially each object is a cluster. Then the most similar clusters are aggregated. The aggregation is repeated until having a single cluster containing all objects.

4.3 Density Based Methods

These methods characterize classes as homogeneous high density regions, separated by regions of low density. Unlike partitioning methods, methods based on density are able to discover classes of concave forms and do not take into account outliers as they are removed during the process. There are two approaches for this type of methods. The first approach is based on the density connectivity. It takes two input parameters, the neighborhood radius *Eps* which represents the maximum distance between points and the density threshold *MinPts* which represents the minimum number of points in the neighborhood. The best known algorithms are DBSCAN and GDBSCA-N [2]. The second approach is based on sound mathematical principles and uses a density function in its process. The best known algorithm is DENCLUE [10].

5   Tests and Results

In this section, we will give a summary of the different tests performed. We will start by presenting the result of k-means using the two proposed distances d_{Dyn1} and d_{Dyn2}, respectively named K-means$_{d_{Dyn1}}$ and K-means$_{d_{Dyn2}}$. K-means with preprocessing using KNN is called K-means$_{static}$. Then we will compare those results to DBSCAN and CAH algorithms. These tests were executed on an unsupervised spatial benchmark Boston[1], represented in Fig. 1. Benchmark Boston contains 506 objects and represents Boston housing and neighborhood data. Each object is represented by 18 non spatial features and longitude and latitude (spatial attributes). The different features describe the type of area of the different housing such as: hot district characterized by high criminality level, rich district, poor one... One can notice that this benchmark is very dense.

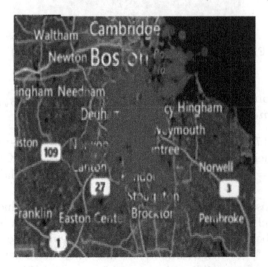

Fig. 1. Representation of benchmark Boston.

[1] https://geodacenter.asu.edu.

5.1   Dynamic vs Static K-means

To compare the dynamic versions of K-means (K-means$_{d_{Dyn1}}$ and K-means$_{d_{Dyn2}}$) to the static one we conducted several tests. For all the tested version of K-means, the number of iterations was set to 100. As this benchmark is unsupervised one, and to compare the results of the three versions of k-means, we executed classical k-means ($k = 3$) on the non spatial attributes of the benchmark to better visualize the objects distribution and their meaning. Moreover, we executed K-means$_{d_{Dyn2}}$ on spatial attributes($\alpha = 0$), obtaining a regionalization. The results of both executions are given in the following Figures. In Fig. 2, the blue color represents north districts, the yellow one the center, and the red one south districts. In Fig. 3, the blue color represents the wealthy neighborhood, the yellow one the middle class neighborhood and the red color the poor neighborhood.

Fig. 2. Regionalization. (Color figure online)

Fig. 3. Classical K-means. (Color figure online)

5.2   Comparison of $K - means_{static}$, $K - means_{d_{Dyn1}}$ and $K - means_{d_{Dyn2}}$

Figures 4, 5, 6 and 7 show the results of visualization of K-means$_{static}$ (with the number of neighbors for KNN being to 2 and 4), K-means$_{d_{Dyn1}}$ and K-means$_{d_{Dyn2}}$ ($\alpha = \beta = 0.5$). For these tests, the number of classes for K-means was set to 3.

If we take a closer look at Figs. 4 and 5, we can see the influence of the KNN parameter on the clustering. For a *number of neighbors* $= 2$, a certain number of objects (the ones between the red class and the yellow one) are considered belonging to the cluster "poor neighborhood". While when more neighbors are taken into consideration when building the smoothing matrix, this number decreases. One notes that some objects classified in the blue cluster in Fig. 4 are classified in the red cluster in Fig. 5.

Fig. 4. Static K-means with number of neighbors = 2. (Color figure online)

Fig. 5. Static K-means with number of neighbors = 4. (Color figure online)

Fig. 6. K-means Dyn1. (Color figure online)

Fig. 7. K-means Dyn2 ($\alpha = \beta = 0.5$). (Color figure online)

If we compare the results of classical K-means and those of K-means$_{static}$ (Fig. 5), we notice that the two classifications can be considered as close. We do not see a clear impact of the spatial attributes on the clustering. One notices that the boundaries between classes are clearly defined for K-means$_{d_{Dyn1}}$ (Fig. 6) when compared to K-means$_{static}$ (Figs. 4 and 7). Classes do not overlap as for K-means$_{static}$. This may be due to the distance used which gives the same importance to both types of attributes. K-means$_{d_{Dyn2}}$ gives a different partitioning of the objects. A close look at Fig. 7 shows that two classes of K-means$_{d_{Dyn1}}$ (the poor neighborhood in red and the middle class neighborhood in yellow, Fig. 6) are merged to form the poor neighborhood (in red). While the rich neighborhood of Fig. 6 is split into two clusters: yellow one for middle class neighborhood

Fig. 8. Results' visualization of K-means static (with 4 neighbors and K = 5). (Color figure online)

and blue one for rich neighborhood. To try to explain this, we executed the K-means$_{static}$ with 4 neighbors for KNN, setting the number of classes to 5. We obtained the following results (Fig. 8): The obtained classes of this execution are described as follows: The pink cluster represents the rich neighborhood, the green one the middle class one, the yellow one the residential area, the blue the economical one and the red one the poor neighborhood. Given this type of information, we can conclude that the clustering of K-means$_{d_{Dyn2}}$ is interesting as it merged the economical and poor neighborhood, found by K-means$_{static}$ and K-means$_{d_{Dyn1}}$, into one class (Red one in Fig. 7). It also, separated the rich class found by K-means$_{static}$ and K-means$_{d_{Dyn1}}$ into two classes: Middle class and rich neighborhood. Now, if we compare the different approaches according to the intra cluster inertia and execution time (Figs. 9 and 10), we note that as the number of classes increases, the inertia decreases. This is due to the fact that as the size of clusters decreases, clusters will have more similar objects. For 3 clusters, the best inertia is given by K-means$_{static}$ with 2 neighbors, followed by K-means$_{d_{Dyn2}}$. However, the dynamic version of K-means is faster than the preprocessing k-means (K-means$_{static}$). That was the goal of using dynamic K-means.

5.3   Dynamic K-means vs DBSCAN and CAH

In this section, we will compare K-means$_{d_{Dyn1}}$ and K-means$_{d_{Dyn1}}$ to DBSCAN and CAH. Figures 11, 12, 13 and 14 display the results of respectively K-means$_{d_{Dyn1}}$, K-means$_{d_{Dyn2}}$, DBSCAN and CAH.

We can notice from these Figures, that the four methods give different clustering. DBSCAN (Fig. 13) give more compact classes. One cluster (rich neighborhood in blue) is bigger, in terms of size than the other two. K-means$_{d_{Dyn2}}$ divide this class into two (rich and middle classe neighborhood). CAH (Fig. 14) does the same; however, its rich neighborhood is smaller and is located south. Both K-means$_{d_{Dyn2}}$ and CAH define the same poor neighborhood.

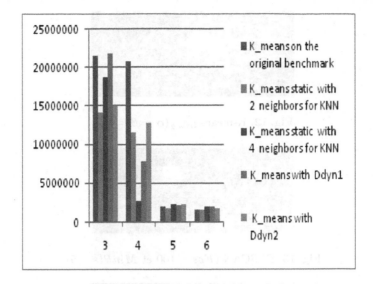

Fig. 9. Intra cluster inertia of the different k-means versions when varying k.

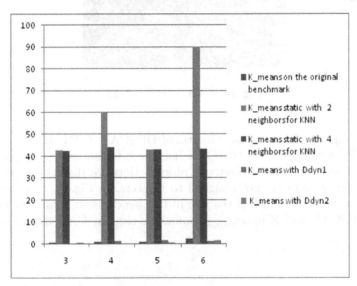

Fig. 10. Execution time of the different versions of K-means when varying K.

Fig. 11. $K - means_{d_{Dyn1}}$.

Fig. 12. K-means$_{d_{Dyn2}}$ ($\alpha = \beta = 0.5$).

Fig. 13. DBSCAN ($Eps = 100$ et $MinPts = 5$).

Fig. 14. CAH ($\alpha = \beta = 0.5$).

If we compare the intra cluster inertia of DBSCAN, CAH and K-means$_{d_{Dyn2}}$ (see Fig. 15), we notice that, DBSCAN has the best followed by K-means$_{d_{Dyn2}}$ and CAH. However, for this method, objects that are distant or isolated are considered noises and are not assigned to clusters. In Fig. 13, these points are in black and represent 3.4% of the size of the benchmark. While the execution time of DBSCAN and K-means$_{d_{Dyn2}}$ somehow similar, CAH takes much more time (Fig. 15).

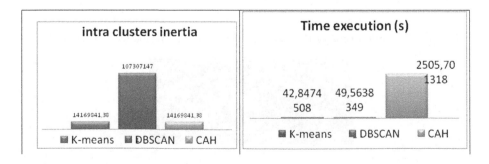

Fig. 15. Comparison between DBSCAN, CAH and K-means$_{d_{Dyn2}}$.

6   Conclusions

We focused, in our work, on the main descriptive task of SDM, spatial clustering. The main goal of our study was to compare two spatial clustering approaches namely: the spatial data preprocessing approach (static) and dynamic approach. For the second approach, we proposed to take into consideration the spatial component in the similarity measure. We proposed two distances, d_{Dyn1} and d_{Dyn2}. The various tests performed on the benchmark Boston, showed that the approach proposed $(K - means_{d_{Dyn1}}, K - means_{d_{Dyn2}})$ gives better results than the preprocessing approach in terms of execution time. This was the main goal as the preprocessing approach takes too much time. The results obtained with $K - means_{d_{Dyn1}}$ seems similar to those of the preprocessing approach but with more precise boundaries. This is due to the fact that $K - means_{d_{Dyn1}}$ takes into consideration both spatial and non spatial attributes. $K - means_{d_{Dyn2}}$ is more efficient in terms of intra-class inertia, it makes a good description in terms of regionalization and characterization of these obtained regions.

References

1. Andritsos, P.: Data clustering techniques. Technical report CSRG-443, University of Toronto (2002)
2. Ankerst, M., Breunig, M., Kriegel, H., Sander, J.: OPTICS: ordering points to identify the clustering structure. In: ACM SIGMOD, International Conference on Management of Data (1999)
3. Anselin, L.: Spatial Econometrics: Methods and Models. Springer, Dordrecht (1988). https://doi.org/10.1007/978-94-015-7799-1
4. Bogorny, V., Martins, E., Alvares, L.O.: A reuse-based spatial data preparation framework for data mining. In: Proceedings of the 17th International Conference on Software Engineering and Knowledge Engineering, pp. 649–652 (2005)
5. Boubou, M.: Contribution aux méthodes de classification non supervisee via des approches prétopologiques et d'aggrégation d'opinions. Thèse de doctorat, Université Claude Bernard - Lyon I (2007)
6. Cleuziou, G.: Une méthode de classification non-supervisée pour l'apprentissage de règles et la recherche d'information. Thèse de doctorat, Université d'Orléans (2004)
7. Ester, M., Kriegel, H.-P., Sander, J.: Spatial data mining: a database approach. In: Scholl, M., Voisard, A. (eds.) SSD 1997. LNCS, vol. 1262, pp. 47–66. Springer, Heidelberg (1997). https://doi.org/10.1007/3-540-63238-7_24
8. Ester, M., Kriegel, H.P., Sander, J.: Knowledge discovery in spatial databases. In: Institute for Computer Science, University of Munich, Invited Paper at 23rd German Conference on Artificial Intelligence (KI 1999), Bonn, Germany (1999)
9. Geary, R.C.: The contiguity ratio and statistical mapping. Incorporated Stat. **5**(3), 115–145 (1954)
10. Han, J., Kamber, M.: Data Mining: Concepts and Techniques. Elsevier Inc. (2006)
11. Jain, A.K., Murty, M.N., Flynn, P.J.: Data clustering: a review. ACM Comput. Surv. **31**(3), 264–323 (1999)
12. Klosgen, W., Zytkow, J.M.: Knowledge discovery in databases: the purpose, necessity, and challenges. In: Handbook of Data Mining and Knowledge Discovery. Oxford University Press, Inc. L.Y (2002)

13. Lebart, L.: Contiguity analysis and classification. In: Gaul, W., Opitz, O., Schader, M. (eds.) Data Analysis. Studies in Classification, Data Analysis, and Knowledge Organization, pp. 233–243. Springer, Heidelberg (2000). https://doi.org/10.1007/978-3-642-58250-9_19

14. Maguire, D.J.: An overview and definition of GIS. In: Geographic Information Systems: Principles and Applications, 1st edn., vols. 1, 2, pp. 9–20 (1991)

15. Mennis, J., Guo, D.: Spatial data mining and geographic knowledge discovery an introduction. Comput. Environ. Urban Syst. **33**(6), 403–408 (2009)

16. Moran, P.A.P.: The interpretation of statistical maps. Biometrika **35**, 255–260 (1948)

17. Rinzivillo, S., Turini, F., Bogorny, V., Körner, C., Kuijpers, B., May, M.: Knowledge discovery from geographical data. In: Giannotti, F., Pedreschi, D. (eds.) Mobility, Data Mining and Privacy, pp. 243–265. Springer, Heidelberg (2008). https://doi.org/10.1007/978-3-540-75177-9_10

18. Tobler, W.R.: Cellular geography. In: Gale, S., Olsson, G. (eds.) Phylosophy in Geography, pp. 379–386. Reidel, Dortrecht (1979)

19. Zeitouni, K., Yeh, L.: Le data mining spatial et les bases de données spatiales. Revue internationale de géomatique, Numéro spécial sur le Data mining spatial **9**(4) (2000)

Computational Ontologies for a Semantic Representation of the Islamic Knowledge

Bendjamaa Fairouz[✉] and Taleb Nora

Computer science Department, University of Badji Mokhtar,
Annaba, Algeria
bendjamaafairouz@gmail.com, talebnr@hotmail.fr

Abstract. In spite of the efforts made in the Arabic language on the syntactic and semantic level, it remains very restricted, even those on the Arabic Sacred Book are few and very limited, due to its difficulties and peculiarities. In this paper we tried to shed the light on some of the recent works that have been conducted to present a semantic representation and manipulation of the Islamic texts to define the problems, limitations and the possible future works that need our intention to improve the semantic support in the Arabic religious texts. Furthermore, we intent to briefly present our project that aims to help us reading, understanding, and interpreting the Islamic legislative sources. The goal of this project is divided into two main tasks which are the creation of an ontology representing the Islamic knowledge and the development of a system which can analyze this knowledge. The ultimate goal is to assist the muftis and facilitate their job.

1 Introduction

Any religion in the word has a number of guides and directives to be followed by its believers, and Islam has no exception. The Islamic laws are derived from the legislative sources which are: Quran (the holy book), Hadith (the words and acts of the Prophet Mohamed peace be upon him), Ijtihad (consensus of the companion) and Quias (the analogical deduction).

Quran is the main religious source of the Islamic rulings or jurisprudence. One of its miracles is its unique style written in classic Arabic. Linguistically and in terms of perfection; it is considered the best Arabic scripture with its expressions and words, in that great meanings were expressed in two or three words and sometimes the same meaning was rehearsed in different and various ways. This holy book contains a huge amount of knowledge in various subjects like: the pillars of Islam, faith, general and political relations, science and art, holy Quran, organising of financial relations, human and social relations, Al-Jihad, religions, judicial relations, working, stories and history, human and ethical relations, trade, agriculture and industry and call for Allah (Daawah).

Lately, Islamic data knew a huge explosion with the growth of Information technology. One of the most used sources of information to search the

© IFIP International Federation for Information Processing 2018
Published by Springer International Publishing AG 2018. All Rights Reserved
A. Amine et al. (Eds.): CIIA 2018, IFIP AICT 522, pp. 37–46, 2018.
https://doi.org/10.1007/978-3-319-89743-1_4

Islamic knowledge is the web. However, this knowledge is sparsely spread all over the web and it is not well organized which make it hard to access, process or reuse it.

Moreover, the majority of software and websites for searching on the Islamic texts are keyword based. And the absence of the semantic support affects the precision of the results, especially when dealing with the ambiguity of the Qur'anic text and the Arabic language in general.

As most Islamic resources are represented as a plain text or image documents, achieving machine interoperability was a major problem. After the introduction of the semantic web, Ontologies were very popular in different communities such as knowledge management, natural language processing, and information retrieval. Ontologies were defined for the first time in computer science by Gruber "an explicit specification of a conceptualization" [1]. As an effect, a lot of works have been done to take advantage of the ontologies for a machine-readable representation to semantically model the Islamic domain.

This paper is laid out as follows: Sect. 2 gives background information on Arabic and Qur'anic texts difficulties. Section 3 is dedicated to outlining the different main sources of the Islamic religion. Section 4 gives a brief overview of some relevant works on Islamic texts. Section 5 briefly introduces our in-progress project. Section 6 presents our recent results. Section 7 shows some possible future works. Finally, we conclude this paper within the Sect. 8.

2   Arabic Linguistics

The Arabic script-based languages share in different degrees an explosion of homograph and word sense ambiguity. Dealing with such a problem represents a real challenge to NLP systems. Resolving ambiguity in NLP requires representation not only of linguistic and contextual knowledge but also of domain knowledge. Ambiguity in Arabic is enormous at every level: lexical, morphological and syntactic. Another serious problem is tokenization and it is extremely common in Arabic to find a token that can function as an entire sentence in English (Fig. 1).

In order to facilitate the comprehension of the Islamic sources structure, the Arabic grammar rules are used to define the meaning which is very important since no decisions can be deduced except when the content of these sources are well understood [2].

It is commonly known that Arabic is one of the most difficult languages. In fact, each language has its problems and limitations. In Arabic for instance, it can be the agglutination because as Arabic native speakers we are able to read any text automatically without any agglutination signs, but it can be more challenging for automatic processing systems or non native-speakers as shows in the Fig. 2. At the other hand, Arabic language has a very strict grammar rules which can be helpful in limiting the problems of the automatic processing of

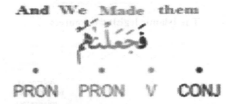

Fig. 1. Example of the Arabic language difficulties: the tokenization problem.

Fig. 2. Examples of the Arabic language difficulties: the problem of agglutination.

Arabic texts. So the problem in this case is the lake of research and works done on the language rather than the difficulty of the language itself. For exemple, Arabic is spoken by more than 300 million people in over 22 countries, but the works made regarding the automatic processing of Arabic or ontologies are almost non-existent, and a big part of these works are very limited especially compared to the evolutions of other languages. Among these works we can quote the works presented in [3–8].

3   Arabic Legislative Sources

The four legislative sources: the holy Qur'an, Sunnah of the prophet, consensus of the companion (Ijama) and the analogical deduction (Quias) (presented in the Fig. 3). Also, we have Fiqh which is the science of having the knowledge of decisions of all Islamic laws which are extracted from the four Islamic legislative sources. In the other hand, we have the foundation of jurisprudence (Usul EL-Fiqh) which it is the theoretical bases relating to the methodology which contains indications and methods used to extract Islamic judgments from the four Islamic legislative sources (Fig. 3).

4   Islamic Ontologies

4.1   Quranic Ontologies

Many applications based on Quranic text have been built to facilitate information retrieval and knowledge sharing. Some works used the Qur'an in its original

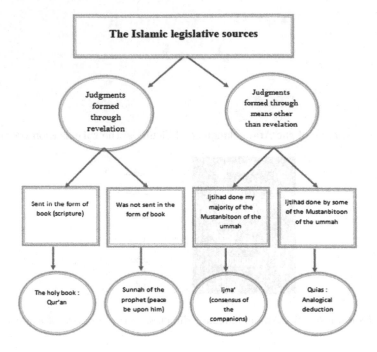

Fig. 3. The Islamic legislative sources.

standard Arabic format; others used a translated version like English, Malay...
etc. Following are some of the recent studies:

Mustapha [9] proposed a dialogue-based visualization system called AQILAH
to facilitate navigating and learning the Qur'anic text. The source of knowledgebase used was an English version of the Qur'anic text. This prototype was able
to answer user's query by listing the related verses.

Fouzi et al. [10] based their work on statistic and linguistic methods. They
applied a linguistic pattern-based approach to extract the concepts and association rules to extract the conceptual relationships from the Qur'anic text related
to the historical stories of the prophets.

Al-Yahya et al. [11] worked on designing and implementing an ontological
model capable of representing the Arabic language lexicons. The application was
applied on "Time" vocabulary in the Holy Qur'an. This ontology only contains
18 concepts where the temporal sequencing of time cannot be captured.

Aliyu et al. in [12,13] proposed a framework capable of responding to complex
natural language queries related to historical concepts mentioned in the Qur'an.
They based their work on a knowledgebase containing an annotated ontology in
RDF, where user's queries are reformulated to match the knowledgebase representation for concept retrieving.

Abbas [14] proposed a bilingual search website for the abstract and concrete concepts in the Qur'anic text. This website is based on a syntactic search using keywords in the user's query to retrieve the answer according to their occurrence in verses. The limit of such systems is the use of "keyword-based algorithms", and thus they do not provide any semantic or contextual information. To enhance the results she used eight English translations of the Qur'an and extended the search by using lemmas rather than the exact words of the queries.

Lamraoui [15] proposed a research model incorporating Dukes' ontology in various levels of the information search system. He integrated the ontology to represent the Qur'an and to reformulate the user's queries.

Zaidi in [16] worked on the implementation of a new process for building ontologies from Arabic texts, and its application on the Qur'an. She proposed a hybrid method for extracting simple and complex terms, as well as the semantic relations. She used the "The Quranic Arabic Corpus" proposed by Dukes and the "Al-Sulaiti corpus" proposed by Eric Atwell.

Sharaf et al. [17] designed a new knowledge representation for the Qur'anic text. They were able to propose a FrameNet frames for the Arabic verbs mentioned in the Qur'an. To manage the ambiguity of some cases, several English translations of Qur'an as well as books of Tafsir were used.

Among the prominent works on Qur'an, we find the works of Dukes, who created a morphological and syntactic annotation for the Qur'an [18]. This project was conducted at Leeds University and their website is world-wide collaboration. Also, he proposed a Qur'anic ontology in [19] that uses knowledge representation to define the key concepts in the Qur'an, and shows the relationships between these concepts using predicate logic.

Saad and Al [20–22], which uses an approach based on a combination of NLP techniques, information extraction and text meaning technologies, to create an ontology for Islamic concepts in the Qur'an. But the work presented in [20]; covers only 63 verses related to the obligatory prayers.

Hakkoum and Raghay [23], their ontology covers the following subjects: Qur'anic chapters and verses, each word of the Qur'an and its root and lemma to facilitate the keyword search, it does not cover words morphology search but they stated that they will add links to QVOC ontology later on.

4.2 Hadith Ontologies

Abdelhamid et al. [24] developed a tool that helps in compiling all the authentic Hadith from the Malay translation of the six books containing the authentic collection of Hadith text (Bukhari, Muslim, Abu Dawud, Tirmidhi, Nasai, Ibn Madja). The final tool was a well-structured relational database with a user interface.

Yehya et al. [25] proposed a decision support system to judge Hadith Isnad using Ontologies. Their work is based on the methodology used by Hadith scholars.

Mohammed [26] proposed an Ontology-based approach to enhance the process of information retrieval from Al-Shamelah digital library. This work presents a method to support semantic search with complex queries by proposing a new ontology to model concepts from Al-Shamela digital library (ADL). For the evaluation process, they compared the results obtained from their system to the results obtained by the ADL. This system was applied to Hadiths covering the Prophetic medicine domain presented in the ADL.

4.3   Islamic Legal Rulings Ontologies

The Islamic legal ruling represents the divine law revealed in the Quran and the Sunnah and developed by the consensus of companions (Ijma) and the analogical deduction (Quias). Despite the importance of the Islamic legal rulings, they are not semantically represented. As to our knowledge, no work has been done in this area.

5   Our Project

Our project is to develop a new system which aims to extract Islamic judgments with the related evidence texts from Islamic legislative sources. The users can ask any question that requires a deep reasoning using complex natural language. The final application could be used by Muslims, non-Muslims and by the decision-makers in the field of El-Fatwa too.

This work is based on Ontologies representing the Islamic knowledge scattered on the four legislative sources. Nevertheless, the knowledge modeling techniques from an Arabic corpus and the technical analysis of knowledge contained in ontologies are sparsely studied, which requires a deep epistemological research. Even more, the application of such studies on Qur'anic texts is very limited. A new work in this area certainly brings benefits to the Islamic world and the Arabic world in general, as well as a huge support for the progress of the modern science in all its fields. The purpose of this project is to present an interdisciplinary approach, which allows us to correctly read, understand and interpret the Islamic legislative sources.

The work of our project is divided into two major axes:

- The first axe consists of the construction of the ontological model representing the four legislative sources of Islam. Diverse problems are put at this stage, but since many works (some presented in this paper) are done in this domain, they give us the methodological frames for this process. During this phase we create a knowledge base combining different available Qur'anic ontologies like the Ontology developed at Leeds University. Whereas for the rest of the legislative texts we aim to create our own ontological representation.

- The second axe consists of the analysis phase of the ontology built to supply an answer to a given question. Indeed, the absence of a system which can analyze on ontology and supply a result leads to design a complementary tool for Protégé 3.2 to reach the aimed goal.

The final result of this project is a dialogue-based system where users can enquire the system in Arabic. Where needed, the system can ask the user for more details about its query and at the end of the dialogue, the system will generate an answer containing the Islamic judgment with the related verses and evidence texts.

6   Results

At is stage we aim collecting different existing ontologies to better describe the Islamic knowledge. Until now, we have collected different ontologies and we quote: The Qur'anic annotation of Dukes [19], The Solat Ontology of Saad et al. [20], the domain Ontology of Hakkoum [23] and our Hadith Ontology.

6.1   Our Ontological Model

The automatic processing of texts in Arabic is not very fruitful, due to the complexity of the Arabic language and the lack of tools that allow proper treatment of the language. This has led us to the manual construction of our ontology. By analyzing the texts of the corpus of the Hadith and with the help of the most relevant terms extracted with RapidMiner, we were able to conceive the dictionary of concepts. Then we extracted the different relationships between these concepts, which allowed us to conceive the conceptual model of our ontology. And finally, we have implemented this ontology with the ontology editor Protégé (Fig. 4), which we evaluated with some SPARQL queries (Fig. 5).

For our ontology we used Volume 1 of "Sahih Al-Bukhari" which contains the following books: book of Revelation, The belief, Ablutions, Menstrual periods, Prayers, The times of prayer, etc. We implemented our ontology with Protégé.

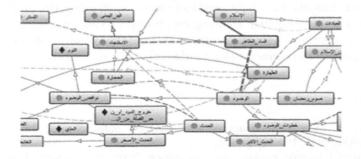

Fig. 4. A sample of the hadith ontology graph in Protégé.

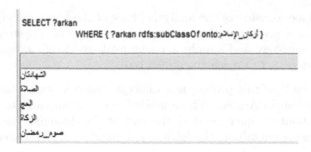

Fig. 5. Result of Sparql query for concepts subsumed by أركان الإســــلام.

7   Future Work

- In Quran or Islamic sources in general, we can find different words or phrases used to refer to the same thing. For this reason, it is preferable to use a large collection of synonyms to help understanding the meaning and avoiding the ambiguity.
- Creating Ontology for Tafsir to facilitate the comprehension of the Islamic knowledge.
- Creating a domain ontologies for the Islamic knowledge
- Using Protégé with other programming languages and tools to enhance the results.

8   Conclusion

In this paper, we presented some works previously made on the Islamic texts. The extensive efforts made in this field were toward ontology building process. Even though the majority of these works were done on Qur'anic text only and they remain very limited, they either focus on certain Surah or verses related to a given topic with some simple queries or struggle dealing with the ambiguity of natural language.

To summarize, there is a huge need to create domain ontologies for the Islamic legislative sources rather than focusing our efforts and time on sample ontologies. Moreover, presenting a semantic representation of the Islamic rulings is a critical need to guaranty the correct understanding the Islamic religion.

References

1. Gruber, T.: A translation approach to portable ontology specifications. Know. Acquisition **5**(2), 199–220 (1993)
2. Institute of Islamic banking and insurance. http://www.islamic-banking.com/islamic-jurisprudence.aspx. Accessed 06 May 2016
3. Nizar, H.: Arabic Morphological Representations for Machine Translation, Center for Computational Learning Systems Columbia University

4. Al-Agha, I.: Using linguistic Analysis to translate Arabic natural language queries to SPARQ, Faculty of Information Technology, The Islamic University of Gaza, Gaza Strip, Palestine (2015)

5. Jarrar, M.: Arabic ontology engineering-challenges and opportunities. Birzeit university, Palestine (2011)

6. Jarrar, M., Meersman, R.: Ontology Engineering -The DOGMA Approach. Vrije Universiteit Brussel

7. Boulaknadel, S.: Traitement Automatique des Langues et Recherche d'Infor-mation en langue arabe dans un domaine de spécialité: Apport des connaissances mor-phologiques et syntaxiques pour l'indexation (2010)

8. Zaidi, S., Laskri, M., Abdelali, A.: Arabic collocations extraction using Gate. In: Proceedings of the 2010 International Conference on Machine and Web Intelligence (ICMWI 2010) (2010)

9. Mustapha, A.: Dialogue-based visualization for Qu'ranic text. Eur. J. Sci. Res. **37**, 36–40 (2009)

10. Fouzi, H., Abdullah, A.N., Abdullah, A.M., Rayan, A., Abdulmalik, S.: Using Association Rules for Ontology Extraction From a Qu'ran Corpus (2014)

11. Al-Yahya, M., Al-Khalifa, H., Bahanshal, A., Al-Odah, I., Al-Helwa, N.: An onto-logical model for representing semantic lexicons: an application on time nouns in the Holy Qur'an. Arab. J. Sci. Eng. **35**, 21 (2010)

12. Aliyu, R.Y., Rabiah, A.K., Azreen, A., Masrah, A.M.: Qu'ranic verse extraction base on concepts using OWL-DL ontology. Res. J. Appl. Sci. Eng. Technol. **6**, 4492–4498 (2013)

13. Aliyu, R.Y., Rabiah, A.K., Azreen, A., Masrah, A.M.: Semantic web application for historical concepts search in Al-Quran. Int. J. Islamic Appl. Comput. Sci. Technol. **2**(2), 1–7 (2014)

14. Abbes, N.H.: Qu'ran Search for a concept tool and website. University of Leeds School of Computing, School of Computing (2009)

15. Lamraoui, Y.: Recherche intelligente des informations dans le coran. Badji Mokhtar university Algeria (2011)

16. Zaidi, S.: Une plateforme pour la construction d'ontologie en arabe: Extraction des termes et des relations à partir de textes. (Application sur le Saint Coran), Badji Mokhtar University Algeria (2013)

17. Sharaf, A.B., Atwell, E.: A corpus-based computational model for knowledge rep-resentation of the Qur'an. University of Leeds (2009)

18. Dukes, K., Atwell, E., Sharaf, A.B: Syntactic annotation guidelines for the Qu'ranic Arabic dependency treebank. In: Proceedings of the Language Resources and Eval-uation Conference (LREC) (2010)

19. Dukes, K.: Ontology of Qu'ran Concepts (2012). http://corpus.Quran.com/ontology.jsp. Accessed 04 May 2016

20. Saad, S., Salim, N., Zainal, H., Muda, Z.: A process for building domain ontology: an experience in developing Solat ontology. In: Proceedings of the International Conference on Electrical Engineering and Informatics, Bandung, Indonesia (2011)

21. Saad, S., Salim, N.: Methodology of Ontology Extraction for Islamic Knowledge Text (2008)

22. Saad, S., Salim, N., Zainal, H.: Islamic knowledge ontology creation. In: Proceed-ings of the Internet Technology and Secured Transactions (2009)

23. Hakkoum, A., Raghay, S.: Ontological approach for semantic modeling and query-ing the Qur'an. In: Proceedings of the International Conference on Islamic Appli-cations in Computer Science And Technology (2015)

24. Abdelhamid, Y., Mahmoud, M., El-Sakka, T.M.: Towards enhancing the compilation of Al-Hathdith text in Malay. Quran and Hadith Research Group, University of Malaya
25. Yehya, M.D.: An Ontology-Based Approach to Support the Process of Judging Hadith Isnad, Master thesis, Islamic University of Gaza (2013)
26. Ghazi, A.M: An Ontology Based Approach to Enhance Information Retrieval from Al-Shamelah Digital Library, Master thesis, Islamic University of Gaza (2015)

A Parallel Implementation of GHB Tree

Zineddine Kouahla[1]([✉]) and Adeel Anjum[2]([✉])

[1] LABSTIC, University of 08 Mai 1945 Guelma, Guelma, Algeria
kouahlazinedine@yahoo.fr
[2] Department of Computer Science,
COMSATS Institute of Information Technology, Islamabad, Pakistan
adeel.anjum@comsats.edu.pk

Abstract. Searching in a dataset remains a fundamental problem for many applications. The general purpose of many similarity measures is to focus the search on as few elements as possible to find the answer. The current indexing techniques divides the target dataset into subsets. However, in large amounts of data, the volume of these regions explodes, which will affect search algorithms. The research tends to degenerate into a complete analysis of the data set. In this paper, we proposed a new indexing technique called GHB-tree. The first idea, is to limit the volume of the space. The goal is to eliminate some objects without the need to compute their relative distances to a query object. Peer-to-peer networks (P2P) are superimposed networks that connect independent computers (also known as nodes or peers). GHB-tree has been optimized for secondary memory in peer-to-peer networks. We proposed a parallel search algorithm on a set of real machine. We also discussed the effectiveness of construction and search algorithms, as well as the quality of the index.

1 Introduction

Efficient indexing is an increasingly important area in computer science. Indexing techniques have been improved to deal with searches on large collections of data. However, it has been found that the indexing processes become more difficult. It is difficult to compare these techniques [1–3], their effectiveness depend on different factors (type of data, quality of the computing machine, etc.). Formally, a metric space is defined for a family of elements that are comparable through a given distance. The distance function measures the dissimilarity between two elements from a given database, in such a way that smaller distances correspond to more similar elements. Let $\mathcal{O}$ be a set of elements. Let $d : \mathcal{O} \times \mathcal{O} \to \mathbb{R}^+$ be a distance function, which verifies: (i) non-negativity: $\forall(x,y) \in \mathcal{O}^2, d(x,y) \geq 0$, (ii) reflexivity: $\forall x \in \mathcal{O}, d(x,x) = 0$, (iii) symmetry: $\forall(x,y) \in \mathcal{O}^2, d(x,y) = d(y,x)$, and (iv) triangle inequality: $\forall(x,y,z) \in \mathcal{O}^3, d(x,y) + d(y,z) \leq d(x,z)$. The concept of metric space is rather simple and leads to a limited number of possibilities for querying an actual database of such elements. These are called similarity queries and several variants exist. We consider k nearest neighbor

© IFIP International Federation for Information Processing 2018
Published by Springer International Publishing AG 2018. All Rights Reserved
A. Amine et al. (Eds.): CIIA 2018, IFIP AICT 522, pp. 47–55, 2018.
https://doi.org/10.1007/978-3-319-89743-1_5

(kNN) searches, i.e., searching for the k closest objects with respect to a query object. There are two main types of similarity queries: the range and the k-nearest neighbor queries.

Let $q \in \mathcal{O}$ be a query point and $k \in \mathbb{N}$ be the expected number of answers.

Then $(\mathcal{O}, d, q, k)$ defines a kNN query, the value of which is $S \subseteq \mathcal{O}$ such that $|S| = k$ (unless $|\mathcal{O}| < k$) and $\forall (s, o) \in S \times \mathcal{O}, d(q, s) \leq d(q, o)$.

The main factor that influence the efficiency of search algorithms, when the dimension increases is called the "dimensionality-curse problem". The current methods have proven to be unreliable, it becomes hard to store, manage, and analyze this amount of data. This problem is caused by inherent deficiencies of space partitioning, and also, the overlap factor between regions that will influence subsequent performance search algorithms. So the problem is still open. An efficient structure is based on a better grouping of similar objects in compact clusters. In a previous work [4], we led our researches on indexing via tree structure. It is based on the successive division of the space with the spheres. It is a technique that leads to simpler data structures, and therefore simple algorithms.

Moreover, on large scale, the regions of balls become very large, which could degenerate the index. This subsequently reflects on the search algorithm. Distributed (P2P) systems, which are framed by similarly advantaged hubs interfacing with each other in a self-sorting out way, have been a standout amongst the most critical models for information sharing. The main difficulty that one faces when searching is a generalized version of the so-called "multidimensional curse problem". When distances tend to be close to each other, the objects become almost indistinguishable, they cannot be grouped into clearly separated clusters and, as reported by several authors, searches tend to degenerate into full scans of the whole data set. These remarks open two possible directions (that can possibly be combined as we discuss below): improve sequential scans; provide parallel algorithms.

Our proposed system is based on the use of the most efficient indexing structure in the peer-to-peer (P2P) network, which are formed by equally privileged nodes connecting to each other in a self-organizing way, and have been one of the most important architectures for data sharing. While P2P networks are well known for their efficiency, scalability and robustness.

The rest of the paper is organized as follows. Section 2 of the paper provides an overview of Bag of features and adaboost algorithm. Section 3 deals with the proposed algorithm and Sect. 4 deals with the experimental analysis and Sect. 5 concludes the paper.

2   Background

Based on these two partitioning techniques, the first class does not enforce a partitioning of the space. The M-tree [5] builds a balanced index, allows incremental updates. On the context of the reorganization of objects in compact clusters, Almeida [6] proposed a new structure but just for an approximate search, called Divisive-Agglomerative Hierarchical Clustering or DAHC-tree. In [7], the authors proposed an extension of Slim-Tree named Slim*-tree, that exploits the best properties from ball and the BST as a hash function to search within a bucket file. The problem has not been resolved and the reinsertion of objects remains costly on a large scale. A novel clustering based dynamic indexing and retrieval approach is proposed, termed as CD-Tree [3], updates the structure with constant insertion of data. The nodes in the CD-Tree are fitted by Gaussian Mixture Models. In our opinion, the problem is not totally solved because the update of construction phase remains slow, and becomes costly on a large scale. The second class is based on the partitioning of the space. There have been a number of longitudinal studies [8,9]. Two sub-approaches are included: the first uses ball partitioning, like VP-tree [10–12]. In this method, the choice of the pivots plays a very important role on the index structure, that is why, Yianilos proposes the VP-tree [10], it is based on finding the median element of a set of objects. The mVP-tree is a generalization of the VP-tree, the nodes are divided into quantiles. This principle of partitioning eliminates the problem of overlapping between shapes. However, in this type of approach, a problem arises in cases where a demand point is close to the border between two regions; it is necessary to visit all the neighboring regions which makes the index less efficient. Combine two trees to improve the search time, an idea that has been proposed by Curtin [13], it uses the kd-tree and ball-tree to take advantage of both information. Several difficulties were cited by the authors. The main problem is that the efficiency decreases if the dimension is greater than 10. Other techniques [14] have been proposed in the last two years trying to index large-scale data but does not meet the exact but approximate queries, and other try to compress the index [15]. This leaves the door open to other proposals in the future.

In our previous work [4], we led our researches on indexing via tree structure. It is based on the successive division of the space with the spheres. Moreover, with the large amount of current data, the region of balls become very large, which could degenerate the index. This subsequently reflects on the search algorithm. This problem is caused by inherent deficiencies of space partitioning, and also, the overlap factor between regions. This is one of the major problems in this type of work. The Parallelism can and should certainly be part of a solution. We believe that no technical indexing can achieve a logarithmic search time. We know that a logarithmic response time is achievable with parallel implementations. On sequential version, a general sort is $O(n.log(n))$, where n is the number of objects. On a parallel machine, a sort can be implemented in $O(log(n))$ time and $O(n)$ on the surface, namely the number of processors.

3   The Proposed GHB-tree

The partitioning of space is a technique that leads to simpler data structures - hence algorithms. Moreover, the problem of exponentially increasing volumes in large spaces argues in favor of techniques that would otherwise reduce or at least limit volumes, or even control their occupancy. We introduce a new structure called GHB-tree (*Generalised Hyper-plane Bucketed*) [16], inspired from

---

Algorithm 1. Insertion in GHB-tree

$$\text{Insert-GHB} \begin{pmatrix} o \in \mathcal{O}, \\ N \in \mathcal{N}, \\ c_{\max} \in \mathbb{N}^*, \end{pmatrix} \in \mathcal{N}$$

$$\triangleq \begin{cases} (o, \bot, \bot, \bot) & \text{if } N = \bot \\ (p_1, o, \bot, \bot) & \text{if } N = (p_1, \bot, \bot, \bot) \\ (p_1, p_2, \text{Insert}(o, d, c_{\max}, G), D) & \text{if } N = (p_1, p_2, L, R) \wedge \\ & \quad d(p_1, o) \leq d(p_2, o) \\ (p_1, p_2, L, \text{Insert}(o, d, c_{\max}, D)) & \text{if } N = (p_1, p_2, L, R) \wedge \\ & \quad d(p_1, o) > d(p_2, o) \end{cases}$$

GH-tree. The Fig. 1 illustrates the development of a tree. At each stage of the recursive process of constructing the tree, two pivots are chosen from a subset of elements $c_{\max}$, they are chosen as the two objects furthest apart from each other. First, a node $Nodes_{GHB}$ - or only $\mathcal{N}$ - consists of two elements and two children:

$$(p_1, p_2, L, R) \in E \times E \times \mathcal{N}_{GHB} \times \mathcal{N}_{GHB}. \tag{1}$$

or: p_1, p_2 are two non-confused elements, $d(p_1, p_2) > 0$, called "pivots", they thus define a hyper-plane; L and R are the subtrees associated with the elements respectively in the "left" parts. A (sub) tree can be empty, which is denoted by $\bot$.

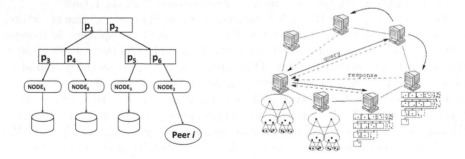

Fig. 1. Parallel version of GHB-tree

Construction of a GHB-tree. Building a GHB-tree is realised incrementally. The insertion is done in a top-down way. Algorithm 1 describes formally the incremental insertion process. When the cardinal limit is reached, a leaf is replaced by an inner node. Besides, the tree tends to be rather balanced, hence inserting a new object is a logarithmic operation, in amortised cost. This algorithm is implemented in the order of balancing network peer loads.

Algorithm 2. Search kNN in GHB-tree

$$\text{kNN-GHB} \begin{pmatrix} N \in \mathcal{N}, \\ q \in \mathbb{R}^n, \\ k \in \mathbb{N}^*, \\ d \in \mathcal{O} \times \mathcal{O} \to \mathbb{R}^+, \\ r_q \in \mathbb{R}^+ = +\infty, \\ A \in (\mathbb{R}^+ \times \mathcal{O})^\mathbb{N} = \emptyset \end{pmatrix} \in (\mathbb{R}^+ \times \mathcal{O})^\mathbb{N}$$

with :

- $A^L = \text{kNN-GHB}(L, q, k, d, r_q, k\text{-insertion}(A, ((d(p_1, q), p))));$
- $A^R = \text{kNN-GHB}(R, q, k, d, r_q, A);$
- $r_q^L = \max\{d : (d, o) \in A^G\}$ if $|A^L| = k$ else r_q;
- $r_q^R = \max\{d : (d, o) \in A^D\}$ if $|A^R| = k$ else r_q;
- $A^{LR} = \text{kNN-GHB}(O, q, k, d, r_q^G, A^G);$
- A^{RL};

$$\triangleq \begin{cases} A & \text{if } N = \bot \\ A^L & \text{if } N = (p, r, L, R) \wedge \\ & \quad d(q, p_1) - r_q^L < d(q, p_2) \wedge \\ & \quad d(q, p_1) < d(q, p_2) \\ A^R & \text{if } N = (p, r, L, R) \wedge \\ & \quad d(q, p_1) - r_q^R > d(q, p_2) \wedge \\ & \quad d(q, p_1) > d(q, p_2) \\ A^{LR} & \text{if } N = (p, r, L, R) \wedge \\ & \quad d(q, p_1) - r_q^L \leq d(q, p_2) + r_q^L \wedge \\ & \quad d(q, p_1) < d(q, p_2) \\ A^{RL} & \text{if } N = (p, r, L, R) \wedge \\ & \quad d(q, p_1) + r_q^R \geq d(q, p_2) - r_q^R \wedge \\ & \quad d(q, p_1) > d(q, p_2) \end{cases}$$

We have considered putting in place strategies to try to balance the tree, such as choosing two elements furthest apart from each other. However, we are careful not to use a function of more than linear complexity, otherwise the algorithm will exceed a complexity in $O(n. \log n)$ which is the one it has in this version.

kNN Search in GHB-tree. The Algorithm 2, which formally describes the search kNN in a GHB-tree, is also quite complex. The searches are made from balls while the space has been partitioned. The search is done by calculating the distance between the query point and the two pivots, while descending into the tree. Not counting the case of the empty tree, we can meet four cases when passing through a tree node:

- The first case is where the search result is located entirely in the left subtree. In other words, the search ball lies entirely in the left half-plane. Similarly, the second case is where the search result is fully present in the right subtree.

– The third and fourth cases are those where the search must *a priori* be continued in the two subtrees because the search ball overlaps the two half-spaces. What distinguishes the third of the fourth case is the position of the center. If the center is in the left hyperplane, then the search will continue first in the left son. Only if the search has not sufficiently reduced the radius of the search ball will the pursuit take place in the right son. The search can be modified *a posteriori* to finally get back to the first case. Obviously, the fourth case is where the search in both threads is reversed.

Note that this algorithm is the same on all the stations of the network. It is on this logical network that the query q is broadcast. In each time the indexes are browsed, the value of the query radius r_q decreases, which actually corresponds to the distance to the k^e object in the ordered list A.

The leaf nodes contain a subset of the indexed data with a maximum cardinal equal to c_{max}. At the leaf level the procedure is quite simple. In order to find the k closest neighbours of a leaf, just sort them according to their increasing distances to the q request object. Then we return *at most* the first k sorted items. Note that a real sort is not necessary; there is a variant, called ≪k-sort≫, which is only in: $O(c_{max}.\log_2 k)$. Note that c_{max} being either a constant, a logarithm of the size of the collection, or its square root, the complexity of the operation on a sheet is very fast, or even constant. The r_q query radius plays the essential role for search optimization (the minimum possible is a maximum of pruning). It is initially set to $+\infty$ by default, but we *hopefully* see it dwindle with each move on an internal node.

Note, again, that this step does not really require sorting, but only a sequence of mergers. The complexity is "constant", that is to say in:

$$O(2.k) \tag{2}$$

rather than:

$$O(2.k.\log_2 k). \tag{3}$$

4   Experiments and Comparison

In this section we provide experimental results on the performance of GHB-tree on real data sets, in order to test and compare its effectiveness. We used tow datasets. We started with the cities of France, which have a low dimensional. We turned to the complex objects, good example is multimedia descriptors, we used a subset of the MPEG-7 Dominant Color Descriptor (KDD), it can be found at http://kdd.ics.uci.edu. We run our structure with same datasets on a workstations computers with the configuration Intel(R) Xeon(R) CPUs, and 8 GB of main memory. All index files were stored on a network partition.

We arrange the size of each tree node to be equal to the size of a disk page. We compared ourselves to the MM-tree [17] its extension onion-tree, as well as slim* [7], an improved version of the M-tree [5], and IM-tree [4]. We used the library C++ "GBDI Arboretum" which implements these methods and we adapt them to be executable in a P2P environment.[1] In Fig. 2, we see that our proposal is the most effective compared to others with a difference of over 30% with the onion-tree in the three collections (average and with the two values of c_{max}), and with more than 40% compared to the Slim*-tree. The difference from MM and onion-trees is easily explainable. The reason is the absence of the respective "semi-balancing algorithm" and "keep-small" that require a number of additional operations. In slim*-tree "slim-down algorithm" also has a significant cost which was noted by its own authors [7]. Our approach, GHB-tree is simple in the insertion of new objects, (which was one of the initial objectives with respect to the complexity $O(nlogn)$ reasonable) and provides an incremental index competitive.

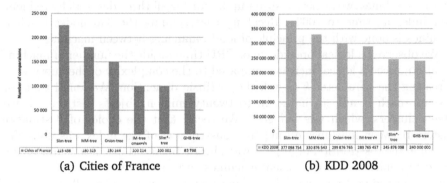

(a) Cities of France   (b) KDD 2008

Fig. 2. Performance statistics of construction algorithms in GHB-tree

We vary in different ways the kNN searches. Firstly, for building the index, we run with different values of c_{max} parameter which was chosen either as the square root or as the logarithm of the size of the collection. Next, we run kNN searches with k between 5 and 100. We note also that the difference between the perfect version which is the most effective, and sequential versions is not negligible. This allows us to say that there is a possibility of minimizing this gap using techniques to find as soon as possible nearest neighbor with a minimum of energy.

This proves that the creation of the index has been beneficial by the creation of dense, even at large scale and also in large amount of data. For the parameter k, we observe that if we increase its value of k, the performances decreases but with a gap between less than 1% and to less than 2% when k = 50. So the value of k has no major influence on the performance of the search algorithm. Figure 3 shows the elapsed time for building indexes for each of the three collections.

As shown from the figure, the elapsed time gradually decreases as the number of cores increases from 1 to 15. When using two machines, the time to build index

---

[1] GBDI Arboretum is a library C++ that implements different metric access methods (MAM) (cf. http://www.gbdi.icmc.usp.br/old/arboretum).

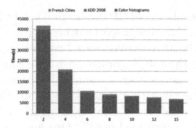

Fig. 3. Elapsed time for building indexes for each of the three collections

for the datasets, is 40,144 s, 41,847 s and 42,547 s, respectively. In comparison with 15 machines, we achieve a speed-up factor for all three datasets by reducing their indexing time to 7000 s. Recalling that sending the leaf nodes to client machines is done with the principle of load balancing between machines.

We observed a logical breakdown of CPU time beside the number of machine. We also noticed a logical increase compared to the complexity of the query while increasing the parameter k as well the intrinsic dimension. We found that this new approach is able to index up to twenty million objects distributed over fifteen clusters, which was our goal. We recall that the choice of destination clusters between machines during the construction of the index was done in a way that the distribution of objects was almost balanced on all machines. Note that communication between client machines and also the exchanges of responses plays a very important role in improving response time, so the effectiveness of our index.

5   Conclusion

In this paper, we have clarified some methods of indexing in metric spaces. Everything is put on a taxonomy of most existing indexing techniques in the literature. Afterwards, we presented a study (GHB-tree), a proposition that was inspired from GH-tree. This technique is incremental, not dependent on a defined data type, and especially easy to construct the index. GHB-tree is a peer-to-peer system supporting similarity search in metric spaces. Compared with the available state-of-the-art, our method significantly improves the query retrieval process. Extensive experimental results show that this improvement, for kNN queries, increases directly proportional with the size of the network, adding ground to our scalability claims.

References

1. Chen, L., Gao, Y., Li, X., Jensen, C.S., Chen, G.: Efficient metric indexing for similarity search and similarity joins. IEEE Trans. Knowl. Data Eng. **29**, 556–571 (2017). Print ISSN 1041-4347

2. Gonzaga, A.S., Cordeiro, R.L.F.: A new division operator to handle complex objects in very large relational datasets. In: EDBT (2017)
3. Wan, Y., Liu, X., Wu, Y.: CD-Tree: a clustering-based dynamic indexing and retrieval approach. Intell. Data Anal. **21**, 243–261 (2017)
4. Kouahla, Z., Martinez, J.: A new intersection tree for content-based image retrieval. In: 10th International Workshop on Content-Based Multimedia Indexing, CBMI 2012, Annecy, France, 27–29 June 2012, pp. 1–6 (2012). https://doi.org/10.1109/CBMI.2012.6269793
5. Ciaccia, P., Patella, M., Zezula, P.: M-tree: an efficient access method for similarity search in metric spaces. In: Proceedings of the 23rd VLDB International Conference, pp. 426–435 (1997)
6. Almeida, J., Valle, E., Torres, R.S., Leite, N.J.: DAHC-tree: an effective index for approximate search in high-dimensional metric spaces. J. Inf. Data Manag. **1**(3), 375–390 (2010)
7. Pola, I.R.V., Traina, A.J.M., Traina Jr., C., Kaster, D.S.: Improving metric access methods with bucket files. In: Amato, G., Connor, R., Falchi, F., Gennaro, C. (eds.) SISAP 2015. LNCS, vol. 9371, pp. 65–76. Springer, Cham (2015). https://doi.org/10.1007/978-3-319-25087-8_6
8. Ooi, B.C.: Spatial kd-tree: a data structure for geographic database. In: Schek, H.J., Schlageter, G. (eds.) Datenbanksysteme in Büro, Technik und Wissenschaft. Informatik-Fachberichte, vol. 136, pp. 247–258. Springer, Heidelberg (1987). https://doi.org/10.1007/978-3-642-72617-0_17
9. Burkhard, W.A., Keller, R.M.: Some approaches to best-match file searching. Commun. ACM **16**(4), 230–236 (1973)
10. Yianilos, P.N.: Data structures and algorithms for nearest neighbor search in general metric spaces. In: Proceedings of the 4th Annual in ACM-SIAM Symposium on Discrete Algorithms, pp. 311–321 (1993)
11. Nielsen, F.: Bregman vantage point trees for efficient nearest neighbor queries. In: Proceedings of Multimedia and Exp (ICME). IEEE (2009)
12. Fu, A.W., Chan, P.M., Cheung, Y.-L., Moon, Y.S.: Dynamic vp-tree indexing for n-nearest neighbor search given pair-wise distances. VLDB J. Very Large Data Bases **9**, 154–173 (2012)
13. Curtin, R.R.: Faster dual-tree traversal for nearest neighbor search. In: Amato, G., Connor, R., Falchi, F., Gennaro, C. (eds.) SISAP 2015. LNCS, vol. 9371, pp. 77–89. Springer, Cham (2015). https://doi.org/10.1007/978-3-319-25087-8_7
14. Arroyuelo, D.: A dynamic pivoting algorithm based on spatial approximation indexes. In: Traina, A.J.M., Traina, C., Cordeiro, R.L.F. (eds.) SISAP 2014. LNCS, vol. 8821, pp. 70–81. Springer, Cham (2014). https://doi.org/10.1007/978-3-319-11988-5_7
15. Pagh, R., Silvestri, F., Sivertsen, J., Skala, M.: Approximate furthest neighbor in high dimensions. In: Amato, G., Connor, R., Falchi, F., Gennaro, C. (eds.) SISAP 2015. LNCS, vol. 9371, pp. 3–14. Springer, Cham (2015). https://doi.org/10.1007/978-3-319-25087-8_1
16. Ulhmann, J.K.: Satisfying general proximity/similarity queries with metric trees. Inf. Process. Lett. **40**, 175–179 (1991)
17. Carélo, C.C.M., Pola, I.R.V., Ciferri, R.R., Traina, A.J.M., Traina Jr., C., de Aguiar Ciferri, C.D.: Slicing the metric space to provide quick indexing of complex data in the main memory. Inf. Syst. **36**, 79–98 (2011)

Leveraging Web Intelligence
for Information Cascade Detection
in Social Streams

Mohamed Cherif Nait-Hamoud[1,2]($\boxtimes$), Fedoua Didi[1], and Abdelatif Ennaji[3]

[1] Department of Science Computing,
University of Abou Bekr Belkaid, 13000 Tlemcen, Algeria
fedouadidi@yahoo.fr
[2] Department of Mathematics and Science Computing,
University of Larbi Tebessi, 12000 Tebessa, Algeria
mc_naithamoud@hotmail.com
[3] LITIS Lab, University of Rouen, Rouen, France
ennaji@univ-rouen.fr

Abstract. In this paper, we present an approach for investigating information cascades in social and collaborative networks. The proposed approach seeks to improve methods limited to the detection of paths through which merely exact content-tokens are propagated. For this sake, we adopt to leverage web intelligence to the purpose of discovering paths that convey exact content-tokens cascades, as well as paths that convey concepts or topics related to these content-tokens. Indeed, we mine sequence of actors involved in cascades of keywords and topics extracted from their posts, using simple to use restful APIs available on the web. For the evaluation of the approach, we conduct experiments based on assimilating a scientific collaborative network to a social network. Our findings reveal the detection of missed information when using merely exact content propagation. Moreover, we noted that the vocabulary of actors is preserved mostly in short cascades, where topics become a better alternative in long cascades.

Keywords: Information cascade · Information diffusion analysis
Social and collaborative networks · Web intelligence
Information flow paths

1 Introduction

Information diffusion in online social networks (OSN) is a research field that aims at addressing concerns about what governs information spread in these networks. According to [1], information diffusion is categorized into four types: *herd behavior*, *information cascades*, *diffusion of innovation* and *epidemics*. Information cascades occur in OSN when actors adopt the behavior of their followees

© IFIP International Federation for Information Processing 2018
Published by Springer International Publishing AG 2018. All Rights Reserved
A. Amine et al. (Eds.): CIIA 2018, IFIP AICT 522, pp. 56–65, 2018.
https://doi.org/10.1007/978-3-319-89743-1_6

or friends due to their actions influence. The taxonomy of information diffusion proposed in [2] reveals a set of challenges for the effective extraction and prediction of valuable information from the exchanged huge amount of data. The authors, in [2] have identified three main research axes: (1) interesting topics detection, (2) information diffusion modeling, and (3) influential spreaders identification. As an improvement for influential actors detection, the authors have suggested to take topics into account. Since that study, many works are dedicated to investigate the detection of influencers using contents of interactions taking into accounts the underlying network structure. Particularly, in [4] and later in [3] authors have studied the problem of information cascades, their work included mining paths that convey more frequent information and influence detection in the context of information flow in networks. In that work, both content of interactions and underlying network structure are considered. In addition, authors have designed an algorithm to detect information flow patterns in social networks called *InFlowMine*. Specifically, they first targeted the detection of paths (sequence of linked nodes of the social graph) through which exact content-tokens are propagated more frequently. They referred to these paths as *frequent paths*, considering the chronological order of actors posted information. Afterwards, they computed a score using the mentioned frequent paths to discover influencers in social and collaborative networks. The content generated by the different actors of the network was considered as a social stream of text content. Each element is a tuple that consists of a unit of information called *content-token* (hash tags, URLs, text in Twitter) and its originating actor. Formally, a social stream was defined as all couples (U_j, a_i) where U_j is the posted token and a_i is the originating actor.

In this paper, we propose to leverage web intelligence to extend the work of authors in [3,4] to the purpose of discovering paths that convey exact content-tokens cascades, as well as paths through which content-tokens referring concepts are propagated. Specifically, we mine sequence of actors involved in cascades of keywords and topics extracted from posts of social or collaborative networks actors, using simple to use restful APIs available on the web. Our proposal allows to reveal eventual missed paths that may not be detected when tracking merely cascades of exact contents.

The rest of this paper is organized as follows; in Sect. 2 we present necessary definitions for the clarity of the paper and the problem formulation; in Sect. 3 we introduce the details of our proposal; in Sect. 4 we present experiments and we discuss the obtained results. Finally, we conclude in Sect. 5 with some comments.

2   Problem Formulation

To make this paper self-contained and for the sake of clarity, we introduce below the basic concepts used in [3,4] to mine frequent paths and influencers in social networks. Some definitions were adapted to the purpose of the problem reformulation.

Definition 1 *(Valid flow path)* [3,4]. *Let $G(N,E)$ be a social or collaborative network where N is a set of nodes and E a set of edges, a valid flow path is an ordered sequence of distinct nodes $n_1 n_2 \ldots n_k, n_i \in \mathbb{N}$, such that for each t ranging from 1 to $k-1$ an edge exists between nodes n_t and n_{t+1}.*

Definition 2 *(Information flow frequency)* [3,4]. *The information flow frequency of actors $n_1 n_2 \ldots n_k$ is the number f of content-tokens $U_1 \ldots U_f$ given that for each U_i the following conditions hold:*

– *Each actor from the sequence $n_1 n_2 \ldots n_k$ has posted U_i*
– *Each U_i was posted by the actors in the order $n_1 n_2 \ldots n_k$.*

Definition 3 *(Frequent path)* [3,4]. *A sequence of actors $n_1 n_2 \ldots n_k$ is defined as a frequent path of a frequency f, if the following two conditions hold:*

– *The sequence $n_1 n_2 \ldots n_k$ is a valid flow path.*
– *The frequency of the actor sequence $n_1 n_2 \ldots n_k$ is at least f.*

To the sake of reformulating mining frequent paths problem considered in this paper, we reformulate Definition 2 as follows:

Definition 4 *(Information flow frequency - reformulated). Information flow frequency of actors $n_1 n_2 \ldots n_k$ is the number f of content-tokens $U'_1, U'_2, \ldots, U'_f$ representing the keywords and concepts extracted from actors posts using simple to use restful APIs available on the web, given that for each U'_i the following conditions hold:*

– *Each actor from the sequence $n_1 n_2 \ldots n_k$ has posted a token U'_i*
– *Each U'_i was posted by the actors in the order of the sequence $n_1 n_2 \ldots n_k$*

Problem 1 (Information flow paths mining - Extended). Given a graph G, a stream of content propagation and a frequency f, the problem of information flow path mining is to find the frequent paths in the underlying graph G using keywords and concepts extracted from actor posts instead of using content-tokens. Keywords and concepts are obtained using available restful APIs.

3   Proposed Approach

The content-tokens considered in [3,4] may represent the results of the tokenization of each actor post (i.e.; text message) after the removal of stop words. To focus only on important words or composed words of the post, we propose to consider the keywords extracted from the post as content-tokens when mining frequent paths of exact content cascades. Indeed, the gain is substantial in the extent that only the important part of posts is considered; this leads to the extraction of more accurate frequent paths.

As an attempt to capture semantic content of posts, authors in [3,4] proposed to use a vocabulary for all content-tokens of a given topic or content-specific flow mining; this way, the occurred content-tokens in posts are treated as regular tokens (i.e.; U_r). In order to seek for all topics and not only specific ones while keeping track of the propagation of important exact content, we propose to leverage web intelligence to extract concepts and keywords from actors posts that will replace the original content-tokens of each actor's post.

Afterwards, we propose to use the *InFlowMine* algorithm [3,4] to mine frequent information flow paths. For this sake, as in [3,4], we use a hash table to track the sequence of actors for each extracted concept C_i and keyword U'_i from the post containing the original set of content-tokens $\{U_i\}$. Each slot of the hash table $h(C_i)$ corresponds to an ordered list of actors ordered chronologically on the basis of posting the concept C_i. Similarly, each slot of the hash table $h(U'_i)$ corresponds to an ordered list of actors in chronological order of posting the keyword U'_i. Figure 1 depicts the proposed approach.

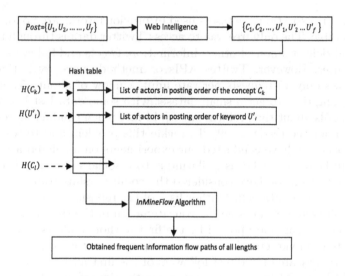

Fig. 1. Diagram chart of the proposed approach

The web intelligence phase mentioned in Fig. 1 and depicted in Fig. 2 is crucial; it consists of using IBM Watson Natural Language Understanding service (NLU) [5] that offers text analytics through a simple to use restful API framework. The IBM Watson NLU allows developers to leverage natural language processing techniques such as: analyzing plain text, URL or HTML and extracting meta-data from unstructured data contents for instance concepts, entities and keywords. Note that, it is not obvious extracting concepts and keywords from one single content-token U_i or a low content string, unless the content-token is itself a keyword. In this case, a solution is inspired by the works in [6,7], a context may be created using Google search restful API [8]. This way, the titles and the snippets of the k-top results of search are used to build an enhanced text.

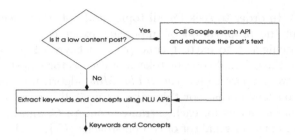

Fig. 2. Web intelligence phase

4   Experiments and Results

Due to Twitter restrictions, datasets used in previous works are no longer available nor sharable. The only free way to access Twitter data is through their restful APIs that deliver amongst others information; tweets and followers or friends of a given user. However, Twitter APIs or methods have restrictions, every method allows only 15 requests per rate limit window of about 15 min. Hence, it will take long time to get a small dataset with some data lost between calls. For more details about the problem one could face when conducting research on Twitter we refer the reader to [9]. To tackle this problem and to test our proposed approach, we have conducted our experiments on a collaborative dataset namely Core Dataset [10,11] assimilating it to a social network. Effectively, as an underlying graph, we have considered the co-authorship graph that we have built using the meta-data field "author list" of the repository records. Moreover, we have considered as a message the concatenation of the title and the abstract of each paper; messages are posted by the first author to all its co-authors that are assimilated to followers. Note that co-authors are not considered as followers of each other, they are only direct followers of the first author of a given paper.

It should be noted that, an important motivation of our choice to conduct experiments on a collaborative network, is the fact that in this kind of networks the co-authors share more likely the same topics.

Core dataset is a collection of open access research outputs from repositories of journals worldwide. It allows free and unrestricted access to research papers. The Core Dataset offers two datasets, (1) a meta-data file of 23.9 million items and a content dataset of 4 million items. The former contains meta-data on scientific papers in JSON format structured in 645 repositories, and the latter contains full articles. We carried out our experiments using *repository 2* of the last dump of the Core dataset (i.e., dump of October 2016). Figure 3 depicts an example of a meta-data dataset item.   As a first step, we imported the repository 2 file as a collection to the NoSQL database *MongoDB* [12]. Secondly, we sorted the information about the papers in ascending order by publication date ("dc:date" as shown in Fig. 3) to conserve the chronological order of posting. Afterwards, we used the IBM Watson NLU APIs to generate keywords and concepts; then, we built the hash table as a python dictionary. Finally, we used

{"Identifier":30474512,
 "ep:Repository":2,
 "dc:type":0,
 "bibo:shortTitle":"Learning from triads: training undergraduates in counselling skills",
 "bibo:abstract":"Background:\\ud\nResearch has shown that counselling skills training....
 ....tutors must act proactively to ensure a safe learning environment",
 "bibo:AuthorList":("Smith, Kate"),
 "dc:date":"2015-05-06",
 "doi":"10.1002/capr.12056",
 "bibo:cites":0,
 "bibo:citedBy":0,"
 similarities":0}

Fig. 3. Example of a Core dataset meta-data item

the *InFlowMine* algorithm to extract frequent paths from the hash table built earlier.

4.1 Results and Discussion

As first experimental tests of our approach, we have set frequency f to 10 (ten content-tokens were diffused through all the detected paths) and obtained all paths of length 1, 2, 3 and 4 as maximal possible length extracted from the repository 2 of the Core dataset. Figure 4 depicts some of the detected length-3 paths. A line of the results below represents a path of length-3, with each author separated by a semi-colon.

> Havard, Catriona;Memon, Amina;Gabbert, Fiona;
> Williams, P. K. G.;Stark, Craig R.;Helling, Ch.;
> George, Keith P.;Grant, Marie Clare;Baker, Julien S.;
> Ivanova, Iva;Pickering, Martin J.;McLean, Janet F.;
> Tummala, Hemanth;Khalil, Hilal S.;Mitev, Vanio;
> Clifford, Brian R.;Havard, Catriona;Memon, Amina;
> Clifford, Brian R.;Memon, Amina;Gabbert, Fiona;

Fig. 4. Example of length-3 paths extracted from the repository 2 of the Core dataset

The path "Havard, Catriona; Memon, Amina; Gabbert, Fiona;" shown in Fig. 4 means that the author "Havard, Catriona" was the first author of the paper that she co-authored with "Memon, Amina". In her turn, "Memon, Amina" was the first author of another paper that she co-authored with "Gabbert, Fiona". The order of the positions of the authors in the path indicates the chronological order of posting (i.e., who first has emitted this content-token) of the propagated content-token. This result means that ten different content-tokens were cascaded trough this path with respect to the order of appearance of authors in the path.

Unlike the work proposed in [3, 4] that uses a vocabulary to capture semantic aspects of posts, our proposal permits, in addition, the extraction of vocabularies. The concepts and the keywords collected in the hash table could be used to build a vocabulary of the whole analyzed texts. The resulted vocabularies allow, among others, the detection of topical similarities between different posts of social networks or research papers of collaborative networks.

To assess the improvement gained by our approach, we started by considering, as a first experiment, the extracted keywords from the text with frequency set to 10. As a second experiment, we considered concepts keeping the same frequency as in the first experiment. Table 1 below shows early results in terms of number of mined frequent paths.

Table 1. Early results of our proposed approach

Path length	# mined frequent paths - frequency = 10	
	Using keywords	Using concepts
2	438	634
3	20	23
4	1	2

Basically, it fall in common sense that actors in a social or a collaborative network may use their own vocabulary while keeping cascading the same topic. We expected that the length and the number of mined frequent paths may increase if concepts or topics are used for mining instead of considering keywords. However, we noted that with low frequencies the numbers of mined frequent paths of short length are better if keywords are used as shown in Table 2. We explain these findings by the fact that the vocabulary (keywords) of actors tend to be conserved in their neighborhood. Moreover, with high frequencies, concepts give better results in terms of path numbers than keywords. This is due to the fact that the extracted keywords represent only a subset of representative terms of a given concept.

Table 2. Results with low frequencies

Path length	# mined frequent paths - frequency = 2	
	Using keywords	Using concepts
2	2756	2113
3	112	93
4	7	10

Table 3 shows a comparison of the results obtained in case of maximal length frequent paths. These results reveal that concepts performs well than keywords in case of long cascades. Hence, in the case of longer cascades the vocabulary vanishes and only concepts (topics) persist. In spite of the aforementioned cases, considering topics in frequent paths detection reveals valuable information, that could be missed when using merely the exact content.

Table 3. Mined paths of maximal detected length with different frequencies

Frequency	# mined frequent paths of maximal length	
	Using keywords	Using concepts
1	9	10
2	7	10
3	6	8
4	4	7
5	4	6
6	4	6
7	3	5
8	2	2
9	1	2
10	1	2

Figure 5 depicts all the extracted frequent paths of maximal length and their respective propagated concepts with a frequency set to 3. We have noted when analyzing the abstracts and the short titles that in the case of the 8th path, the propagated concepts are somewhat generic which is due to the employed technique of concepts extraction.

```
1.  Tinlin, Rowan M.;Watkins, Christopher D.;DeBruine, Lisa M.;Jones, Benedict C.;
2.  Quist, Michelle C.;Watkins, Christopher D.;DeBruine, Lisa M.;Jones, Benedict C.;
3.  Clifford, Brian R.;Havard, Catriona;Memon, Amina;Gabbert, Fiona;
4.  Davies, J. W.;Butler, D.;Jefferies, Christopher;Duffy, A.;
5.  Simpson, Edward;Gilmour, Daniel J.;Blackwood, David J.;Isaacs, John P.;
6.  Phillips, P. J.;Jamison, S. P.;Berden, G.;van der Meer, A. F. G.;
7.  Phillips, P. J.;Jamison, S. P.;Berden, G.;MacLeod, Allan M.;
8.  Scott-Brown, Kenneth C.;Gilmour, Daniel J.;Blackwood, David J.;Isaacs, John P.;
```

Respective propagated Concepts:

```
1.  Anorexia nervosa; Nutrition; Obesity;
2.  Characteristic; Histrionic personality disorder; Evidence;
3.  Immune system; Pharmacist; Mycelium;
4.  Evidence; Critical thinking; Major;
5.  Human resources; Management; Project;
6.  Holography; Fundamental physics concepts; Optics;
7.  Optics; Systems of measurement; Measurement;
8.  Management; Higher education; Learning;
```

Fig. 5. Extracted paths and their respective propagated concepts- frequency set to 3

In this specific case there are no propagated keywords; what explains that the path was not detected when using keywords as shown in Fig. 6.

```
1. Isaacs, John P.;Blackwood, David J.;Gilmour, Daniel J.;Falconer, Ruth E.;
2. Tinlin, Rowan M.;Watkins, Christopher D.;DeBruine, Lisa M.;Jones, Benedict C.;
3. Quist, Michelle C.;Watkins, Christopher D.;DeBruine, Lisa M.;Jones, Benedict C.;
4. Simpson, Edward;Gilmour, Daniel J.;Blackwood, David J.;Isaacs, John P.;
5. Stojanovic, V.;Blackwood, David J.;Gilmour, Daniel J.;Falconer, Ruth E.;
6. Clifford, Brian R.;Havard, Catriona;Memon, Amina;Gabbert, Fiona;
```

Respective propagated keywords

```
1. rural development project; rural planning projects; inclusive decision making;
2. Positive correlations; facial characteristics; facial attractiveness propose;
3. dominance questionnaire; average facial characteristics; facial masculinity;
4. post project review; formal learning context; large redevelopment project;
5. GPU shader programs; rendering methods; Short paper version;
6. TP line-up; video identification parade; TP line-ups;
```

Fig. 6. Extracted paths and their respective propagated keywords- frequency set to 3

5   Conclusion

In this paper we have presented an approach to extend previous works on information flow frequent paths detection. The main objective is to capture in addition semantics of propagated information in a given social stream, considering in that both underlying graph structure and the content of interactions. Our experiments were conducted on a collaborative dataset that we have assimilated to a social network; the results showed an improvement that might reveal useful information missed when considering only cascades of exact content. As findings, we have noted that the vocabulary (keywords) of actors tend to be reused in their neighborhood, but vanishes in the case of long cascades and considering concepts remain the best alternative.

References

1. Zafarani, R., Abbasi, M., Liu, H.: Social Media Mining. Cambridge University Press, Cambridge (2014)
2. Guilles, A., Hacid, H., Fabre, C., Zighed, D.A.: Information diffusion in online social networks: a survey. ACM SIGMOD **42**(2), 17–28 (2013)
3. Subbian, K., Aggarwal, C., Srivastava, J.: Content-centric flow mining for influence analysis in social streams. In: Proceedings of the 22nd ACM International Conference on Information & Knowledge Management, pp. 841–846 (2013)
4. Subbian, K., Aggarwal, C., Srivastava, J.: Mining influencers using information flow in social streams. ACM Trans. Knowl. Discov. Data **10**(3), Article 26 (2016)
5. IBM NLU API. https://www.ibm.com/watson/developercloud/natural-language-understanding/api/v1
6. Singhal, A.: Leveraging Open Source Web Resources to Improve Retrieval of Low Text Content Items, Ph.D. thesis in Department of Computer science, university of Minnesota, Minneapolis, MN, 144 (2014)
7. Singhal, A., Kasturi, R., Sivakumar, V., et al.: Leveraging web intelligence for finding interesting research datasets. In: IEEE/WIC/ACM Proceedings of the International Joint Conferences on Web Intelligence (WI) and Intelligent Agent Technologies (IAT), vol. 1, pp. 321–328. IEEE (2013)
8. Google search API. https://developers.google.com/custom-search/json-api/v1/overview

9. Kelly, P.G., Sleeper, M., Cranshaw, J.: Conducting research on twitter: a call for guidelines and metrics. In: CSCW Measuring Networked Social Privacy Workshop (2013)
10. Knoth, P., Zdrahal, Z.: CORE: three access levels to underpin open access. D-Lib Mag. **18**(11/12) (2012). Corporation for National Research Initiatives
11. Knoth, P.: From Open Access Metadata to Open Access Content: Two Principles for Increased Visibility of Open Access Content. Open Repositories 2013, Charlottetown, Prince Edward Island, Canada (2013)
12. Chodorow, K.: MongoDB The Definitive Guide. O'Reilly Media Inc., Sebastopol (2013)

Understanding User's Intention in Semantic Based Image Retrieval: Combining Positive and Negative Examples

Meriem Korichi[1](✉), Mohamed Lamine Kherfi[2], Mohamed Batouche[1], Zineb Kaoudja[2], and Hadjer Bencheikh[2]

[1] Computer Science Department, University of Constantine 2 - Abdelhamid Mehri, 25 000 Constantine, Algeria
meriemkorichi@gmail.com, mcbatouche@gmail.com
[2] Department of Computer Science and Information Technology, University of Ouargla - Kasdi Merbah, 30000 Ouargla, Algeria
Mohammedlamine.Kherfi@uqtr.ca, zinebkaoudja@gmail.com, bencheikh1991@gmail.com

Abstract. Understanding user's intention is at the core of an effective images retrieval systems. It still a significant challenge for current systems, especially in situations where user's needs are ambiguous. It is in this perspective that fits our study.

In this paper, we address the challenge of grasping user's intention in semantic based images retrieval. We propose an algorithm that performs a thorough analysis of the semantic concepts presented in user's query. The proposed algorithm is based on an ontology and takes into account the combination of positive and negative examples. The positive examples are used to perform generalization and the negative examples are used to perform specialization which considerably decrease the two famous problems of image retrieval: noise and miss.

Our algorithm processed in two steps: in the first step, we deal only with the positive examples where we will generalize the query from the explicit concepts to infer the others hidden concepts desired by the user. whereas the second step deal with the negative examples to refine results obtained in the first step. We created an image retrieval system based on the proposed algorithm. Experimental results show that our algorithm could well capture user's intention and improve significantly precision and recall.

Keywords: Image retrieval · Grasping user's intention
Positive examples · Negative examples · Ontology

© IFIP International Federation for Information Processing 2018
Published by Springer International Publishing AG 2018. All Rights Reserved
A. Amine et al. (Eds.): CIIA 2018, IFIP AICT 522, pp. 66–77, 2018.
https://doi.org/10.1007/978-3-319-89743-1_7

1 Introduction

Image retrieval is a growing field. Recently, it has witnessed an explosion of personal and professional image collections especially with the development of new technologies that allow users share their images via internet. There was an urgent need for automatic tools that organizing these large quantities of images. Such tools can help users to find the desired images within a reasonable time. These tools are called image retrieval engines.

Image retrieval systems can be classified into two main categories: the first category called Content-Based Image Retrieval (CBIR). In this category, the user is generally asked to formulate a visual query (image for example) that correspond to what he is looking for. The research is performed by measuring similarities between the low levels features of the query and the images of the collection. Various visual features, including both global features [1–3] (e.g., color, texture, and shape) and local features (e.g., SIFT keypoints [4–6]) have been studied for CBIR.

The second category which exploits the semantic concepts associated with images is called Semantic-Based Image Retrieval: SBIR. The user formulates his/her query with textual terms that express his/her needs. In this case, the search is performed by comparing between these terms and textual annotations that represent the images in the collection by the use of semantic tools. Among the tools that are often used to represent the semantics, there are "ontologies" [7].

There is another secondary category where the user formulates his/her query with visual example like CBIR but the research is carried out on the basis of the semantics associated with those images instead of the low levels features [8]. This category try to overcome the drawbacks of the two previous categories by avoiding user to identify keywords and letting the images speak for themselves.

Whatever the category of the image retrieval system, Their main objective is to understand user's needs and answer him/her accurately. However, these needs may vary from one person to another, and even from a given user at different times; therefore, the success of an image retrieval system depends on how accurate the system understands the intention of the user when he/she formulates a query [9].

Creating the query is a difficult problem. It arises two important challenges for both the user and the retrieval system [9]: Firstly, for the user the challenge is how to express his/her needs accurately. Secondly, for the system the challenge is how to understand what the user wants based on the query that he/she formulated.

The difficulty in dealing with these two challenges lead to many problems in image retrieval, such as noise and miss. Noise can be defined as the set of retrieved images which do not correspond to what the user wants [10] whereas, miss is the set of images corresponding to what the user wants which have not been retrieved.

In this paper, our objective is to improve the efficiency of image retrieval by decrease noise and miss via understanding user's intention. We will be interested by the systems of the last category where the user formulates a visual query using images and the retrieval system exploits the semantic concepts associated with these images.

We propose an algorithm that performs a thorough analysis of the semantic concepts presented in the user's query. Our algorithm is processed in two steps: in the first step, we will perform a generalization from the explicit concepts to infer the other hidden concepts that the user can't express them which decrease the miss problem. In the second step, we ask user to select negative examples to refine results obtained in the first step. In this case the images similar to those of the negative examples will be rejected, which reduces the noise. At the same time, the rejected images are replaced by others correspond to what the user wants, so miss will be decrease also.

The rest of the paper is organized as follows: Sect. 2 briefly reviews previous related work. Section 3 present the principle of our algorithm, then, we will present our data and our working hypotheses in Sect. 4. Section 5 details our algorithm. An experimental evaluation of the performance of our algorithm is presented in Sect. 6. Finally, Sect. 7 presents a conclusion and suggests some possibilities for future research.

2   Related Works

User is at the centre of any image retrieval system. One of the primary challenges for these systems is to grasp user intent given the limited amount of data from the input query [11]. In order to solve this challenge, several methods has been proposed. One way is keyword expansion where the user's original query is augmented by new concepts with a similar meaning to decrease the miss problem.

Keyword expansion was implemented by different techniques. Some of them [12–14] are typically based on dictionaries, thesauri, or other similar knowledge models such as ontology. It is noted by many researchers that the use of an ontology for query expansion is advantageous especially if the query words are disambiguated [11,15]. Note that query expansion is very effective, provided that initial query perfectly matches the user's need [16], However, this is not always the case, especially in situation where the user has a difficulty to express them. In this case, the extensions of the query can lead to more inappropriate results than the initial query because of irrelevant concepts which leads to significantly increase noise [17].

Another way to grasp user's intention is relevance feedback [10,18–20] and pseudo relevance feedback. In Relevance feedback technique, the user choose from the returned images those that he/she finds relevants (positive examples) and those that he/she finds not relevants (negative examples) then a query-specific similarity metric was learned from the selected examples.

In pseudo relevance feedback [21,22], the query is expanding by taking the top N images visually most similar to the query image as positive examples.

For many researchers [23–25], The task of learning concepts from user's query to understand his/her intent has seen as a supervised learning problem in which the classifiers search through a hypothesis space to find an adequate hypothesis that will make good predictions. Several algorithms of supervised learning were proposed [26–29] such us: Multi-Class Support Vector Machines (SVM), Naive Bayes Classifier, Maximum Entropy (MaxEnt) classifier, Multi-Layer Perceptron (MLP).

Recent search [25], proposed an ensembling multiple classifiers for detecting user's intention instead of focusing on single classifiers.

This problem has an analogue in the cognitive sciences which called human generalization behavior and inductive inference.
Bayesian models of generalization [30–33] are models from cognitive science that focus on understanding the phenomena of how to generalize from few positive examples. They has been remarkably successful at explaining human generalization behavior in a wide range of domains [33], however their success is largely depend on the reliability of the initial query.
The main contribution of our work is to propose an algorithm to perform the challenge of grasping user's intention and consequently decrease miss and noise. Our algorithm refines the initial query formulated by the user to infer the relevant concepts. It uses an ontology and exploits its semantic richness which allowed us to extract the hidden concepts that the user could not express them explicitly. Then we can improve the obtained results through the relevance feedback with negative examples.

3   Principle of Our Algorithm

In medicine, doctors apply a characteristic rule to summarize the symptoms of a specific disease, whereas, to distinguish one disease from others, they apply a discrimination rule to summarize the symptoms that discriminate this disease from others. This principle is exactly what we will use in our algorithm to understand user's intention from a given query. We perform generalization using positive examples and specialization using negative examples. This principle is applied in content-based image retrieval [10] as follows:

- For positive examples, a concentrated characteristic is important and a scattered characteristic is not important.
- For positive and negative examples, a discriminated characteristic is important and a non-discriminated characteristic is not important.

We carry out a similar work with semantic concepts as follows:

1. For positive examples: with a query that contains several images:
   (a) A concept which is salient in the majority of query's images is important.

(b) A concept that appears in some images and does not appear in the others, in this case, we cannot judge immediately that it is not important. However, it can be classified into three types:

   i. Either, it has concepts that reinforces it in the other images via direct or indirect relationships (for example: apple - fruit - banana), and so it is important, or at least the common concept (here it is the ancestor Fruit) is important.

   ii. Either it has concepts that contradicts it in the other images, directly or indirectly, and therefore it is not important.

   iii. Either there is no concept that reinforces it or a concept that contradicts it in the other images, in this case, it can either be considered with moderate importance, or neglected.

2. For positive and negative examples:

(a) The common concepts between the two sets (positive and negative) are to be neglected. common concept mean either the same concept or concepts that have a positive relations between them as it is illustrated in the case of positive example only.

(b) Discriminant concepts (directly or indirectly through relationships) between the two sets are important. If the concept is present in the positive set, it must be sought; and if it is present in the negative set, it must be dismissed.

4 Data and Hypothesis

4.1 Query Formulation

We ask user to select from displayed images the maximum number of images that correspond to what he/she is looking for. It is the set of concepts that annotate the selected images that will constitute our initial query.

4.2 Ontology Creation

Our algorithm use an ontology to exploit its semantic richness and apply the principle describe above. We have chosen the domain of "nature" for our ontology. The first step of the ontology creation is to choose a collection of images representative of "Nature" domain. The ontology will then be used to annotate the images in this collection. Relationships in our ontology are classified into three categories:

1. Positive relationships (reinforcement relationship): means that the presence of such concept implies the presence of such other concept, for example: presence of concept "Dune" implies the presence of concept "Desert".

2. Negative relationships (contradiction relationships): means that the presence of such concept generally implies the absence of such other concept, for example: presence of the concept "Snow" implies the absence of concept "Desert".

3. Neutral relationships: means that the relations between concepts are null (ie neither positive nor negative).

The other steps of the ontology creation are the following:

1. Identify the different concepts presents in the images collection: for example: desert, camel, dune, snow, etc.
2. Calculate the implications weights between concepts that illustrate the degree of semantic correlation between these concepts.
   For each concept, the probability of having the other concepts was calculated as follows:
$$P(A|B) = \frac{P(A \cap B)}{P(B)}$$
   Where $P(A \mid B)$ means the probability of having a concept A knowing that we are in the presence of the concept B.
   Because our challenge is not the ontology creation, we have calculated this probability manually. Firstly, we started the search with the concept B using the retrieval engine "Google". The obtained images represent P(B). In this collection, we calculate the number of images that represents the concept A which gives us $P(A \cap B)$.
3. Represent the ontology by a value oriented graph where concepts represent the nodes and the arcs are the semantic relationships. Our graph can be seen as a probabilistic graph.
4. Represent the relationships between concepts with a relevance matrix. This latter is a square matrix whose lines and columns are concepts and the value in box M [i, j] represents the probability of the relation between the concept i and the concept j.

A weight between two concepts models a type of semantic relationship between them.

* The weights with the value 0: mean that the two concepts can never be found in the same image so this weight models a negative relationship;
* [0,01 0, 49]: meant that there are times when the two concepts can be found in the same image and sometimes it cannot be found. Which models a neutral relationship;
* [0.5 1]: means that the two concepts are often found in the same image so this weight models a positive relationship.

5   Our Algorithm

Our algorithm is processed in two steps. We called the first step: "Generalization algorithm" (see Algorithm 1) and the second step "Specialization algorithm" (see Algorithm 2).

Algorithm 1. Generalization

Background: a set of an annotated images D
Preprocess: Extract the list of semantic concepts (LCpt) that annotate the positive examples
Input: list of concept LCpt =C1, C2, ... Cn, relevance matrix M, ontology G
Output: L1 list of relevant concepts (explicit and hidden)
  for all C ∈ LCpt **do**
    compute weight (C): the number of occurrences of C in Q
    compute P(C) whith
    W ← weight (C)
    NB ← Number of images in Q

$$P(C) = \frac{W}{NB}$$

    if P(C) ≥ 0.5 **then**
      add (C) to L1 (the list of accepted concepts)
    else
      add (C) to L2 (the list of concepts to be verified)
    end if
  end for
  for all C1 ∈ L2 **do**
    for all C2 ∈ LCpt **and** C1 ≠ C2 **do**
      Sum = Sum + M[C1,C2]

$$Avg = \frac{Sum}{length(LCpt) - 1}$$

    if Avg ≥ 0.5 **then**
      add (C1) to L1
    end if
    end for
  end for
  for all C1 ∈ LCpt **do**
    for all C2 ∈ G **do**
      if M[C1,C2] ≥ 0.5 **then**
        add (C2) to Hid (the list of hidden concepts)
      end if
    end for
  end for
  for all C ∈ Hid **do**
    compute wght (C):the number of occurrences of C in Hid
    compute Moy(C) whith
    Wh ← wght (C)
    Nb ← length (LCpt)

$$Moy(C) = \frac{Wh}{Nb}$$

    if Moy(C) ≥ 0.5 **then**
      add (C) to L1
    end if
  end for

Algorithm 2. Specialization

Background: a set of an annotated images D (search results using L1 as query)

Preprocess: Extract the list of semantic concepts (LNE) that annotate the negative examples

Input: list of relevant concepts L1, list of negative concepts LNE, relevance matrix M, ontology G

Output: L3 list of undesired concepts.

Apply the Generalization algorithm to the list LNE, as results we obtain a an initial list of undesired concepts (L3).

  for all C $\in$ L1 **do**
    if C $\in$ L3 **then**
      delete C from L3
    end if
  end for
  for all Ci $\in$ L3 **do**
    for all Cj $\in$ L1 **do**
      Som = Som + M[Ci,Cj]

$$Moyn = \frac{Som}{length(L1)}$$

    if Moyn $\geq$ 0.5 **then**
      delete Ci from L3
    end if
    end for
  end for

Display results by deleting all images where the concepts of l3 are appeared

---

6 Experimental Evaluation

6.1 Images Collection

To be able to apply our algorithm, we need an ontology and an images collection annotated with the concepts of this ontology. We have chosen the domain of "nature" for our ontology. Then, we have used Google images as an image retrieval engine to collect images for each concept in the ontology. The top 200 returned images of each query are crawled and manually labeled to construct the experimental data set.

6.2 Experiments

We implemented an image retrieval system that integrate our algorithm. We invited thirty users to participate in our experiments. We displayed them a set of images randomly selected from the images collection. Users can ask to see more images if the presented images do not reflect to what they wants.

To evaluate the performance of our algorithm, we carried out a comparison between: our algorithm, the Bayesian model of generalization used in [33] and the standard matching method (without generalization). As performance criteria, we use the precision and the recall measures.

First Experiment. The aim of this experiment is to evaluate the effectiveness of the first step of our algorithm (Generalization algorithm) to increase recall compared to the Bayesian model of generalization used in [33] and the standard matching method (without generalization). We asked users to make different queries, and then we calculated the recall of the three models. Fig. 1 illustrate the obtained results from 16 queries.

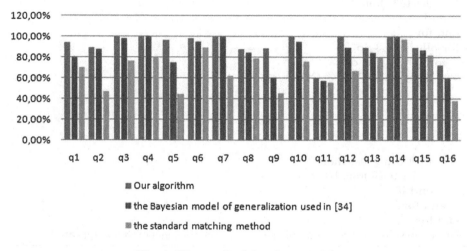

Fig. 1. The recall of the three models

We can see that the recall is considerably increased with our approach and then consequently the miss problem is decreased. Compared to our algorithm, the success of the Bayesian model of generalization is largely depend on the reliability of the examples present in the query to good reflect the user's intention which is not always the case because sometimes the images of the query may composed of many objects, however the user is interest only with some of them. Our algorithm overcomes this problem by refining the initial query of the user.

Second Experiment. The aim of this experiment is to illustrate the effect of the negative examples to decease noise by improving the accuracy. We asked users to formulate queries using the Generalization algorithm and we calculated the accuracy then we asked him to select negative images and restart the search. the obtained results are illustrated in Fig. 2.

It is clear that the use of negative examples increase the accuracy. Indeed, after obtaining the results of a given query, the user can keep the positive images and enrich the query by including some undesired images as a negative examples.

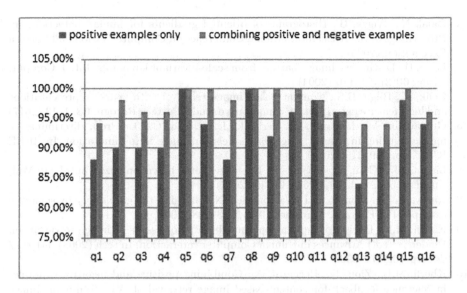

Fig. 2. The effect of negative examples to increase accuracy.

This implies that images similar to those of the negative example will be rejected, which reduces the noise. At the same time, the rejected images are replaced by others corresponding to what the user wants, so silence will be diminished.

7   Conclusion

In this paper, we proposed an algorithm that process in two steps: we called the first step "Generalization algorithm" and the second step "specialization algorithm". The objective of this algorithm is to improve the semantic image retrieval quality by reducing the famous problems of image retrieval: noise and miss via understanding user's intention. Our algorithm is based on the use of an ontology which allows us to exploits it's semantic richness. We implemented an image retrieval system that integrate the proposed algorithm. Experimental evaluation verify the efficiency of our approach in comparison to the others proposed in the state-of-the-art. We have shown that the combination of positive and negative examples improve significantly precision and recall. The perspective of our work is to automate the ontology creation.

References

1. Lowe, D.G.: Object recognition from local scale-invariant features. In: IEEE ICCV (1999)
2. Torralba, A., Murphy, K., Freeman, W., Rubin, M.: Context-based vision system for place and object recognition. In: Proceedings of the International Conference on Computer Vision (2003)

3. Dalal, N., Triggs, B.: Histograms of oriented gradients for human detection. In: Proceedings of the IEEE International Conference on Computer Vision and Pattern Recognition (2005)

4. Lowe, D.: Distinctive image features from scale-invariant keypoints. Int. J. Comput. Vision **60**(2), 91–110 (2004)

5. Celik, C., Bilge, H.S.: Content based image retrieval with sparse representations and local feature descriptors: a comparative study. Pattern Recogn. **68**, 1–13 (2017)

6. Guillaumin, M., Mensink, T., Verbeek, J., Schmid, C.: TagProp: discriminative metric learning in nearest neighbor models for image auto-annotation. In: IEEE ICCV, pp. 309–316 (2009)

7. Liaqat, M., Khan, S., Majid, M.: Image retrieval based on fuzzy ontology. Multimed. Tools Appl. **76**, 22623–22645 (2017)

8. Rasiwasia, N., Moreno, P.J., Vasconcelos, N.: Bridging the gap: query by semantic example. IEEE Trans. Multimed. **9**(5), 923–938 (2007)

9. Kherfi, M.L.: Review of human-computer interaction issues in image retrieval. In: Pinder, S. (ed.) Advances in Human Computer Interaction. InTech (2008). https://doi.org/10.5772/5929

10. Kherfi, M.L., Ziou, D., Bernardi, A.: Combining positive and negative examples in relevance feedback for content-based image retrieval. J. Vis. Commun. Image Represent. **14**(4), 428–457 (2003)

11. Tang, X., Liu, K., Cui, J., Wen, F., Wang, X.: Intentsearch: capturing user intention for one-click internet image search. IEEE Trans. Pattern Anal. Mach. Intell. **34**(7), 1342–1353 (2012)

12. Bilotti, M., Katz, B., Lin, J.: What works better for question answering: stemming or morphological query expansion? In: Proceedings of the Information Retrieval for Question Answering (IR4QA) Workshop at SIGIR 2004 (2004)

13. Navigli, R.: Word sense disambiguation: a survey. ACM Comput. Surv. **41**(2), 1–69 (2009)

14. Azizan, A., Bakar, Z.A., Noah, S.A.: Analysis of retrieval result on ontology-based query reformulation. In: Proceedings of the 1st International Conference on Computer, Communication and Control Technology, I4CT 2014, pp. 244–248 (2014)

15. Zha, Z.-J., Yang, L., Mei, T., Wang, M., Wang, Z.: Visual query suggestion. In: Proceedings of the 17th ACM International Conference on Multimedia, MM 2009, vol. 6, no. 3, p. 15 (2009)

16. Fergus, R., Perona, P., Zisserman, A.: A visual category filter for google images. In: Pajdla, T., Matas, J. (eds.) ECCV 2004. LNCS, vol. 3021, pp. 242–256. Springer, Heidelberg (2004). https://doi.org/10.1007/978-3-540-24670-1_19

17. Park, G., Baek, Y., Lee, H.-K.: Majority based ranking approach in web image retrieval. In: Bakker, E.M., Lew, M.S., Huang, T.S., Sebe, N., Zhou, X.S. (eds.) CIVR 2003. LNCS, vol. 2728, pp. 111–120. Springer, Heidelberg (2003). https://doi.org/10.1007/3-540-45113-7_12

18. Deng, J., Berg, A.C., Fei-Fei, L.: Hierarchical semantic indexing for large scale image retrieval. In: Proceedings of the IEEE International Conference on Computer Vision and Pattern Recognition (2011)

19. Glowacka, D., Shawe-Taylor, J.: Image Retrieval with a Bayesian Model of Relevance Feedback (2016)

20. Tao, D., Tang, X., Li, X., Wu, X.: Asymmetric bagging and random subspace for support vector machines-based relevance feedback in image retrieval. IEEE Trans. Pattern Anal. Mach. Intell. **28**, 1088–1099 (2006)

21. Torjmen, M., Pinel-Sauvagnat, K., Boughanem, M.: Using pseudo-relevance feedback to improve image retrieval results. In: Peters, C., Jijkoun, V., Mandl, T., Müller, H., Oard, D.W., Peñas, A., Petras, V., Santos, D. (eds.) CLEF 2007. LNCS, vol. 5152, pp. 665–673. Springer, Heidelberg (2008). https://doi.org/10.1007/978-3-540-85760-0_85

22. Yan, R., Hauptmann, A., Jin, R.: Multimedia search with pseudo-relevance feedback. In: Bakker, E.M., Lew, M.S., Huang, T.S., Sebe, N., Zhou, X.S. (eds.) CIVR 2003. LNCS, vol. 2728, pp. 238–247. Springer, Heidelberg (2003). https://doi.org/10.1007/3-540-45113-7_24

23. Kang, I.-H., Kim, G.: Query type classification for web document retrieval. In: Proceedings of the 26th Annual International ACM SIGIR Conference on Research and Development in Informaion Retrieval, SIGIR 2003, pp. 64–71. ACM, New York (2003)

24. Lee, U., Liu, Z., Cho, J.: Automatic identification of user goals in web search. In: Proceedings of the 14th International Conference on World Wide Web, WWW 2005, pp. 391–400. ACM, New York (2005)

25. Figueroa, A., Atkinson, J.: Ensembling classifiers for detecting user intentions behind web queries. IEEE Internet Comput. **20**(2), 8–16 (2016)

26. Krammer, K., Singer, Y.: On the algorithmic implementation of multi-class svms. Proc. J. Mach. Learn. Res. **2**, 265–292 (2001)

27. Hsu, C.-W., Lin, C.-J.: A comparison of methods for multiclass support vector machines. IEEE Trans. Neural Networks **13**(2), 415–425 (2002)

28. Nigam, K., Lafferty, J., Mccallum, A.: Using maximum entropy for text classification

29. Foody, G.M.: Hard and soft classifications by a neural network with a non-exhaustively defined set of classes. Int. J. Remote Sens. **23**(18), 3853–3864 (2002)

30. Tenenbaum, J.B., Griffiths, T.L.: Generalization, similarity, and Bayesian inference. Behav. Brain Sci. **24**, 629–630 (2001)

31. Tenenbaum, J.B., Xu, F.: Word learning as Bayesian inference. In: Proceedings of the 22nd Annual Conference of the Cognitive Science Society (2000)

32. Abbott, J.: Constructing a hypothesis space from the Web for large-scale Bayesian word learning. Proc. Annu. Meet. Cogn. Sci. Soc. **34**, 54–59 (2012)

33. Jia, Y., Abbott, J., Austerweil, J.L., Griffiths, T.L., Darrell, T.: Visual concept learning: combining machine vision and Bayesian generalization on concept hierarchies. In: Advances in Neural Information Processing Systems 27 (NIPS 2013), vol. 1, no. 1, pp. 1–9 (2013)

Exploring Graph Bushy Paths to Improve Statistical Multilingual Automatic Text Summarization

Abdelkrime Aries[(✉)] , Djamel Eddine Zegour, and Walid Khaled Hidouci

École nationale Supérieure d'Informatique (ESI, ex. INI),
BP 68M, 16309 Oued Smar, El-Harrach, Algiers, Algeria
{ab_aries,d_zegour,w_hidouci}@esi.dz

Abstract. Statistical extractive summarization is one of the most exploited approach in automatic text summarization due to its generation speed, implementation easiness and multilingual property. We want to improve statistical sentence scoring by exploring a simple, yet powerful, property of graphs called bushy paths represented by the number of node's neighbors. A graph of similarities is constructed in order to select candidate sentences. Statistical features such as sentence position, sentence length, term frequency and sentences similarities are used to get a primary score for each candidate sentence. The graph is used again to enhance the primary score by using bushy paths property. Also, we tried to exploit the graph in order to enhance summary's coherence. We experimented our method using MultiLing'15 workshop's corpora for multilingual single document summarization. Using graph properties can improve statistical scoring without loosing the multilingualism of the method.

1 Introduction

Summarization is an effective way to handle huge amount of information, especially in this century. Its main purpose is to keep the most important information and get rid of redundant one. Therefore, a summary must be short, representative, readable and not redundant. A summary has to present the main points and key details of the input document. Information redundancy is not a good aspect of a summary, since it prevents adding new useful details especially when that summary's size is limited. Readability is another important aspect of a summary; this later must be coherent and grammatically correct. Manual summarization is a very expensive task either in time or cost, this is why researches have been conducted since 1958 [15] to find a perfect automatic text summarization (ATS) system.

Recently, multilingual content started to grow, leading to a huge need for more adaptable methods for languages other than English. As a result, multilingual ATS began to receive more attention from research community [9,10,12].

A truly multilingual method must not rely on any language-dependent resource, either toolkits or corpora. ATS methods categorized in linguistic approach or machine learning need a lot of changes to handle another language. The problem with linguistic approach is its heavy dependency on languages and the lack of NLP toolkits. In case of machine learning (ML), using a corpus for training can cause the system to be more dependent on the corpus's language. The two remaining approaches are statistical and graph-based. Beside their multilingualism, statistical methods are fast and easy to implement. Even if they do not use deep linguistic techniques, they still give good results. As for graph-based approach, it is not as widely used as the statistical one. Its main factor is the coherence relations between sentences, where the score of a sentence is calculated based on its neighbors. In our work, we want to exploit the graph bushy paths property in order to improve statistical scoring in multilingual ATS. To place our work among others of the same domain, here is a summary of our contribution:

- Proposing a multilingual method based on statistical features avoiding their weights attribution when combining them linearly.
- Investigating the impact of using graph properties (here, bushy paths) to improve the original scores.
- Using the graph to select candidate sentences relying on the assumption that a sentence without either much or strong relations with its neighbors will not be considered for scoring. This will help decrease the number of sentences to be processed, resulting in a faster generation.
- Testing the impact of extraction methods on the summarization quality.

The remainder of this article is divided as follows. Section 2 discusses related works, namely those based on statistical or/and graph-based approaches. Section 3 provides the details of our method starting from preprocessing, through graph construction and scoring, ending with sentences extraction and summary construction. Section 4 presents the experimental evaluation of our method in the context of multilingual single document summarization. Finally, Sect. 5 concludes the work carried in this research and provides future insights.

2   Related Works

Statistical approach was the first exploited for ATS and still widely used nowadays. It is based on statistical features which are used to judge the importance of a sentence. The first feature ever used for ATS is term frequency (tf) [15] which scores a sentence based on its terms distribution in the original document. Very high frequency terms can be related to the domain and not the subject, this is why the author fixes a high threshold to eliminate such terms. Another solution is by combining tf with inverse document frequency to get $tf - idf$ [19]. Position is another feature allowing to capture the importance of a sentence, either by its words positions [15], by its position in the original document [4, 6] or by its position in a paragraph [8]. Sentence position in the document is based on the assumption that first and last sentences tend to be more important than others [6]. To score a sentence using this feature, researchers tested with many different formulas [8,17]. Sentence length feature is used to penalize short sentences with a size less than a given threshold [13]. Other length-based formulas can be found in [8,17]. Mostly, these features are combined linearly into one score representing how pertinent a sentence is. The weight of each feature is fixed manually, or estimated using optimization or machine learning.

Graph-based approach transforms document sentences into a graph using a similarity measure. Then, this structure is used to calculate the importance of each sentence. In [20], a graph of similarities is constructed from document paragraphs, where similarities below a given threshold are ignored. The authors define a feature called bushiness which is the number of a node's connections. The most scored paragraphs in term of bushiness are extracted to form a summary. Node's bushiness can be used to score sentences rather than paragraphs, to be combined with other statistical features scores [8]. Likewise, we use this feature to score sentences, but we use it to enhance statistical scoring rather than consider it as one. Iterative scoring using graphs is very popular among graph-based ATS methods; TextRank [16] and LexRank [7] are the most known ones. They use a modified version of PageRank [5] in order to score sentences. LexRank incorporates $tf - idf$ into similarity calculation, while LexRank uses simple cosine similarity. When multiple documents are summarized, it is preferable to include temporal information to the score favoring recent documents [21]. To capture the impact of documents on sentences in multi-document ATS, sentence to document dimension can be used [22]. The author incorporates sentence position into the score to distinguish sentences from each other when they belong to the same document. Another work aiming to fuse statistical features with graph-based approach is iSpreadRank [23]. The method is based on activation theory [18] explaining the cognitive process of human comprehension. Each sentence is scored using some features (centroid, position, and first-sentence overlap), then these scores are spread from a sentence to its neighbors iteratively until equilibrium. The ranking in iterative graph-based ATS may never converge. Also, the ranking convergence depends on the initial scores and the damping factor.

3   Method Overview

3.1   Preprocessing

Preprocessing phase consists of: sentence segmentation, words tokenization, stop words removal, and words stemming. Sentence segmentation is the task of splitting a text into several sentences using punctuation. Sometimes it is difficult to detect sentence boundaries due to punctuation use in abbreviations (mr., ms., etc.). Words tokenization divides a text into words, which is not simple in some languages such as Japanese where the words are attached. Stop words removal aims to get rid of subject-independent words such as pronouns, prepositions, etc. Finally, stemming is the task of prefixes, suffixes and infixes removal; which helps associating variants of the same term to a common root form. In our work, we use an open source tool (LangPi[1]) which affords these tasks for 40 languages.

3.2   Candidate Sentences Generation

The task of finding candidate sentences comes to find those less probable to be included in the summary and exclude them. That is, the stronger a sentence connection to other sentences in the document, the more it is important to be kept. This connection is represented, in our case, by cosine similarity between sentences. A graph of similarities $G(V, E)$ is constructed where each sentence is represented by a node, and arcs represent the similarities between the sentences. This graph can be reduced to keep just significant nodes and arcs.

Using a given similarity threshold, we decide the most important neighbors $MImpN(v_i)$ to a node v_i. We define an insignificant node v_i as the one with a sum of similarities less than $\frac{1}{|MImpN(v_i)|}$ where $|MImpN(v_i)|$ is the number of its important neighbors. When the number of important neighbors increases, the chance of a node to be kept as a candidate sentence will be higher. In this case, many factors determine if a node is good enough to be included in the summary. The first factor is the similarity threshold's value; When it is high, the number of important neighbors will be lower and thus a low chance. The second one is the number of connections; When it has a lot of connections, the sum of similarities (even if they are low) has a great chance to exceed the significance threshold. The third one is the similarities strength; If a sentence has high similarities with its neighbors, their sum can be higher enough to make it significant.

Once we have deleted the insignificant nodes, we want to remove weak arcs as well. We define a weak arc (s_i, s_j) from a node s_i as the one with a value (similarity) less than $\frac{threshold}{|MImpN(v_i)|}$. In here, we try to distribute the threshold uniformly between the different arcs joined to each node. This will result in a very small value, especially when the node has many neighbors. We should point out that, using this method, a node can consider another as a neighbor while the second one does not.

3.3   Sentence Scoring

We want to attribute to each candidate sentence a score based on statistical features reflecting its pertinence towards the main topic of the input document. Our intention is to come up with a method totally independent from the input document's language and domain. This is why we choose four features: its similarity to other candidates, the terms it contains, its size and its position in the document. These features can be calculated easily for any language without the need of affording language-dependent resources except the preprocessing task which is simple, mostly.

A sentence is more likely to be included in the summary if it can reflect the input document's content. In our method, we just calculate the similarity between

---

[1] LangPi: https://github.com/kariminf/langpi.

a candidate sentence s_i and the remaining candidates $C \backslash s_i$ as shown in (1).

$$Score_{sim}(s_i) = sim(s_i, C \backslash s_i) \qquad (1)$$

Term frequency has been used since the first days of ATS [15]. It is a very good sign of sentence importance, especially when it is combined with inversed document frequency (idf) [19]. Unfortunately, we can't use idf since we want a domain-independent method. But, we can use a similar concept which is inverse sentence frequency isf, first used in [1]. Our $tf - isf$ based sentence score follows the same one used in [17] represented by the Euclidean normalization of $tf - isf$ values, but using sentence terms instead of document terms. Equation 2 represents the $tf - isf$ score of a sentence s_i given its words (terms) w_{ik}.

$$Score_{tf-isf}(s_i) = \sqrt{\sum_{w_{ik} \in s_i} tf - isf(w_{ik})^2} \qquad (2)$$

Sentence size (length) is another feature used to calculate its importance in a document. Mostly, short sentences do not carry so much information, this is why they are omitted [13]. Also, we believe that long sentences are not a good fit to be included in a summary. This is justified by the fact that a summary must be short, informative and without redundancy. Pushing long sentences into a summary reduces the number of ideas that can be extracted from the original document. Furthermore, they can contain redundant or unimportant information reduced just by another type of language-dependent summarization approach called sentence compression. So, in our case, we prefer the shortest ones; The shortest they are, the most scored they will be. The two previous scores, namely similarity and $tf - isf$, will favor sentences similar to the main topic. This score (size), in the other hand, will help them select pertinent sentences but with the least possible size. It is calculated by, simply, dividing 1 on the sentence's size, as shown in (3).

$$Score_{size}(s_i) = \frac{1}{|s_i|} \qquad (3)$$

The last score is sentence position which is a good feature, since important sentences tend to occur at first or at last [4,6]. Equation 4 represents the position score which is proposed among others in [17], where $|D|$ is the number of sentences in the input document.

$$Score_{pos}(s_i) = \max(\frac{1}{i}, \frac{1}{|D| - i + 1}) \qquad (4)$$

The previous scores can be considered as probabilities that a sentence belongs to the summary using a given feature, after normalizing them of course. In our case, we won't normalize these scores even if they exceed 1 since they will be used to reorder sentences and the denominator will be the same in all sentences thus a constant. So, the overall statistical score $(Score_{stat})$, which is the probability that a sentence belongs to the summary using all features, can be calculated as the multiplication of all these scores assuming features independence.

A big number of neighbors is a good sign that a sentence is representative and discusses topics covered by many other sentences [20], so its score must be high. In contrast of [20], which scores a sentence by the number of its neighbors, our bushy paths score is based on amplifying the statistical score of a sentence by the number of its neighbors plus one. A sentence with a little statistical score can be given another chance to be included in the summary when it has many neighbors. Pertinence with the main topics is not the only aspect of sentence importance; Coherence with other sentences has proven to have very large impact on summary's quality. To combine these two aspects, we simply amplify the statistical score by the sentence's neighbors number as presented in (5).

$$Score_{bushy}(s_i) = Score_{stat}(s_i) * (1 + |\{(s_i, s_j) \in E\}|) \tag{5}$$

3.4   Extraction

After the sentences are scored, they are extracted to form a summary. Most works extract the most scored ones without any further processing. One of the problems resulting from that, is information redundancy; If two sentences are highly scored and similar, they risk to be included in the summary though they almost have the same information. Some works try to handle redundancy after scoring, by exploring the similarity between the candidate sentence and the last added one [3]. Similarly, we try to investigate the extraction phase impact on the generated summary's quality. This is why we use three simple extraction methods.

The first one (e_0) is used by most works: extract the first scored sentences till reaching summary's maximum size. This method does not manage redundancy and therefore it is expected to give the least precision score among the others. It, also, does not consider presenting the summary in a more coherent way. We tried to formulate this in (6), where *next* represents the sentence to be selected next; C is the list of candidate sentences; S is the list of summary sentences.

$$next_{e0} = \arg\max_i score(s_i) \text{ where } s_i \in C \backslash S \tag{6}$$

The second variant (e_1) is used in [3] by adding the most scored sentences which are not similar to the summary. The idea is to reorder sentences using their scores then each time we want to add a sentence to the summary, we must compare it to the last added one. This way, we will have a summary with diverse information, but no coherence is considered. The extraction method is illustrated in (7), where *sim* is the similarity function (cosine similarity, in our case); *last* is the last sentence added to the summary S; Th is a given similarity threshold.

$$next_{e1} = \arg\max_i score(s_i) \text{ where } s_i \in C \backslash S \text{ and } sim(s_i, last_{e1}) < Th \tag{7}$$

In the third one (e_2), we want to incorporate the graph into the extraction task. The graph affords the information about which sentence is similar to the other.

We believe consecutive sentences must have some similarity to preserve coherence. When a sentence discusses an idea, the next one must have some shared terms with it. In this case, we extract the highest scored sentence, then the highest scored one among its neighbors. This will ensure some shared information between a sentence and the next, but does not prevent redundancy. Equation 8 is a formulation of this variant, where E is the set of graph's arcs.

$$next_{e2} = \arg\max_i score(s_i) \text{ where } (last_{e2}, s_i) \in E \tag{8}$$

4   Experiments

4.1   Metrics

To evaluate our method's performance, we apply ROUGE method which is used widely in ATS workshops. ROUGE (Recall-Oriented Understudy for Gisting Evaluation) [14] is an automatic evaluation method for ATS systems. Its principle is to compare automatic text summaries against human made abstracts based on words grams. It includes five measures: ROUGE-N, ROUGE-L, ROUGE-W, ROUGE-S and ROUGE-SU. In our evaluation, we use two metrics of ROUGE: ROUGE-1 and ROUGE-2, which are used in most ATS workshops. The original ROUGE package supports just English, this is why we use another implementation afforded by LangPi project, used in our preprocessing phase. This implementation supports all MultiLing'15 languages.

4.2   Corpora

MultiLing 2015 Multilingual Single-document Summarization (MSS) corpora contains two sets: training and testing. Each of them affords 30 documents extracted from Wikipedia featured documents for every language out of 38. The documents are UTF-8 encoded and packed with their respective human-generated summaries. For each summary, the number of characters which the generated summaries must not exceed is afforded.

4.3   Baseline System

We choose to use AllSummarizer[2] system as a baseline for these reasons:

- It is an available open source system.
- It is a multilingual system.
- It participated on MultiLing'15 for all the proposed languages.

---

[2] AllSummarizer: https://github.com/kariminf/allsummarizer.

AllSummarizer_TCC (threshold clustering and classification) method [2,3] starts by detecting the different topics in the input document using a simple clustering algorithm based on cosine similarity and a threshold. Then, using Bayes classification and a set of features, it learns the characteristics of each cluster. Each sentence is scored based on its probability to represent all these clusters. In our experiments, we use the optimal parameters fixed for each language in [3]. Also, we use testing corpus since the training one was used to fix these parameters and using it will favor AllSummarizer_TCC method.

4.4 Tests and Discussion

To generate summaries, first, we have to fix similarity threshold which is an important parameter in our method. If the threshold is too low, the number of candidate sentences will grow and also the number of arcs connecting a sentence to another. If it is too high, we will have a few number of candidate sentences which can be selected without even scoring. Also, the number of arcs connected to a sentence may drop to one and hence, the graph will not be necessary since its score remains the same. To avoid all of these scenarios, we use the mean between non null similarities as a threshold.

We generated summaries using the two scoring schemes (just statistical or bushy paths) and the different extraction methods ($e0$, $e1$ et $e2$) for the 38 languages of test corpus. Figure 1 represents ROUGE-1 recall for the 38 languages using Multiling'15 test corpus, compared to AllSummarizer_TCC method.

Lets discuss how much did the introduction of bushy paths improve statistical scoring for ATS. Applying bushy paths surpasses statistical method in 20 languages using extraction method $e0$, in 24 languages using $e1$, and in 25 languages using $e2$. We can say that introducing inter-sentences coherence (graph-based approach) has the potential to improve the quantity of information in summaries. Also, exploring graph structure to help extracting sentences (the case of $e2$) can help increase the quantity of information, although the method $e2$ is intended to increase the coherence between the extracted sentences and not the quantity of information.

Regarding the impact of extraction methods on the generated summary's recall using bushy paths, the application of $e1$ improves 24 languages summaries over $e0$. Using $e2$ improves 29 languages over $e0$ and 18 languages over $e1$. Also, extraction method $e1$ with bushy paths proves to be better compared to Allsumarizer_TCC which uses the same method (overcomes it in 20 languages). While using $e2$, bushy paths beats Allsumarizer_TCC in 18 languages, and beats statistical method with $e0$ in 36 languages except for Hungarian and Malay.

It is also important to point out that summaries quality varies from language to another, even if we use different methods. Languages such as Arabic and Hebrew, which are from the semitic family, show low recall when the automatic generated summaries are compared against human abstracts. This is probably due to the assumption that summaries from these languages are made by abstraction rather than extracting some parts of the original documents. In the contrary, languages such as English and French tend to have better recall.

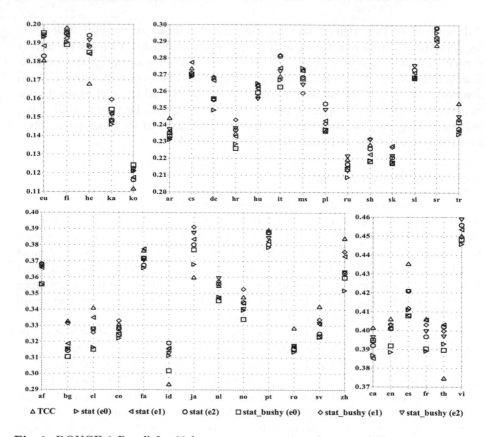

Fig. 1. ROUGE-1 Recall for 38 languages: comparison between AllSummarizer_TCC and different parameters of our method.

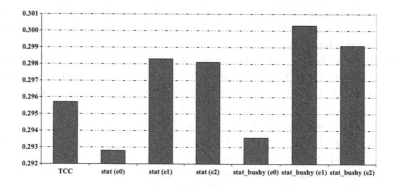

Fig. 2. ROUGE-1 Recall scores average for 38 languages.

Table 1. ROUGE-1 and ROUGE-2 of AllSummarizer_TCC and our method for English testing set.

Peer	ROUGE-1			ROUGE-2		
	R	P	F1	R	P	F1
TCC (allsummarizer)	**0.40619**	0.40866	0.40709	0.10053	0.10115	0.10077
score_stat (e0)	0.38848	0.40101	0.39431	0.10141	0.10478	0.10298
score_stat (e1)	0.40113	0.40504	0.40284	0.10447	0.10539	0.10487
score_stat (e2)	0.40104	0.41323	0.40658	**0.11030**	**0.11351**	**0.11176**
score_bushy (e0)	0.39187	0.40658	0.39874	0.10514	0.10881	0.10686
score_bushy (e1)	0.40314	0.41215	0.40703	0.10974	0.11202	0.11072
score_bushy (e2)	0.40267	**0.41380**	**0.40788**	0.10818	0.11097	0.10948

So, professional summarizers of these languages use parts of the original document to construct final summaries. In fact, it is shown in [11] that 78% of summaries sentences come from the original document while half of them have been compressed.

The average ROUGE-1 recall scores of every method over all 38 languages are represented in Fig. 2. It is clear that using statistical features with bushy paths property and extraction method *e*1 gives better average recall. Overall, bushy paths improves the recall score of summaries.

In English language, our method did not perform as well as the baseline system in term of ROUGE-1 recall score. In the contrary, looking to ROUGE-2 recall scores (see Table 1), it did a good job. Also, in term of precision, our method performs better which means it affords less redundant summaries.

5   Conclusion

We tried to improve statistical scores by exploiting the bushy paths property of graphs. A graph of similarities between sentences is constructed. First, the graph is minimized by deleting unimportant nodes and weak arcs. This can help decreasing the number of sentences to be considered for scoring, especially when there is a big document to be processed. To score each candidate sentence, statistical are used to calculate some sort of probabilities of a sentence being relevant to the main topic of the input document. Then, using the minimized graph, each sentence is re-scored based on its initial statistical score and the number of its neighbors. We want to increase the amount of information a summary can express compared to an abstract made by humans, which is known as summary's informativeness. But, it is not the only issue of ATS; There are also redundancy and readability. These three properties are often inconsistent; i.e. we can not enhance one without affect the quality of another. So, we tried to use an extraction method which prefers non redundant sentences from the high scored ones. Also, we tried to propose a testing version of an extraction method which aims to exploit graph's links when generating sentences in order to have some coherence.

To test our method, we used the testing corpus from MultiLing'15 workshop and one of its participant systems as a baseline. Our method shows improvements in recall score with many languages. In term of precision, using graph's bushy paths improves the original scoring method. Also, trying to extract coherent sentences can cause a little drop in summaries recall score. This is why we have to reformulate this method in order to find a trade-off between informativeness, non redundancy and coherence. We should, also, point out that we fixed our similarity threshold to be the mean of sentences similarities. The value of this threshold can affect our method's performance greatly, which means it has to be investigated in future works. We may, as well, consider iterative graphs with some conditions to prevent non convergence.

References

1. Allan, J., Wade, C., Bolivar, A.: Retrieval and novelty detection at the sentence level. In: Proceedings of the 26th Annual International ACM SIGIR Conference on Research and Development in Informaion Retrieval, pp. 314–321. SIGIR 2003. ACM, New York (2003). https://doi.org/10.1145/860435.860493
2. Aries, A., Oufaida, H., Nouali, O.: Using clustering and a modified classification algorithm for automatic text summarization. In: Proceedings of SPIE, vol. 8658, pp. 865811–865811-9 (2013). https://doi.org/10.1117/12.2004001
3. Aries, A., Zegour, E.D., Hidouci, W.K.: AllSummarizer system at MultiLing 2015: multilingual single and multi-document summarization. In: Proceedings of the 16th Annual Meeting of the Special Interest Group on Discourse and Dialogue, pp. 237–244. Association for Computational Linguistics (2015). http://aclweb.org/anthology/W15-4634
4. Baxendale, P.B.: Machine-made index for technical literature: an experiment. IBM J. Res. Dev. 2(4), 354–361 (1958). https://doi.org/10.1147/rd.24.0354
5. Brin, S., Page, L.: The anatomy of a large-scale hypertextual web search engine. In: Seventh International World-Wide Web Conference (WWW 1998) (1998). http://ilpubs.stanford.edu:8090/361/
6. Edmundson, H.P.: New methods in automatic extracting. J. ACM 16(2), 264–285 (1969). https://doi.org/10.1145/321510.321519
7. Erkan, G., Radev, D.R.: LexRank: graph-based lexical centrality as salience in text summarization. J. Artif. Int. Res. 22(1), 457–479 (2004). http://dl.acm.org/citation.cfm?id=1622487.1622501
8. Fattah, M.A., Ren, F.: GA, MR, FFNN, PNN and GMM based models for automatic text summarization. Comput. Speech Lang. 23(1), 126–144 (2009). https://doi.org/10.1016/j.csl.2008.04.002
9. Giannakopoulos, G., El-Haj, M., Favre, B., Litvak, M., Steinberger, J., Varma, V.: TAC 2011 MultiLing pilot overview. In: Proceedings of the Fourth Text Analysis Conference (TAC 2011)-MultiLing Pilot Track, Gaithersburg, Maryland, USA (2011)
10. Giannakopoulos, G.: Multi-document multilingual summarization and evaluation tracks in ACL 2013 MultiLing workshop. In: Proceedings of the MultiLing 2013 Workshop on Multilingual Multi-document Summarization, pp. 20–28. Association for Computational Linguistics, Sofia, August 2013. http://www.aclweb.org/anthology/W13-3103

11. Jing, H., McKeown, K.R.: The decomposition of human-written summary sentences. In: Proceedings of the 22nd Annual International ACM SIGIR Conference on Research and Development in Information Retrieval, SIGIR 1999, pp. 129–136. ACM, New York (1999). https://doi.org/10.1145/312624.312666

12. Kubina, J., Conroy, J., Schlesinger, J.: ACL 2013 MultiLing pilot overview. In: Proceedings of the MultiLing 2013 Workshop on Multilingual Multi-document Summarization, pp. 29–38. Association for Computational Linguistics, Sofia, August 2013. http://www.aclweb.org/anthology/W13-3104

13. Kupiec, J., Pedersen, J., Chen, F.: A trainable document summarizer. In: Proceedings of the 18th Annual International ACM SIGIR Conference on Research and Development in Information Retrieval, SIGIR 1995, pp. 68–73. ACM, New York (1995). https://doi.org/10.1145/215206.215333

14. Lin, C.Y.: ROUGE: a package for automatic evaluation of summaries. In: Proceedings of ACL Workshop on Text Summarization Branches Out, p. 10 (2004). http://research.microsoft.com/~cyl/download/papers/WAS2004.pdf

15. Luhn, H.P.: The automatic creation of literature abstracts. IBM J. Res. Dev. **2**(2), 159–165 (1958). https://doi.org/10.1147/rd.22.0159

16. Mihalcea, R., Tarau, P.: Textrank: bringing order into texts. In: Lin, D., Wu, D. (eds.) Proceedings of EMNLP 2004, pp. 404–411. Association for Computational Linguistics, Barcelona, July 2004. http://www.aclweb.org/anthology/W04-3252

17. Nobata, C., Sekine, S.: CRL/NYU summarization system at DUC-2004. In: DUC (2004)

18. Quillian, M.R.: Semantic Memory, pp. 227–270. The MIT Press, Cambridge (1968)

19. Salton, G., Yang, C.S.: On the specification of term values in automatic indexing. J. Documentation **29**(4), 351–372 (1973)

20. Salton, G., Singhal, A., Mitra, M., Buckley, C.: Automatic text structuring and summarization. Inf. Process. Manage. **33**(2), 193–207 (1997). https://doi.org/10.1016/S0306-4573(96)00062-3

21. Wan, X.: TimedTextRank: adding the temporal dimension to multi-document summarization. In: Proceedings of the 30th Annual International ACM SIGIR Conference on Research and Development in Information Retrieval, SIGIR 2007, pp. 867–868. ACM, New York (2007). https://doi.org/10.1145/1277741.1277949

22. Wan, X.: An exploration of document impact on graph-based multi-document summarization. In: Proceedings of the Conference on Empirical Methods in Natural Language Processing, EMNLP 2008, pp. 755–762. Association for Computational Linguistics, Stroudsburg (2008). http://dl.acm.org/citation.cfm?id=1613715.1613811

23. Yeh, J.Y., Ke, H.R., Yang, W.P.: iSpreadRank: ranking sentences for extraction-based summarization using feature weight propagation in the sentence similarity network. Expert Syst. Appl. **35**(3), 1451–1462 (2008). http://www.sciencedirect.com/science/article/pii/S0957417407003612

Evolutionary Computation

Hybrid Artificial Bees Colony and Particle Swarm on Feature Selection

Hayet Djellali[1,2(✉)], Akila Djebbar[1,3], Nacira Ghoualmi Zine[1,2], and Nabiha Azizi[1]

[1] Computer Science Department, Badji Mokhtar University, Annaba, Algeria
hayetdjellali@yahoo.fr, ghoualmi@yahoo.fr,
aki_djebbar@yahoo.fr
[2] LRS Laboratory, Badji Mokhtar University, Annaba, Algeria
[3] LRI Laboratory, Badji Mokhtar University, Annaba, Algeria

Abstract. This paper investigates feature selection method using two hybrid approaches based on artificial Bee colony ABC with Particle Swarm PSO algorithm (ABC-PSO) and ABC with genetic algorithm (ABC-GA). To achieve balance between exploration and exploitation a novel improvement is integrated in ABC algorithm. In this work, particle swarm PSO contribute in ABC during employed bees, and GA mutation operators are applied in Onlooker phase and Scout phase. It has been found that the proposed method hybrid ABC-GA method is competitive than exiting methods (GA, PSO, ABC) for finding minimal number of features and classifying WDBC, colon, hepatitis, DLBCL, lung cancer dataset. Experimental results are carried out on UCI data repository and show the effectiveness of mutation operators in term of accuracy and particle swarm for less size of features.

Keywords: Feature selection · Artificial Bees colony · Particle swarm
Genetic algorithm · Hybrid ABC-GA · Hybrid ABC-PSO

1 Introduction

Classification problem can be well resolved using machine learning algorithm and efficient feature selection (FS) methods. FS algorithms act as filter of redundant and irrelevant features and are categorized on three types: filter, wrapper and embedded methods. In filters approach, feature relevance are evaluated by applying statistical methods, wrapper approach find the best subset of features based on machine learning algorithms evaluations[1–3]. Embedded methods incorporate filter inside wrapper.

Existing feature selection FS methods combine various evaluation measures and search strategies for selecting optimal feature subset. Large number of FS algorithms used but, none of them can outperform all others FS methods with all types of data. There is a need to find better algorithm with less computational cost.

In the category of swarm intelligence, several methods are introduced such as Ant Colony Optimization (ACO) [4], Artificial Bee Colony (ABC) [5], particle swarm optimization (PSO) [6].

© IFIP International Federation for Information Processing 2018
Published by Springer International Publishing AG 2018. All Rights Reserved
A. Amine et al. (Eds.): CIIA 2018, IFIP AICT 522, pp. 93–105, 2018.
https://doi.org/10.1007/978-3-319-89743-1_9

The Artificial Bee Colony (ABC) algorithm [5] proposed by Karaboga is one of the most recent swarm intelligence based optimization technique, which simulates the foraging behavior of a bee swarm. Because of ABC is easy to implement, converge fastly, it has been often applied in numerical optimization domain [7–9] and feature selection [3, 10]. However, there are still some drawbacks. ABC is computationally inefficient if it is trapped in the local optima [11].

The exploration and the exploitation abilities of the population based algorithms are both necessary characteristics. The exploration acts to search the various unknown regions to find the global optimum, while the exploitation attempts to apply the knowledge of the previous good solutions to find better solutions. The exploration and exploitation abilities should be well balanced to achieve good performance. However, research shows that the solution search equation of ABC algorithm is good at exploration but poor at exploitation [7, 8, 12].

Many variants of ABC algorithm were developed to enhance the exploitation ability, Zhu and Kwong [7] presented a global best solution guided ABC (GABC) algorithm by incorporating the information of global best into the solution search equation. Wu et al. [13] proposed improvement of global swarm optimization (GSO) by hybridizing it with ABC and PSO. They used neighborhood solution generation scheme of ABC and accepted new solution only when it was better than previous one to improve GSO performance.

Yan et al. [14] proposed for clustering a hybrid artificial bee colony (HABC) algorithm and embedding the crossover operator of GA to ABC in information exchange phase between bees. Gao et al. [15] develop a hybrid ABC using the modified search equation and orthogonal learning strategies to achieve efficiency.

Applying machine learning algorithms as evaluation function is time consuming especially for high dimensional data. So, it is necessary to improve any lack of FS methods to produce best selected subset.

In this paper, a hybridized swarm intelligence ABC with PSO and meta-heuristic approach like GA for optimal feature selection is presented to improve the balance between exploration and exploitation. To achieve that, particles swarm PSO contribute in ABC during employed bees, in addition, GA mutation operators are applied in Onlooker and Scout phase.

The main contributions are:

- Reduce the features space to the relevant characteristics achieving the highest accuracy.
- Improve the exploitation process by mutating bad solutions of onlookers.
- Replace abandoned solutions of scout by mutating worst source foods and also random good solutions of onlookers.
- Compare the behavior of ABC-PSO, ABC-GA with PSO, GA and ABC.
- Validate the efficiency of the exploitation and exploration of the proposed solutions.

The manuscript is organized as follows. Section 2 describes features selection methods: ABC, Particle swarm and Genetic algorithms and related work. Section 3 gives a detailed description of the proposed FS methods: ABC-PSO and ABC- Genetic Algorithm. Experimental results and discussions are presented in Sects. 4 and 5. finally conclusion in Sect. 6.

2 Background and Related Work

2.1 Artificial Bees Colony

Artificial bee colony (ABC) was proposed by Karaboga in 2005 [5] and mimics the foraging behaviors of honey bee colony. In optimization problem, the food sources represent candidate solutions and the nectar amounts correspond to fitness values. The ABC has three types of bees: employed, onlooker, scout. Employed bees exploit their food sources and convey the information concerning quality and position of food sources to onlooker bees via waggle dance. Onlooker bees select food source to be exploited with the help of the employed bees. Scout bees search a new food source. [9, 16]. ABC algorithm has slower convergence speed in some cases because of the lack of powerful local exploitation capacity [17].

Algorithm 1. Pseudo-code ABC algorithm [18].

1: Initialize the population of solutions Xi $\forall$ i, i =1, ..., NB.
2: Evaluate the population Xi using fitness function, $\forall$ i, i = 1, ..., NB.
3: for cycle = 1 to Maximum Iteration do
4: **For each Employed Bees** i
Produce and evaluate new solutions Vi using Eq. (1).

$$V(i,j) = X(i,j) + phi(i,j) * (X(i,j) - X(k, j)) \tag{1}$$

5: Apply the greedy selection process.
Endfor
6: **For each Onlooker** Bee i
Calculate the probability values Pi for the solution Xi by Eq. (2).

$$P(i) = fit (i) / \sum fit(k) , k=1..NB. \tag{2}$$

7: Produce and evaluate new solutions Vi for the solution Xi depending on Pi.
8: Apply the greedy selection process.
9: **Scout Bees phase**:
Replace the abandoned solutions with a new one Xi by using Eq. (3).

$$X(I,j) = X(min, j) + rand(0,1) * (X(max,j) - X(min, j)) \tag{3}$$

10: Memorize the best solution achieved so far. cycle = cycle + 1
12: end for

2.2 Particle Swarm Optimization

Particle swarm optimization PSO [6] inspired by the social behavior of flocks of birds. Feature selection has intrinsic discrete binary search spaces. The original Binary BPSO was proposed by Kennedy and Eberhart [19] to allow PSO to operate in binary problem spaces. In this version, particles could only fly in a binary search space by taking values of 0 or 1 for their position vectors.

PSO method presents the advantage of high convergence rate and low computational cost. PSO adjusts only a few parameters. It is also efficient. PSO has much better global optimization ability and lower computational time than GA and has been widely used in feature selection [20]. Unfortunately, falling into local optimum is the principal drawback of particles swarms.

$$V_i^{t+1} = w * V_i^t + c_1 * rand * (pbest - X_i^t)$$
$$c_2 * rand * (gbest - X_i^t)$$

(4)

The position is:

$$X_i^{t+1} = X_i^t + V_i^{t+1}$$

(5)

Where c1, c2: learning coefficients; w: inertia weight; pbest: personal best; gbest: global best.

PSO Pseudo Code

Input   Data, class, NBfeatures

Output   finalSubset, accuracy;

Initialize PSO particles

For I =1: MaxIteration   for k =1: NBparticles

Compute velocity with equation (4) and position with equation (5);

Calculate fitness with SVM classifier;

Update personal best;   Update global best;

End; End;

End.

2.3   Genetic Algorithm

Genetic algorithm (GA) is an evolutionary algorithm developed by John Holland and follow Darwin's principle of survival of the fittest [21]. GA is a successful wrapper method where each individual is a chromosome (candidate solution) and the set of chromosomes constitutes a population. However, GA is computationally expensive and tends to prematurely converge into local optima if the diversity mechanism is not well adapted. GA phases start with population initialization and generate new individuals by using successive genetic operators (selection, crossover, mutation) during every iteration, the best chromosomes of them are preserved.

GA Pseudo Code

Input Data, class, NBfeatures

Output finalSubset, accuracy;

Initialize GA chromosomes

For i=1: MaxIteration   for k=1: NBchromosomes /2 * CrossRate

  Selection of two chromosomes parents;

  Apply crossover with Crossover rate and evaluate fitness

Apply mutation and calculate fitness;

Rank all chromosomes according to fitness value and keep N top chromosomes;

End;End;End.

2.4   Related Work

Hancer et al. [3] propose a binary artificial Bees colony (ABC) for feature selection. Evolutionary based similarity search mechanism is introduced in binary ABC. This method is compared to PSO, GA, discrete ABC on 10 datasets. The results indicate that the performance is higher in term of accuracy and eliminate effectively redundant features.

In this paper [11], a hybrid approach based on the life cycle for the artificial bee colony algorithm is achieved, to ensure appropriate balance between exploration and exploitation. Each individual can reproduce or die dynamically throughout the searching process and population size can dynamically vary. The bees incorporate the information of global best solution into the search equation for exploration. Powell's search enables the bees deeply to exploit around the candidate area. Experiments based on the CEC 2014 benchmarks compared the performance of Hybrid ABC against other bio-mimetic algorithms and proved the strength of this algorithm.

Gao et al. [15] combined ABC with Differential evolution DE. An eliminative rule and the new search strategy are introduced into the iteration of ABC to improve the convergence rate. Then, to stabilize the population diversity, DE simulates evolution and all individuals are taken into account in each generation.

Authors [17] proposed a hybrid algorithm called PS–ABC, which combines the local search phase in PSO with two global search phases in ABC for the global optimum. In the iteration process, the algorithm checked the aging degree of pbest for each individual to decide which type of search phase to adopt. The proposed PS–ABC algorithm is validated on 13 high-dimensional benchmark functions from the IEEE-CEC 2014 competition problems, and compared with ABC, PSO. Results show that the PS–ABC algorithm is efficient, fast converging and robust optimization method for solving high dimensional optimization problems.

In this paper [18], authors described a method to perform a crop classification based on the Gray Level Co-Occurrence Matrix (GLCM) and the artificial bee colony (ABC) algorithm. The proposed methodology selects the set of features from the GLCM that allow classify the crops with a good accuracy using the ABC algorithm in terms of a distance classifier.

A hybrid PSO local search HPSO-LS [20] has been implemented for feature selection. They embedded local search in PSO to select low and salient feature subset. Local search guides the search process of the PSO to select distinct features by considering their correlation information. HPSO-LS was compared with GA, PSO, ACO, simulated annealing SA on 13 benchmark and prove the superiority of this approach.

Djellali et al. [22] propose two hybrid FS methods: multivariate filter fast correlation based filter FCBF and PSO (FCBF-PSO) and FCBF with genetic algorithm (FCBF-GA), FCBF-PSO combines in two phases particles swarm PSO and In first stage, FCBF filter is applied and provide a subset of features (relevant and no redundant), in the second stage, PSO is performed with previous selected features and executed on five UCI datasets. The experiments show that both FCBF-PSO and FCBF-GA (genetic algorithm replace PSO) are competitive.

3   Proposed Hybrid Feature Selection Method

In this study ABC, mutation operators and PSO are combined in ABC phases. This combination is called ABC-PSO mutate in Fig. 1. The reason is the fact that the ABC shows poor balance between exploitation and exploration.

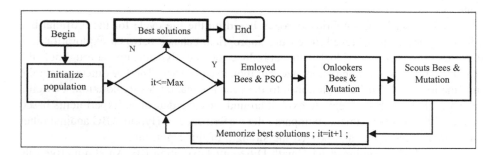

Fig. 1. Hybrid ABC-PSO mutate Algorithm

This scheme attempts to improve the search capability of Bees when their behavior do not provide good source foods, So, mutation is introduced in both Onlookers and Scout Bees to produce better solutions and PSO particles are compared to Employed Bees to take advantage of them. To establish the efficiency of the proposed algorithm ABC-PSO-mutate, we first compare two hybrids algorithms ABC-PSO and ABC-GA.

In ABC-PSO, the PSO particles are incorporated in employed Bees like described in Modified employed Bees code. Scout and onlookers keep the same code as standard ABC. In ABC-GA, we preserve standard employed bees unchanged and insert mutation in onlooker and scouts bees phases.

Proposed Hybrid ABC-PSO-mutate algorithm is built. Firstly, ABC employed Bees is modified by using PSO velocity and comparison between fitness of Employed Bees solutions and Particle solution. The best is saved.

Secondly, GA mutation is used in onlooker phase. After enough iteration, when probability of food source (i) is not improved, this food source solution is mutated and compared its fitness with previous value, in case where the new solution is better, we replace the current food source with new food source.

Scout Bees search new solutions when employed and onlookers are stuck. So, we propose to modify scouts behavior by mutating both 'the worst solution' and selected two 'random good solutions'. The new food source is the best of them.

3.1 Hybrid ABC-PSO Algorithm

ABC-PSO is proposed to stimulate competitive behavior between onlooker bees and PSO particles and applied to achieve better balance between exploration and exploitation. First, all initial employed Bees are copied into swarm PSO particles with their fitness. Second, both particles and employed Bees cooperate, particles compute velocity and position; Employed Bees calculate new source food. Each solution is evaluated and compared, then the best is affected to Employed Bees.

Hybrid ABC-PSO Algorithm

Input Data, class, NBfeatures, PSO parameters,
Output finalSubset, accuracy;
1: Initialize the population of solutions Xi $\forall$ i, i = 1, ..., NB.
2: Evaluate the population Xi, $\forall$ i, i = 1, ..., NB.
3: Copy the population Xi and its fitness in PSO initialization population.
4: for cycle = 1 to Maximum Cycle Number do
  5: Modified Employed Bees();
  6: Onlooker Bee phase;
  7: Scout Bees phase
  8: Memorize the best solution achieved; cycle = cycle + 1
End for

Modified Employed Bees()

---
1. **For** each Employed Bees i
2. Produce new solutions Vi using Eq. (1) and evaluate it.

$$V(i,j) = X(i,j) + phi(i,j) * (X(i,j) - X(k, j)) \tag{6}$$

  J is a feature index 1..D.

3. Calculate new velocity Vel(i+1) PSO equation (4) and new position X(i+1) equation (5).

4. Compare fitness X(i+1) and new source food V(i). keep the Best (Vi, Xi+1).
5. Update pbest and gbest and apply the greedy selection process.
6. If fitness(V(I,:) > fitness (X(I,:)) then X(I,:) = V(I, :)

Endfor

3.2   Hybrid ABC-GA Algorithm

In ABC-GA, we introduce mutation operators first on scout bees where worst source food are mutated an compared with random mutated good source food. Second, Onlookers bees are also modified with mutation operator. The code is presented below:

Hybrid ABC-GA Algorithm

Input Data, class, NBfeatures, CrossRate, MuRate, PSO parameters,
Output finalSubset, accuracy;

1: Initialize the population of solutions Xi $\forall$ i, i = 1, ..., NB.
2: Evaluate the population Xi, $\forall$ i, i = 1, ..., NB.
3: for cycle = 1 to Maximum Cycle Number MCN do
    4: Employed Bees;
    5: **Modified Onlooker Bee and mutation**
    6: Scout Bees phase: **ImprovedLazyBees();**
    7: Memorize the best solution achieved. cycle = cycle + 1

End for

Modified Onlookers Bees and Mutation

New source food solution Vi is selected depending on probability Pi. In fact, onlookers will not choose the worst solutions Vi from employed Bees only if in the next iteration the solutions are improved. Vi is selected and improved depending on Pi.

In the modified onlookers code, we mutate source foods (i) = Vi only after few iterations (20% of MaxIterations) and if (prob(i) < rand(0,1)) is true, then, the new mutated source is compared to source Vi. If the new source is better, we update Vi and its fitness.

Modified Onlookers()

Input Fitness, Foods,data, label;
Output Fitness, Foods;
For each Onlooker Bees i
  Calculate the probability values Pi for the solution Xi by Eq.(7).

$$P(i) = fit\ (i)\ /\ \sum fit(k)\ , k=1..NB \tag{7}$$

-   **If prob(i)>rand(0,1)**
    Produce new solutions Vi for the solution Xi selected depending on Pi.
    NewSubFeat = indices of selected value of Vi (equal to 1);
    Evaluate new source:
      errorNew = SVM_CostFitness (data(:,NewSubFeat), label);
    If errorNew < previousError(Foods(i))
      Foods(i) = V(i); Fitness(i) = errorNew; counter=0 **Else**   Counter++;
    End
    Else // added new code for ABC
      If nbIteration >= 20%*maxIteration
      NewFood = Mutate_method(Foods(i)); Evaluate new solution NewFood;
      Get selected features (where NewFood is set to 1).
      ErrorNew = SVM_CostFitness (data(:,NewSubFood), label);
      If errorNew < previousError(Foods(i))
        Foods(i) = V(i); Fitness(i) = errorNew; counter=0; **Else** Counter++;
    End. End.

Scout Bees phase

Scout Bees search new solutions using ImprovedLazyBees function. In original ABC code, Eq. (3) initialized new solutions, but there is no guarantee that these new solutions provide better results. In this regard, mutate both previous worst ABC solutions and random good ABC solution is required to guide scout bees towards better solutions.

Mutation operators help to diversify solutions whatever the source issue, from bad or good solution. Mutating good solutions can deliver also good solutions and improve the search process, consequently, increase the search ability of hybridized ABC. The solution is to replace the abandoned solutions (counter > limit) with a new one Xi by applying this code:

ImprovedLazyBees()

Input: Foods,Fitness
Output: NewFoods, newFitness;
For each abandoned solution j
Choose randomly two good food sources V(k1), V(k2) where k1 and k2 <> j ; and
K1, k2 are indices of good food sources;
- Mutate V(k1) , V(k2) and also current bad solution.
- Evaluate the three food sources and keep the best as new source of food.

---

4 Experimental Results

In this section, we first describe the experimental settings, then, we evaluate the performance of the proposed feature selection method ABC-PSO and ABC-GA on Support Vector Machine (SVM) [23]. We also compare the proposed algorithms with FS methods: ABC, GA, PSO.

4.1 Databases

To evaluate the performance of the proposed feature selection algorithms, we choose the dataset from UCI [24] machine learning repository. The dataset tested are Wisconsin Breast cancer Dataset WBCD, hepatitis, colon cancer, DLBCL. The datasets properties are summarized in Table 1.

Table 1. Dataset tested in the experimental study

Dataset	Number of features	Class	Number of instances
Hepatitis	19	2	80
Colon Tumor	2000	2	62
DLBCL	4026	2	45
WDBC	30	2	569

4.2   Parameters Setting of FS Algorithms

We describe the parameters setting of GA, PSO, hybrid ABCGA, ABC-PSO; all algorithms are implemented by using MATLAB 7.0. Each dataset is divided into training set (70%) and test set (30%). Classification accuracies of algorithms are measured by the standard 10 fold cross-validation methods on training data. To provide the test classification accuracy, the selected features are assessed on test set [25]. The average accuracy of the algorithms are computed with 20 independent runs. The parameters setting of algorithms are described in Table 2.

Table 2. Setting parameters

Algorithms	Setting parameters
ABC-PSO	Maximal iteration = 100; Population size = 80; mutation rate = 0.1; Crossover rate = 0.8; Fitness function : Accuracy of SVM classifier
PSO	Population size = 100; inertia weight = 0.7; personal learning coefficient $c1 = 1.49$; global learning coefficient $c2 = 1.5$; max iteration = 100; fitness function: Accuracy of SVM classifier
GA	Maximal iteration = 100; Population size = 50; mutation rate = 0.1; Fitness function: Accuracy of SVM classifier

Support Vector machine is an efficient machine learning technique that uses kernel to construct linear classification boundaries in higher dimensional space [6, 23]. SVMs have a good generalization performance in various applications.

5   Discussions

The experiments on WDBC data in Table 3 show that ABC-GA provides best accuracy (99.41%) and minimal number of features involved (13). ABC-PSO (accuracy = 96.71 and 16 features selected) accomplishes less efficiency in both accuracy and minimal number of features. ABC is better than PSO and Genetic algorithm (accuracy = 99.41% 15 features).

Table 3. WDBC performance

FS methods	SVM accuracy %	Numbers of features
GA	99.38	21
PSO	96.80	17
ABC	99.41	15
ABCPSO	96.71	16
ABCGA	**99.41**	**13**

The best results on hepatitis are achieved with hybrid ABC-GA (accuracy = 91.67 % with 9 features) in Table 4 but ABC-PSO provide less number of features (accuracy = 87.50% with 4 features).

Table 4. Hepatitis performance

FS methods	SVM accuracy %	Numbers of features
ABCPSO	87.50	4
GA	88.86	6
PSO	84.33	9
ABC	87.50	12
ABCGA	**91.67**	**9**

Colon cancer shows in Table 5 that the GA algorithm realizes the best SVM accuracy (96.34%) with 1001 features. ABC-PSO achieves minimal number of features (599) but less accuracy 79.17.

Table 5. Colon performance

FS methods	SVM accuracy %	Numbers of features
ABCPSO	79.17	599
GA	**96.34**	**1001**
PSO	82.44	1023
ABC	88.89	981
ABCGA	81.67	1027

The results on DLBCL data in Table 6 demonstrate that ABC-GA achieves the highest accuracy (91.45%) but more features involved (1943). ABC-PSO algorithm outperforms all FS methods for minimal number of features (1383) but less accuracy 90.00%. PSO provides (90.43% accuracy but more features 2068).

Table 6. DLBCL performance

FS methods	SVM accuracy %	Numbers of features
ABC-PSO	90.00	**1383**
GA	89.31	1991
PSO	90.43	2068
ABC	90.00	1541
ABCGA	**91.45**	1943

It was observed that the combination of ABC-GA gives the best results for hepatitis and WDBC. ABC-PSO provides low feature subset size than ABC-GA on all dataset except for WDBC data.

6 Conclusions

In this paper two hybrid approaches based on artificial bees colony and particles swarm ABC-PSO and ABC with Genetic algorithm ABC-GA are proposed. It has been found that the proposed method ABC-GA outperforms the proposed ABC-PSO method and exiting methods (ABC, GA, PSO) on hepatitis, WDBC and DLBCL. Despite the lowest accuracy of ABC-PSO, this hybrid method tends to produce less number of features. These observations lead us to make assumption that adding PSO in employed phase and mutations in onlookers and scout phase in ABC will maximize accuracy and minimize the feature subset size.

Adapting the behaviors of scout and onlooker bees lead to obtain the best result with hybrid ABC-GA and consolidate the first assumption concerning increasing the numbers of possible solutions in every iteration and consequently improve the exploitation process. We propose ABC-PSO mutate scheme and to establish efficiency of mutations alone we tested ABC-GA, to validate the performance of particle swarm on ABC alone, we introduce particles in employed bees phase. Both mutations and particles swarm are competitive, which means incorporating all these modification in ABC-PSO mutate should provide better results. We have to test all these modification in ABCPSOGA in future work.

References

1. Yu, L., Liu, H.: Feature selection for high-dimensional data: a fast correlation based filter solution. In: Proceedings of International Conference on Machine Learning, pp. 856–863 (2003)
2. Djellali, H., Zine, N.G., Azizi, N.: Two stages feature selection based on filter ranking methods and SVMRFE on medical applications. In: Chikhi, S., Amine, A., Chaoui, A., Kholladi, M.K., Saidouni, D.E. (eds.) Modelling and Implementation of Complex Systems. LNNS, vol. 1, pp. 281–293. Springer, Cham (2016). https://doi.org/10.1007/978-3-319-33410-3_20
3. Hancer, E., Xue, B., Karaboga, D., Zhang, M.: A binary ABC algorithm based on advanced similarity scheme for feature selection. J. Appl. Soft. Comput. 36, 334–348 (2015)
4. Dorigo, M., Maniezzo, V., Colorni, A.: Ant System: optimization by a colony of cooperating agents. IEEE Trans. Syst. Man Cybern. 26(1), 29–41 (1996)
5. Karaboga, D.: An idea based on honey bee swarm for numerical optimization, Technical report TR06, Erciyes University Engineering Faculty (2005)
6. Kennedy, J., Eberhart, R.C.: Particle swarm optimization. In: IEEE International Conference on Neural Networks, vol. 4, pp. 1942–1948 (1995)
7. Zhu, G., Kwong, S.: Gbest-guided artificial bee colony algorithm for numerical function optimization. J. Appl. Math. Comput. 217(7), 3166–3173 (2010)
8. Li, G., Niu, P., Xiao, X.: Development and investigation of efficient artificial bee colony algorithm for numerical function optimization. J. Appl. Soft Comput. 12(1), 320–332 (2012)
9. Karaboga, D., Basturk, B.: A Powerful and efficient algorithm for numerical function optimization: artificial bee colony (ABC) algorithm. J. Glob. Optim. 39(3), 459–471 (2007)
10. Schiezaro, M., Pedrini, H.: Data feature selection based on artificial bee colony algorithm. J. Image Video Process. 47, 1–8 (2013)

11. Ma, L., Zhu, Y., Zhang, D., Niu, B.: A hybrid approach to artificial bee colony algorithm. J. Neural Comput. Appl. **27**, 387–409 (2016)
12. Murugan, R., Mohan, M.: Artificial bee colony optimization for the combined heat and power economic dispatch problem. ARPN J. Eng. Appl. Sci. **7**(5), 597–604 (2012)
13. Wu, B., Qian, C., Ni, W., Fan, S.: The improvement of glowworm swarm optimization for the combined heat and power economic dispatch problem. ARPN J. Eng. Appl. Sci. **7**(5), 597–604 (2012)
14. Yan, X., Zhu, Y., Zou, W., Wang, L.: A new approach for data clustering using hybrid artificial bee colony algorithm. J. Neurocomputing **97**(15), 241–250 (2012)
15. Gao, W., Liu, S., Jiang, F., Zhang, J.: Hybrid artificial bee colony algorithm. J. Syst. Eng. Electron. **33**(5), 1167–1170 (2011)
16. Das, S., Biswas, S., Kundu, S.: Synergizing fitness learning with proximity-based food source selection in artificial bee colony algorithm for numerical optimization. J. Appl. Soft Comput. **13**(12), 4676–4694 (2013)
17. Li, Z., Wang, W., Yan, Y., Li, Z.: PS–ABC: a hybrid algorithm based on particle swarm and artificial bee colony for high-dimensional optimization problems. J. Expert. Syst. Appl. **42**(22), 8881–8895 (2015)
18. Vazquez, R.A., Garro, B.A: Crop classification using artificial bee colony (ABC) algorithm, advances in swarm intelligence (2016)
19. Kennedy, J., Eberhart, R.: A discrete binary version of the particle swarm algorithm. In: Proceedings of IEEE International Conference on Systems and Man, Cybernetics, Computational Cybernetics and Simulation, vol. 5, pp. 4104–4108 (1997)
20. Moradi, P., Gholampour, M.: A hybrid particle swarm optimization for feature subset selection by integrating a novel local search strategy. J. Appl. Soft Comput. **43**, 1–14 (2016)
21. Holland, J.H.: Adaptation in Natural and Artificial Systems. University of Michigan Press, Ann Arbor (1975)
22. Djellali, H., Guessoum, S., Ghoualmi-Zine, N., Layachi, S.: Fast correlation based filter combined with genetic algorithm and particle swarm on feature selection. In: 5th International Conference on Electrical Engineering - Boumerdes (ICEE-B), pp. 1–6 (2017). https://doi.org/10.1109/icee-b.2017.8192090
23. Vapnick, V.: Statistical Learning Theory. Wiley, Hoboken (1998)
24. Machine Learning Repository UCI https://archive.ics.uci.edu/ml/datasets.html
25. Kohavi, R., John, J.H.: wrappers for feature selection. J. Artif. Intell. **97**(1/2), 273 324 (1997)

An Efficiency Fuzzy Logic Controller Power Management for Light Electric Vehicle Under Different Speed Variation

Nouria Nair[✉], Ibrahim Gasbaoui[✉], and Abd El Kader Ghazouani[✉]

Laboratory of Smart Grids and Renewable Energies (SGRE),
Department of Electrical Engineering, University Tahri Mohamed of Bechar,
BP 417, 08000 Bechar, Algeria
nouia0479@gmail.com

Abstract. Light electric vehicle LEV autonomous present a major important problem for modern commercialized Electric Vehicle propulsion system. To improve the perfomance of LEV an efficiency fuzzy logic controller power management are proposed. The proton exchange mebran fuel hybrid system considered in this paper consists of fuel cells, lithium-ion batteries, and supercapacitors. The LEV is moving in the Algerian Saharan region, exactly in Bechar city. The aim objective of this work is to study the comportment of 2WDLEV based direct torque control supplied by differents sources of energy under diffrents speed varaiation. The performances of the proposed strategy controller give a satisfactory simulation results. The proposed control law increases the utility of LEV autonomous under several speed variations. Moreover, the future industrial's vehicle must take into considerations the hybrid power management choice into design steps. ...

Keywords: LEV · Buck boost · DC-DC converter
Fuzzy logic controller · PEMFC · Power management

1 Introduction

As known that the hydrocarbons sector is the backbone of the Algerian economy witch the fall in oil and natural gas prices, has led Algeria government to adopt modest austerity measures and increased the pressure for structural and institutional economic reform. In this way the Electric Vehicles are proposed in this paper. LEV present several advantages, no emission of hydrocarbons, fumes or particles, no consumption during idling phases, batteries are recharged during deceleration phases, reduced maintenance costs and the engine is perfectly silent. The battery is capable of storing sufficient energy, offer high energy efficiency, high current discharge, and good charge acceptance from regenerative braking, high cycle time and calendar life and abuse tolerant capability. It should

A. Amine et al. (Eds.): CIIA 2018, IFIP AICT 522, pp. 106–118, 2018.
https://doi.org/10.1007/978-3-319-89743-1_10

also meet the necessary temperature and safety requisites. Nickel metal hydride (NiMH) batteries have dominated the automotive application since 1990's due to their overall performance and best available combination of energy and power densities, thermal performance and cycle life. They do not need maintenance, require simple and inexpensive charging and electronic control and are made of environmentally acceptable recyclable materials. [1–3] This paper deals with the behavior light electric vehicle (LEV) moving in the hot region under different speed variation. The LEV is equipped with two induction motors providing each one 3,3 HP. In order to minimize the ripple of current, flux and the electromagnetic torque of both induction motors an artificial neural network based direct torque control is proposed. To evaluate the performance of the proposed system, a simulation test is realized in all operating system conditions.

2   Light Electric Vehicle Description

According to Fig. 1 the opposition forces acting to the light electric vehicle motion are: the rolling resistance force Ftire due to the friction of the vehicle tires on the road; the aerodynamic drag force Faero caused by the friction on the body moving through the air; and the climbing force Fslope that depends on the road slope [1]. The total resistive force is equal to and is the sum of the resistance forces, as in (1).

$$F_r = F_{tire} + F_{aero} + F_{slope} \tag{1}$$

The rolling resistance force is defined by:

$$F_{tire} = mgf_r \tag{2}$$

The aerodynamic resistance torque is defined as follows:

$$F_{aero} = \frac{1}{2}\rho_{air}A_fC_d\nu^2 \tag{3}$$

The rolling resistance force is usually modeled as:

$$F_{tire} = mgf_r \tag{4}$$

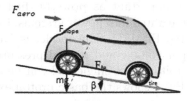

Fig. 1. The forces acting on a vehicle moving along a slope [2].

where is the tire radius, m is the vehicle total mass, is the rolling resistance force constant, the gravity acceleration, is Air density, is the aerodynamic drag coefficient, is the frontal surface area of the vehicle, is the vehicle speed, is the road slope angle. Values for these parameters are shown in Table 1. The Light Electric Vehicle considered in this work is two wheels drive destined to urban transportation. Two induction motors are used forlight electric vehicle. The energy source of the electric motors comes from PEMFC hybrid system composed by Fuel cell, lithium ion battery and supercapacitor Lithium-ion battery controller by Buck boost DC-DC converter and boost converter the energy management are assured by fuzzy logic controller [3].

Table 1. Parameters of the electric vehicle model

r	0.82 m	AF	0.80 m^2
m	400 Kg	Cd	0.32
fr	0.01	pair	1.2 Kg/m^3

3   Fuel Cell Static Model

Hydrogen PEM fuel cells transform chemical energy into electrical and thermal energy by the simple chemical reaction [4–6].

$$H_2 + \frac{1}{2}O_{t2} \Longrightarrow H_{2o} + heat + electricalenergy \tag{5}$$

In order to get an electric current out of this reaction, hydrogen oxidation and oxygen reduction are separated by a membrane, which is conducting protons from the anode to the cathode side. The semi reactions on both electrodes are:

$$H_2 \Longrightarrow 2H + 2e^- anode \tag{6}$$

$$O_2 + 4e^- \Longrightarrow 2O_2^- cathode \tag{7}$$

While the protons are transported through the membrane, electrons are carried by an electric circuit in which their energy can be used. Modelling of fuel cells is getting more and more important as powerful fuel cell stacks are getting available and have to be integrated into power systems. In [7] Jeferson M. Corrêa introduced a model for the PEMFC. The model is based on simulating the relationship between output voltage and partial pressure of hydrogen, oxygen, and currant. The output voltage of a single cell can be defined as the result of the following expression, Fig. 2 shows the basic proton exchange membrane fuel cell scheme:

$$V_{FC} = E_{Nenst} - V_{act} - V_{ohm} - V_{con} \tag{8}$$

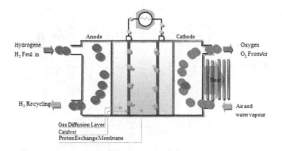

Fig. 2. Basic Proton exchange membrane fuel cell scheme [3,6].

4   Direct Torque Control Strategy Based Space Vector Modulation (SVM-DTC)

In this technique there are two proportional integral (PI) type controllers instead of hysteresis band regulating the torque and the magnitude of flux. Refereeing to Fig. 3, two proportional integral (PI) type controllers regulate the flux amplitude and the torque, respectively. Therefore, both the torque and the magnitude of flux are under control, thereby generating the voltage command for inverter control. Noting that no decoupling mechanism is required as the flux magnitude and the torque can be regulated by the PI controllers. Due to the structure of the inverter, the DC bus voltage is fixed, therefore the speed of voltage space vectors are not controllable, but we can adjust the speed by means of inserting the zero voltage vectors to control the electromagnetic torque generated by the induction motor. The selection of vectors is also changed. It is not based on the region of the flux linkage, but on the error vector between the expected and the estimated flux linkage [8–11]. The induction motor stator flux can be estimated by:

$$\Phi_{ds} = \int_0^t (V_{ds} - R_s i_{ds})dt \tag{9}$$

$$\Phi_{qs} = \int_0^t (V_{qs} - R_s i_{qs})dt \tag{10}$$

$$|\Phi_s| = \sqrt{\Phi_{d_s}^2 + \Phi_{q_s}^2} \tag{11}$$

$$\theta_s = tan^{-1}(\frac{\Phi_{qs}}{\Phi_{ds}}) \tag{12}$$

The electromagnetic torque Tem can be written as follow:

$$T_{em} = \frac{3}{2}p(\Phi_{ds}i_{qs} - \Phi_{qs}i_{ds}) \tag{13}$$

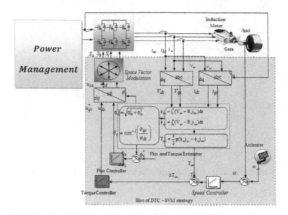

Fig. 3. Bloc diagram for DTC strategy based space vector modulation [2].

The SVM principle is based on the switching between two adjacent active vectors and two zero vectors during one switching period. It uses the space vector concept to compute the duty cycle of the switches.

5   Energy Management Strategies

The energy management system is required to ensure the following:

- low hydrogen consumption;
- high overall system efficiency;
- narrow scope of the battery/supercapacitor SOC;
- long life cycle [11].

This is achieved by controlling the power response of each energy source with load demand through their associated converters, using a given EMS. For this paper, fuzzy logic controller state-of-the-art EMSs are considered and designed based on the requirements given in Table 2.

6   The Rule Based Fuzzy Logic Strategy

This scheme has a faster response to load change compared to state machine control and is more robust to measurement imprecisions. The fuel cell power is obtained based on the load power and SOC membership functions and the set of if-then rules. The scheme is shown in Fig. 4. The design is made following an approach similar [12] to where trapezoidal membership functions are used as shown in Figs. 5, 6 and 7. The fuzzy logic rules are derived from the state machine control decisions as shown in Table 2. The Mamdani's fuzzy inference approach is used along with the centroid method for defuzzification. The fuzzy

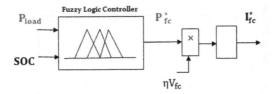

Fig. 4. Basic fuzzy logic controller

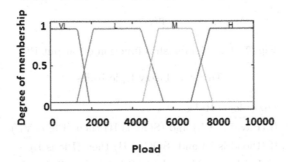

Fig. 5. The Membership function of input Pload

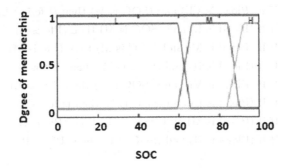

Fig. 6. The Membership function of input SOC

logic control surface obtained is shown in Fig. 8. The linguistic variables are defines as H, M, L, VL meaning high, medium, low and very low respectively, and the membership function is illustrated in the Figs. 5, 6 and 7. Using the settings given in Table 2 the fuzzy controllers were obtained and are given in Fig. 8 [10].

The rule based fuzzy logics strategy is implemented in SPS using a Simulink Fuzzy Logic Controller block from the Fuzzy logic Toolbox. The design of this Fuzzy Logic controller is made with the help of the FIS (Fuzzy Inference System) Editor GUI (Graphical user interface) tool of Matlab. This tool allows to create input/output variables, membership functions and rules in a very convenient fashion, without having to develop complicated fuzzy logic system code. The twelve Fuzzy tuning rules are descript in Table 2 below.

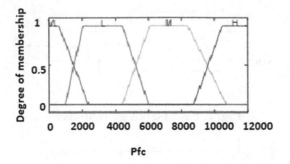

Fig. 7. The Membership function of output Pfc

Table 2. Fuzzy logic Rules

N	Fuzzy logic Rules
1	If (Pload is VL) and (SOC is H) then (Pfc is VL)
2	If (Pload is L) and (SOC is H) then (Pfc is L)
3	If If (Pload is M) and (SOC is H) then (Pfc is M)
4	If If (Pload is H) and (SOC is H) then (Pfc is H)
5	If If (Pload is VL) and (SOC is M) then (Pfc is VL)
6	If If (Pload is L) and (SOC is M) then (Pfc is L)
7	If If (Pload is M) and (SOC is M) then (Pfc is M)
8	If If (Pload is H) and (SOC is M) then (Pfc is H)
9	If If (Pload is VL) and (SOC is L) then (Pfc is L)
10	If If (Pload is L) and (SOC is L) then (Pfc is M)
11	If If (Pload is M) and (SOC is L) then (Pfc is H)
12	If If (Pload is H) and (SOC is L) then (Pfc is H)

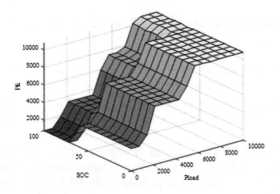

Fig. 8. Fuzzy logic control surface.

7   Simulation Results

In order to characterize the behavior of the driving wheel system, simulations were carried out using the model in Fig. 9. The following results were simulated in MATLAB/SIMULINK, and it's divided into two phases, the first one represent a performance test of 2WDES controlled by DTC-SVM in various speed and second phase shows the behavior of 2WDES energy management schemes based fuzzy logic controller for the fuel-cell hybrid system during the different scenarios consideration. The fuel-cell hybrid system are composed by fuel cells, lithium-ion batteries, and supercapacitor.

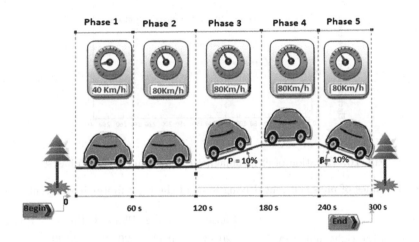

Fig. 9. Specify driving route topology

7.1   Direct Torque Control Scheme with Space Vector Modulation

The topology studied in this present work consists of five phases: the first one represent straight road with 40 km/h, the second phase symbolize straight road with 80 km/h, the tired phase a 2WDES is moving up the slopped road of 10% under 80 km/h, fourth cases represent directly road and finally the proposed system are moving the inverse sloped road with 80 km/h, the speed road constraints are described in the Table 3.

Table 4 explains the variation of phase current and driving force respectively. In the first step and to reach 80 km/h. The 2WDEV demand a current of 20.43 A for each motor which explained with driving force of 128.70 N. The third phases explain the effect of sloped road. The driving wheels forces increase and the current demand undergo double of the current braking phases the PEMFC use 41.87% of his power to satisfy the motorization demand under the slopped road condition which can interpreted physically the augmentation of the globally vehicle resistive torque illustrate in Fig. 10. In the other hand the linear speeds of

Table 3. Specified driving route topology

Phases	Event information	Vehicle Speed [km/h]
Phase 1	Straight road	40
Phase 2	Straight road	80
Phase 3	climbing a slope 10%	80
Phase 4	Straight road	80
Phase 5	climbing a inverse slope 10%	80

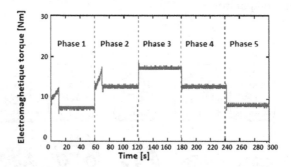

Fig. 10. Variation of electromagnetic torque in different phases.

Table 4. Values of phase current driving force of the right motor in different phases.

Phases	Phase 1	Phase 2	Phase 3	Phase 4	Phase 5
Current of the right motor [A]	20.43	20.43	20.43	20.43	18.50
Driving force of the right motor [N]	128.70	128.70	161.10	128.70	96.28

the two induction motors stay the same and the road drop does not influence the torque control of each wheels. In the fifth phases the current and driving forces demand decrees by means that the vehicle is in recharging phase's which explained with the decreasing of current demand and developed driving forces shown in Fig. 10. The results are listed in Table 4. According to the formulas (1), (2), (3) and (4) and Table 5, the variation of vehicle torques in different cases as depicted in Fig. 12, the vehicle resistive torque was 127.60 N.m in the first case (curved road), in the sloped road (phases 3), the front and rear driving wheels develop more and more efforts to satisfy the traction chain demand which impose an resistive torque equal to 168.00 N.m. In the last phase (breaking phase) the resistive vehicle torque are equal to 86.00 N.m. The result prove that the traction chain under sloped road demand develop the double effort comparing with the breaking phase case's by means that the vehicle needs the half of its energy in the inverse slop phase's compared with the sloped road one's as it specified in Table 4 and Fig. 11.

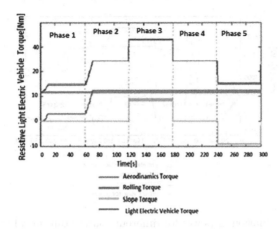

Fig. 11. Evaluation of resistive vehicle torque in different phases.

Table 5. Variation of vehicle torque in different Phases.

Phases	Phase 1	Phase 2	Phase 3	Phase 4	Phase 5
The Vehicle resistive torque [N.m]	127.60	127.60	168.40	127.60	86.00
The globally vehicle resistive torque Percent compared with nominal motor torque of 392.46 Nm	32.51%	32.51%	42.90%	32.51%	24.34%

7.2 Fuzzy Logic Controller Power Management for 2WDES

The PEMFC hybrid system considered in this paper consists of PEMFC, lithium-ion batteries, and supercapacitors must be able to supply sufficient power to the 2WDES in different phases, which means that the peak power of the PEM fuel cell supply must be greater than or at least equal to the peak power of the two electric motors. The PEMFC must store sufficient energy to maintain their Fuel at a reasonable level during driving, Fig. 12, describe the evaluation of power for different sources during all trajectory.

It is very interesting to describe the power of different sources distribution in the electrical traction under several speed variation as it described in Fig. 12. The PEMFC power provides about 2.07 Kw in the first phase in order to reach the electronic differential reference speed of 40 km/h. At t = 70 s. At this time the extra load power required is instantly supplied by the supercapacitor due to its fast dynamics, while the fuel cell power increases slowly. In the second phases and exactly at t = 70 s, the peak of 2WDEC is assured by the lithium ion battery power, in the third phases (sloped phase's) the power demand are assured by the PEMFC power provides about 8.80 Kw and battery power (1.8 Kw). In the fourth phases the globally nominal power PEMFC (7.02 Kw). The power of the battery is negative that mean the lithium ion is recharged via PEMFC and the supercapacitor provide their power to satisfied the power demand. And finally

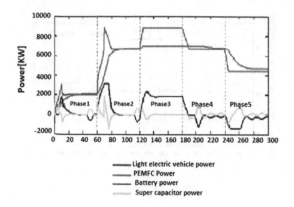

Fig. 12. Evaluation of power for different sources durant all trajectory.

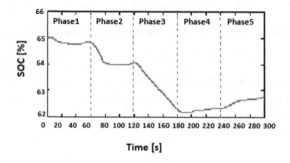

Fig. 13. Variation of state of charge during all trajectory.

the PEMFC power is about 4.7 Kw. Of the demanded power battery increased about 13.00 Kw that present 16.10% produced power is equal to 6.70 Kw under inverse slopped road state.

Figure 13 explains how SOC in the Lithium-ion battery changes during the driving cycle, it seems that the SOC decreases rapidly at the third phases (sloped road 10%), At t = 192 s, the SOC battery becomes less than 62.18% (it was initialized at 65% at the beginning of the simulation), so the battery must be recharged. The PEMFC and super capacitor shares its power between the lithium ion battery and the 2WDES you can see that the battery power becomes negative. This means that the battery receives power from the PEMFC and super capacitor (Table 6).

Table 6. State of charge in different scenarios.

Time [s]	0	60	120	180	240	300
State of charge[%]	65	64.73	64.32	64.37	62.32	62.73

The relationship between SOC and time in different phases are defined by the flowing linear fitting formula:

$$
\begin{aligned}
SOC[\%] = & -3.6864e^{-21}t^{10} + 4.7848e^{-186}t^9 - 2.4794e^{-15}t^8 + 6.3089e^{-13}t^7 \\
& - 7.0985e^{-11}t^6 - 9.6612e^{-10}t^5 - 1.1196e^{-6}t^4 - 0.0001144t^3 \\
& + 0.0047067t^2 - 0.078718t + 65.241
\end{aligned}
\tag{14}
$$

$$
\begin{aligned}
FulConsumption[g] = & -2.0377e - 06t^3 + 0.0010537t^2 - -0.031357t \\
& + 0.40936
\end{aligned}
\tag{15}
$$

8   Conclusion

The vehicle energy policy outlined in this paper has demonstrated that the PEMFC behavior controlled by buck boost DC-DC converter for utility 2WDEV which utilize two rear and front deriving wheel for motion can be improved using direct torque control strategy based on space vector modulation when the PEMFC developed power depend on the speed reference of the driver. The several topologies road do not affect the performances of the buck boost DC-DC converter output voltage and the control strategy gives good dynamic characteristics of the 2WDEV propulsion system. This paper proposes novel fitting formulas which give the relationship between the PEMFC voltage and distance traveled and others formulas that give more efficiency to different propulsion systems paths. This study enables the prediction of PEMFC dynamic behavior under different road topologies conditions, which is considered as a foundation for control and power management for 2WDEV.

References

1. Gasbaoui, B., Chaker, A., Laoufi, A., Allaoua, B., Nasri, A.: The efficiency of direct torque control for electric vehicle behavior improvement. Serb. J. Electr. Eng. 8(2), 127–146 (2011)
2. Gasbaoui, B., Nasri, A., Laoufi, A., Mouloudi, Y.: 4 WD urban electric vehicle motion studies based on MIMO fuzzy logic speed controller. Int. J. Control Autom. 6(1), 1–14 (2013). Faculty of the sciences and technology, Bechar University
3. Brahim, G.: Commande direct du couple d'un vehicule electrique à deux roues motrices. These p. 47, 97, 99 (2012)
4. Colella, W.: Cleaning the air with fuel cell vehicles: net impact on emissions and energy use of replacing conventional internal combustion engine vehicles with hydrogen fuel cell vehicles. In: The First European Fuel Cell Technology and Applications Conference, ASME (2005)
5. Boettner, E., Paganelli, G., Guezennec, Y.G., Rizzoni, G., Moran, M.J.: Proton exchange membrane fuel cell system model for automotive vehicle simulation and control. ASME J. Energy Res. Technol. 124(1), 20–27 (2002)

6. Hirschenhofer, J.H., Stauffer, D.B., Engleman, R.R., Klett, M.G.: Fuel Cell Handbook, Seventh Edition. FETC (2004)
7. Corrêa, J.M.: Simulation of fuel-cell stacks using a computer-controlled power rectifier with the purposes of actual high-power injection applications. IEEE Trans. Ind. Appl. **39**(4), 1136–1142 (2003)
8. Jafarboland, M., Zarchi, H.A.: Efficiency-optimized variable structure direct torque control for synchronous reluctance motor drives. J. Electr. Syst. **8**(1), 95–107 (2012)
9. Zhang, Z., Tang, R., Bai, B., Xie, D.: Novel direct torque control based on space vector modulation with adaptive stator flux observer for induction. IEEE Trans. Magn. **46**(8), 3133–3136 (2010)
10. Belkacem, S., Naceri, F., Abdessemed, R.: Improvement in DTC-SVM of AC drives using a new robust adaptive control algorithm. Int. J. Control Autom. Syst. IJCAS **9**(2), 267–275 (2011)
11. Motapon, S.N., Dessaint, L.-A., Al-Haddad, K.: A comparative study of energy management schemes for a fuel-cell hybrid emergency power system of more-electric aircraft. IEEE Trans. Industr. Electron. **61**(3), 1320–1334 (2014)
12. Caux, L.S., Hankache, W., Fadel, M., Hissel, D.: On-line fuzzy energy management for hybrid fuel cell systems. Int. J. Hydrogen Energy **35**(5), 2134–2143 (2010)

A New Handwritten Signature Verification System Based on the Histogram of Templates Feature and the Joint Use of the Artificial Immune System with SVM

Yasmine Serdouk, Hassiba Nemmour[✉][iD], and Youcef Chibani

Laboratoire d'Ingénierie des Systèmes Intelligents et Communicants (LISIC),
Faculty of Electronic and Computer Sciences, University of Sciences
and Technology Houari Boumediene (USTHB), 16111 Algiers, Algeria
{yserdouk,hnemmour,ychibani}@usthb.dz

Abstract. Verifying the authenticity of handwritten signatures is required in various current life domains, notably with official contracts, banking or financial transactions. Therefore, in this paper a novel histogram-based descriptor and an improved classification of the bio-inspired Artificial Immune Recognition System (AIRS) are proposed for handwritten signature verification. Precisely, the Histogram Of Templates (HOT) is introduced to characterize the most widespread orientations of local strokes in handwritten signatures, while the combination of AIRS and SVM is proposed to achieve the verification task. Usually, using the k Nearest Neighbor rule, a questioned signature is classified by computing dissimilarities with respect to all AIRS outputs. In this work, using these dissimilarities, a second round of training is achieved by the SVM classifier to further improve the discrimination power. In comparison with existing methods, the experiments on two widely-used datasets show the potential and the effectiveness of the proposed system.

Keywords: Artificial Immune Recognition System
Handwritten signature verification · Histogram Of Templates · SVM

1 Introduction

Handwritten signature is a biometric feature unique to each person. As it depends on physical and psychological conditions of the writer, researchers have to deal with the intra variability of the signer in order to develop robust systems for signature verification. One can mention two verification approaches: on-line and off-line. The on-line verification, in which signatures are acquired via an electronic device, considers dynamic information of signatures. In the off-line approach, signatures are written on a sheet of paper. In this case, features are

© IFIP International Federation for Information Processing 2018
Published by Springer International Publishing AG 2018. All Rights Reserved
A. Amine et al. (Eds.): CIIA 2018, IFIP AICT 522, pp. 119–127, 2018.
https://doi.org/10.1007/978-3-319-89743-1_11

calculated from the signature shape. Furthermore, the verification can be carried out according to two strategies: writer-dependent or writer-independent [1]. To authenticate genuine and forged signatures, the writer-dependent strategy develops a specific system adapted to each person's style while only one generic system is developed for all persons in the writer-independent framework.

To characterize effectively signature images, several global and local descriptors were employed during the past years. For instance, typical global features are the mathematical transforms, such as Wavelets, Ridgelets and Contourlets [2–4]. Nevertheless, local features are much preferred since they describe specific parts of signature images, which makes them robust to global shape variations [5]. In this respect, we note topological features, such as pixel density and pixel distribution, curvature features, orientation features and gradient features [1,5,6].

Moreover, various methods were developed to achieve the verification task, such as dynamic time warping, neural networks, hidden Markov models and SVM [7]. Currently, SVM is the most commonly used classifier since it can significantly outperform the others [8]. Nevertheless, the scores reported in literature are not optimal and still need improvements. Recently, many interesting mechanisms inspired from the natural immune system allowed the development of Artificial Immune Recognition Systems (AIRS) that tackle with various pattern recognition applications, such as thyroid diagnosis [9] and fault detection [10]. AIRS classification adopts a supervised learning process to create new representative data for each class, called Memory Cells (MC). Then, the k Nearest Neighbor (kNN) rule is performed over the established MC to classify test data. In [6,11,12], the authors successfully employed the AIRS classifier for off-line signature verification. However, experiments showed that the user-defined parameters must be carefully tuned for each writer's characteristics in order to achieve a competitive performance. Also, since the kNN decision depends only on the pertinence of the produced MC, a more powerful decision is conceivable.

Presently, we propose a robust system for off-line signature verification. A novel descriptor using the Histogram Of Templates (HOT) is introduced to characterize stroke orientations in signatures. For the verification step, we jointly use AIRS and SVM to overcome the shortcomings of the conventional AIRS classifier. Precisely, after the AIRS training, a set of dissimilarities is calculated between original data and the evolved MC from the AIRS training. Then, these dissimilarities are used to train a SVM to develop an automatic decision about questioned signatures. The rest of this paper is organized as follows: Sect. 2 introduces the proposed signature verification system. Experiments are presented and discussed in Sect. 3, followed by the main conclusions in the last Section.

2   Proposed Signature Verification System

The proposed Signature Verification System (SVS) is composed of a feature generation module that is based on the Histogram Of Templates and a verification module, which combines AIRS with SVM. The verification task is achieved according to the writer-dependent strategy. So, for each writer, an SVS is developed to discriminate between genuine signatures and skilled forgeries. In this section, a detailed explanation of the feature extractor "Histogram Of Templates" is carried out. Then, a brief overview of the AIRS theory is made, followed by the details of how the combination of AIRS and SVM is performed.

2.1   Histogram Of Templates

The Histogram Of Templates (HOT) is proposed for highlighting local stroke orientations by using a set of templates. As shown in Fig. 1, sliding windows covering (3×3) pixels are applied on a signature image to count the number of pixels that fit each template [13]. The resulting counts constitute the histogram of templates. So, if we consider twenty templates, the histogram will have 20 bins. Each bin corresponds to the number of pixels P matching a template k. Presently, HOT feature is computed by considering both pixel information and gradient information. This leads to a histogram of 40 bins that combines pixel and gradient information vectors.

Pixel Information-Based HOT. For each template, if the gray value $I(P)$ of a pixel P is greater than the gray value of the two adjacent pixels, P matches the template.

$$I(P) > I(P1) \quad \&\& \quad I(P) > I(P2). \tag{1}$$

Gradient Information-Based HOT. For each template, if the gradient magnitude $Mag(P)$ of a pixel P is greater than the gradient magnitudes of the two adjacent pixels, P matches the template.

$$Mag(P) > Mag(P1) \quad \&\& \quad Mag(P) > Mag(P2). \tag{2}$$

2.2   Artificial Immune Recognition System

The Artificial Immune Recognition System (AIRS) is a bio-inspired classifier that was introduced by Watkins in [14]. Through mutations and resources competition processes, the training of AIRS generates new data that are called antibodies (or Memory Cells, MC) in order to represent variability within the classes of interest. The training algorithm considers each training signature as an antigen. Also, each generated antibody with its associated class label (genuine class or forged class) is called Artificial Recognition Ball (ARB). Note that ARBs are provisional MC that will be used during the training process to produce the final established MC. Before the beginning of the AIRS training, an initialization of the MC set is carried out by selecting randomly one training sample from each class.

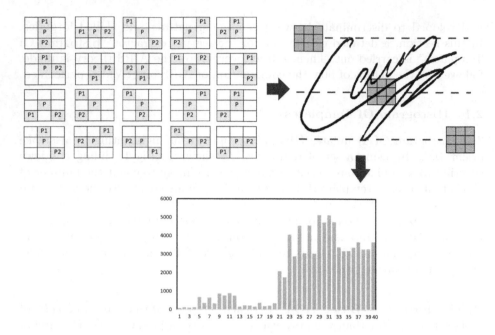

Fig. 1. HOT calculation for a signature sample.

Training Process. The training of each antigen (i.e. a training signature image) is a one-shot process that is described in what follows:

- **MC-match selection:** MC-match represents the highest stimulated MC. The stimulation ST is calculated between each MC and the actual antigen as:

$$ST(ag_i, MC) = 1 - affinity(ag_i, MC). \tag{3}$$

  While *affinity* is the Euclidian distance and ag_i is the ith antigen. Then, using the selected MC-match a set of randomly mutated clones (ARBs) is generated.
- **Resources competition:** a competition between the generated clones (ARBs) is carried according to their stimulation level in order to ensure the development of more representative cells (i.e. the most stimulated ARBs that allow the recognition of antigens).
- **MC-candidate selection and MC pool update:** the ARB having the highest stimulation is selected as being the MC-candidate. Then, based on a comparison with MC-match, MC-candidate will replace MC-match or will be added to the memory cells population.

Classification. A questioned signature is classified as being genuine or forged according to its k Nearest Neighbors within the MC population.

AIRS Shortcomings. Because of the writer-dependent protocol, the AIRS training employs several set-up parameters that must be tuned according each writer's characteristics. This leads to several tests to find the optimal combination of parameters [6]. Mainly, AIRS parameters are described as follows:

- **Mutation rate:** a real taken between [0.002–0.01] that represents the mutation probability of an ARB.
- **Clonal rate:** an integer in the range [10–200] that controls the number of generated mutated clones.
- **Affinity threshold scalar:** a real taken in the range [0.1–1] used within the MC-candidate and MC-match comparison.
- **Stimulation threshold:** a real ranged between [0.1–1] used as a stopping criterion in the training routine of an antigen.
- **Resources number:** an integer between [100–700] that limits the number of mutated clones (ARBs) allowed in the system.

Furthermore, as the decision of the classical AIRS depends only on the pertinence of produced MC, we propose a hybrid verification system, in which the kNN classification is substituted by a support vector decision. This implementation allows us to globally tune the AIRS parameters for all writers while improving greatly the verification performance.

2.3 The Joint Use AIRS-SVM

The joint use AIRS-SVM as a verification system is achieved according to the following steps (see Fig. 2).

- Train AIRS according to the steps reported in Subsect. 2.2.
- Develop new training and testing sets by substituting signature features by their dissimilarities with respect to all memory cells in the MC Pool.
- Train SVM with the new training dissimilarity set to separate genuine dissimilarities from forged dissimilarities.
- Incorporate the dissimilarity vector of each questioned signature into the support vector decision to decide if it is a genuine or a forged signature.

3   Experimental Results

Our experimental study is conducted on two widely-used datasets: MCYT-75 and GPDS-300. MCYT-75[1] contains off-line signatures of 75 writers with 15 genuine and 15 skilled forgeries each. While the GPDS-300 corpus[2] contains off-line signatures of 300 writers represented by 24 genuine signatures and 30 skilled forgeries for each. Performance evaluation is based on the False Rejection Rate (FRR), the False Acceptance Rate (FAR) and the Average Error Rate (AER). This latter represents the average value between FAR and FRR.

---

[1] MCYT dataset is available on: http://atvs.ii.uam.es/databases.jsp.
[2] GPDS dataset is available on: http://www.gpds.ulpgc.es/download/.

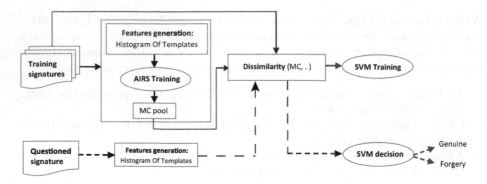

Fig. 2. Proposed signature verification system (continuous lines indicate the training flowchart while dashed lines indicate the verification flowchart).

Following the protocol reported in [15], for each dataset, the training stage utilizes 10 genuine and 10 forged signatures that are randomly selected while the remaining signatures are used to test the verification performance. In this work, AIRS parameters take the same values for all writers to facilitate its implementation. So, as a first experiment, we tried to select the optimal k^{th} neighbor allowing the best AER on training data. As shown in Fig. 3 for both datasets, the best accuracy is obtained when considering one neighbor with an AER about 31% and 18% for MCYT-75 and GPDS-300, respectively. From these outcomes, we deduce that the kNN classification cannot deal with the variability of information offered by the evolved memory cells. Consequently, to improve the AIRS classifier, the proposed system performs a second round of training to take more advantage from both training set and memory cells to achieve a more robust verification. Table 1 reports error rates as well as the verification time using AIRS,

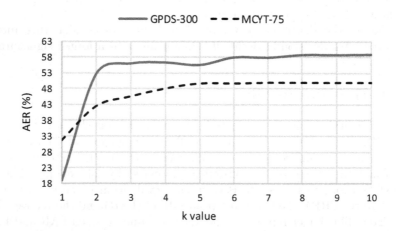

Fig. 3. AER variations according to the k values for kNN-based AIRS verification.

SVM and the proposed joint use of AIRS-SVM. For the proposed system, the verification time includes the HOT calculation, the dissimilarity computation, the SVM training and the decision time. In the conventional AIRS, the verification time corresponds to the calculation of HOT features and the computation of the KNN decision for a questioned signature while in SVM, it corresponds to the HOT computation plus the support vector decision time.

The results show that the proposed joint use AIRS-SVM outperforms both the classical AIRS and SVM performances. Indeed, the AIRS-SVM combination allows a significant improvement in AER values with at least a gain 6% for MCYT-75 and 0.7% for GPDS-300. Moreover, thanks to the support vector decision, the proposed combined verification system provides lower FAR than FRR, which reflects its ability to favor the reduction of false accepted signatures. The comparison of the verification time required to treat a questioned signature reveals that the proposed combination requires approximately the same duration as AIRS or as SVM. In addition, compared to the state-of-the-art results reported in Tables 2 and 3, the joint use of AIRS with SVM provides competitive outcomes.

Table 1. Signature verification results obtained for AIRS and the proposed implementation.

Dataset	Classifier	Verification time (s)	FAR (%)	FRR (%)	AER (%)
MCYT-75	AIRS	3.13	50.40	12.53	31.47
	SVM	3.13	17.06	17.06	17.07
	AIRS-SVM	3.15	**08.80**	**13.33**	**11.07**
GPDS-300	AIRS	1.42	22.43	13.76	18.87
	SVM	1.42	12.85	10.80	12.04
	AIRS-SVM	1.44	**07.61**	**16.66**	**11.35**

Table 2. MCYT-75 state-of-the-art.

References	Features	Classifier	#Genuine signatures	AER (%)
[16]	Slant measure	Variability measure	10	22.13
[17]	Geometric centroids	Degree of authenticity	9	21.61
[18]	LDP	LS-SVM	10	11.54
Proposed system	**HOT**	**AIRS-SVM**	**10**	**11.07**

Table 3. GPDS-300 state-of-the-art.

References	Features	Classifier	#Genuine signatures	AER (%)
[15]	LDP	SVM	10	15.35
[19]	Surroundedness	MLP	24	13.76
[20]	Gradient + equimass pyramid	Adaptive feature thresholding	16	14.01
Proposed system	**HOT**	**AIRS-SVM**	**10**	**11.35**

4   Conclusion

This paper aimed to introduce a novel histogram-based descriptor to characterize off-line signatures and proposes to verify the authenticity of these signatures using a joint use of Artificial Immune Recognition System with SVM. Specifically the kNN decision, which is commonly associated with the conventional AIRS is substituted by a support vector decision. Experiments conducted on two public datasets demonstrated the effectiveness of the proposed algorithm despite of using the same parameters selection for all writers. Precisely, a gain of 20.4% for MCYT-75 and of 7.52% for GPDS-300 in the AER is achieved over the classical AIRS performance. In order to further improve the verification accuracies, the histogram of templates descriptor could be implemented in different local parts of signature images for extracting more accurate information.

References

1. Bertolini, D., Oliveira, L.S., Justino, E., Sabourin, R.: Reducing forgeries in writer-independent off-line signature verification through ensemble of classifiers. Pattern Recogn. **43**, 387–396 (2010)
2. Deng, P.S., Liao, H.M., Ho, C.W., Tyan, H.: Wavelet-based off-line handwritten signature verification. Comput. Vis. Image Underst. **76**, 173–190 (1999)
3. Nemmour, H., Chibani, Y.: Off-line signature verification using artificial immune recognition system. In: 10th International Conference on Electronics Computer and Computation (ICECCO), pp. 164–167 (2013)
4. Pourshahabi, M.R., Sigari, M.H., Pourreza, H.R.: Offline handwritten signature identification and verification using contourlet transform. In: International Conference of Soft Computing and Pattern Recognition (SoCPaR), pp. 670–673 (2009)
5. Yilmaz, M.B., Yanikoglu, B., Tirkaz, C., Kholmatov, A.: Offline signature verification using classifier combination of HOG and LBP features. In: International Joint Conference on Biometrics (IJCB), pp. 1–7 (2011)
6. Serdouk, Y., Nemmour, H., Chibani, Y.: New off-line handwritten signature verification method based on artificial immune recognition system. Expert Syst. Appl. **51**, 186–194 (2016)
7. Impedovo, D., Modugno, R., Pirlo, G., Stasolla, E.: Handwritten signature verification by multiple reference sets. In: 11th International Conference on Frontiers in Handwriting Recognition (ICFHR), pp. 19–21 (2008)

8. Justino, E.J.R., Bortolozzi, F., Sabourin, R.: A comparison of SVM and HMM classifiers in off-line signature verification. Pattern Recogn. Lett. **26**, 1377–1385 (2005)
9. Kodaz, H., Ozsen, S., Arslan, A., Gunes, S.: Medical application of information gain based artificial immune recognition system (AIRS). Expert Syst. Appl. **36**, 3086–3092 (2009)
10. Laurentys, C.A., Palhares, R.M., Caminhas, W.M.: A novel artificial immune system for fault behavior detection. Expert Syst. Appl. **38**, 6957–6966 (2011)
11. Serdouk, Y., Nemmour, H., Chibani, Y.: Topological and textural features for off-line signature verification based on artificial immune algorithm. In: 6th International Conference on Soft Computing and Pattern Recognition (SoCPaR), pp. 118–122 (2014)
12. Serdouk, Y., Nemmour, H., Chibani, Y.: An improved artificial immune recognition system for off-line handwritten signature verification. In: 13th International Conference on Document Analysis and Recognition (ICDAR), pp. 196–200 (2015)
13. Tang, S., Goto, S.: Histogram of template for human detection. In: International Conference on Acoustics Speech and Signal Processing (ICASSP), pp. 2186–2189 (2010)
14. Watkins, A.B.: AIRS, A resource limited artificial immune classifier. Master's thesis, Faculty of Mississippi state university, USA (2001)
15. Ferrer, M.A., Vargas, J.F., Morales, A., Ordonez, A.: Robustness of off-line signature verification based on gray level features. IEEE Trans. Inf. Forensics Secur. **7**, 966–977 (2012)
16. Alonso-Fernandez, F., Fairhurst, M.C., Fierrez, J., Ortega-Garcia, J.: Automatic measures for predicting performance in off-line signature. In: International Conference on Image Processing (ICIP), pp. 369–372 (2007)
17. Prakash, H.N., Guru, D.S.: Geometric centroids and their relative distances for off-line signature verification. In: 10th International Conference on Document Analysis and Recognition (ICDAR), pp. 121–125 (2009)
18. Grupo de Procesado Digital de Senales. http://www.gpds.ulpgc.es/download/index.htm
19. Kumar, R., Sharma, J.D., Chanda, B.: Writer-independent off-line signature verification using surroundedness feature. Pattern Recogn. Lett. **33**, 301–308 (2012)
20. Larkins, R., Mayo, M.: Adaptive feature thresholding for off-line signature verification. In: 23rd International Conference on Image and Vision Computing, pp. 1–6 (2008)

Improved Quantum Chaotic Animal Migration Optimization Algorithm for QoS Multicast Routing Problem

Mohammed Mahseur[1(✉)], Abdelmadjid Boukra[1], and Yassine Meraihi[2]

[1] Department of Informatics, Faculty of Electronics and Informatics,
University of Sciences and Technology Houari Boumediene,
16025 El Alia Bab Ezzouar, Algiers, Algeria
mahseur.mohammed@gmail.com, aboukra@usthb.dz
[2] Automation Department, University of MHamed Bougara Boumerdes,
Avenue of Independence, 35000 Boumerdes, Algeria
yassine.meraihi@yahoo.fr

Abstract. In recent years, we are witnessing the spread of many and various modern real-time applications implemented on computer networks such as video conferencing, distance education, online games, and video streaming. These applications require the high quality of different network resources such as bandwidth, delay, jitter, and packet loss rate. In this paper, we propose an improved quantum chaotic animal migration optimization algorithm to solve the multicast routing problem (Multi-Constrained Least Cost MCLC). We used a quantum representation of the solutions that allow the use of the original AMO version without discretization, as well as improving AMO by introducing chaotic map to determine the random numbers. These two contributions improve the diversification and intensification of the algorithm. The simulation results show that our proposed algorithm has a good scalability and efficiency compared with other existing algorithms in the literature.

Keywords: Animal migration optimization
Quantum representation · Chaotic maps · Quality of Service (QoS)
Multicast routing

1 Introduction

The evolution of computer networks has been accompanied by the emergence of a new range of real-time multimedia network applications that exchange and share data, images, videos, sound files, etc. These applications require multicast transmissions with strict QoS guarantees as essential capabilities, where the same data streams are delivered from the source node to a selected group of nodes simultaneously. The most important requirements of QoS parameters are cost, delay, delay jitter, bandwidth, and packet loss rate. Therefore, it is essential

© IFIP International Federation for Information Processing 2018
Published by Springer International Publishing AG 2018. All Rights Reserved
A. Amine et al. (Eds.): CIIA 2018, IFIP AICT 522, pp. 128–139, 2018.
https://doi.org/10.1007/978-3-319-89743-1_12

to use the different network resources in a rational way. The QoS multicast routing problem (MCLC) is a nonlinear combinatorial optimization problem for transmission in the areas of networks. It is proved to be NP-Hard [1] and has recently attracted an increasing attention from researchers. Its main objective consists in finding the optimal multicast routing tree by minimizing the cost under multiple constraints such as delay, delay-jitter, bandwidth, and packet loss rate. It was approached by various algorithms based on meta-heuristics such as Genetic Algorithm (GA), Tabu Search (TS), Particle Swarm Optimization (PSO), Bat Algorithm (BA), and Optimization Based on Biogeography (BBO).

Hwang et al. [2] proposed a GA-based algorithm for multicast routing problem without constraints, the optimization of the multicast tree is achieved through a serial path selection, by crossover and mutation operation. Results showed that the proposed algorithm is able to find a better solution than the RSR algorithm. Haghighat et al. [3] proposed a novel QoS-based multicast routing algorithm based on the GA to find a multicast tree with least cost satisfying the constraints of delay and bandwidth. Simulation results showed that the proposed algorithm outperformed other algorithms.

Youssef et al. [4] present a TS algorithm to build the near optimal solution, their algorithm starts with an initial feasible solution and builds a tree of the sink for each destination with the shortest path of Dijkstra algorithm, in each iteration the algorithm refines the tree at a lower cost. Experimental results showed that TS is able to find better solutions that other reported multicast algorithms. Ghaboosi and Haghighat [5] used the PRA, which is a progressive method. They construct solutions by the combination of elements of a couple of solutions chosen randomly (an initial solution and a guide one) using systematic and deterministic rules.

Liu et al. [6] proposed a PSO algorithm to solve the QoS multicast routing problem by means of serial path selection to obtain a feasible multicast tree by exchanging paths in the vector. Experimental results indicated that the proposed algorithm can converge to the optimal or near the optimal solution with lower computational cost. It also appeared that PSO outperformed GA on QoS the tested multicast routing problem. Qu et al. [7] developed a PSO algorithm based on the jump (JPSO), such that each particle moves either based on its position, either based on a chosen attractor, once the particle reaches his new position, it performs a local search to improve its position.

Mahseur and Boukra [8] proposed two approaches to solve the MCLC Problem, the first one is based on BBO and the second on BA, the results of their experimentation are encouraging in comparison with other existing algorithms. Meraihi et al. [9] proposed an improved chaotic binary bat algorithm to solve the QoS multicast routing problem. Simulation results reveal the efficiency of the proposed algorithm compared with some other existing algorithms in the literature.

In this paper, we have applied a new meta-heuristic called Animal Migration Optimization (AMO) to resolve the problem of MCLC. To improve diversification and intensification, we have used the quantum representation of solutions and integrated the chaotic maps in some characteristic equations of AMO.

The rest of the paper is organized as follows. Section 2 presents the mathematical multicast routing problem formulation followed by Sect. 3 which describes the AMO algorithm. In Sect. 4, we propose the improved quantum chaotic animal migration optimization algorithm. In Sect. 5 we study three scenarios, we investigate the performance of our algorithm in first one, the convergence and scalability in the second one, and the robustness and speed in the last one. Finally Sect. 6 summarizes the findings.

2   Multicast Routing Problem Formulation

Throughout this work, the communication network is modeled as a directed weighted graph $G = (V, E)$ where V denotes the set of nodes, and E denotes the set of the edges representing wireless links between nodes. $|V| = n$ is the number of nodes and $|E| = l$ is the number of links in the network. Each link $e = (i, j) \in E$ that connects node i with node j is associated with link cost $C(e) : E \to R^+$, link delay $D(e) : E \to R^+$, link delay jitter $J(e) : E \to R^+$, link bandwidth $B(e) : E \to R^+$, and link packet loss rate $PL(e) : E \to R^+$, where R^+ is the set of all nonnegative real numbers. We assume that $s \in V$ represents the source node and $M \subseteq \{V - \{s\}\}$ represents a set of multicast destination nodes such that s and M construct a multicast tree $T(s, M)$.

The multicast tree has the following parameters [10,11]:
The total cost of the multicast tree $T(s, M)$, denoted by $Cost(T(s, M))$, is defined as the sum of the costs of all links in that tree. It can be given by:

$$Cost(T(s, M)) = \sum_{e \in T(s,M)} C(e) \tag{1}$$

The total delay of the path $P_T(s, m)$, denoted by $Delay(P_T(s, m))$, is simply the sum of the delays of all links along the path:

$$Delay(P_T(s, m)) = \sum_{e \in P_T(s,m)} D(e) \tag{2}$$

where $P_T(s, m)$ denotes the routing path of the multicast tree $T(s, M)$ that connects the source node s with the destination node $m \in M$ (clearly $P_T(s, m)$ is a subset of $T(s, M)$).

The delay jitter of the path $P_T(s, m)$, denoted by $Jitter(P_T(s, m))$, is defined as the sum of the delays-jitter of all links along the path and can be given by:

$$Jitter(P_T(s, m)) = \sum_{e \in P_T(s,m)} J(e) \tag{3}$$

The bandwidth of the path $P_T(s, m)$, denoted by $B(P_T(s, m))$, is defined as the minimum required bandwidth at any link along $P_T(s, m)$:

$$B(P_T(s, m)) = \min_{e \in P_T(s,m)} (B(e)) \tag{4}$$

The Packet loss rate of the path from source node s to destination node m, denoted by $PL(P_T(s,m))$ is given by:

$$PL(P_T(s,m)) = 1 - \prod_{e \in P_T(s,m)} (1 - PL(e)) \tag{5}$$

The MCLC problem can be formulated as follows:

$$Minimize\ Cost(T(s,M)) \tag{6}$$

Subject to:

$$Delay(P_T(s,m)) \leq D_{max} \tag{7}$$

$$Jitter(P_T(s,m)) \leq J_{max} \tag{8}$$

$$Min(B(P_T(s,m))) \geq B_{min} \tag{9}$$

$$PL(P_T(s,m)) \leq PL_{max} \tag{10}$$

3   Animal Migration Optimization AMO

Animal Migration Optimization Algorithm (AMO), developed by Li et al. in 2014 [12], is a novel interesting nature-inspired metaheuristic optimization algorithm used to solve various optimization problems. This optimization algorithm is inspired by the migration behavior of animals in groups to ensure their survival. These animals can be found in several families such as mammals, birds, fish, reptiles, amphibians, insects, and crustaceans. These families are leaving their current habitats and migrating to other geographically distant habitats that can provide better living conditions. There are several reasons for migration such as food shortage, drought, reproduction, and seasonal change. For simplicity, AMO uses the following two idealized rules [12]:

1. The leading animal with high position quality will be kept for the next generation;
2. The number of animals in the group is fixed, and the animal leaving the group will be replaced by a new individual with a probability Pa.

AMO algorithm works based on two processes. In the first process, the algorithm simulates how the groups of animals move from the current position to the new position. In the second process, the algorithm simulates how some animals leave the group and some join the group during the migration.

Each individual i changes its current position during the iteration $G + 1$ by changing the components of its vector X_i according to one of the neighbors according to the following formula:

$$X_{i,G+1} = X_{i,G} + \delta.(X_{neighborhood,G} - X_{i,G}) \tag{11}$$

where δ is a parameter defined according to the problem treated, and the neighboring individual is randomly selected from the neighbors. Remember that the topology and the size of the population are fixed. An animal is asked to leave the group with probability Pa and replaced by another obtained by the following formula:

$$X_{i,G+1} = X_{r1,G} + rand.(X_{best,G} - X_{i,G}) + rand.(X_{r2,G} - X_{i,G}) \tag{12}$$

4   IQCAMO for QoS Multicast Routing

4.1   Improved Quantum Chaotic AMO Algorithm

In order to improve the effectiveness and the robustness of the AMO algorithm, we have proposed two modification methods. The first modification method introduces the quantum representation of the solutions to allow the use of the original AMO version without discretization.

The second modification method adopts the chaotic map to overcome the problem of choosing the right value of the two random numbers used in Eq. 12. There are several chaotic maps like logistic map, circle map, tent map, piecewise map, iterative map, Chebyshev map, sine map, sinusoidal map etc. Logistic map is the most representative chaotic map with simple operations and well dynamic randomness [13]. So this chaotic behavior map is adopted in this paper to generate the two random numbers at each iteration. The random number generated by a logistic map is given by:

$$CM_{t+1} = aCM_t(1 - CM_t), t = 0, 1, 2 \ldots \tag{13}$$

Where CM_t is the chaotic map value at the t-th iteration, it is in $[0, 1]$. Here a is a control parameter and $0 < a < 4$. The chaotic system is very sensitive to the initial values, to make the logistic map in a complete chaos, we adopt the parameter $a = 4$ and CM_t does not belong to $\{0, 0.25, 0.5, 0.75, 1\}$ otherwise the logistic equation does not show chaotic behavior [14]. In our experiments, we take $CM_0 = 0.7$. Algorithm 1 describes the pseudo-code of the IQCAMO algorithm for multicast routing problem.

4.2   Representation of the Solution

The solution representation of the i-th Qbit individual is defined by a real vector $Q_i^t = (q_{i1}^t, q_{i2}^t, \ldots, q_{im}^t), j = 1, 2 \ldots, m$, where m represents the length of the vector which is equal to the total number of links in the network.

The Qbit vector Qi is defined as follows:

$$\begin{bmatrix} \alpha_{i1}^t & \alpha_{i2}^t & \cdots & \alpha_{im}^t \\ \beta_{i1}^t & \beta_{i2}^t & \cdots & \beta_{im}^t \end{bmatrix} \tag{14}$$

The binary representation of the solution which is a multicast tree is a binary vector $X_i^t = (x_{i1}^t, x_{i2}^t, \ldots, x_{im}^t), j = 1, 2 \ldots, m$, where $x_{ij}^t = 0 \ or \ 1$, which indicates

Algorithm 1. The pseudo-code of the IQCAMO Algorithm

1: Introduce the problem data
2: Optimize the graph
3: Initialize QPopulation
4: Initialize generation number $G = 0$
5: **while** $(G < G_{max})$ **do**
6:     **for** $i = 1$ to NP **do**
7:         $\delta = L_i/L_{max}$
8:         **for** $j = 1$ to $|E|$ **do**
9:             $Q_{i,j,G+1} = Q_{i,j,G} + \delta.(Q_{neighborhood\ of\ i,j,G} - Q_{i,j,G})$
10:             **if** $Q_{i,j,G+1} < 0$ **then**
11:                 $Q_{i,j,G+1} = 0$
12:             **end if**
13:             **if** $Q_{i,j,G+1} > \Pi/2$ **then**
14:                 $Q_{i,j,G+1} = \Pi/2$
15:             **end if**
16:         **end for**
17:     **end for**
18:     **for** $i = 1$ to NP **do**
19:         Repair infeasible solutions
20:         Evaluate Fitness($Q_{i,G+1}$)
21:         **if** $Fitness(Q_{i,G+1}) < Fitness(Q_{i,G})$ **then**
22:             $Q_i = Q_{i,G+1}$
23:         **end if**
24:     **end for**
25:     **for** $i = 1$ to NP **do**
26:         Calculate Pa_i
27:         **if** $rand < Pa_i$ **then**
28:             **for** $j = 1$ to $|E|$ **do**
29:                 $Q_{i,j,G+1} = Q_{r1,j,G} + CM1(G + 1).(Q_{best,j,G} - Q_{i,j,G}) + CM2(G + 1).(Q_{r2,j,G} - Q_{i,j,G})$
30:                 **if** $Q_{i,j,G+1} < 0$ **then**
31:                     $Q_{i,j,G+1} = 0$
32:                 **end if**
33:                 **if** $Q_{i,j,G+1} > \Pi/2$ **then**
34:                     $Q_{i,j,G+1} = \Pi/2$
35:                 **end if**
36:             **end for**
37:         **end if**
38:     **end for**
39:     **for** $i = 1$ to NP **do**
40:         Repair infeasible solutions
41:         Evaluate Fitness($Q_{i,G+1}$)
42:         **if** $Fitness(Q_{i,G+1}) < Fitness(Q_{i,G})$ **then**
43:             $Q_i = Q_{i,G+1}$
44:         **end if**
45:     **end for**
46:     Store Qbest
47:     $G = G + 1$
48: **end while**

whether the corresponding link i is selected or not to construct the multicast tree. The binary vector X_i^t is defined as follows:

$$x_{ij} = \begin{cases} 1; \text{ if } rand(0,1) > |\alpha_{ij}^t|^2 \text{ (i is selected to construct the multicast tree).} \\ \\ 0; \hspace{4cm} \text{otherwise.} \end{cases}$$

$$(15)$$

4.3   Initial Population

Before the generation of the initial population, we delete all link with bandwidths less than the bandwidth constraint B_{min}. The initial population is produced by the Depth First Search (DFS) method, we take the source node and connect it randomly to one of its successors and compare it with the set of destination nodes M, if it is a destination so it is a valid path, otherwise we redo the same approach until obtaining a solution in natural form. If the path is saturated (blocked road), we restart the operation from the source node. Each connection of a node, a check is made to avoid the construction of loops.

5   Simulation Results

The performances of the proposed approach are studied and compared with the implemented algorithms: GA [2] and JPSO [7]. The three algorithms were implemented using Visual C++ on a PC Intel Core 2 Duo 2.0 GHz. The nodes are randomly distributed in a rectangle of 4000 km × 2400 km, links are generated according to the topology model Waxman [15]. In this model, the existence of a link between two nodes u and v depends on a probability $p(u,v)$ given by the following formula: $P(u,v) = B \exp(-l(u,v)/AL)$ Where: $l(u,v)$ represents the Euclidean distance between two nodes u and v. L is the maximum distance between any two points in the network. B is a parameter that controls the number of links in the network. A is a parameter that controls the number of short links in the network. A and B are set to 0.8 and 0.7 respectively like in [7]. The cost, delay, packet loss rate and bandwidth of the links are taken randomly in the interval [1, 100], [1 ms, 30], [0.0001, 0.01] and [2 Mbps, 10] respectively. The thresholds constraints Δ, ξ, and δ are set to 120 ms, 60 ms and 0.05 respectively. The bandwidth required by the flow to send β is randomly generated in the interval [2 Mbps, 10] [15]. The source and destination nodes are selected randomly from the network. We perform tree scenarios:

1. In the first scenario, and in order to evaluate the performance of each algorithm, we evaluate the cost of the multicast tree while varying the number of nodes in the network with 10% of destinations;
2. In the second scenario, we observe the cost of the best solution for each algorithm according to the number of iterations and the size of network to study the convergence and scalability of algorithms;
3. In the third scenario, we have varied the number of destination nodes to evaluate for each algorithm the cost of the optimal solution and the convergence time in order to study the robustness and speed of algorithms.

5.1   Results of the First Scenario

In the first scenario, we varied the size of the network from 20 to 120 nodes. Each time, the source node and the destinations (10% of the nodes) are chosen randomly. The results are represented in Fig. 1.

Figure 1 shows the multicast tree cost of GA, JPSO, and IQCAMO, according to the network size. First, we observe that the multicast tree cost increases with the increase of the network size. Second, we observe that the multicast tree cost generated by IQCAMO is less than the other multicast tree costs.

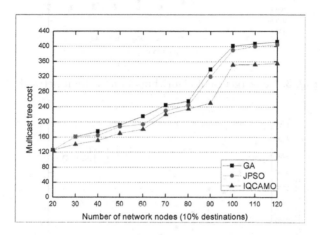

Fig. 1. The cost of the multicast tree while varying the number of nodes in the network.

5.2   Results of the Second Scenario

In the second scenario, we have study three kinds of networks: small network(50 nodes), medium network (200 nodes), and large network (500 nodes) with 10%

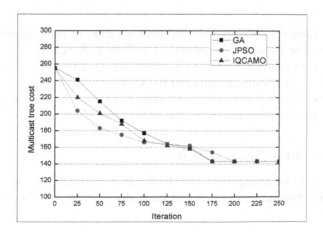

Fig. 2. Convergence of algorithms in a small network.

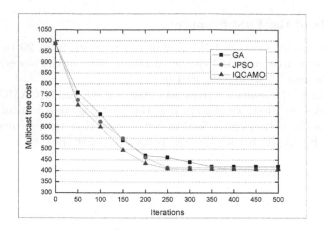

Fig. 3. Convergence of algorithms in a medium network.

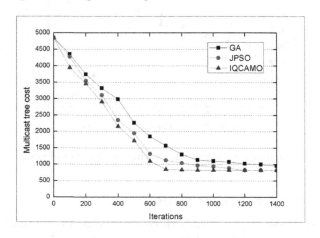

Fig. 4. Convergence of algorithms in large network.

destinations, then we evaluated the optimal solution of each algorithm during the iterations. The results are represented in Figs. 2, 3, and 4.

Figure 2 shows solutions obtained by the three algorithms in a small network, we observe that the solutions proposed by JPSO are the best.

We observe in Figs. 3 and 4 that our algorithm has the best speed of convergence; this is provided by the use of quantum principles.

5.3   Results of the Third Scenario

In the third test, we study the cost of multicast routing and the convergence time by varying the number of destination nodes. The results are represented in Figs. 5 and 6.

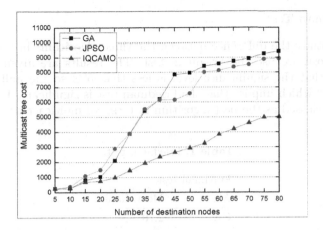

Fig. 5. The cost of the multicast tree while varying the number of destination nodes.

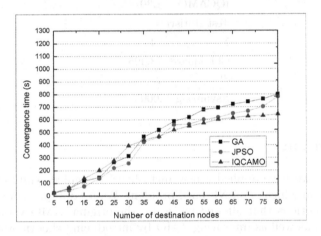

Fig. 6. The convergence time while varying the number of destination nodes.

Figure 5 shows the multicast tree cost of GA, JPSO, and IQCAMO while varying the number of destination nodes. We observe that IQCAMO gives the best routing cost. This is due to the use of the quantum representation and the chaotic maps which are more efficient in exploring large search space. Figure 6 shows that the convergence time increases while increasing the number of destination nodes and we observe that IQCAMO gives the best convergence time. Thus, the proposed approach is more robust and fast. This is due to the use of the quantum mechanism and the chaotic maps which are more efficient in exploring large search space.

5.4   Friedman Test

In order to study the difference between the three algorithms in terms of multicast tree cost, we use the Friedman test. The results are given in Table 1. We observe that the significance level is less than 5%, so the null hypothesis H_0 is rejected which implies that the Friedman test is significant. Furthermore, IQCAMO is classed as the best algorithm in terms of multicast tree cost.

Table 1. Friedman test.

Friedman test	
Algo	Mean Rank
GA	2.70
JPSO	2.01
IQCAMO	1.30
Test statistics	
N	64
Chi-Square	73.495
df	2
Asymp.sig	.000

6   Conclusion

In this paper, we have applied a new meta-heuristic called AMO on multicast routing problem which is an NP-hard problem. We used a quantum representation of the solutions that allowed us to use the original AMO version without discretization, as well as improving AMO by introducing chaotic Maps into the calculation of some equations. These two contributions improve the diversification and intensification of the algorithm. We performed an optimization on the graph representing the network and another on the initial population which accelerated the exploration of the search space. To evaluate the performance of our solution we tested three comparative scenarios with two other algorithms: in the first scenario, we compared the multicast tree cost with varying the network size and using 10% as destinations. We notice that IQCAMO produces the minimum cost, In the second scenario, we study the convergence of the three algorithms; we observe that IQCAMO is the best. In the last scenario, we evaluate the cost and the convergence time while varying the size of the set of destination nodes. We found that our approach has an ability to deliver the best results. The experimental results lead to conclude that our algorithm increases the overall research capacity. As future work, we intend to investigate the multi-objective aspect of the problem. Indeed, to have a better resolution of the problem, other objectives must be considered, like delay, bandwidth, jitter etc.

References

1. Wang, Z., Crowcroft, J.: Quality-of-service routing for supporting multimedia applications. IEEE J. Sel. Areas Commun. **14**(7), 1228–1234 (1996)
2. Hwang, R.H., Do, W.Y., Yang, S.C.: Multicast routing based on genetic algorithms. J. Inf. Sci. Eng. **16**(6), 885–901 (2000)
3. Haghighat, A.T., Faez, K., Dehghan, M., Mowlaei, A., Ghahremani, Y.: GA-based heuristic algorithms for QoS based multicast routing. Knowl. Based Syst. **16**(5), 305–312 (2003)
4. Youssef, H., Al-Mulhem, A., Sait, S.M., Tahir, M.A.: QoS-driven multicast tree generation using tabu search. Comput. Commun. **25**(11), 1140–1149 (2002)
5. Ghaboosi, N., Haghighat, A.T.: A path relinking approach for delay-constrained least-cost multicast routing problem. In: 19th IEEE International Conference onTools with Artificial Intelligence, ICTAI 2007, vol. 1, pp. 383–390. IEEE (2007)
6. Liu, J., Sun, J., Xu, W.: QoS multicast routing based on particle swarm optimization. In: Corchado, E., Yin, H., Botti, V., Fyfe, C. (eds.) IDEAL 2006. LNCS, vol. 4224, pp. 936–943. Springer, Heidelberg (2006). https://doi.org/10.1007/11875581_112
7. Qu, R., Xu, Y., Castro, J.P., Landa-Silva, D.: Particle swarm optimization for the Steiner tree in graph and delay-constrained multicast routing problems. J. Heuristics **19**(2), 317–342 (2013)
8. Mahseur, M., Boukra, A.: Using bio-inspired approaches to improve the quality of service in a multicast routing. Int. J. Commun. Netw. Distrib. Syst. **19**(2), 186–213 (2017)
9. Meraihi, Y., Acheli, D., Ramdane-Cherif, A.: An improved chaotic binary bat algorithm for QoS multicast routing. Int. J. Artif. Intell. Tools **25**(04), 1650025 (2016)
10. Sun, J., Fang, W., Wu, X., Xie, Z., Xu, W.: QoS multicast routing using a quantum-behaved particle swarm optimization algorithm. Eng. Appl. Artif. Intell. **24**(1), 123–131 (2011)
11. Abdel-Kader, R.F.: An improved discrete PSO with GA operators for efficient QoS-multicast routing. Int. J. Hybrid Inf. Technol. **4**(2), 23–38 (2011)
12. Li, X., Zhang, J., Yin, M.: Animal migration optimization: an optimization algorithm inspired by animal migration behavior. Neural Comput. Appl. **24**(7–8), 1867–1877 (2014)
13. Kanso, A., Smaoui, N.: Logistic chaotic maps for binary numbers generations. Chaos Solitons Fractals **40**(5), 2557–2568 (2009)
14. dos Santos Coelho, L., Sauer, J.G., Rudek, M.: Differential evolution optimization combined with chaotic sequences for image contrast enhancement. Chaos Solitons Fractals **42**(1), 522–529 (2009)
15. Waxman, B.M.: Routing of multipoint connections. IEEE J. Sel. Areas Commun. **6**(9), 1617–1622 (1988)

Developing a Conceptual Framework for Software Evolution Methods via Architectural Metrics

Nouredine Gasmallah[1,3]($\boxtimes$), Abdelkrim Amirat[1], Mourad Oussalah[2], and Hassina Seridi[3]

[1] Department of Mathematics and Computer Science, University of Souk-Ahras, 41000 Souk Ahras, Algeria
{gasmallahedi,a.amirat}@univ-soukahras.dz
[2] Department of Computer Science, University of Nantes, 44300 Nantes, France
mourad.oussalah@univ-nantes.fr
[3] Department of Computer Science, University of Annaba, 23000 Annaba, Algeria
seridi@labged.net

Abstract. Because of the vital need for software systems to evolve and change over time in order to account for new requirements, software evolution at higher levels of modeling is considered as one of the main foundation within software engineering used to reduce complexity and ensure flexibility, usability and reliability. In similar studies for migration technique and software engineering, presenting a framework do not usually cover the specification of systems based on software architecture. In this paper, we specify a conceptual framework based on six explicit dimensions in respect to an architectural view-point as first class citizen. Indeed, sketching evolution relies upon identifying dimensions on which researchers try to answer while performing a new approach. The proposed model is based on answering What, Where, When, Who, Why and How questions. Analyzing these dimensions could provide a multiple choice to implement classification for architectural techniques. Further and using an example, these dimensions are quantified and then analyzed. This framework aims to provide a blueprint to guide researches to position architectural evolution approaches and maps them according a selected set of dimensions.

1 Introduction

Software evolution has become a major concern for stakeholders involved in the process of designing and modeling computer systems. Because of the rudimentary nature of software systems to evolve and change over time in order to account for new requirements and different needs, software evolution is considered as a vital pillar within the area of software engineering to ensure consistency and maintainability as well as to allow the system to open up for new directions and

A. Amine et al. (Eds.): CIIA 2018, IFIP AICT 522, pp. 140–149, 2018.
https://doi.org/10.1007/978-3-319-89743-1_13

strategic opportunities. Further, software architecture must evolve within existing production systems as it constitutes the favorable blueprint for supporting such evolvability at higher modeling levels [1]. This implies that evolution could be handled at earlier modeling phases where future changes could be anticipated. The area of software evolution has received unprecedented interest during the last two decades where numerous research studies have been published describing various methods and frameworks. In return, few studies have been devoted to present classification for software architecture evolution [2]. However, the main stream of most studies are either specialized in software architecture evolution as knowledge re-usability or addressing software evolution from an overall perspective [3,4]. Buckley et al. (2005) proposed a taxonomy of software changes which characterize the mechanisms of change and their influencing factors. Williams et al. (2010) suggested that the key solution to address software changes is to identify the causes and effects which can be used to illustrate the potential impact of the change. The major drawback of their study is the lack of an explicit framework capable of positioning a given approach based on well-defined criteria or metrics. Introducing an evolution framework would fulfill the need to identify certain specifications according to which approaches can be classified. Because of the dearth of classification procedures that address the architectural evolution techniques with respect to well-defined dimensions, a conceptual framework that addresses the major concerns related to the evolution nature of software architecture is proposed in this research study. A framework can offer a structured and coherent design for examining the organizational aspect in software architecture evolution as well as assist architects to address decisions on new requirements. De facto, introducing a clear and concise architectural evolution framework would fulfill the need: (i) to identify certain specifications according to the major concerns related to the evolution nature of software architecture and, (ii) to provide a common vocabulary to understand, analyze and categorize a given evolution approach. The discussed model is inspired from the work of Buckley (2005) for software change in addition to the work of [5] for the architecture of information systems. The proposed framework is devised to reshape the dimensions reported in both earlier studies with a newly set of proposed metrics for the arena of software architecture evolution. Such metrics should assist to examine and assess evolution approaches through providing answers which revolve around these dimensions: What, Why, Where, Who, When and How the software architecture evolves? Based on these dimensions, the following research questions are addressed during this research study:

RQ1: What are the major metrics that can be used to analyze and compare the different architectural evolution methods?
RQ2: Based on such dimensions, can a classification framework devises to regroup and categorize architectural evolution approaches?

The remainder of the paper is organized as follows: the proposed dimensions defined for positioning architectural evolution and a method for classifying them are explained in the following section. The third section illustrates the results drawn by the application of the proposed framework on a set of well-known frameworks in the field of software evolution. The penultimate section is dedicated to discuss the findings to explore and compare the potentials of the proposed dimensions. Conclusions are drawn within the last section.

2   Proposed Framework

2.1   Framework Dimensions

What? Object of Evolution Dimension - The object of evolution is the subject on which the evolution is operated. It answers the what question and comprises of the following:

- **Artifact:** is an abstraction of any element belonging to the architectural structure. It may be a software architecture, a component, a service and so forth. It can be simple and even concrete description as program codes for example.
- **Process:** Process is as a set of interacting activities, which transforms inputs into outputs. Therefore the process also evolves and presents suitable means for anchoring the benefits pertaining to cost optimization and quality promotion [6], workflows in SOA are one of the best examples of processes.

Where? Hierarchical Level Dimension - Two types of hierarchical levels are commonly identified within the software engineering literature as depicted in Fig. 1, which are modeling and abstraction levels.

- **Modeling levels (M_0 to M_2):** Software systems must be mapped over several levels, from lower-level constructs (code) to higher-level constructs, to ensure convergence [7]. Modeling level refers to one of the four levels (from M_0 to M_3) as defined by the OMG [8].
- **Abstraction levels (a_0 to a_n):** Experts often use abstraction as an approach to address complex problems. This is still very often used for solving problems related to software modeling. Thereby, a solution is performed through a series of model description at a number of abstraction levels referred to in the Fig. 1 as a_0, a_1..., a_n.

When? Time of Evolution - In respect of architectural viewpoint, the time of evolution embraces one of these metrics:

- **Design-time:** Predicting evolution at the earlier stages of design allows improving and extending the system architecture. Model Driven Architecture is a good of methods addressing evolution at design-time, especially co-evolution approaches.
- **Run-time:** The evolution can occur while the system is running [9]. So far, evolving at run-time is considered as a major topic to address architecture adaptation [10]. Noting that, this metric encompasses: compile-time, Load-time and Dynamic-time [10,11].

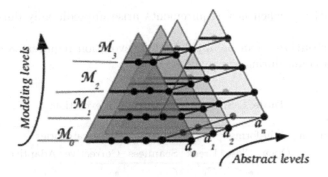

Fig. 1. Modeling levels vs. Abstraction-levels

Why? Type of Evolution Dimension - It is one of the most commonly-used metric when addressing classification issues. This dimension encloses the two following main sub-categories:

Main-categories: Instead of categories based on the architects intentions, main-categories sub-dimension is responding to the objective evidence of architect's activities identifiable from observations and artifacts before and after comparison of the software architecture.

Main-Forms of resolution: Architects can use several strategies for the evolution problem-solving according to different forms. From the literature, the following two major studies discussed the various forms in the area of evolution.

(i) Chaki et al. [12] grouped the different resolution methods into two main strategies:
  - *Open evolution* means from an initial architecture, of a new architecture reflecting a system solution in which a set of invariants and constraints are respected.
  - *Close evolution* means the evolution activity uses induction resolution to find the best permissible sequence of operations to be applied to achieve the desired result.
(ii) Oussalah et al. [13] proposed two other forms:
  - *Break evolution* means that interventions are applied directly to the initial architecture without having the ability to go back on the trace of the evolution.
  - *Seamless evolution* denotes an evolution with trace whereby the architecture keeps a trace of its initial properties and operations performed before each evolving process.

The proposed structuring for deducing the assumed evolution type for a particular method is shown in Table 1 such that the categorization is based hybridly on categories and forms of resolution. Herein, the proposed type of evolution is structured according to two main forms (curative or antici-pative) which are explained as follows:

- **Curative** - when new requirements arise unpredictably during the life cycle.
- **Anticipative** - can be applied when evolution requirements are taken into account during the analysis.

Table 1. Structure of the type of evolution.

Type of evolution	Main-forms				Main categories		
	Open	Close	Break	Seamless	Corrective	Adaptive	Perfective
Curative	○	•	⊙	○	•	○	○
Anticipative	•	⊙	⊙	○	○	•	⊙

Symbol legend: • *higher* ⊙ *Medium* ○ *Weak.*

The two main forms (*Curative* and *Anticipative*) can be set either to *Close* or *Open*. In the same way, the latter sub-forms can be set to *Break* or *Seamless*.

Who? Stakeholders - Stakeholders contribute in architecture enhancement in multiple roles regarding the responsibility they assume and the range of challenges they face [14]. For instance, this dimension covers only those stakeholders that operate evolution on the software system itself by referring to the development team enclosing *researches, architects, designers, developers, analysts* and *programmers* [10]. This dimension specifies two main knowledge about how are involved and interested by the evolution of software architecture.

How? Operating Mechanism of Evolution - The operating mechanism of evolution (OME) is introduced to refer to the general behavior and employed procedure by taking into account solely the described hierarchical levels. Mainly, evolution solving approaches commonly adhere to one of the two main following methods:

- **Reduce:** For the reductionist evolution, the process gets through a predefined evolution path; until the solution is satisfied (evolved model). Brooks [15] 25 defines the reductionist approach as classical approach to problem solving to which the overall resolution task is decomposed into subtasks.
- **Emergence:** On the contrary during an emergentist evolution approach, the process builds the path to a solution. The emergence exposes a passage between the activity of micro-level and that of macro-level.

3   Case Study

3.1   Metric Calculations

In order to clarify applicability of the proposed framework, two studies of architectural evolution are analyzed and compared. These studies are only cited as an example for evolution approaches among many other relevant approaches.

The first study by Oussalah et al. [16] presents a software architecture evolution model (SAEV) which aims to describe and manage the evolution through the architectural elements of a given software architecture. SAEV considers software architecture elements (component, connector, interface and configuration) as first class entities. Managing these elements is conceived independently of any architecture language level by considering the different levels of modeling (meta-level, architectural level and application level). The second by Barnes et al. [17] is an automatic approach provided to assist architects by planning alternative evolution paths for evolving software architecture. An evolution path is expressed in terms of intermediate architecture generated from the initial state. If the architect's goal is clear, this approach aims to assist architects to find the optimal path that meets the evolution requirements. The evolution path acts explicitly on software architecture and applies evolution from a higher to lower level to reach the desired architecture. This approach helps architects to find the optimal path for the evolved architecture. In order to apply the conceptual framework for the two approaches, all the discussed dimensions are quantified using three values 1, 0.5 and 0 which reflect respectively explicit, implicit and not mentioned. Conventionally, a dimension, which is explicitly mentioned, is assigned the value one. However, if it is shown in implicit fashion, 0.5 is assigned. Otherwise, the estimated value is zero. The values obtained from Table 2 shows that the two approaches have considered the six dimensions. For the where dimension for example, the first approach has explicitly expressed evolution at different modeling levels and implicitly shows the consideration of the abstraction levels which leads to a value of 1.5 from two of the whole dimension expressiveness which gives 75% against 25% for the second approach. These percentages serve as evaluation and comparison for determining the approach focus regarding the proposed dimensions. In similar fashion, the other dimensions are analyzed and compared accordingly. In summary, the first approach focuses on modeling levels of the architecture whereas the second is based on abstraction levels mainly used for reducing evolution complexity. It is noteworthy that the two approaches show identical values of object, type, stakeholder and time of evolution and both support the 'reduce' operating mechanism of evolution.

3.2 Framework Assessment

Three prominent existing studies have been selected and surveyed to further evaluate the conceptual framework using the proposed metrics on their set of surveyed papers as follows:

The first considered survey study that involves 32 research papers was conducted by Ahmad [18] which focuses on approaches wherein changes impact the architectural level when analyzing and improving software evolvability. The investigation of the existing methods or techniques, either for systematic application or for empirical acquisition of architectural knowledge, categorizes evolution reuse knowledge into six broad themes: (i) evolution styles, (ii) change patterns, (iii) adaptation strategies and policies, (iv) pattern discovery, (v) architecture

Table 2. Comparison of two evolution approaches using the framework

Model		Oussalah (2006)		Barnes (2014)	
Object	Artifact	1	50%	1	50%
	Process	0		0	
Levels	Modeling	1	75%	0	25%
	Abstraction	0.5		0.5	
Stakeholders	Architects/designers	1	100%	1	100%
Time	Run	0	50%	0	50%
	Design	1		1	
OME	Reduce	1	50%	0.5	25%
	Emergence	0		0	
Type	Main-forms	1	12.50%	0.5	12.50%
		0		0.5	
		0		0	
		0		0	
	Main-categories	1	16.67%	0.5	16.67%
		0		0.5	
		0		0	

configuration analysis, and (vi) evolution and maintenance prediction. The proposed thematic classification focuses on both time of evolution and type of evolution for reuse knowledge.

The second study by Breivold et al. [2] where five main categories of themes are identified based substantially on research topics through an investigation of 82 research papers: (i) techniques supporting quality consideration during software architecture design, (ii) architectural quality evaluation, (iii) economic valuation, (iv) architectural knowledge management, and (v) modeling techniques. A set of specific characteristics is provided with a view to refine each category to subcategories reflecting a common specification in terms of research focus, research concepts and context.

The third by Chaki et al. [12] recommends three classes of common type for architectural evolution:

- Maintenance focused evolution - these are works concerned with the use of correction decisions and architectural modifiability to address architects' concerns. These works often require intervention at the lowest modeling level.
- Open evolution - refers to all software architectural evolution works whose final architecture is not known a priori. These approaches infer, from an initial architecture, one or more solutions in respect of a context known beforehand.
- Closed evolution - Categorizes works in which the target architecture of the model is known before proceeding to the evolution of the initial architecture. It is a question of finding a sequence of operations that may guide an initial architecture to a desired one.

Table 3. Coverage percentages of the proposed dimensions.

Dimensions	Studies	Ahmad 32	Breivold 82	Chaki 5	Weighted average %
Object	Artifact	66.67	74.60	20	70.17
	Process	33.33	25.40	80	29.83
Levels	Modeling	86.67	88.23	64.29	86.80
	Abstraction	13.33	11.77	35.71	13.20
Stakeholders	Architects/designers	100	100	100	100
Time	Run	34.62	24.44	22.22	27.08
	Design	65.38	75.56	77.78	72.92
OME	Reduce	96.78	94.29	100	95.20
	Emergence	3.22	5.71	0	4.80
Type	Main-forms	10.59	41.86	100	35.89
	Main-categories	89.41	58.14	0	64.11

$Weighted\ average = \frac{\sum percentage_i \times n_i}{\sum n_i}$ where n_i is the number of the surveyed papers.

Table 4. Details of form and category percentages.

Studies	Nbr	Main-forms %				Main-categories %		
		Open Break	Open Seam	Close Break	Close Seam	Corr.	Perf.	Adap.
Ahmad	32	04.71	03.53	2.35	0	37.65	28.24	23.53
Breiold	82	36.05	04.65	1.16	0	24.42	26.74	6.98
Chaki	5	80.00	00.00	0.00	20	-	-	-

It is worth emphasizing that the first step was to devise a set of projections for the conceptual benchmarks and afterwards for the proposed taxonomy. These projections aim to reveal the coverage of each of the benchmarks that appear most frequently across the range of architectural evolution studies. Table 3 summarizes the quantification using weighted results of the six dimensions. The percentage of each dimension is calculated by dividing the sum of expressed studies by the total number of investigated studies. Detailed illustration for the provided results in Table 3 relating to the *Type* dimension (including *Forms* and *Categories*) are further shown in Table 4. The results presented in both tables reveal the just proportion (or disproportion) between weights granted for each proposed dimension.

4   Discussion

Significantly, Table 3 shows that over three-quarters of the selected studies have explicitly fostered modeling levels for all works, which explains these studies' relevance to the architecture evolution topic. Further, percentages show that a wide range of studies are committed to artifact as an object of evolution, except for [12] which focuses on process evolution studies. This is due to the fact that both existing evolution support formalism (SAEV [16], Query/View/Transformation-based [19], ...) and architecture description languages (ADL, UML, xADL, ...) for evolution models are all artifact oriented. It is worth noting that styles and patterns are actually more suitable for evolving architecture elements (artifacts) but contribute little to address the evolution issue of the styles themselves [20]. In addition, the modeling level has attracted major interest from the evolution community for the reason that it overlaps with the abstraction level concept. In the same way, an overwhelming percentage deal with reducing evolution from higher to lower level, i.e. top-down hierarchy modeling fashion. In our view, this preponderance is mostly due to a deductive reasoning influence which promotes the top-down method (i.e. reduce OME). However, this reductionism may deprive software architecture systems to introduce new solution opportunities according to the current and/or future environment assumptions. Regarding the temporal dimension, values are systematically in favor of the design-time which in our opinion reflects the interest to the anticipation activity to deal with evolution. It is noted that all these studies specify the importance of stakeholders requirements when dealing with evolution. The percentages in Table 4, show that the main categories sub-dimension attracted an explicit interest from all the studies apart from Chaki et al. (2009).

5   Conclusion

In this paper, a conceptual framework is presented for modeling the classification of software architecture evolution approaches. This model is based on evaluating six proposed dimensions on a given approach. The proposal is part of a wider strategy to evaluate tendency of existing research within software architecture evolution according to well-defined dimensions. Thus, the proposed model provides architects with the capacity for understanding and analyzing the evolution before making technical choices. Considering the impact of evolution process through the hierarchical levels, an evolution framework of six different dimensions is devised. This would help to sketch the interior design of the architectural evolution solution space. Based on an experimental study of three well known classifications, the investigation of the coverage in software architecture evolution has drawn a number of conclusions on opportunities, strength and weakness of existing approaches. The obtained results highlight the lack of emergence techniques, which could be a very promising track. The proposed conceptual framework enjoys the merits of (i) compare different architectural evolution approaches, (ii) determine the appropriate evolution approach according to the dimensions that seems the most relevant to the application class and (iii) to guide thinking within research teams on new architectural evolution approaches.

References

1. Jazayeri, M.: Species evolve, individuals age. In: Eighth International Workshop on Principles of Software Evolution, pp. 3–9. IEEE (2005)
2. Breivold, H.P., Crnkovic, I., Larsson, M.: A systematic review of software architecture evolution research. Inf. Softw. Technol. **54**(1), 16–40 (2012)
3. Williams, B.J., Carver, J.C.: Characterizing software architecture changes: a systematic review. Inf. Softw. Technol. **52**(1), 31–51 (2010)
4. Buckley, J., Mens, T., Zenger, M., Rashid, A., Kniesel, G.: Towards a taxonomy of software change. J. Softw. Maint. Evol. Res. Pract. **17**(5), 309–332 (2005)
5. Zachman, J.A.: A framework for information systems architecture. IBM Syst. J. **26**(3), 276–292 (1987)
6. Pigoski, T.M.: Practical Software Maintenance: Best Practices for Managing Your Software Investment. Wiley, New York (1996)
7. Shaw, M., DeLine, R., Klein, D.V., Ross, T.L., Young, D.M., Zelesnik, G.: Abstractions for software architecture and tools to support them. IEEE Trans. Softw. Eng. **21**(4), 314–335 (1995)
8. Bézivin, J.: La transformation de modèles, p. 13. INRIA-ATLAS & Université de Nantes (2003)
9. Oreizy, P., Taylor, R.N.: On the role of software architectures in runtime system reconfiguration. IEE Proc. Softw. **145**(5), 137–145 (1998)
10. Mens, T., Wermelinger, M., Ducasse, S., Demeyer, S., Hirschfeld, R., Jazayeri, M.: Challenges in software evolution. In: Eighth International Workshop on Principles of Software Evolution, pp. 13–22. IEEE (2005)
11. Kniesel, G., Costanza, P., Austermann, M.: Jmangler-a framework for load-time transformation of Java class files. In: Proceedings of the First IEEE International Workshop on Source Code Analysis and Manipulation, pp. 98–108. IEEE (2001)
12. Chaki, S., Diaz-Pace, A., Garlan, D., Gurfinkel, A., Ozkaya, I.: Towards engineered architecture evolution. In: Proceedings of the 2009 ICSE Workshop on Modeling in Software Engineering, pp. 1–6. IEEE Computer Society (2009)
13. Oussalah, M., et al.: Génie objet: analyse et conception de l'évolution. Hermès Science Publications (1999)
14. Terho, H., Suonsyrjä, S., Systä, K., Mikkonen, T.: Understanding the relations between iterative cycles in software engineering. In: Proceedings of the 50th Hawaii International Conference on System Sciences (2017)
15. Brooks, R.A.: A robot that walks; emergent behaviors from a carefully evolved network. Neural comput. **1**(2), 253–262 (1989)
16. Oussalah, M., Sadou, N., Tamzalit, D.: SAEV: a model to face evolution problem in software architecture. In: Proceedings of the International ERCIM Workshop on Software Evolution, pp. 137–146 (2006)
17. Barnes, J.M., Garlan, D., Schmerl, B.: Evolution styles: foundations and models for software architecture evolution. Softw. Syst. Model. **13**(2), 649–678 (2014)
18. Ahmad, A., Jamshidi, P., Pahl, C.: Classification and comparison of architecture evolution reuse knowledgea systematic review. J. Softw. Evol. Process **26**(7), 654–691 (2014)
19. Mens, T., Van Gorp, P.: A taxonomy of model transformation. Electron. Notes Theor. Comput. Sci. **152**, 125–142 (2006)
20. Hassan, A., Oussalah, M.: Meta-evolution style for software architecture evolution. In: Freivalds, R.M., Engels, G., Catania, B. (eds.) SOFSEM 2016. LNCS, vol. 9587, pp. 478–489. Springer, Heidelberg (2016). https://doi.org/10.1007/978-3-662-49192-8_39

A Study on Self-adaptation in the Evolutionary Strategy Algorithm

Noureddine Boukhari[1]($\boxtimes$) (iD), Fatima Debbat[1]($\boxtimes$),
Nicolas Monmarché[2]($\boxtimes$), and Mohamed Slimane[2]($\boxtimes$)

[1] Department of Computer Science, Mascara University, Mascara, Algeria
boukhari.noureddine@gmail.com, debbat_fati@yahoo.fr
[2] Laboratoire Informatique (EA6300), Université François Rabelais Tours, 64,
avenue Jean Portalis, 37200 Tours, France
{nicolas.monmarche,mohamed.slimane}@univ-tours.fr

Abstract. Nature-inspired algorithms attract many researchers worldwide for solving the hardest optimization problems. One of the well-known members of this extensive family is the evolutionary strategy ES algorithm. To date, many variants of this algorithm have emerged for solving continuous as well as combinatorial problems. One of the more promising variants, a self-adaptive evolutionary algorithm, has recently been proposed that enables a self-adaptation of its control parameters. In this paper, we discuss and evaluate popular common and self-adaptive evolutionary strategy (ES) algorithms. In particular, we present an empirical comparison between three self-adaptive ES variants and common ES methods. In order to assure a fair comparison, we test the methods by using a number of well-known unimodal and multimodal, separable and non-separable, benchmark optimization problems for different dimensions and population size. The results of this experiments study were promising and have encouraged us to invest more efforts into developing in this direction.

Keywords: Meta-heuristics · Evolutionary algorithms · Evolution strategy
Parameter control · Self-adaptation

1 Introduction

The most successful methods in global optimization are based on stochastic components, which allow escaping from local optima and overcome premature stagnation. A famous class of global optimization methods is ES. They are exceptionally successful in continuous solution spaces. ES belong to the most famous evolutionary methods for black-box optimization, i.e., for optimization scenarios, where no functional expressions are explicitly given and no derivatives can be computed [1, 2]. ES imitate the biological principle of evolution [3] and can serve as an excellent introduction to learning and optimization. They are based on three main mechanisms oriented to the process of Darwinian evolution, which led to the development of all species. Evolutionary concepts are translated into algorithmic operators, i.e., recombination, mutation, and selection. This paper focuses on the self-adaptation in evolution

A. Amine et al. (Eds.): CIIA 2018, IFIP AICT 522, pp. 150–160, 2018.
https://doi.org/10.1007/978-3-319-89743-1_14

strategies (ES). Primarily, the self-adaptation has been gaining popularity due to the flexibility in adaptation to different fitness landscapes [8].

This method enables an implicit learning of mutation strengths in the real-valued search spaces. Self-adaptation bases on the mutation operator that modifies the problem variables using the strategy parameters to search the problem and parameter spaces simultaneously [8]. In line with this, the best values of problem variables and strategy parameters survive during the evolutionary process.

Evolutionary algorithms (EAs) are intrinsically dynamic, adaptive processes [19]. Therefore, setting the strategy parameters, controlling the behavior of these algorithms, fixed during the run is in contrast with an idea of evolution on which bases an evolutionary computation (EC). In line with this, the evolution of evolution has been developed by Rechengerg [10] and Schwefel [6], where the strategy parameters are put into a representation of individuals and undergo operations of the variation operators. As a result, the values of strategy parameters those modify the problem variables, which are the most adapted to the fitness landscape, as determined by the fitness function of the problem to be solved during the evolutionary search process. However, the process of the simultaneously evolution of strategy parameters together with the problem variables is also named a self-adaptation in EC.

2 Evolution Strategies

Evolution strategies derive inspiration from principles of biological evolution. We assume a population, P, of so-called individuals. Each individual consists of a solution or object parameter vector $x \in R\ n$ (the visible traits) and further endogenous parameters, s (the hidden traits), and an associated fitness value, f(x). In some cases the population contains only one individual. Individuals are also denoted as parents or offspring, depending on the context. In a generational procedure [4]. First, one or several parents are picked from the population (mating selection) and new off- spring are generated by duplication and recombination of these parents; then, the new offspring undergo mutation and become new members of the population; finally, Environmental selection reduces the population to its original size.

2.1 Pseudo Code for Evolutionary Algorithms

Using these ideas a computer algorithm can be developed to analyze a problem and its data to achieve an optimal solution to that problem, as shown in Fig. 1. First, an initial population P(t) is generated randomly and evaluated. Second a mutation technique is applied to adjust the children to a new set P'(t). The fitness of those children is evaluated and then children are chosen from the set of parents P(t) and children P'(t) to form the new parent set P(t + 1). The loop terminates if either the maximum number of iterations are reached or the desired solution is found. As outlined here by Castro: [16].

Pseudo code For Evolutionary Algorithms

```
Initialize P (t)
 Evaluate P (t)
 While not Terminate do
          P' (t) = mutate P(t)
          Evaluate P'(t)
          P(t+1) = P'(t)
          or best of {P'(t) U P(t)}
 t = t + 1
 Loop
```

Fig. 1. Pseudo Code for Evolutionary Algorithms

3 Mutation and Parameter Control

Mutation introduces small, random and unbiased changes to an individual. These changes typically affect all variables. The average size of these changes depends on endogenous parameters that change over time. These parameters are also called control parameters, or endogenous strategy parameters, and define the notion of "small", for example via the step-size σ. In contrast, exogenous strategy parameters are fixed once and for all, for example parent number μ. Parameter control is not always directly inspired by biological evolution, but is an indispensable and central feature of evolution strategies [4]. Control parameters are encoded into chromosomes and undergo actions by the variation operators (e.g., crossover and mutation) using self-adaptive parameter control. The better values of parameter variables and control parameters have more chances to survive and reproduce their genetic material into the next generations. This phenomenon makes EAs more flexible and closer to natural evolution [5]. This feature was firstly introduced in ES by Schweffel [6]. Parameter control addresses only one side of the parameter setting, where the strategic parameters are changed during the run.

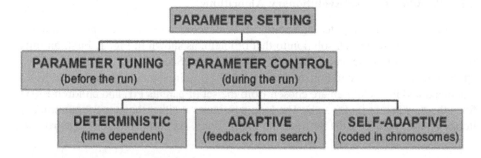

Fig. 2. Parameter setting in CI algorithms

In contrast, when the parameters are fixed during the run, an optimal parameter setting needs to be found by an algorithm's developer. Typically, these optimal parameters are searched during a tuning. In general, the taxonomy of parameter setting according to Eiben and Smith [7] is as illustrated in Fig. 2 Sample Heading (Third Level). Only two levels of headings should be numbered. Lower level headings remain unnumbered; they are formatted as run-in headings.

3.1 Deterministic Parameter Control

Deterministic parameter control [21] means that the parameters are adjusted according to a fixed time scheme, explicitly depending on the number of generations t. It may be useful to reduce the mutation strengths during the evolutionary search, in order to allow convergence of the population. There are many examples where parameters are controlled deterministically. Mezura-Montes et al. [20] make use of deterministic parameter control for differential evolution based constraint handling.

3.2 Adaptive Parameter Control

Adaptive parameter control methods make use of rules the practitioner defines for the heuristic. A feedback from the search determines magnitude and direction of the parameter change. An example for an adaptive control of endogenous strategy parameters is the 1/5-th success rule for the mutation strengths by Rechenberg [10]. Rchenberg's rule adapts the mutation strengths in the following way: the whole population makes use of a global mutation strength r for all individuals. If the ratio of successful candidate solutions is greater than 1/5-th, the step size should be increased, because bigger steps towards the optimum are possible, while small steps would be a waste of time. If the success ratio is less than 1/5-th the step size should be decreased. This rule is applied every g generations [3].

3.3 Self-adaptation

This article focuses on the self-adaptation in evolution strategies (ES). Primarily, the self-adaptation has been gaining popularity due to the flexibility in adaptation to different fitness landscapes [8]. This method enables an implicit learning of mutation strengths in the real-valued search spaces. Self-adaptation bases on the mutation operator that modifies the problem variables using the strategy parameters to search the problem and parameter spaces simultaneously [9]. In line with this, the best values of problem variables and strategy parameters survive during the evolutionary process.

There are different strategy parameters in self-adaptive EAs, e.g., probability of mutation pc, probability of crossover pm, population size Np, etc. In order to better adapt the parameter setting to the fitness landscape of the problem, the self-adaptation of strategy parameters has been emerged that is tightly connected with a development of so named evolution strategies (SA-ES) [9–11]. SA-ES were especially useful for solving the continuous optimization problems [12].

As each EAs, the SA-ES also consists of the following components [7]:

- representation of individuals,
- evaluation function,
- mutation,
- crossover,
- Survivor selection.

4 Next Generation Selection Strategies

Two different selection processes were used to pick the next generation of data to be sent through the algorithm. In standard evolutionary strategy literature "given μ parents generating λ offspring ($\lambda >= \mu$)" [16] The simplest way is to use the (μ, λ)-ES technique where the new set of children is used as the parents for the next iteration, the algorithm in Fig. 1 would use P(t + 1) = P'(t). A second technique is to use ($\mu + \lambda$)-ES technique [16], by looking at the fitness of the parents and the fitness of the new children and sorting their resulting finesses from best to worst. A next generation set is created from the top μ best parents and children, the algorithm in Fig. 1 would use P (t + 1) = best of {P' (t) U P (t)}.

5 Mutation Operators

In classical ES, three types of mutation operators have been applied: an uncorrelated mutation with one step size, an uncorrelated mutation with n-step sizes and correlated mutation using a covariance matrix [7]. In the first mutation, the same distribution is used to mutate each problem variable, in the second, different step sizes are used for different dimensions, while in the last mutation, not only the magnitude in each dimension but also the direction (i.e., rotation angle) are taken into account [13, 14]. Self-adaptation of control parameters is realized in classical evolution strategies (ES) using the appropriate mutation operators controlled by strategy parameters (i.e. mutation strengths) that are embedded into representation of individuals. The mutation strengths determine the direction and the magnitude of the changes on the basis of the new position of the individuals in the search space is determined. This article analyzes the characteristics of classical mutation operators, like uncorrelated mutation with one step size and uncorrelated mutation with n step sizes [15], and the comparison with newest approaches and improvements of self-adaptation strategies like the correlation and hybridization in the recent literature.

As schematically illustrated in Table 1, parameters that yield good solutions with high probability are selected as survivors. It is quite reasonable, at least for local optimization. Concerning selection of the set of parameters to be adapted, we are chosen in our experiments three options from Back (1996).

5.1   Uncorrelated Mutation with One Step Size (Type 1)

As this algorithm loops through each iteration, all parents are being mutated by the same single step size. Using the same standard deviation σ to create all children will have the result that "lines of equal probability density of the normal distribution are hyper-spheres in an l-dimensional space" [16].

- Chromosomes: $\langle x_1, \ldots, x_n, \sigma \rangle$
- $\sigma' = \sigma \bullet \exp(\tau \bullet N(0, 1))$
- $x'_i = x_i + \sigma' \bullet N(0, 1)$
- Typically the "learning rate" $\tau \propto 1/n^{1/2}$

5.2   Uncorrelated Mutation with Individual Step Size (Type 2)

As this algorithm loops through each iteration, each parent is being mutated by an individual unique step size. Using the different standard deviation σi to create each child will have the result that "lines of equal probability density of the normal distribution are hyper-ellipsoids" [16].

- Chromosomes: $\langle x_1, \ldots, x_n, \sigma_1, \ldots, \sigma_n \rangle$
- $\sigma'_i = \sigma_i \bullet \exp(\tau' \bullet N(0, 1) + \tau \bullet N_i(0, 1))$
- Two learning rate parmeters: τ' overall learning rate and τ coordinate wise learning rate
- $\tau \propto 1/(2\,n)^{1/2}$ and $\tau \propto 1/(2\,n^{1/2})^{1/2}$

5.3   Correlated Mutation (Type 3)

As this algorithm loops through each iteration, each parent is being mutated by an individual unique step size. Just like the previous uncorrelated mutation with individual step sizes, the formulas use a different standard deviation σi to create each child. The correlated mutation formulas add an additional step of introducing rotation angles to "describe the coordinate rotations necessary to transform the uncorrelated mutation vector to a correlated mutation vector. Now the previous hyper-ellipsoids can be rotated randomly at angle α (t) to give the data more freedom of movement through the plane" [16].

- Chromosomes: $\langle x_1, \ldots, x_n, \sigma_1, \ldots, \sigma_n, \alpha_1, \ldots, \alpha_k \rangle$ where $k = n \bullet (n - 1)/2$
- and the covariance matrix C is defined as:
- $c_{ii} = \sigma_i^2$
- $c_{ij} = 0$ if i and j are not correlated
- $c_{ij} = 1/2 \bullet (\sigma_i^2 - \sigma_j^2) \bullet \tan(2\alpha_{ij})$ if i and j are correlated

The mutation mechanism is then:

- $\sigma'_i = \sigma_i \bullet \exp(\tau' \bullet N(0, 1) + \tau \bullet N_i(0, 1))$
- $\alpha'_j = \alpha_j + \beta \bullet N(0, 1)$
- $x' = x + N(0, C')$

- *C'* is the covariance matrix *C* after mutation of the α values
- $\tau \propto 1/(2\,n)^{1/2}$ and $\tau \propto 1/(2\,n^{1/2})^{1/2}$ and $\beta \approx 5°$

6 Numerical and Equations Experiments

As discussed above, the ES achieve self-adaptive search in different manners. To understand similarities and differences of both mechanisms, first a comparison of them through numerical experiments has been carried out. In the experiments, the ES types are applied to several test functions picked from [18]. Usually, a comparative study of different optimization methods is performed by using a set of test functions from the literature. In this paper, five well-known benchmark functions are used. We have chosen two unimodal and three multimodal functions; the functions can also be grouped into separable or non-separable. These are the Sphere function, the generalized Rosenbrock's function, the generalized Rastrigin's function, the Ackley's function, and the Griewank's function. The above functions are defined respectively as:

$$f_{Sphere}(x) = \sum_{j=0}^{D-1} x_j^2, |x_j| \leq 5 \; and \; f_{Sphere}(0,0,\ldots0) = 0 \tag{1}$$

$$f_{Rosenbrock}(x) = \sum_{j=0}^{D-2} \left(100\left(x_{j+1} - x_j^2\right)^2 + (x_j - 1)^2 \right),$$
$$|x_j| \leq 5 \; and \; f_{Rosenbrock}(1,1,\ldots1) = 0 \tag{2}$$

$$f_{Rastrigin}(x) = \sum_{j=0}^{D-1} \left(x_j^2 - 10\cos\left(2\pi x_j\right) + 10 \right), |x_j| \leq 5$$
$$and \; f_{Rastrigin}(0,0,\ldots0) = 0 \tag{3}$$

$$f_{Ackley}(x) = -20\exp\left(-\frac{1}{5}\sqrt{\frac{1}{D}\sum_{j=0}^{D-1} x_j^2} \right) - \exp\left(\frac{1}{D}\sum_{j=0}^{D-1}\cos\left(2\pi x_j\right) \right) + 20 +$$
$$\exp(1), |x_j| \leq 5 \; and \; f_{Ackley}(0,0,\ldots0) = 0 \tag{4}$$

$$f_{Griewank}(x) = \frac{1}{4000}\sum_{j=1}^{D} x_j^2 - \prod_{j=1}^{D}\cos\left(\frac{x_{ij}}{\sqrt{j}}\right), |x_j| \leq 5$$
$$and \; f_{Griewank}(0,0,\ldots0) = 0 \tag{5}$$

The sphere function is one of the simplest benchmarks. It is a continuous, unimodal and separable problem. The generalized Rosenbrock's global optimum lies inside a parabolic shaped flat valley. It is easy to find the valley but convergence to the global optimum is difficult. This problem is unimodal and non-separable. The generalized Rastrigin function is a complex multimodal separable problem with many local optima. The Ackley's function is a multimodal non-separable problem and has many local optima and a narrow global optimum. The fifth function is highly multimodal and non-separable. It has many widespread local minima, which are regularly distributed.

6.1 Algorithm Parameters Used for Experiments

The ES has many options that should be selected considering both the difficulty of the problems to be solved and the resources available for computation. Based on the recommendation by Back (1996), the following options are adopted:

- The comma strategy with 15 parents and 100 children, i.e., (15,100)-ES is used.
- The discrete recombination is used for the decision variables, and the panmictic intermediate recombination for the standard deviations and the rotation angles.
- Recommended values for parameters τ, τ 0, and β are used.
- All the initial values of α_{ij} are set at $\pi/4$.
- We used experiments with 500 and 2000 number of iterations.
- The results of the all experiments are averaged over 20 independent runs.

Table 1. Best normalized optimization results in five benchmark functions. The values shown are the minimum objective function values found by each algorithm with 500 iterations, averaged over 20 Monte Carlo simulations.

Benchmark function	D	Type 1 One step size	Type 2 N step size	Type 3 Correlated	Adaptive 1/5 rule	Adaptive cumulative
Sphere	5	4.0546e−114	**4.9457e−178**	2.6385e−108	9.9912e−54	9.5627e−91
	10	3.4808e−85	**1.5605e−106**	9.6233e−42	1.378e−15	4.9752e−54
	20	**1.531e−61**	3.2016e−40	0.0028815	0.21719	1.1019e−37
Rosenbrock	5	0.21096	0.025719	**1.9657e−06**	0.10707	0.055967
	10	4.8468	**0.021611**	0.058623	2.5602	1.3705
	20	15.6707	**9.9448**	42.3292	25.2971	12.6153
Rastrigin	5	**0**	**0**	**0**	**0**	**0**
	10	0.99496	1.9899	0.99496	**0**	**0**
	20	3.9798	7.9597	**2.3175**	9.6734	3.9798
Ackely	5	**0**	**0**	**0**	**0**	**0**
	10	3.5527e−15	**0**	3.5527e−15	3.5909e−10	**0**
	20	**3.5527e−15**	**3.5527e−15**	0.032326	0.22965	**3.5527e−15**
Griewank	5	**0**	**0**	**0**	**0**	**0**
	10	**0**	**0**	0.007396	**0**	**0**
	20	**0**	**0**	0.019238	1.5486e−07	**0**

Table 2. Best normalized optimization results in five benchmark functions. The values shown are the minimum objective function values found by each algorithm with 2000 iterations, averaged over 20 Monte Carlo simulations.

Benchmark function	D	Type 1 One step size	Type 2 N step size	Type 3 Correlated	Adaptive 1/5 rule	Adaptive cumulative
Sphere	5	**0**	**0**	**0**	5.2256e−204	**0**
	10	**0**	**0**	**0**	9.4191e−51	1.6795e−219
	20	**2.097e−248**	1.7093e−147	1.0321e−15	0.0005046	1.0846e−150
Rosenbrock	5	0.021332	0.00014261	**1.1442e−05**	0.0044214	0.00059227
	10	0.023147	4.5627	**0.0023164**	0.10187	0.045193
	20	12.3151	3.2985	**2.0112**	6.4291	3.8319
Rastrigin	5	**0**	**0**	**0**	**0**	**0**
	10	**0**	0.99496	0.99496	9.9496	0.99496
	20	2.9849	5.9698	**2.3175**	5.9721	2.9849
Ackely	5	**0**	**0**	**0**	**0**	**0**
	10	3.5527e−15	**0**	**0**	**0**	**0**
	20	**3.5527e−15**	**3.5527e−15**	**3.5527e−15**	0.0020593	**3.5527e−15**
Griewank	5	**0**	**0**	**0**	**0**	**0**
	10	**0**	**0**	**0**	**0**	**0**
	20	**0**	**0**	**0**	**0**	**0**

From Tables 1 and 2 we summarize how many times each algorithm get best results (Table 3):

Table 3. Statistical results

ES algorithms	Type 1 One step size	Type 2 N step size	Type 3 Correlated	Adaptive 1/5 rule	Adaptive cumulative
Best times	16	20	18	11	16

6.2   Behavior in Higher-Dimensional Search Spaces

For problems in the five-dimensional search space, all the algorithms succeeded in finding good solutions. Especially, the ES Type 3 and Type 2, show good performances and the auto-adaptation process outperform the adaptation one. Increasing the dimension of the search space, performances of the ES algorithms get poorer because the non-separability of the test function becomes serious. The results show clearly that the ES type 2 and 3 is more sufficient and perform better in comparing Sphere and Rosenbrock functions. Hence, using the rotated angle recommendation give the ES forces to reach a good solutions as well as the strategy parameter with individual step sizes help ES to improve solutions as shown between ES type 1 and type 2.

6.3   Behaviors for Multi-modal Objective Functions

The results show that the auto-adaptation ES mostly get hard in finding the optimum. The best solutions found by the ESs often stayed at points that are not even local optima. This result suggests that the mechanism of the self-adaptive mutation may have some difficulty in multi-modal objective functions. Since the ES has many options and parameters, and the above experiment uses only a single test function and the recommended options and parameters in literature (Back, 1996), more efficient search may be achieved by the ES by adjusting the options and parameters. It is a subject of further study as well as more detailed examination of the behavior of the self-adaptive mechanism.

7   Conclusion and Future Work

The self-adaptation of strategy parameters is the most advanced mechanism in ES, where the mutation strength strategy parameters control the magnitude and the direction of changes generated by the mutation operator. There are three mutation operators in classical ES, as follows: the uncorrelated mutation with one step size, the uncorrelated mutation with n step sizes and the correlated mutations.

This paper analyses the characteristics of these three mutation operators. In last decades several studies and papers focus at the self-adaptation as tools to improves evolutionary strategy algorithm performance, for that many techniques are used, either mathematics such including geometric function to improve solutions in multimodal functions [17], incorporating machine learning technique to expect suitable parameters which leads to best solutions [2]. Although the research domain of the ES seems to be already explored and, by this flood of the new nature-inspired algorithms every day, unattractive for the developers, the results of our experiments showed the opposite. Additionally, hybridization adds to a classical self-adapting ES the new value. As the first step of our future work, however, the next step is to extend this comparative study with another well-known meta-heuristics such as PSO, DE, HS in order to recognize the weakness of the self-adaptation ES for eventual future work improvement by hybridization with other local search techniques.

Regardless of the problems mentioned, self-adaptation is a state-of-the-art adaptation technique with a high degree of robustness, especially in real-coded search spaces and in environments with uncertain or noisy fitness information.

References

1. Fister, I., Fong, S., Fister Jr., I., Brest, J.: A novel hybrid self-adaptive bat algorithm. Sci. World J. **2014**, 1–12 (2014). Article ID 709738
2. Kramer, O.: Machine Learning for Evolution Strategies. Springer, Heidelberg (2016). https://doi.org/10.1007/978-3-319-33383-0
3. Kramer, O., Ciaurri, D.E., Koziel, S.: Derivative-free optimization. In: Koziel, S., Yang, X. S. (eds.) Computational Optimization and Applications in Engineering and Industry. Springer, Heidelberg (2011). https://doi.org/10.1007/978-3-642-20859-1_4

4. Hansen, N., Arnold, D., Auger, A.: Evolution strategies. In: Kacprzyk, J., Pedrycz, W. (eds.) Handbook of Computational Intelligence, pp. 871–898. Springer, Heidelberg (2015). https://doi.org/10.1007/978-3-662-43505-2_44

5. Deb, K.: Multi-objective Optimization Using Evolutionary Algorithms. Wiley, New York (2001)

6. Schwefel, H.P.: Numerische Optimierung von Computer–Modellen mittels der Evolutions strategie. Birkhäuser, Basel (1977)

7. Eiben, A., Smith, J.: Introduction to Evolutionary Computing. Springer, Berlin (2003). https://doi.org/10.1007/978-3-662-44874-8

8. Back, T., Hammel, U., Schwefel, H.-P.: Evolutionary computation: comments on the history and current state. IEEE Trans. Evol. Comput. 1(1), 3–17 (1997)

9. Beyer, H.-G., Deb, K.: On self-adaptive features in real-parameter evolutionary algorithms. IEEE Trans. Evol. Comput. 5(3), 250–270 (2001)

10. Rechenberg, I.: Evolutionsstrategie Optimierung technischer Systeme nach Prinzipien der biologischen Evolution. Frommann-Holzboog, Stuttgart (1973)

11. Schwefel, H.P.: Numerische Optimierung von Computer-Modellen mittels der Evolutions strategie. Birkhäuser, Basel (1977)

12. Fister, I., Mernik, M., Filipič, B.: Graph 3-coloring with a hybrid self-adaptive evolutionary algorithm. Comp. Opt. Appl. 54(3), 741–770 (2013)

13. Hansen, N.: The CMA evolution strategy: a tutorial. Vu le 29 (2005)

14. Igel, C., Hansen, N., Roth, S.: Covariance matrix adaptation for multi-objective optimization. Evol. Comput. 15(1), 1–28 (2007)

15. Fister, I., Fister, I.: On the mutation operators in evolution strategies. In: Fister, I., Fister Jr., I. (eds.) Adaptation and Hybridization in Computational Intelligence. ALO, vol. 18, pp. 69–89. Springer, Cham (2015). https://doi.org/10.1007/978-3-319-14400-9_3

16. Castro, L.N.: Fundamentals of Natural Computing, Chapter 3. Taylor and Francis Group, LLC, New York (2006)

17. DeBruyne, S., Kaur, D.: Comparison of uncorrelated and correlated evolutionary strategies with proposed additional geometric translations. In: Proceedings of the International Conference on Genetic and Evolutionary Computing, GEM 13, held at WORLDCOMP 2013 Congress, Las Vegas, USA (2013)

18. Kita, H.: A comparison study of self-adaptation in evolution strategies and realcoded genetic algorithms. Evol. Comput. 9(2), 223–241 (2001)

19. Eiben, A.E., Hinterding, R., Michalewicz, Z.: Parameter control in evolutionary algorithms. Trans. Evol. Comp. 3(2), 124–141 (1999)

20. Mezura-Montes, E., Palomeque-Ortiz, A.G.: Self-adaptive and deterministic parameter control in differential evolution for constrained optimization. Constraint-Handling Evol. Optim. 189, 95–120 (2009)

21. Kramer, O.: Evolutionary self-adaptation: a survey of operators and strategy parameters. Evol. Intell. 3, 51–65 (2010)

Swarm Intelligence Algorithm for Microwave Filter Optimization

Erredir Chahrazad[(⊠)], Emir Bouarroudj, and Mohamed Lahdi Riabi

Laboratory of Electromagnetic and Telecommunication,
University Brothers Mentouri Constantine 1, Constantine, Algeria
cerredir@yahoo.fr

Abstract. In this paper, three recent swarm intelligence algorithms (spider monkey optimization (SMO), social spider optimization (SSO) and teaching learning based optimization (TLBO)) are proposed to the optimization of microwave filter (H-plane three-cavity filter). The results of convergence and optimization use of these algorithms are compared with the results of the most popular swarm intelligences algorithm, namely particle swarm optimization (PSO) for different common parameters (population size and maximum number of iteration). The results showed validation of the proposed algorithms.

Keywords: Swarm intelligence · Spider monkey optimization
Social spider optimization · Teaching learning based optimization
Particle swarm optimization · Microwave filter

1 Introduction

Swarm intelligence (SI) algorithms are natural inspired techniques that involve the study of collective behavior of decentralized, self-organized systems [1, 2]. The swarm intelligence systems contain a set of particles (or agents) that interact locally with one another and with their environment. Swarm Intelligence techniques can be used in several engineering applications where the SI algorithms have been successfully applied to solve complex optimization problems including continuous optimization, constrained optimization and combinatorial optimization.

To date, several swarm intelligence models based on different natural swarm systems have been proposed in the literature, and successfully applied in many real-life applications. Examples of swarm intelligence models are: Ant Colony Optimization [3], Particle Swarm Optimization [4], Artificial Bee Colony [5], Bacterial Foraging [6], Cat Swarm Optimization [7], Artificial Immune System [8], and Glowworm Swarm Optimization [9].

In this paper, we will primarily focus on three of the recent swarm intelligences models, namely, spider monkey optimization (SMO), social spider optimization (SSO) and teaching learning based optimization (TLBO) to optimize the waveguide microwave filter (H-plane three-cavity filter). The results of optimization obtained are validated by comparing them with those obtained using the optimization algorithm available in the literature Particle swarm optimization (PSO). The details of each algorithm are presented in the next section.

© IFIP International Federation for Information Processing 2018
Published by Springer International Publishing AG 2018. All Rights Reserved
A. Amine et al. (Eds.): CIIA 2018, IFIP AICT 522, pp. 161–172, 2018.
https://doi.org/10.1007/978-3-319-89743-1_15

2   Swarm Intelligence Algorithm

2.1   Particle Swarm Optimization

Particle swarm optimization (PSO) is a stochastic method of optimization based on the reproduction of a social behavior. It was invented by Eberhart and Kennedy [4] in 1995. They tried to simulate the ability of animal societies that don't have any leader in their group or swarm (bird flocking and fish schooling) to move synchronously and their ability to change direction suddenly while remaining in an optimal formation (food source). PSO consists of a swarm of particles, where particle represent a potential solution. The particles of the swarm fly through hyperspace and have two essential reasoning capabilities: their memory of their own best position local best (LB) and knowledge of the global or their neighborhood's best global best (GB). The essential steps of particle swarm optimization are presented by the following algorithm:

> **Step 1.** Initialize the optimization parameters (population size, number of generations, design variables of the optimization problem and the specific parameters of algorithm) and define the optimization problem (minimization or maximization of fitness function).
> **Step 2.** Generate a random population (position and velocities), according to the population size and the limits of the design variables.
> **Step 3.** Evaluate each initialized particle's fitness value, then calculate LB the positions of the current particles, and GB the best position of the particles.
> **Step 4.** the best particle of the current particles is stored. The positions and velocities of all the particles are updated according to (1) and (2), then a group of new particles are generated.

$$X_i(t+1) = X_i(t) + V_i(t+1) \tag{1}$$

$$V_i(t+1) = w * V_i(t) + c_1 r_1 [LB_i(t) - X_i(t)] + c_2 r_2 [GB(t) - X_i(t)] \tag{2}$$

$V_i(t)$, $X_i(t)$ are the velocity and the position for particle i at time t. w is the inertia weight, at each iteration update with the following equation [10].

$$w(t) = w_{max} - \frac{(w_{max} - w_{max}) * t}{maxit} \tag{3}$$

The parameters w_{max}, w_{min}, $maxit$, c_1 and c_2 are constant coefficients determined by the user. r_1 and r_2 are random numbers between 0 and 1.
> **Step 5.** Repeat the procedure from step 3 until the maximal iteration is met.

2.2   Spider Monkey Optimization Algorithm

Spider Monkey Optimization (SMO) algorithm is a new swarm intelligence algorithm based on the foraging behavior of spider monkeys, proposed by Bansal et al. in [11]. There are four important control parameters necessary for this algorithm: perturbation rate (Pr), local leader limit (LLL), global leader limit (GLL) and maximum number of groups (MG). The SMO process consists of six phases:

Local Leader Phase (LLP). In this phase, Spider Monkey SM of each group updates its position based on the experience of the local leader and local group members as following expression:

$$SM_{ij} = SM_{ij} + r_1\left(LL_{Kj} - SM_{ij}\right) + r_2\left(SM_{rj} - SM_{ij}\right) \tag{4}$$

Where SM_{ij} is the jth dimension of the ith SM, LL_{kj} represents the jth dimension of the kth local group leader position. SM_{rj} is the jth dimension of the rth SM which is chosen randomly within k^{th} group such that $r \neq i$, r_1 is a random number between 0 and 1 and r_2 is a random number between -1 and 1.

Global Leader Phase (GLP). In GLP phase, all the SM's update their position using experience of global leader and local group member's experience. The position update equation for this phase is as follows:

$$SM_{ij} = SM_{ij} + r_1\left(GL_j - SM_{ij}\right) + r_2\left(SM_{rj} - SM_{ij}\right) \tag{5}$$

Where GL_j represents the jth dimension of the global leader position and $j \in \{1, 2 \dots D\}$ is the randomly chosen index; D is the number of design variables. The positions are updated based upon some probability given by the following formula.

$$P_i = 0.9\frac{F_i}{max\_F} + 0.1 \tag{6}$$

Where P_i is the probability, F_i is the fitness of ith SM, and max_F is the maximum fitness of the group.

Local Leader Learning Phase (LLLP). In this phase, the position of the local leader is updated by applying the greedy selection in that group. If the LL's position remains same as before, then the Local Limit Count is increased by 1.

Global Leader Learning Phase (GLLP). In this phase, the position of the global leader is updated by applying the greedy selection in the population. If the GL's position remains same as before, then the Global Limit Count is increased by 1.

Local Leader Decision Phase (LLDP). If a LL position is not updated for a predetermined number of iterations Local Leader Limit (LLL), then the positions of the spider monkeys are updated either by random initialization or by using information from both LL and GL through Eq. (7) based on the perturbation rate (pr)

$$SM_{ij} = SM_{ij} + r_1\left(GL_j - SM_{ij}\right) + r_1\left(SM_{ij} - LL_{Kj}\right) \tag{7}$$

Global Leader Decision Phase (GLDP). If the position of GL is not updated in predetermined number of iterations Global Leader Limit, then the population is split into subgroups. The groups are split till the number of groups reaches to maximum allowed groups (MG), then they are combined to form a single group again.

The details of each step of SMO implementation are explained below:

Step 1. Initialize the optimization parameters and define the optimization problem. Control parameters necessary for this phases; perturbation rate (Pr), local leader limit (LLL), global leader limit (GLL) and maximum number of groups (MG). Some settings of control parameters are suggested as follows:

- MG = N/10, i.e., it is chosen such that minimum number of SM's in a group should be 10.
- Global Leader Limit ∈ [N/2, 2 × N].
- Local Leader Limit should be D × N.

$$Pr(t+1) = Pr(t) + \frac{(0.4 - 0.1)}{maxit}, \quad Pr(1) = 0.1 \tag{8}$$

Step 2. Initialize the population and evaluate the corresponding objective function value.

Step 3. Locate global and local leaders.

Step 4. The local leader phase starts by update the position of all group members' Eq. (4). Accept of a new solution if it gives better function value. All the accepted function values at the end of this phase are maintained and these values become the input to the global leader phase.

Step 5. Produce new positions for all the group members, selected by probability (P_i), by using self experience, global leader experience and group members' experiences Eq. (5).

Step 6. Update the position of local and global leaders, by applying the greedy selection process on all the groups (see. LLLP, GLLP).

Step 7. If any Local group leader is not updating her position after a specified number of times (Local Leader Limit) then redirect all members of that particular group for foraging (LLDP).

Step 8. If Global Leader is not updating her position for a specified number of times (Global Leader Limit) then she divides the group into smaller groups (see. GLDP).

Step 9. Repeat the procedure from step 3 to until the termination criterion is met.

2.3 Social Spider Optimization Algorithm

Social-spiders optimization (SSO) [12, 13] is a new proposed swarm optimization algorithm; it is based on the natural spider's colony behavior. An interesting characteristic of social-spiders is the highly female-biased populations, where the number of females N_f is randomly selected within the range of 65–90% of the entire population NP and the rest is the number of male Nm. Therefore, N_f and N_m are calculated by the following equations:

$$N_f = floor((0.9 - rand * 0.25) * NP) \tag{9}$$

$$N_m = NP - N_f \tag{10}$$

Where *floor* rounds each element to the nearest integer, and *rand* is a random number in the unitary range [0, 1].

After the initialization process the algorithm starts the searching loop that only ends when the maximum number of function evaluations or the target function value is reached. The first step in the searching loop is to calculate the spider's weight. This calculation is done according to:

$$W_i = \frac{(Worst - f(x_i))}{(Worst - Best)} \qquad (11)$$

Where W_i is the weight of the ith spider, $f(x_i)$ is the fitness value of the spider x_i. The values *Worst* and *Best* are defined as follows (considering a minimization problem):

$$Best = min_{i=1...NP} f(x_i) \qquad (12)$$

$$Worst = max_{i=1...NP} f(x_i) \qquad (13)$$

In the colony, the spiders communicate with each other directly by mating or indirect by a small vibration to determine the potential direction of a food source, this vibration depend on the weight and distance of the spider which has generated them.

$$Vib_{ij} = w_j * \exp\left(-\left(d_{ij}\right)^2\right) \qquad (14)$$

Where w_j indicates the weight of the jth spider, and d_{ij} is the euclidean distance between i^{th} and j^{th} spiders. Every spider is able to consider three vibrations from other spiders as:

- Vibrations Vib_{ci} are perceived by the individual i (X_i) as a result of the information transmitted by the member c (X_c) who is an individual that has two important characteristics: it is the nearest member to i and possesses a higher weight in comparison to i ($W_c > W_i$).
- The vibrations Vib_{hi} perceived by the individual i as a result of the information transmitted by the member b (X_b), with b being the individual holding the best weight (best fitness value) of the entire population NP, such that:

$$Wb = max_{k=1...N}(W_k). \qquad (15)$$

- The vibrations Vib_{fi} perceived by the individual i (X_i) as a result of the information transmitted by the member f (X_f), with f being the nearest female individual to i.

Depending on gender, each individual is updating their position according to three operations (female operation, male operation and mating operation). In female operation, the female individuals are updating as follow equation:

$$X_i = X_i + \alpha\, Vib_{ci}(X_c - X_i) + \beta\, Vib_{bi}(X_b - X_i) \\ + \delta(rand - 0.5) \text{ with probability } PF \qquad (16)$$

$$X_i = X_i + \alpha \, Vib_{ci}(X_c - X_i) + \beta \, Vib_{bi}(X_b - X_i)$$
$$+ \delta(rand - 0.5) \text{ with probability } 1 - PF \tag{17}$$

Where PF is threshold parameter, α, β, δ and *rand* are random numbers between [0, 1].

The male spiders are divided into two different groups (dominant members D and non-dominant members ND) according to their position with regard to the median member. According to this, change of positions for the male spider can be modeled as follows:

$$X_i = X_i + \alpha \left(\frac{\sum_{h=1}^{N_m} X_h W_{N_f + h}}{\sum_{h=1}^{N_m} W_{N_f + h}} - X_i \right), \text{ Male D} \tag{18}$$

$$Xi = Xi + \alpha \, Vibfi(Xf - Xi) + \delta.(rand - 0.5), \text{ Male ND} \tag{19}$$

Where the individual (X_f) represent the nearest female individual to the male member.

After all males and females spiders are update, the last operator is representing the mating behavior where only dominant males will participate with females who are within a certain radius called mating radius given by

$$R = \frac{\sum_{d=1...n} (X_d^h - X_d^l)}{2D} \tag{20}$$

Where X^h and X^l are respectively the upper and lower bound for a given dimension and n is the problem dimension. Males and females which are under the mating radius generate new candidate spiders according to the roulette method. Each candidate spider is evaluated in the objective function and the result is tested against all the actual population members. If any member is worse than a new candidate, the new candidate will take the actual individual position assuming actual individual's gender.

2.4 Teaching Learning Based Optimization

Rao et al. [14–16] proposed an algorithm, called Teaching-Learning-Based Optimization (TLBO), based on the traditional teaching learning phenomenon of a classroom. TLBO is a population based algorithm, where a group of students (i.e. learner) is considered as population and the different subjects offered to the learners are analogous with the different design variables of the optimization problem. The results of the learner are analogous to the fitness value of the optimization problem. The best solution in the entire population is considered as the teacher. Teacher and learners are the two vital components of the algorithm, so there are two modes of learning; through the teacher (known as the teacher phase) and interacting with other learners (known as the learner phase).

Teacher Phase. In this part, learners take their knowledge directly through the teacher, where a teacher tries to increase the mean result value of the classroom to another value, which is better than, depending on his or her capability. This follows a random process depending on many factors. In this work, the value of solution is represented as

$X_{j,k,i}$, where j means the jth design variable (i.e. subject taken by the learners), $j = 1, 2, \ldots, m$; k represents the kth population member (i.e. learner), $k = 1, 2, \ldots, N$; and i represents the ith iteration, $i = 1, 2, \ldots, maxit$, where $maxit$ is the number of maximum generations (iterations). The existing solution is updated according to the following expression

$$X'_{j,k,i} = X_{j,k,i} + DM_{j,k,i} \tag{21}$$

$DM_{j,k,i}$ the difference between the existing mean and the new mean of each subject is given by

$$DM_{j,k,i} = r * \left(X_{j,kbest,i} - TF * M_{j,i}\right) \tag{22}$$

$M_{j,i}$ the mean result of the learners in a particular subject j, $X_{j,kbest,i}$ the new mean and is the result of the best learner (i.e. teacher) in subject j. r is the random number in the range [0, 1]. TF The teaching factor is generated randomly during the algorithm in the range of [1, 2], in which 1 corresponds to no increase in the knowledge level and 2 corresponds to complete transfer of knowledge. The in between values indicates the amount of transfer level of knowledge. The value of TF is not given as an input to the algorithm its value is randomly decided by the algorithm

$$TF = round[1 + rand(0, 1)\{2 - 1\}] \tag{23}$$

Learner Phase. In this part, learners increase their knowledge by interaction among themselves. A learner interacts randomly with other learners for enhancing his or her knowledge. A learner learns new things if the other learner has more knowledge than him or her. At any iteration i, each learner is compared with the other learners randomly. For comparison, randomly select two learners P and Q such that $X'_{P,i} \neq X'_{Q,i}$ (where $X'_{P,i}$ and $X'_{Q,i}$ are the updated values at the end of the teacher phase).

$$X''_{j,P,i} = X'_{j,P,i} + r * \left(X'_{j,P,i} + X'_{j,Q,i}\right), f\left(X'_{P,i}\right) < f\left(X'_{Q,i}\right) \tag{24}$$

$$X''_{j,P,i} = X'_{j,P,i} + r * \left(X'_{j,Q,i} + X'_{j,P,i}\right), f\left(X'_{P,i}\right) > f\left(X'_{Q,i}\right) \tag{25}$$

Accept $X''_{j,P,i}$ if it gives a better function value.

3 Comparison of Optimization Techniques

The swarm intelligence algorithms have been widely used to solve complex optimization problems. These methods are more powerful than conventional methods based on formal logic or mathematical programming. In terms of comparison of intelligence algorithms of the swarm we have:

- The four algorithms studied in this paper are population-based techniques that implement a group of solutions to achieve the optimal solution.
- The PSO and TLBO algorithms use the best solution of the iteration to change the existing solution in the population, which increases the rate of convergence.
- TLBO and PSO do not divide the population unlike SSO and SMO.
- TLBO, SSO and SMO implement greed to accept the right solution.
- Each method requires parameters that affect the performance of the algorithm.
  - PSO requires coefficients of confidence and inertia.
  - SSO requires the threshold setting.
  - SMO requires the perturbation rate (Pr), the local leader limit (LLL), the global leader limit (LGL) and the maximum number of groups (MG).
  - In contrary, TLBO does not require any parameters, which simplifies the implementation of TLBO.

4   Application Example and Results

In this section, the application of a proposed algorithm is presented for the optimization of rectangular waveguide H-plane three-cavity filter [17] Fig. 1. When the main guide is WR28, four parameters are to be optimized W_1 and W_2 (the opening of the iris), l_1 and l_2 (the distance between the iris). The thicknesses of the iris are fixed to $t_1 = 1.45$ mm, $t_2 = 1.1$ mm. Table 1 contains the geometric variables of the structure and the corresponding ranges. As for the frequency range, it was chosen to be $f \in (34, 35.5 \text{ GHz})$.

The objective is to minimize the fitness function in the frequency range, where the fitness functions is the mean value of the coefficient of reflection S11.

$$fitness = \frac{\sum_{f_1}^{f_2} S_{11}(f)}{PT} \tag{26}$$

With PT is the number of points in the interval $[f_1, f_2]$.

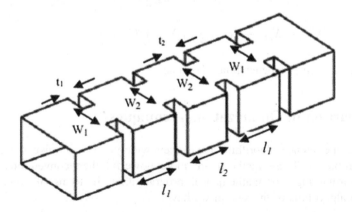

Fig. 1. Rectangular waveguide H-plane three-cavity filter

Table 1. Geometric variables of the structure and the corresponding ranges

Variables	Min	Max
W_1(mm)	1.80	5.40
W_2(mm)	1.20	3.60
l_1(mm)	2.07	6.22
l_2(mm)	2.35	7.05

The convergence of the fitness functions of each algorithm (Best, Worst, Mean) is presented in Table 2 with the population size of 50 and the maximum number of iterations take the following values (30, 50, 300). Table 3 shows the convergence of the fitness functions for the number of iterations is 50 and the population size (30, 70 and 100). Every algorithm is run 10 independent times. The other specific parameters of algorithms are given below:

Table 2. The fitness functions for population size = 50

Maxit	Algorithm	Best	Worst	Mean
30	PSO	0.0193	0.1588	0.0907
	SSO	0.0191	0.1405	0.0680
	TLBO	0.0082	0.0100	0.0089
	SMO	0.0082	0.0092	0.0088
50	PSO	0.0126	0.0945	0.0791
	SSO	0.0110	0.0353	0.0190
	TLBO	0.0081	0.0089	0.0082
	SMO	0.0081	0.0081	0.0081
300	PSO	0.0084	0.0210	0.0108
	SSO	0.0081	0.0121	0.0091
	TLBO	0.0081	0.0081	0.0081
	SMO	0.0081	0.0081	0.0081

-   PSO Settings: c_1 and c_2 are constant coefficients $c_1 = c_2 = 2$, the inertia weight decreased linearly from 0.9 to 0.2.
-   SSO Settings: the threshold parameter PF = 0.7.
-   TLBO Settings: for TLBO there is no such constant to set.
-   SMO Settings: the parameter of SMO depends on the population size where: (N = 30: MG = 3, GLL = 30 and LLL = 120); (N = 70: MG = 7, GLL = 70 and LLL = 280); (N = 100: MG = 10, GLL = 100 and LLL = 400).

Figure 2 shows the convergence of the fitness function of the best individual of each algorithm. The results of the optimization are presented in the Table 4.

Table 3. The fitness functions for Maxit = 50

Population	Algorithm	Best	Worst	Mean
30	PSO	0.0197	0.0849	0.0411
	SSO	0.0120	0.0536	0.0409
	TLBO	0.0081	0.0140	0.0089
	SMO	0.0081	0.0081	0.0081
70	PSO	0.0126	0.0634	0.0326
	SSO	0.0119	0.0501	0.0272
	TLBO	0.0081	0.0082	0.0081
	SMO	0.0081	0.0081	0.0081
100	PSO	0.0120	0.0352	0.0255
	SSO	0.0103	0.0311	0.0162
	TLBO	0.0081	0.0081	0.0081
	SMO	0.0081	0.0081	0.0081

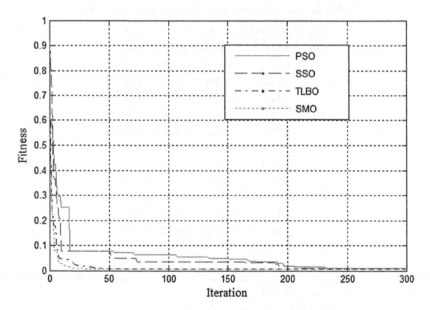

Fig. 2. The convergences of the fitness function of the best individual of each algorithm.

Table 4. The geometrical parameters optimized

Variables	[17]	PSO	SSO	TLBO	SMO
W_1(mm)	3.60	4.705	4.694	4.709	4.078
W_2(mm)	2.40	3.598	3.585	3.600	3.599
l_1(mm)	4.15	2.936	2.917	2.915	2.910
l_2(mm)	4.70	3.633	3.613	3.613	3.608

It is observed from Tables 2, 3 and Fig. 2 that, the SMO and TLBO algorithms performs better in terms of convergence than the PSO and SSO algorithms. In which SMO and TLBO algorithms converge to the minimum optimal for the first's iterations maxit = 30, on the other hand PSO and SSO algorithms converge to the minimum optimal in 300 iterations.

5 Conclusion

In this works, the study and application of the three recent swarm intelligence algorithms in literature (spider monkey optimization (SMO), social spider optimization (SSO) and teaching learning based optimization (TLBO)) are presented. These three algorithms are used for the optimization of microwave filter (H-plane three-cavity filter). The results of convergence and optimization are compared with the results of the most popular swarm intelligences algorithm, particle swarm optimization (PSO). The results showed validation of the proposed algorithms.

References

1. Yang, X.S.: Nature-Inspired Metaheuristic Algorithms. Luniver Press, Beckington (2008)
2. Lim, C.P., Jain, L.C., Dehuri, S.: Innovations in Swarm Intelligence. Springer, Berlin (2009). https://doi.org/10.1007/978-3-642-04225-6
3. Dorigo, M., Stützle, T.: Ant Colony Optimization. MIT Press, Cambridge (2004)
4. Kennedy, J., Eberhart, R.C.: Particle swarm optimization. In: Proceedings of IEEE International Conference on Neural Networks, Perth, Australia, pp. 1942–1948 (1995)
5. Karaboga, D.: An idea based on honey bee swarm for numerical optimization. Technical report-TR06, Erciyes University, Engineering Faculty, Computer Engineering Department (2005)
6. Passino, K.M.: Biomimicry of bacteria foraging for distributed optimization and control. IEEE Control Syst. Mag. 22(3), 52–67 (2002)
7. Chu, S.-C., Tsai, P.-W., Pan, J.-S.: Cat swarm optimization. In: Yang, Q., Webb, G. (eds.) PRICAI 2006. LNCS (LNAI), vol. 4099, pp. 854–858. Springer, Heidelberg (2006). https://doi.org/10.1007/978-3-540-36668-3_94
8. Bakhouya, M., Gaber, J.: An immune inspired-based optimization algorithm: application to the Traveling Salesman problem. Adv. Model. Optim. 9(1), 105–116 (2007)
9. Krishnanand, K.N., Ghose, D.: Glowworm swarm optimization for searching higher dimensional spaces. In: Lim, C.P., Jain, L.C., Dehuri, S. (eds.) Innovations in Swarm Intelligence. SCI, vol. 248, pp. 61–75. Springer, Heidelberg (2009). https://doi.org/10.1007/978-3-642-04225-6_4
10. Singh, N., Singh, S.B.: A new modified approach of mean particle swarm optimization algorithm. In: IEEE 5th International Conference on Computational Intelligence and Communication Networks, Mathura, India, pp. 296–300 (2013)
11. Bansal, J.C., Sharma, H., Jadon, S.S., Clerc, M.: Spider monkey optimization algorithm for numerical optimization. Memet. Comput. 6(1), 31–47 (2014)
12. Cuevas, E., Cienfuegos, M., Zaldívar, D., Cisneros, M.P.: A swarm optimization algorithm inspired in the behavior of the social-spider. Expert Syst. Appl. 40(16), 6374–6384 (2013)

13. Cuevas, E., Cienfuegos, M.: A new algorithm inspired in the behavior of the social-spider for constrained optimization. Expert Syst. Appl. **41**(2), 412–425 (2014)
14. Rao, R.V., Savsani, V.J., Vakharia, D.P.: Teaching–learning-based optimization: a novel method for constrained mechanical design optimization problems. Comput. Aided Des. **43**(3), 303–315 (2011)
15. Rao, R.V., Savsani, V.J., Vakharia, D.P.: Teaching–learning-based optimization: an optimization method for continuous non-linear large scale problems. Inf. Sci. **183**(1), 1–15 (2012)
16. Rao, R.V., Rai, D.P.: Optimization of fused deposition modeling process using teaching-learning-based optimization algorithm. Eng. Sci. Technol. Int. J. **19**, 587–603 (2016)
17. Yang, R., Omar, A.S.: Investigation of multiple rectangular aperture irises in rectangular waveguide using TE modes. IEEE Trans. Microw. Theory Tech. **41**(8), 1369–1374 (1993)

Innovation Diffusion in Social Networks:
A Survey

Somia Chikouche[1(✉)], Abderraouf Bouziane[1],
Salah Eddine Bouhouita-Guermech[2], Messaoud Mostefai[1],
and Mourad Gouffi[2]

[1] Faculty of Mathematics and Informatics, University of Bordj Bou Arreridj,
34030 El-Anasser, Algeria
chikouchesoumia@yahoo.fr,
bouziane.abderraouf@gmail.com, mostefaimess@gmail.com
[2] University of Mohamed Boudiaf, BP 166, 28000 M'sila, Algeria
bouhouita@usa.com, gofimorad@gmail.com

Abstract. The innovations diffusion is a process of communication by which a new idea spreads among a population. A successful propagating leads researchers to seek the elements, which approve and consequently contribute to its spread, or otherwise, find the elements that prevent it. In fact, many efforts were made to models this social phenomenon; however, each one had its strengths and weaknesses. Therefore, this paper aims (1) to discuss and to analyze existing models showing their utility and limitation, (2) highlighting the detail of their applications, and (3) suggesting a taxonomy, which resumes the state of art. Then, to address the current social networks models, it seems necessary to begin by presenting the diffusion of innovation theory, then to detail models, which do not incorporate the social structure, more precisely the mathematical models.

Keywords: Diffusion of innovation · Mathematical models
Social networks models

1 Introduction

The advent of the Internet, and its universally emerging use, by its Democratic policy, allows any person to be connected to the network, and therefore anyone can benefit, distributes information, and proposes its own services or product as a marketing process. As a matter of fact, this free exchange of information involves unrestricted spread. It has also encouraged academics to reflect, what are the reasons that encourage the propagation or the diffusion, in a specific case, if it carries a certain innovation.

In any social system, the diffusion notion takes a chief place, as its influences the individuals' behavior [1]. Thus, this process is the heart of the society evolution. Studying the innovations diffusion is analyzing and identifying mechanisms that allow the social system to change [2]. It is also to recognize the way that encourages a population to adopt an innovation whose advantage is not immediately observable [3].

A. Amine et al. (Eds.): CIIA 2018, IFIP AICT 522, pp. 173–184, 2018.
https://doi.org/10.1007/978-3-319-89743-1_16

In other words, the research in diffusion is mainly to provide more possibility to comprehend how and why this phenomenon do well [3–8]. The study of the innovation diffusion affords an insight to the adoption prediction. To expound and analyze this social phenomenon, the diffusion of innovation theory was introduced as a general framework. This theory is the ultimate prominent behavioral application of network investigation [7, 9].

Coming back in history, the psychologist used the mathematical models as a behavior quantitative theory [10]. After a duration, an answer is wanted, how much could mathematical models support social science? [11]. Epstein [12] maintains that these models may possibly explain the social phenomenon but have not the possibility to predict its future. Therefore, Silverman and Bryden [13] recommend simulation because it offers an optimistic solution to social science. The current development of technologies affords the simulation with more opening and capability to study complex systems comparing with mathematical models [14]. Axelrod [15] explains that the fact that in social science, the simulation helps to take into account the interaction and its influence. In other word, the simulation takes into account the social network structure. Models and simulation approaches provide a better understanding of this social process and give convincing answers to these questions [16]. Geroski submits *"We use models to help illuminate phenomena that we find difficult to understand or to solve problems which are too difficult to think through. These benefits come because models simplify reality"* [17, p. 621].

The richness of literature prompted the study to propose a taxonomy that summarizes the state of the art on this subject. The objective is to provide analysis around the social networks models. To correctly place this problem, some basic concepts will be first defined, then in Sect. 2; the mathematical models will be explained and analyzed. Moreover, Sect. 3 will dealt the social networks models by presenting the necessary details. Finally, Sect. 4 will conclude the work's findings.

2 The Diffusion of Innovation

Rogers [6] defines the innovation diffusion as *"the process by which an (1) innovation is communicated through certain (2) channels over (3) time among the members of (4) a social system"*.

(1) **The innovation** is anything perceived as new by the potential adopters.
(2) **The communication channels** are the medium whereby the message is transferred and exchanged.
(3) **The time** is key factors, because whatever the innovation contribution, it diffuses slowly in the population. Thus, it requires time.
(4) **The social system** is a set of interdependent units that are got together to achieve a common objective.

The following definition and assumption are important to the understanding of the subsequent models.

2.1 Adoption Rate

It presents the level of innovation adoption by the population as a function of time. Initially, the diffusion rate goes up pretty much slowly. Then, rises over time, which leads to a period of rapid adoption from the first period, until saturation. It draws an "S" curve, which should be guaranteed by the diffusion models [18].

2.2 Cycle of Innovation Diffusion

Rogers [4] in the book of 1962 defined different types of attitudes towards the innovation. Its idea is to associate the different types corresponding adopters to the different phases of adoption of a novelty where some people are more open to newness than others do. These stages and rates draw a curve in "bell".

Innovators (2.5%). Courageous people, aim change. They are individuals with a strong ambition towards new ideas or technologies "the taste of adventure", they prefer to be the first to possess them. They present innovation to their social system, they have very significant communication mechanisms.

Early Adopters (13.5%). They follow very closely the innovators. They try new ideas, but in a careful way. They are educated and more integrated into the social system than innovators, their opinions matter and can facilitate the diffusion process. They are the opinion leaders and with whom other members get information and seek advice on innovation.

Early majority (34%). They are thoughtful, pragmatic and do not seek changes but developments that improve life. They have many social contacts. They accept change more quickly than others do. They adopt innovation as a group with the accumulation of positive opinions. It represent alone a third of individuals. The one by which the innovation can be considered to be in diffusion on a large scale. Because it serves as a transmission link between members who relatively adopted the innovation early, and those whose decision is tardy. Therefore, it is the privileged target of companies because of its quantity and its ease to convince.

Late Majority (34%). They are conservative skeptics; they accept it only when the majority has already adopted it. They are the ones who undergo social pressure before deciding. Indeed, the late majority adopts innovation more by social pressures than by actual predisposition. They also represent 1/3 of the members of the system. If it resembles the early majority, it differs from them in the motivations for its adoption of innovation.

Laggards (16%). They are traditional people, very skeptical, they like to be glued to "old ways", and will admit them only if the new idea has become the mainstream or even the tradition. Sometimes they are critical about new ideas and they expect a great reliability of innovation, they adopt only by constraint or absolute necessity. They are volunteers who refuse to innovate.

3 Mathematical Models

Several mathematical models have been suggested to model the spread of information among the population. In this section, three well-known models are detailed:

3.1 Logistic Model

It is based on the cumulative distribution patterns. It follows a growth model, which has proven to be a consistent pattern by empirical research on the diffusion of innovation [18]. It is approximated by a logistic function and described by the following equation

$$F(t) = P/r * e^{-st} \qquad (1)$$

Where t is the time P and s are positive real numbers, r any real. This function was proved by Verhulst in 1845 [19]. He proposed as a model of evolution of the non-exponential population with s steepness and capacity P.

3.2 Gompertz Model

It is a mathematical model, where the development begins and finish slowly. This sigmoid function is come close gradually to α value. This function refers to Gompertz [20]. It is an exceptional case of the logistic function. Gompertz function is described by

$$F(t) = \alpha e^{-\beta e^{\lambda t}} \qquad (2)$$

Where α, β and λ are positive real numbers.

3.3 Bass Model

A remarkable expansion took place in the marketing literature on the diffusion 1970s and the largest enhancement to this scientific explosion is a model for predicting of the spread of new consumer products proposed by Bass [21] in 1969. Bass prediction model has become so important in the field of marketing because it offers some plausible answers to the uncertainty linked to the introduction of a new product in the market. Some of the largest US companies have used the Bass model [3].

To better describe the distribution, particularly to take into account the effect of social pressure Bass defines two main forces driving adoption: innovation and imitation. These explanatory factors reflect the fact that part of individuals adopts by social pressure of individuals who have already adopted, while some adopt by interpersonal persuasion. Diffusion speed is proportionate to (a) the population that has already adopted the service, denoted by f(t), and (b) the remaining market potential represented by M F(t).

From the 19th century, the sociologist Tarde [22] has well highlighted the importance of social influence and the way it is at the base of the formation of a value of the invention, based on its diffusion [2]. He used the word "imitation" to describe the adoption; in his book, he called "the laws of imitation" which published it in 1903.

The Bass model is a predictive model of the future adoption of a product. To apply, you must first define the market size M, then the coefficients p innovation and imitation q, and it can estimate the number of adoptions f(t) in any specified time. The equation is

$$f(t) = M\,p + (q-p)\,F(t) - q/M\,F(t)^2 \tag{3}$$

3.4 Discussion

For the scientist who attempts to predict the future, there is always a need for simple models that describing events. Logistic and Gompertz models could be used to compare the growth rates for different innovations [23]. Each such model should be based upon easily understood assumptions; that support the forecaster in his efforts to make the future what he wants it to be. The models aforementioned are simply based on clearly coefficient. The best-known diffusion models used for technology diffusion purposes are the Bass model, the logistic family models [24], as well as the Gompertz model [20].

For several decades, great attention turned to the Bass diffusion model [3]. The estimation of accurate parameters was the most important point for getting a proximate forecasting [25]. To deal with this problem, the evolutionary approach is a practical solution. For example, Venkatesan and Kumar [26] joined the diffusion model of Bass with the evolutionary technique precisely the genetic algorithms (GA). They aim to predict future sales in the telecommunications sector. Three main genetic operations used: reproduction, crossover, and mutation. By using cellular phone adoption data, they found out that the integration of genetic algorithm gives more robustness and produces more consistency comparable to non-linear least squares(NLS), ordinary least squares(OLS), and time series model. A further illustration is a work of Wang et al. [25], where a hybrid approach that combines the genetic algorithms with the particle swarm optimization (PSO) were proposed. By comparison with NLS, GA and PSO, the consequence of this conjunction provides better predictions, and more accurate parameters for forecasting the notebook shipment.

A significant number of existing models based on the use of parameters that determine the process of adoption of innovation and simple mathematical functions concentrated on observation and description of patterns of diffusion. These models allow more explicit diffusion pattern, but require the estimation of diffusion coefficients, obtained from historical data or time series. This raises the problems of the application in contexts where there is no data or the data are insufficient [27].

The problem with these mathematical models that is impossible to reproduce the failure of the diffusion of innovation, except by artificially placing the potential adopters at 0, while potential adopters may exist but not be reached during the propagation. The aggregative approach itself does not take into account network effects, geographical boundaries, or the heterogeneity of people. Precisely, the difference between Rogers's categories of adopters is overlooked.

Modeling at this macro level, however, is imprecise because it assumes the high-connected social network, in which each person interacts with all world [23, 28]. Pelc [29] aimed at encouraging researchers to integrate the social network where interactivity level is so important. Epstein [12] addressed this issue, stating different reasons for building social models: mathematical models could explain (not predict) the emergence of collective phenomena, or capture qualitative behavior of phenomena. Such criticism strongly limits the description of the mathematical models. To take into account the heterogeneity of the population and the network effect, the models are moving towards approaches called individual-oriented models where the social structure and the communication channels are taken into account. The following section probes into the social networks models.

4   Social Networks Models

4.1   Threshold Model

The way that the social network influences the diffusion process, is the principal focus of threshold model [30]. This model has been assumed to understand the process of innovation diffusion [23]. The threshold is the fraction of adopters in a system needed for an individual to be an adopter. Valente [30] applied this concept as an extension of Granovetter's [28] work, where, the individual's decision depends on the behavior of others in the group or the system. For the reason that the individuals are more effected by facts received from personal network. Valente [30] proposes using the direct communication network rather than a system to which the individual belongs. The exposure E_i is the proportion of the adopters in the personal networks. It is measured by

$$E_i = \sum w_{ij} \times y_i / \sum w_i \qquad (4)$$

The principle of threshold models is as follows: an individual changes its state if a sufficiently large proportion of his neighbors are in this state. This proportion constitutes the threshold of the individual. The diversity of these thresholds, as well as the position of the individual on the social network, create a heterogeneity in the population. The linear threshold model is derived from threshold model.

Linear Threshold Model. Suggested by Kempe et al. [31]. LTM works as follow: the individual is influenced by each neighbor according to a weight. A uniform random threshold from 0 to 1 are assigned to each individual I. This threshold is imperative to I to became an adopter. For each step of simulation, an activated sequence of individuals activate similarly their neighbors' and so on. The propagation ends when there are no further inactive person.

4.2   Epidemic Approach

It is easy to observe the relation between epidemic disease and the diffusion of information. The Influenza for example, is a contagious disease that spread from person to person. Epidemics can pass through a population, or they can persist over long time periods at low levels. As epidemics can spread between people, ideas and information can also spread from person to person, across similar kinds of networks that connect people. The Diffusion of innovation is usually treated as social contagion [32]. Similar to the spread of disease, as a reaction chain phenomenon. First there are some adopters, while members of their networks adopt, then they move on to their own networks and so on. First slowly, then faster and faster and then slows down again because potential adopters are reduced.

This approach assumes that exposure to information enough to become informed, and potentially transmit the information to someone else. However, we know that all individuals do not have the same propensity to adopt and transmit the information immediately after its receipt. Mathematically, this phenomenon is best described by an S curve, it can be easily constructed from an array of single frequency of adoption time. The epidemic approach assumes that meetings between individuals can appear randomly in the population. To simulate the diffusion of innovation, where the main channel of communication is the word-of-mouth, this approach is the most appropriate for this case [17]. Indeed, it is generally used to describe the transmission of innovations [33].

The SIR Epidemic Model. The state of individuals changes through three possible stages:

Susceptible. The individual is suspected to catch the disease from its neighbors.

Infectious. If the individual catch the disease, he became infectious and has some probability to infect his neighbors.

Removed. After some period of infection, the individual became removed from consideration.

Ba et al. [34] have designed a model that is based on the epidemic approach. To take into account the heterogeneity of the population, which is a real fact, they inspired by the Granovetter threshold models. They simulated the innovation diffusion in Senegalese rural environment by using Netlogo development software. The simulation was done using multi-agent systems. The network on which the diffusion took place was generated using an interaction network generator. This simulation has led to some interesting conclusions: on the first part, the geographical disparity that separates individuals does not in itself constitute a barrier to the wide dissemination of information; on the other part, an important factor is the social structure of the environment. The social network and links on through which the influence of external sources happens is the central interesting for Myers et al. [35]. They investigate how information spread in the network from node to node. They apply the contagion and exposure models. The results confirms that the effects of external sources on the information diffusion. Namely, by quantifying this degree of influence over time, they argue that the information is likely to "jump" across the network.

4.3   Evolutionary Model

Despite the promising outcomes and recent interest in modeling social systems, it was pointed out that in many cases; models lack realism and do not take the realities of the field into account [11]. In recent years, numerous publications have appeared dealing with the incorporation of the evolutionary perspective to the social science, this integration known as evolutionary psychology. This combination gets a chief place within psychological science [36]. The evolutionary psychology is a field that apply the evolutionary concepts in psychology arena. Buss et al. [36] highlight the importance of incorporating the evolutionary perspective to social science. The evolutionary principal base on the notion that the individuals with more effective characteristics have more probability to replicate and live [36]. The evolutionary computation need only a little specific problem knowledge. Thus, they can be applied to deal with a varied range of problems [37].

Nevertheless, a little available literature on innovation diffusion that applies techniques based on evolutionary computation [27]. An example of the evolutionary algorithm is the paper of Sampaio et al. [27]. They recommend the evolutionary algorithm to simulate the diffusion of innovation process. The evolutionary computation is suitable to model the progress of adoption decision process. Their central support is Rogers' innovation diffusion theory and the threshold approach. Another important factor that was taken into account, is the knowledge function of Verhulst [19]. This function is used to simulate the evolution accumulation acquaintance by time. The result was compared to Logistic, Gompertz, and Bass models. The effect of first adopters and the social structure are studied.

In a previous work [38], we proposed a novel evolutionary model based on the human interactions as an evolutionary learning process, because the decision does not occur directly from the first exposition to the innovation, but it follows an evolutionary process affected by the other neighbors already been adopters. We used three social networks type: regular ring network, lattice network on a torus and random network. The simulation proved the capability of the model to produce the S-shaped diffusion pattern, it is built on two associated factors: the individual decision process and the social learning influence. Consequently, the paper may offer a significant outline to better understand the subject of human behavior.

Yavaş and Yücel [39] addressed the way the homophily in the social network can affect the diffusion process. An agent-based simulation implemented model demonstrates that the homophily is self-reinforcing especially in its early increases. It is, in other words, more supporting at the beginning of the diffusion rather than being so later. In addition, by applying the evolutionary homophilous network, the extent of macro high homophily degree gets more influence than the one of individual's local neighborhood preference. Furthermore, Cowana and Jonard [40] focused on the relationship between the network topology and the diffusion process. They examined the effects of incremental innovations diffusion by means of three different network architectures. Additionally, they implemented heterogeneous agents connected exchanging information over time. This interaction involves the knowledge, which is represented as an evolutionary vector.

4.4   Discussion

The elements of the innovation diffusion are differently considered in the process of modeling. Rogers [3] mentioned four important elements that are: (1) the innovation,

(2) the time, (3) the social system and (4) the communication channels. The threshold models showed particular interest on the social structure, making distributed probability to create the social heterogeneity. The question is what about a personal network with a minimum of communication? However, the impact of communication channels was the focus of epidemiology approach and the evolutionary models. Another problem, is our choices depend only on the social pressure? While all the mentioned models ignored the various innovations characteristics, the need of a model that take into account the innovation features' is vital.

5   Summary

The innovation diffusion models converge to similar findings that are the "S" diffusion curve. Figure 1 illustrates a taxonomy that resumes the state of art in the subject of innovation diffusion.

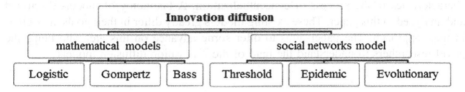

Fig. 1. Represent a taxonomy of the innovation diffusion models.

Table 1 gives a summary of the innovation diffusion elements more or less considered by the abovementioned models.

Table 1. The elements of innovation diffusion theory considered by the mentioned models.

Models	Innovation	Social network	Communication	Time
Logistic	✗	✗	✗	✓
Gompertz	✗	✗	✗	✓
Bass	✗	✗	✗	✓
Threshold	✗	✓	✗	✓
Epidemic	✗	✓	✓	✓
Evolutionary	✗	✓	✓	✓

6   Conclusion

When new idea, activity or object are invented, diffused, adopted or rejected, social changes occur. Research in this field has produced various theoretical models. A survey of mathematical models and social networks models for innovation diffusion was presented. Rogers's theory suggests that the adopters of any innovation can be classified into five categories: innovators, early adopters, the early majority, the late majority, and laggards. This theory describes the mechanisms that drive the diffusion of innovation [41]. The diffusion of innovation leads to the S-shaped curves representations, this object is guaranteed by the most of the models, the logistic and Gompertz models are good examples, which is much more used to demonstrate the cumulative effect. In 1969, inspired by Rogers and other theorists of the domain, Frank Bass developed a quantitative model that was as a reference to quantify the rate of adoption of an innovation in a population [3]. The mathematical models concentrate on the rate of adopters. A time ago, the innovation diffusion was only measured by the rate of adoption.

However, taking into account the interactions between individuals is critical [29], because, they are the main driver of the evolution of individuals [3], which the mathematical models do not take into account. The individual-oriented approach was a solution to integrate the social network. Later, several models have been introduced, in order to better comprehend the process and to incorporate the essential elements. These social network models confirm hypotheses about the causes of success or failure. The designed models still need some improvements like the integration of the innovation characteristics in the process of conceptualization. A taxonomy of models discussed and analyzed in this paper. These existing models greatly differ in their goals, as well as in their approach. The paper aims to offer some advance on this area and helps the novel researchers to understand the field of the innovation diffusion models.

References

1. Steyer, A., Zimmermann, J.-B.: Influence social et diffusion de l'innovation (Social influence and diffusion of innovation). Math. Sci. Hum. **42**(168), 43–57 (2004)
2. Rogers, E.M.: Diffusion of Innovation. The Free Press, A Division of Macmillan Publishing Co., Inc., New York (1983)
3. Rogers, E.M.: Diffusion of Innovations, 5th edn. Free Press, New York (2003)
4. Rogers, E.M.: Diffusion of Innovation, 1st edn. Free Press of Glencoe, New York (1962)
5. Rogers, E.M.: A prospective and retrospective look at the diffusion model. J. Health Commun. **9**(Suppl 1), 13–19 (2004)
6. Rogers, E.M.: Diffusion of Innovations, 4th edn. The Free Press, A Division of Simon & Schuster Inc., New York (1995)
7. Valente, T.W.: Social Networks and Health, p. 172. Oxford University Press, Oxford (2010)
8. Valente, T.W.: Evaluating Health Promotion Programs. Oxford University Press, New York (2002)
9. Valente, T.W., Rogers, E.M.: The origins and development of the diffusion of innovations paradigm as an example of scientific growth. Sci. Commun. **16**(3), 242–273 (1995)
10. Suppes, P.: Models of data. In: Proceedings of the 1960 International Congress on Logic, Methodology, and Philosophy of Science, Palo Alto, CA (1962)
11. Apolloni, A., Channakeshava, K., Durbeck, L., Khan, M., Kuhlman, C., Lewis, B., Swarup, S.: A study of information diffusion over a realistic social network model. In: Proceedings of the 2009 International Conference on Computational Science and Engineering, Washington, DC, USA (2009)
12. Epstein, J.M.: Why model? J. Artif. Soc. Soc. Simul. **11**(4), 12 (2008)

13. Silverman, E., Bryden, J.: From artificial societies to new social science theory. In: Almeida e Costa, F., Rocha, L.M., Costa, E., Harvey, I., Coutinho, A. (eds.) ECAL 2007. LNCS (LNAI), vol. 4648, pp. 565–574. Springer, Heidelberg (2007). https://doi.org/10.1007/978-3-540-74913-4_57

14. De Jong, K.A.: Evolutionary Computation a Unified Approach. The MIT Press, Cambridge (2006)

15. Axelrod, R.: Advancing the art of simulation in the social sciences. In: Conte, R., Hegselmann, R., Terna, P. (eds.) Simulating Social Phenomena. LNE, vol. 456, pp. 21–40. Springer, Heidelberg (1997). https://doi.org/10.1007/978-3-662-03366-1_2

16. Thiriot, S.: Vers une modélisation plus réaliste de la diffusion de l'innovations à l'aide de la simulation multi-agents (Towards more descriptive models of innovation diffusion using multi-agents simulation). University of Pierre and Marie Curie, Paris (2009)

17. Geroski, P.: Models of technology diffusion. Res. Policy 29(4–5), 603–625 (2000)

18. Mahajan, V., Peterson, R.A.: Models for Innovation Diffusion. Quantitative Applications in the Social Sciences. SAGE Publications Inc., Newbury Park (1985)

19. Verhulst, P.F.: Recherches mathématiques sur la loi d'accroissement de la population (mathematical researches into the law of population growth increase), vol. 18, pp. 1–42 (1845)

20. Gompertz, B.: On the nature of the function expressive of the law of human mortality, and on a new mode of determining the value of life contingencies. Philos. Trans. Roy. Soc. Lond. 115, 513–583 (1825)

21. Bass, F.: A new product growth model for consumer durables. Manag. Sci. 15(5), 215–227 (1969)

22. Tarde, G.: The Laws of Imitation, p. 140. H. Holt and Company, New York (1903)

23. Valente, T.W.: Network models and methods for studying the diffusion of innovations. In: Models and Methods in Social Network Analysis, pp. 98–116. Cambridge University Press, New York (2005)

24. Bewley, R., Fiebig, D.: A flexible logistic growth model with applications in telecommunications. Int. J. Forecast. 4(2), 177–192 (1988)

25. Wang, F.-K., Chang, K.-K., Hsiao, Y.-Y.: Implementing a diffusion model optimized by a hybrid evolutionary algorithm to forecast notebook shipments. Appl. Soft Comput. 13(2), 1147–1151 (2013)

26. Venkatesan, R., Kumar, V.: A genetic algorithms approach to growth phase forecasting of wireless subscribers. Int. J. Forecast. 18(4), 625–646 (2002)

27. Sampaio, L., Varajão, J., Pires, E.J.S., de Moura Oliveira, P.B.: Diffusion of innovation simulation using an evolutionary algorithm. In: Gavrilova, M.L., Tan, C.J.K., Abraham, A. (eds.) Transactions on Computational Science XXI. LNCS, vol. 8160, pp. 46–63. Springer, Heidelberg (2013). https://doi.org/10.1007/978-3-642-45318-2_2

28. Granovetter, M.: Threshold models of collective behavior. Am. J. Sociol. 83(6), 1420–1443 (1978)

29. Pelc, K.I.: Diffusion of innovation in social networking. In: Zacher, L.W. (ed.) Technology, Society and Sustainability, pp. 3–13. Springer, Cham (2017). https://doi.org/10.1007/978-3-319-47164-8_1

30. Valente, T.W.: Social network thresholds in the diffusion of innovations. Soc. Netw. 18, 69–89 (1996)

31. Kempe, D., Kleinberg, J., Tardos, E.: Maximizing the spread of influence through a social network. In: Proceedings of the Ninth ACM SIGKDD International Conference on Knowledge Discovery and Data Mining, KDD 2003. ACM, New York (2003)

32. Ronald, S.B.: Social contagion and innovation: cohesion versus structural equivalence. Am. J. Sociol. 92(6), 1287–1335 (1987)

33. Easley, D., Kleinberg, J.: Networks, Crowds, and Markets: Reasoning about a Highly Connected World. Cambridge University Press, Cambridge (2010)
34. Ba, K., Boutet, A., Corenthin, A., Lishou, C.: Étude de la diffusion d'innovations en milieu rural à l'aide de simulations multi-agents, Study of the diffusion of innovations in rural areas using multi-agent simulations. Studia Informatica Universalis **10**(1), 129–154 (2012)
35. Myers, S.A., Zhu, C., Leskovec, J.: Information diffusion and external influence in networks. In: Proceedings of the 18th ACM SIGKDD International Conference on Knowledge Discovery and Data Mining, KDD 2012, New York, NY, USA (2012)
36. Buss, D.M., Haselton, M.G., Shackelford, T.K., Bleske, A.L., Wakefield, J.C.: Adaptations, exaptations, and spandrels. Am. Psychol. **53**(5), 533–548 (1998)
37. Streichert, F.: Introduction to evolutionary algorithms. In: Frankfurt MathFinance Workshop, Frankfurt, Germany, 2–4 April (2002)
38. Chikouche, S., Bouhouita-Guermech, S.E., Bouziane, A., Mostefai, M., Gouffi, M.: Evolutionary knowledge based on human interaction model for the innovation diffusion in social networks. In: 3rd International Conference on Networking and Advanced Systems, Annaba, Algeria (2017)
39. Yavaş, M., Yücel, G.: Impact of homophily on diffusion dynamics over social networks. Soc. Sci. Comput. Rev. **32**(3), 354–372 (2014)
40. Cowana, R., Jonard, N.: Network structure and the diffusion of knowledge. J. Econ. Dyn. Control **28**(8), 1557–1575 (2004)
41. Rogers, E.M.: Rise of the classical diffusion model. Curr. Contents **13**(15), 16 (1991)

An Efficient Cooperative Method to Solve Multiple Sequence Alignment Problem

Lamiche Chaabane[⊠] iD

Computer Science Department, Mohamed Boudiaf University, M'sila, Algeria
Lamiche07@gmail.com

Abstract. In this research work, we propose a cooperative approach called simulated particle swarm optimization (SPSO) which is based on metaheuristics to find an approximate solution for the multiple sequence alignment (MSA) problem. The developed approach uses the particle swarm optimization (PSO) algorithm to discover the search space globally and the simulated annealing (SA) technique to improve the population leader «gbest» quality in order to overcome local optimum problem. Simulation results on BaliBASE benchmarks have shown the potent of the proposed method to produce good quality alignments comparing to those given by other existing methods.

Keywords: Cooperative approach · Multiple sequence alignment
SPSO · PSO · SA · BaliBASE benchmarks

1 Introduction

Multiple sequence alignment (MSA) is a crucial tool in molecular biology and genome analysis. It has been considered as one of the important tasks in bioinformatics [1]. It helps to construct a phylogenetic tree of related DNA sequences, to predict the function and structure of unknown protein sequences, by aligning with other sequences whose function and structure is already known, and to allow comparison of the structural relationships between sequences by simultaneously aligning multiple sequences and constructing connections between the elements in different sequences [2].

Discovering optimal alignment in multiple biological sequence data is known as a NP-complete problem [3]. It has been identified as a combinatorial optimization problem [4], which is solved by using an exact or approximate algorithms. These algorithms lead to exploit various genetic information to determine evolutionary relationships among living beings [3].

Recently, the trend has shifted to the use of iterative algorithms in order to tackle the MSA problem. These approaches are based on the improvement of the given initial alignment through a series of some iterations until a stopping criterion is reached. They include genetic algorithm (GA) [5], simulated annealing algorithm (SA) [6], particle swarm optimization (PSO) [7], GA-ACO algorithm [8], Ant Colony Algorithm [9] and so on. Generally, these metaheuristics are able to find nearly optimal solutions for large instances in a reasonable processing time.

In this study, we propose a hybrid approach called SPSO algorithm to solve the MSA problem. The developed model is the cooperation between both PSO algorithm

A. Amine et al. (Eds.): CIIA 2018, IFIP AICT 522, pp. 185–195, 2018.
https://doi.org/10.1007/978-3-319-89743-1_17

and simulated annealing technique. The remainder of the paper is organized as follows: Sect. 2 presents a brief review of the researches related to the proposed framework. In Sect. 3, both PSO and SA concepts are described. In Sect. 4, our proposed SPSO algorithm is explained in detail. In Sect. 5, the simulation results are provided. Finally, the study is concluded in Sect. 6.

2   Background

A brief review of some related works in the multiple sequence alignment field using iterative methods is presented in this section. Riaz et al. [10] presented a tabu search algorithm to align multiple sequences. The framework of his work consists to implement the adaptive memory features typical of tabu searches in order to obtain multiple sequences alignment where the quality of an alignment is measured by the COFFEE objective function. In [11], the authors proposed a novel approach to multiple sequence alignment based on Particle Swarm Optimization (PSO) to improve a sequence alignment previously obtained using Clustal X.

In Ref. [12], the authors presented an approach to the MSA problem by applying genetic algorithm with a reserve selection mechanism to avoid premature convergence in GA. A better results are obtained compared with those produced by classical GA. The authors in [13] proposed an algorithm based on binary PSO algorithm to address the multiple sequence alignment problem. Simulation results using SP score measure and nine BaliBASE tests case showed that the proposed BPSO algorithm has superior performance when compared to ClustalW and SAGA algorithms.

An artificial bee colony algorithm for solving MSA problem is introduced in [14]. In Ref. [15], Cutello et al. presented an immune inspired algorithm (IMSA) to tackle the multiple sequence alignment problem using ad-hoc mutation operators. Experimental results on BALIBASE v.1.0 show that IMSA is superior to PRRP, CLUS-TALX, SAGA, DIALIGN, PIMA, MULTIALIGN and PILEUP8. In [16], simulated annealing technique was applied to solve MSA problem using a set of DNA benchmarks of HIV virus genes of human and simian.

In Ref. [17], the authors proposed a hybrid algorithm using a GA and cuckoo search algorithm to improve multiple sequence alignment. The obtained results are compared with ClustalW by using five different datasets. Recently, an efficient method by using multi-objective genetic algorithm (MSAGMOGA) to discover optimal alignments is proposed in [18]. Experiments on the BAliBASE 2.0 database confirmed that MSAGMOGA obtained better results than MUSCLE, SAGA and MSA-GA methods.

3   Preliminaries

3.1   Outline of the Particle Swarm Optimization (PSO)

Particle swarm optimization (PSO) is an adapted algorithm developed the first time by Kennedy and Eberhart [19], inspired by bird flocking and fish schooling. PSO used a population of individuals called particles and two primary operators: velocity update and position update. During each generation, each particle moves toward the particles according to its best position and the global best position. In addition, a new velocity value for each particle is calculated based on its current velocity, the distance from its previous best position and the distance from the global best position. The evolution of the swarm is governed by the following equations:

$$V^{(k+1)} = w.V^{(k)} + c_1.rand_1.\left(pbest^{(k)} - X^{(k)}\right) + c_2.rand_2.\left(gbest^{(k)} - X^{(k)}\right). \quad (1)$$

$$X^{(k+1)} = X^{(k)} + V^{(k+1)}. \quad (2)$$

where:
  X is the position of the particle,
  V is the velocity of the particle,
  w is the inertia weight,
  $pbest$ is the best position of the particle,
  $gbest$ is the global best position of the swarm,
  $rand1$, $rand2$ are random values between 0 and 1,
  $c1$, $c2$ are positive constants which determine the impact of the personal best solution and the global best solution on the search process, respectively, k is the iteration number.

Concerning the stopping condition, generally PSO algorithm terminates when a set number of times or until a minimum error is achieved. All parameters of PSO algorithm are fixed experimentally in order to have a good compromise between the convergence time of the algorithm and the final solution quality.

3.2   Outline of the Simulated Annealing (SA)

Simulated annealing (SA) is a general probabilistic local search algorithm proposed by Kirkpatrick et al. [20] to solve difficult optimization problems. It is inspired by the annealing of solids in physics. SA models the slow cooling process of solids to achieve the minimum energy as an analogy, reaching the minimum function value. As a result, it attains an optimal/near-optimal solution by implementing an iterative cooling process from a high temperature, at which solid particles are in the liquid phase. Simulated annealing utilizes a control parameter, temperature T, for the cooling process. The solid is allowed to attain the thermal equilibrium for every T degree that has its energy E probabilistically distributed, as given in Eq. (3), where k_b is the Boltzmann constant.

$$P(E) = e^{\left(\frac{-E}{k_b t}\right)}. \quad (3)$$

In the combinatorial optimization context, if we aim to find a good solution then we move from a solution to one of its neighbors in the search space according to a probabilistic criterion. If the cost decreases then the solution is retained and the move is accepted. Otherwise, the move is accepted only with a probability depending on the cost increase and the temperature parameter T [20].

4 Proposed Method

PSO performs excellently in the case of global search but it is not efficient in local search. It suffers from its weak local search ability and the local minima limit. On the other hand, SA is good in local search while less good in global search. However, it takes advantage of the acceptance of candidate solutions by the use of metropolis criteria and must escape from local optimum to search in other solution space.

In order to construct an intelligent algorithm which can be effectively avoid weaknesses and fully use the advantages of both PSO and SA algorithms, we propose a hybrid approach which combines the PSO with simulated annealing, so the new hybrid algorithm called Simulated Particle Swarm optimization (SPSO) conducts both global search and local search in every iteration. According to this hybridization manner, the probability to obtain better solutions significantly increases. At each iteration, the proposed hybrid SPSO algorithm consists of applying PSO algorithm in order to guide global search, and use SA to improve the *gbest* which helps PSO to escape from local optimum and increase the convergence speed of SPSO algorithm. The flowchart of the proposed SPSO is presented in Fig. 1.

4.1 PSO Components of MSA Problem

Particle Representation. Each particle represents a potential solution to the MSA problem, effectively it corresponds to a sequence alignment. A particle is then represented as a set of vectors, where each vector specifies the positions of the gaps in each one of the sequences to be aligned [21].

Swarm Initialization. The size of the whole swarm is determined by the user. The initial set of particles is generated by adding gaps into each sequence at random position, thus, all the sequences have the same length L in which its value is 1.2 times of the longest sequences [21].

Fitness Evaluation. A parameter to determine which alignment will survive in the next generation is its fitness value. A formal definition of the sum-of-pairs (SPS) of multiple sequence alignment is introduced which is used as a tool to compute fitness. The score assigned to each alignment is the sum of the scores (SP) of the alignment of each pair of sequences. The score of each pair of sequences is the sum of the score assigned to the match of each pair of symbols, which is given by the substitution matrix. The score of a multiple alignment is given as follows:

$$Score(A) = \sum_{i=1}^{k-1} \sum_{j=i+1}^{k} S(A_i, A_j). \tag{4}$$

where the $S(A_i, A_j)$ is the alignment score between two given sequences A_i and A_j.

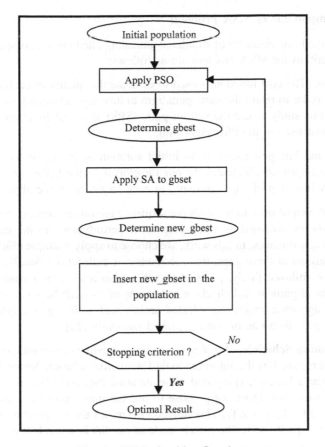

Fig. 1. SPSO algorithm flowchart.

Particle Move. In the PSO algorithm, each particle moves towards the leader at a speed proportional to the distance between the particle and the leader. In this paper, this distance will be measured as the proportion of gaps that do not match in the sequences, according to the formula:

$$Distance = \frac{no\,matching\,gaps}{total\,gaps}. \tag{5}$$

To move particles towards the leader, an operator similar to the crossover operator from genetic algorithms is used [21]. It consists to select a crossover point which divides the alignment into two segments, and then a segment of the particle is replaced with a segment from the leader. This replacement is achieved removing from the particle the gaps that are in the segment, and then adding the gaps from the leader's segment.

4.2   SA Components of MSA Problem

According to the basic elements of simulated annealing cited above, components of our proposed algorithm for MSA problem are as follows:

Cost Function. The cost function is used to evaluate the quality of each alignment in the swarm. In order to retain the same paradigm to measure the quality of the solution, we decide in this study to use the sum-of-pairs (SPS) as a cost function which is the same one of that used in the PSO algorithm.

Initial Solution. The generation of an initial solution is an important step towards getting a final improved alignment. In our developed method, the initial solution is constructed by insertion of gaps randomly in each sequence of the alignment.

Generation of Neighbors. In the multiple sequence alignment context, neighbors of a current solution are obtained by employing some perturbations to the gaps positions into the different sequences. In this work, we choose to apply a simple efficient strategy to change positions of these gaps, this mechanism is called LocalShuffle operator, its main idea is as follows: firstly, picks a random amino acid from a randomly chosen sequence in the alignment and checks whether one of its neighbors is a gap. If this is the case, the algorithm swaps the selected amino acid with a gap neighbor. If both neighbors are gaps then one of them is picked randomly [22].

Choice of Cooling Schedule. An effective cooling schedule is essential to reducing the amount of time required by the algorithm to find an optimal solution. Several temperature decreasing schemes have been proposed in the literature, they include static schedules and adaptive schedules [16]. Here, we propose to use the most common cooling function which is defined by $T_{k+1} = \alpha.T_k$. This function decreases the temperature value by a α factor, where $\alpha \in [0.70, 1.0[$. The pseudo-code of our SA is given below.

Pseudo-code of SA algorithm

```
Fix T_initial, T_final, α
Set T ←T_initial, S_current ←S_initial, score(S_current) ← score(S_initial)
While (T > T_final) do
  S_new ←LocalShuffle(S_current)
  Compute the alculate score of S_new
  d:=score(S_new)- score(S_current)
  If (d >= 0) then
      S_current ← S_new
      If score (S_new) > score(S_better) then
          S_better ← S_new
      end if
  else
  If exp(d/T) > random(0,1) then
      S_current ← S_new
  end if
  end if
  T= α.T
end while
```

After the description of the key components of both PSO and SA algorithms, the pseudo-code of our SPSO procedure is summarized as follows:

SPSO Pseudo-code

---
1. Generate randomly a set of particles
2. Fix ITMAX value
3. Set it ← 1
   4. While (it ≤ ITMAX) do
      4.1. Determine the leader particle *gbest*
      4.2. For each particle in population do
         a) Measure the distance between *gbest* and the particle
         b) Select randomly a crossover point
         c) Apply particle move mechanism
      End for
         // Calculate the new leader particle gbest
      4.3 Update *gbest*
      4.4 *New_gbest* ← SA(*gbest*)   *// Apply SA to improve gbest*
      4.5 Insert *New_gbest* into the current population
      4.6. it ← it + 1
5. End while
// Output the result.
6. Return *gbset* value and its correspondent particle.

---

5   Simulation and Results

The proposed approach is implemented in Java language. All tests have been fulfilled on a PC with an 2.66 GHz Intel Pentium IV processor and 4 GB RAM. We conducted some experiments in order to demonstrate the effectiveness of our SPSO algorithm. For that, a set of benchmark sequences with different identities and lengths are chosen from the BAliBase 1.0 library [23]. Characteristics of the used data and all parameters setting of the algorithm SPSO are summarized in Tables 1 and 2 respectively.

Table 1. Characteristics of benchmark sequences.

Types	Name	N	(Min, Max)	Identity
Short	SH3	5	(80, 49)	<25%
	Cytochrome c	5	(70, 87)	20%–40%
	Serine protease	5	(66, 82)	>35%
Medium	Protein kinase	5	(263, 276)	<25%
	Anthranilate isomerase	4	(247, 259)	20%–40%
	Serine protease	5	(222, 245)	>35%
	Aminotransferase	4	(358, 387)	<25%
Long	Glutamyl-trna synthetase	5	(438, 486)	20%–40%
	Taq DNA polymerase	5	(806, 928)	>35%

Table 2. Parameter settings for experiments.

Parameter	Value
Number of iterations (*ITMAX*)	1000
Swarm size	50
Initial temperature ($T_{initial}$), final temperature (T_{final})	100, 0.0001
Score matrix, gap open	PAM250, 10
c_1, c_2 constants	1.49618

All obtained results by GA, PSO, ABC and our SPSO algorithm using the best, the worst and the average SPS values of 10 experiments are summarized in Tables 3, 4 and 5 respectively.

The results plotted in Tables 3, 4 and 5 show clearly the considerable improvement of scores by the proposed SPSO approach. Indeed, it produces alignments of much better quality than that of the other cited methods in both specified categories.

Table 3. Comparison results for short sequences.

Sequence name	Algorithm	Average	Best	Worst
SH3	GA [14]	0.60.90	0.6829	0.5463
	PSO [14]	0.6444	0.6543	0.6296
	ABC [14]	0.6895	0.7654	0.6543
	Our SPSO	**0.7698**	**0.8712**	**0.8197**
Cytochrome c	GA [14]	0.5141	0.5398	0.4891
	PSO [14]	0.6252	0.6407	0.6116
	ABC [14]	0.6451	0.6854	0.5922
	Our SPSO	**0.7206**	**0.7417**	**0.6806**
Serine protease	GA [14]	0.7455	0.8065	0.6921
	PSO [14]	0.7120	0.7331	0.7008
	ABC [14]	0.7956	0.8168	0.7732
	Our SPSO	**0.8878**	**0.8904**	**0.8505**

Table 4. Comparison results for medium sequences.

Sequence name	Algorithm	Average	Best	Worst
Protein kinase	GA [14]	0.3508	0.3620	0.3453
	PSO [14]	0.4848	0.5065	0.4765
	ABC [14]	0.4951	0.5272	0.4765
	Our SPSO	**0.5277**	**0.5801**	**0.5312**
Anthranilate isomerase	GA [14]	0.3317	0.3440	0.3244
	PSO [14]	0.3813	0.4117	0.3529
	ABC [14]	0.3825	0.4292	0.3682
	Our SPSO	**0.4144**	**0.4549**	**0.4218**

(*continued*)

Table 4. (*continued*)

Sequence name	Algorithm	Average	Best	Worst
Serine protease	GA [14]	0.3792	0.4334	0.3579
	PSO [14]	0.2626	0.2725	0.2507
	ABC [14]	0.4178	0.4480	0.4036
	Our SPSO	**0.4646**	**0.4703**	**0.4209**

Table 5. Comparison results for long sequences

Sequence name	Algorithm	Average	Best	Worst
Aminotransferase	GA [14]	0.3646	0.3719	0.3548
	PSO [14]	0.5819	0.5903	0.5645
	ABC [14]	0.5819	0.5936	0.5741
	Our SPSO	**0.6602**	**0.6378**	**0.6612**
Glutamyl-trna synthetase	GA [14]	0.3738	0.3824	0.3648
	PSO [14]	0.3987	0.4154	0.3828
	ABC [14]	0.4040	0.4254	0.3857
	Our SPSO	**0.4316**	**0.4449**	**0.4207**
Taq DNA polymerase	GA [14]	0.2500	0.2644	0.2441
	PSO [14]	0.2533	0.2588	0.2484
	ABC [14]	0.2569	0.2615	0.2520
	Our SPSO	**0.2603**	**0.2882**	**0.2794**

In order to further assess the potent of our SPSO algorithm, a second experiment using another datasets is performed, where the goal is to compare our SPSO approach with TLPSO-MSA [24] technique. In this experiment, three different sets of protein coming from the BaliBASE 3.0 database [25] are selected. The average SP and TC scores [26] are portrayed in Table 6.

Table 6. Comparison results on the selected test cases.

Sequence set	N	TLPSO-MSA [24]		Our SPSO	
		SP	TC	SP	TC
RV11	38	0.80	0.71	**0.84**	**0.74**
RV12	44	**0.92**	0.80	0.86	**0.81**
RV20	41	0.91	0.64	**0.92**	**0.68**
Overall average		0.87	0.72	**0.87**	**0.74**

From Table 6, it can be seen that our SPSO outperforms clearly TLPSO-MSA method in terms of TC score for all cases. In terms of SP score, it finds results better than TSPSO-MSA for RV11 and RV20 protein families and produces a competitive results on RV12 dataset.

6 Conclusion

In this work, we contributed to the ongoing research by proposing a hybrid model for finding optimized alignments to the MSA problem. The developed approach uses the characteristics of the random search, global convergence of the PSO and the potent of the simulated annealing to update the global optimal solution. The performance of the proposed SPSO algorithm is judged on a set of BaliBase benchmark problems and is favorably compared with other algorithms in the literature. The results demonstrate that the proposed approach is overall more effective to find better alignments in a reasonable processing time.

In the future, we will revise our score function to make this score more realistic, also, we can use another intelligent heuristic to generate the initial whole swarm in order to increase the convergence speed of the algorithm. In addition, we can incorporate other efficient mechanisms for the neighborhood structure to improve the global solution quality. A comparison of the proposed method with some other aligners such as Clustal W, SAGA or MULTALIGN is possible to verify its effectiveness.

References

1. Thompson, J.D., Thierry, J.C., Poch, O.: RASCAL: rapid scanning and correction of multiple sequence alignments. Bioinformatics **19**(9), 1155–1161 (2003)
2. Jiang, T., Wang, L.: On the complexity of multiple sequence alignment. J. Comput. Biol. **1**, 337–378 (1994)
3. Bonizzoni, P., Della Vedova, G.: The complexity of multiple sequence alignment with SP-score that is a metric. Theor. Comput. Sci. **259**, 63–79 (2001)
4. Papadimitriou, C.H., Steiglitz, K.: Combinatorial Optimization: Algorithms and Complexity. Dover Publications, New York (1998)
5. Horng, J.T., Wu, L.C., Lin, C.M., Yang, B.H.: A Genetic algorithm for multiple sequence alignment. Soft Comput. **9**, 407–420 (2005)
6. Hernández-Guía, M., Mulet, R., Rodríguez-Pérez, S.: A new simulated annealing algorithm for the multiple sequence alignment problem. The approach of polymers in a random media. Phys. Rev. E **72**, 1–7 (2005)
7. Lei, C.W., Ruan, J.H.: A particle swarm optimization algorithm for finding DNA sequence motifs. In: Proceedings IEEE, pp. 166–173 (2008)
8. Lee, Z.J., Su, S.F., Chuang, C.C., Liu, K.H.: Genetic algorithm with ant colony optimization (GA-ACO) for multiple sequence alignment. Appl. Soft Comput. **8**, 55–78 (2008)
9. Chen, L., Zou, L., Chen, J.: An efficient ant colony algorithm for multiple sequences alignment. In: Proceedings of the 3rd International Conference on Natural Computation (ICNC 2007), pp. 208–212 (2007)
10. Riaz, T., Wang, Y., Li, K.B.: Multiple sequence alignment using tabu search. In: Proceedings of 2nd Asia-Pacific Bioinformatics Conference (APBC), Dunedin, New Zealand, pp. 223–232 (2004)
11. Xu, F., Chen, Y.: A method for multiple sequence alignment based on particle swarm optimization. In: Huang, D.-S., Jo, K.-H., Lee, H.-H., Kang, H.-J., Bevilacqua, V. (eds.) ICIC 2009. LNCS (LNAI), vol. 5755, pp. 965–973. Springer, Heidelberg (2009). https://doi.org/10.1007/978-3-642-04020-7_104

12. Chen, Y., Hu, J., Hirasawa, K., Yu, S.: Multiple sequence alignment based on genetic algorithm with reserve selection. In: Proceedings of International Conference on Networking, Sensing and Control (ICNSC), pp. 1511–1516 (2008)
13. Long, H.X., Xu, W.B., Sun, J., Ji, W.J.: Multiple sequence alignment based on a binary particle swarm optimization algorithm. In: Proceedings of Fifth International Conference on Natural Computation, pp. 265–269 (2009)
14. Lei, X., Sun, J., Xu, X., Guo, L.: Artificial bee colony algorithm for solving multiple sequence alignment. In: Proceedings of 2010 IEEE Fifth International Conference on BIC-TA, pp. 337–342 (2010)
15. Cutello, V., Nicosia, G., Pavone, M., Prizzi, I.: Protein multiple sequence alignment by hybrid bio-inspired algorithms. Nucleic Acids Res. 39(6), 1980–1992 (2011)
16. Liñán-García, E., Gallegos-Araiza, L.M.: Simulated annealing with previous solutions applied to DNA sequence alignment. ISRN Artif. Intell. 2012, 1–6 (2012)
17. Abu-Srhan, A., Al Daoud, E.: A hybrid algorithm using a genetic algorithm and cuckoo search algorithm to solve the traveling salesman problem and its application to multiple sequence alignment. Int. J. Adv. Sci. Technol. 61, 29–38 (2013)
18. Kayaa, M., Sarhanb, A., Alhajjb, R.: Multiple sequence alignment with affine gap by using multi-objective genetic algorithm. Comput. Meth. Programs Biomed. 114, 38–49 (2014)
19. Kennedy, J., Eberhart, R.: Particle swarm optimization. In: Proceedings of IEEE International Conference on Neural Networks, Perth, vol. 4, pp. 1942–1948, (1995)
20. Kirkpatrick, S., Gelatt, C.D., Vecchi, M.P.: Optimization by simulated annealing. Science 220, 671–680 (1983)
21. Rodriguez, P.F., Nino, L.F., Alonso, O.M.: Multiple sequence alignment using swarm intelligence. Int. J. Comput. Intell. Res. 3(2), 123–130 (2007)
22. Lamiche, C.: An effective alignment technique to solve MSA problem. Asian J. Math. Comput. 17(2), 2395–4213 (2017)
23. Thompson, J.D., Plewniak, F., Poch, O.: BAliBASE: a benchmark alignment database for the evaluation of multiple alignment programs. Bioinformatics 15, 87–88 (1999)
24. Lalwani, S., Kumar, R., Gupta, N.: A novel two-level particle swarm optimization approach for efficient multiple sequence alignment. Memet. Comput. 7(2), 119–133 (2015)
25. Thompson, J.D., Koehl, P., Ripp, R., Poch, O.: BAliBASE 3.0: latest developments of the multiple sequence alignment benchmark. Proteins 61(1), 127–136 (2005)
26. Thompson, J.D., Plewniak, F., Poch, O.: A comprehensive comparison of multiple sequence alignment programs. Nucleic Acids Res. 27(13), 2682–2690 (1999)

11. Chen, X., Hu, J., Hu aney, Z.S., Yu X.: Multiple sequence alignment based on a suequens algorithm with restric. Proceeding of the National Conference on Networking, Sec. ng and Signal (ICNSC), pp. 151–156 (2008).

12. Deng, H.X., Xu, W.B., Sun, J., Liu, Y.J.: Multiple sequence alignment based on a binary particle swarm optimization algorithm. In: Proceedings of Fifth International Conference on Natural Computation, pp. 203–206 (2009).

13. Lei, X., Sun, J., Xu, X., Guo, L.: Artificial bee colony algorithm for solving multiple sequence alignment. In: Proceedings of 2010 IEEE Fifth International Conference on BIC-TA, pp. 337–342 (2010).

14. Needle, W.S., Wunsch, C.D.: A general method applicable to the search for similarities in the amino acid sequence of two proteins. J. Mol. Biol. 48(3), 443–453 (1970).

15. Gotoh, O.: An improved algorithm for matching biological sequences. J. Mol. Biol. 162(3), 705–708 (1982).

16. Carrillo, H., Lipman, D.: The multiple sequence alignment problem in biology. SIAM J. Appl. Math. 48(5), 1073–1082 (1988).

17. Thompson, J.D., Higgins, D.G., Gibson, T.J.: CLUSTAL W: improving the sensitivity of progressive multiple sequence alignment through sequence weighting, position-specific gap penalties and weight matrix choice. Nucleic Acids Res. 22(22), 4673–4680 (1994).

18. Notredame, C., Higgins, D.G., Heringa, J.: T-Coffee: a novel method for fast and accurate multiple sequence alignment. J. Mol. Biol. 302(1), 205–217 (2000).

19. Katoh, K., Misawa, K., Kuma, K., Miyata, T.: MAFFT: a novel method for rapid multiple sequence alignment based on fast Fourier transform. Nucleic Acids Res. 30(14), 3059–3066 (2002).

20. Edgar, R.C.: MUSCLE: multiple sequence alignment with high accuracy and high throughput. Nucleic Acids Res. 32(5), 1792–1797 (2004).

Machine Learning

Automatic Ontology Learning
from Heterogeneous Relational Databases:
Application in Alimentation Risks Field

Aicha Aggoune(✉) [iD]

LabSTIC Laboratory, Department of Computer Science,
University of 8th May 45, Box 401, Guelma, Algeria
aggoune_ai@yahoo.fr

Abstract. In this paper, we propose a semantic approach for automatic ontology learning from heterogeneous relational databases in order to facilitate their integration. The semantic enrichment of heterogeneous databases, which cover the same domain, is essential to integrate them. Our approach is based on Wordnet and Wup's measure for measuring the semantic similarity between elements of these databases. It is described by a detailed process that can allow not only the generation of ontology but also its evolution as the evolution of its databases. We applied our approach in the alimentation risks field that is characterized by a large number of scientific databases. The developed prototype has been compared with similar tools of generation ontology from databases. The result confirms the quality of our prototype that returns the generic ontology from many relational databases.

Keywords: Ontology learning · Ontology evolution · Relational databases
Wup's similarity measure · Wordnet

1 Introduction

Ontologies do not replace relational databases (RDBs) but offer an alternative to RDBs to provide the meaning of the data and so they can facilitate the data integration. Different approaches for ontology building such as ontology building from scratch, ontology learning from text, ontology learning by reusing of existing Ontologies, etc. [8]. The ontology learning has since emerged as an important domain of ontology engineering that consists to generate an ontology from various sources such as text, database, dictionary, etc. [16]. The ontology learning from heterogeneous relational databases is the scope of this paper. We focus on two kinds of heterogeneity: the structural heterogeneity and semantic heterogeneity. The structural heterogeneity is related to different representations of data via tables, the attributes, and the relations, which represent themselves by tables, but the semantic heterogeneity corresponds to the different definitions to describe the same data [1].

In our previous work [1], we have proposed a novel semantic mediation system for solving semantic heterogeneity at both the query and database levels. The proposed system is based on hybrid ontology approach, which consists in attributing to each database its own ontology, called local ontologies describe its knowledge.

© IFIP International Federation for Information Processing 2018
Published by Springer International Publishing AG 2018. All Rights Reserved
A. Amine et al. (Eds.): CIIA 2018, IFIP AICT 522, pp. 199–210, 2018.
https://doi.org/10.1007/978-3-319-89743-1_18

In the context of semantic integration of heterogeneous databases, we attempt to present in this paper a novel approach to automatic learning of domain ontology from relational databases, which cover the same domain rather than built from each database its own ontology. Using one domain ontology of many relational databases allows improving the performance by reducing the execution time and space memory occupied by the ontology schema. To validate our approach, we have chosen the alimentation risks field that is characterized by a large number of scientific databases.

This paper is divided into five sections. Except for the introduction presented in the Sect. 1, the Sect. 2 devoted to the works in the literature, which are related to automatic ontology generation from relational databases. Section 3 gives in detail our semantic approach. Section 4 describes the experiments and results. In Sect. 5, we give conclusion with future work.

2   Related Work

Ontologies represent a primordial semantic web technology that serves as a standard vocabulary for the sharing and reusability of knowledge [3]. They are used to improve data representation by associating well-defined meaning with data. Thus, the ontology is used in different topics related to semantic databases such as semantic enrichment for database [11, 15, 18], disambiguation data [20], semantic data integration [1], etc. Using ontology as a key element for dealing semantic conflicts in databases according to different ways; either for creating database from ontologies [10], ontology learning from database [14, 23], or managing data and making decisions [7]. In this paper, we are interested in the second way in order to propose a new approach for automatic ontology learning from relational databases. In this area, different approaches have been proposed for only generating an ontology from a single database [4, 13]. Upadhyaya et al. [21] proposed an algorithm to transfer extended Entity/Relationship diagram (E/R Diagram) to OWL ontology. The algorithm is semi-automated because that requires a domain expert to aid more meaningful information and obtain a richer ontology. Fahad [9] proposes a framework for transforming the structured analysis and design artifact, E/R diagram, into the OWL ontology. This framework is limited to handle other cases of relationships between entities that are not binary, which require reification.

Recently, Dadjoo and Kheirkhah [5] proposed an approach for automatic ontology construction based on the relational database. This approach is based on graph theory, which leads to product the graph from database, and also with transforming of the graph obtained, final the ontology has been generated. This approach has succeeded to show richer semantics in the target ontology but it is not suitable for ontology construction from heterogeneous databases.

In addition, there are several tools allowing mapping relational databases (RDBs) to ontologies. Some of the most notable tools are DataMaster [19], KAON2 [11] and RDBToOnto [12]. These approaches define the mapping between components of both database and ontology, which an ontology class corresponds to a RDB table, an ontology datatype property corresponds to a table field, an ontology object property corresponds to an RDB attribute, and an ontology class instance corresponds to a RDB record [4].

The aforementioned tools are limited to the homogeneous database and they cannot ensure the ontology evolution as the heterogeneous databases, which used to generate ontology. Thus, they are based on a set of rules and definitions for mapping relational databases to OWL Ontologies.

In this paper, we present a semantic approach for automatic ontology learning from heterogeneous relational databases. Our approach is based on Wordnet for measuring the semantic similarity between components. We apply our proposal to the alimentation risks field in order to integrate easily a set of heterogeneous sources.

3   Semantic Approach for Automatic Ontology Learning

Our purpose is to automatically generate ontology from heterogeneous relational databases using a new method, which based on semantic similarity metric and a Wordnet as a lexical database aided to select the best terms for representing ontology components. We are interested to transform the logical model of database expressed by SQL language to the class hierarchy of ontology, which presents by OWL language. In general, the process of mapping relational databases into OWL structure produces the problem of incompatible between schemata. In this context, we attempt to reduce the gap between logic and ontological models. Our semantic approach provides to generate an ontology from many relational databases in the same domain, especially the alimentation risks field. The recent development of alimentation risks analysis have led to the need for strong health information systems that provide a unified access to various scientific bases, with the aim of discovery, knowing and predicting possible threats to public health [6]. These scientific bases are often heterogeneous databases, which makes their access a complex task [12]. The proposal approach for ontology learning is of type semantic because is based on the use of two important notions related to the semantic are the Wordnet and the similarity measure. The Wordnet is one of the most widely used lexical databases for English [17]. It has been extensively used to improve the quality of data sources with its semantic relations of terms.

Otherwise, the semantic similarity measure is a central issue in different domains of computer science such as natural language processing, information retrieval, word sense disambiguation, text segmentation, question answering and so on [16]. In our work, we focus on one popular similarity measure, Wu and Palmer's similarity (wup) between concepts (C1, C2), which is defined by the function of their distance and the lowest common subsume C to C1 and C2. Thus, it is based on the use of depth (the number of arcs) according to the following formula [22]:

$$SIMwup(C1, C2) = \frac{2 \times \text{depth}(C)}{\text{depth}(C1) + \text{depth}(C2)} \tag{1}$$

Wup similarity measure has the advantage of having good performance and the fast execution time rather than the others [1]. The full process of the proposed approach for automatic ontology learning from heterogeneous relational databases is depicted in Fig. 1.

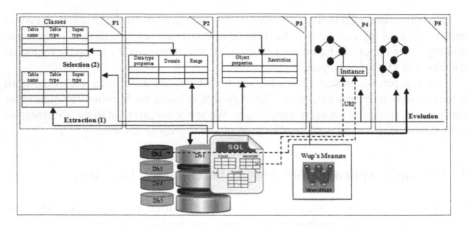

Fig. 1. A full process for automatic ontology learning

The full process for automatic ontology learning is based on four sequential processes (P1 until P4 from Fig. 1) with an additional process for ensuring the evolution of obtained ontology (P5 from Fig. 1). The system starts with the first process called generate classes process in order to extract and to select classes from tables' names of all databases. From the result of this process, we can extract the datatype properties of each class in the generate datatype properties process. The third process aims to generate object properties between obtained classes, and the fourth process allows integrating the records of databases as well as ontology individuals.

Regarding the fifth process, it is based on the algorithm for detecting all modifications of databases, which must be applied to their ontology. All of these five processes need to three essential elements, which are data source (databases), WordNet (a lexical database) and the Wup's similarity measure. In the following sub sections, we are going to present the detail of these processes for automatic ontology learning.

3.1 Generate Classes Process

The first process attempts to generate ontology classes from databases tables using wup's similarity based on Wordnet. Each relational database is created separately, which is heterogeneous according to the different criteria such as structure, semantic, language, format, etc. In this work, we are interested to the two first types of heterogeneity. The structural heterogeneity is defined by the different representations of data but the semantic heterogeneity corresponds to the different definitions to describe the same data [1]. We use five heterogeneous databases in alimentation risks field, which can contain the similar tables, most of the same attributes and records.

It is clear that the input of the Generate classes process is the five relational databases and the output is a set of ontology classes. The first step of this process consists to extract the tables' information of all relational databases and stored it into a new relational table called Class. The latter contains the database's number, the table's name, type using for creating tables, and the super type for representing the inheritance

Table 1. The result of extracting tables' information of databases

Num_db	Table_Name	Table_type	Super_type
01	Food	Food_t	–
01	Microbe	Microbe_t	–
01	To_live	–	–
02	Alimentation	Alimentation_t	–
02	Fruit	Fruit_t	Alimentation_t
02	Vegetable	Vegetable_t	Alimentation_t
03	Microorganism	Microorganism_t	–
03	Mycotoxin	Mycotoxin_t	Microorganism_t
04	Feed	Feed_t	–
05	Nutriment	Nutriment_t	–
05	Virus	Virus_t	–
05	To_Exist	–	–

relation between types (in SQL is presented by the clause Under). The following table presents a part of the class table (Table 1).

From the table presented above, the second step involves to identify the similar tables' names of five relational databases using Wup's measure, which detects the synonymy relation between concepts (tables' names) used by Wordnet. In the first, we are interested to select the similar super tables. The result is presented in the following table (Table 2).

Table 2. The result of similarity measure between tables

Num_db1	Table_Name1	Table_Name2	Num_db2	Wup's similarity
01	Food	Alimentation	02	0.9231
	Food	Microorganism	03	0.4286
	Food	Feed	04	0.9231
	Food	Nutriment	05	0.9231
	Food	Virus	05	0.4615
	Microbe	Alimentation	02	0.3750
	Microbe	*Microorganism*	*03*	*0.9412*
	Microbe	Feed	04	0.3750
	Microbe	Nutriment	05	0.3750
	Microbe	Virus	05	0.8889
02	Alimentation	Microorganism	03	0.4000
	Alimentation	Feed	04	0.8571
	Alimentation	*Nutriment*	*05*	*1*
	Alimentation	Virus	05	0.4286
03	Microorganism	Feed	04	0.4000
	Microorganism	Nutriment	05	0.4000
	Microorganism	Virus	05	0.9412
04	Feed	Nutriment	05	0.8571
	Feed	Virus	05	0.4286

According to the results presented in the table above, we can deduct the similar tables existing in different databases, for example, the Food table form database01 is similar to Alimentation table from database02 and Feed table from database04 and also Nutriment table from database05. When a similarity between two concepts is greater than the threshold, then they are considered similar. It can be clearly seen that the Alimentation and Nutriment tables are semantically closet. According to high value of similarity measure between tables' names, we can obtain the following set of ontology classes: Nutriment and Microbe.

3.2   Generate Datatype Properties Process

In this process, we attempt to identify the datatype properties of ontology classes obtained from the previous process. All attributes of nutriment table from database05 became the datatype properties of nutriment class. Thus, we attempt to select other attributes from similar tables to nutriment table. It is about Food, Alimentation and Feed tables. The selection of additional attributes is based on a similarity measure between datatype properties of nutriment class and all attributes of similar tables. The attributes with a low similarity have been selected as new datatype properties (select the different attributes to the datatype properties). In the case of composite attributes related to the relational table, for example, the diameter of microbe is composed of length and width. The mapping of relational databases to ontology consists to consider composite attribute (for example diameter) as simple datatype properties and its components (length, width) as the sub property of corresponding datatype property.

Datatype properties which stand as a primary key in the relational model, are tagged with a Functional tag for restricting the object to take only one value for a given subject, and also, tagged with inverse-functional which restricts the subject to associate with only one object [13]. In the case of the attributes which stand as a foreign key, the mapping to ontological model is depicted in the third process (see Subsect. 3.3).

Furthermore, the domain and the range mappings of datatype properties are based on both the classes and the domain of corresponding attributes from relational tables. The following algorithm illustrates the domain and the range mappings.

```
Input: C: set of ontology classes, A: set of relational
attributes
Begin
For each attributes of A do
Begin
Rdfs :domain is corresponding class of C;
If  domain of attributes in A is 'Varchar' or 'Char' then
    rdfs:range rdf :resource="string"
else rdfs:range rdf :resource=the domain of attribute;
End;
End.
```

3.3   Generate Object Properties Process

After generating classes ontology and its datatype properties, it's time to identify the semantic relations between classes. That's the purpose of the third process in order to define the object properties between classes. From the subtypes of the relational model, which are presented in table 01, we can deduct the is_a relation and subclasses of the ontological model. So, the subclasses of nutriment class are the subtypes of similar relational tables, which are Fruit and Vegetable. Thus, the microbe class has Mycotoxin as a subclass. In addition, a wup's measure between subtypes of all similar relational tables is necessary for selecting the best subclasses of the ontological model.

About the one-to-many relationship from the relational model, the transformation to the ontological model consists to detect the foreign key from the logical model and generate two object properties with identifying the restriction between classes. For example, the following OWL code interprets the Category relation, which can have one or more records relating to a single record in microbe relation:

```
<owl: class rdf: ID="Microbe">
<rdfs: subclassof>
<owl:Restriction>
<owl:ValuesFrom>   <owl: class rdf: ID="Category">
        <rdfs: subclassof> <owl:Restriction>
        <owl:someValuesFrom rdf:resource="#Microbe/">
</ValuesFrom>
        <owl:onProperty> <owl: Objectproperty
rdf:ID="has-Microbe"/>   </owl:onProperty>
        </owl:Restriction> </rdfs: subclassof>
</owl: class> </ValuesFrom>
<owl:onProperty> <owl: Objectproperty rdf:ID="has-
category"/>   </owl:onProperty>
</owl:Restriction>
 </rdfs: subclassof>
</owl: class>
```

We apply the same rule for presenting many-to-many relationship. In this kind of relation, we use the tag <owl:someValuesFrom> for each restriction related to the classes.

3.4   Generate Instances Process

The final process attempts to generate instances (or individuals) of ontology classes corresponds to an appropriate record from relational tables. Thus, the same instance can be related to many relational tables. In our method, we create a new special attribute in the database that represents a pointer referring to an instance of ontology class [1]. So, each instance is related to its records via this special attribute, which contains the URI (Uniform Resource Identifier) of the corresponding instance from the ontology.

An instance of ontology class is a set of values grant to each datatype property. So, the superfluous attributes from relational table have been eliminated. Form the special attribute; we can manage consistency between the content of the database and obtained ontology (more detail in Subsect. 3.5). From these four processes, the ontology has been created from relational databases.

3.5 Ontology Evolution Process

The ontology evolution process allows the obtained ontology to change according to its relational databases [2]. Each modification of relational databases content involves the same operation in the generated ontology. It is through the special attribute, which contains the URI of the instance of ontology class, we can ensure the auto evolution of ontology. The following algorithm illustrates the evolution of ontology by the evolution of its database.

```
Input : DB : database, Ont: ontology
            Op : {'any', 'insert', 'delete', 'update'}
Begin
For   each i^th table of DB Do
 Begin
   If op= 'any' Then Exit
       ELse if  op= 'Insert' Then
         Begin
                Insert into Ont the new instance according
to the new record;
                Update the SA a special attribute by the
URI of the new instance ;
            End if
   Else if op= 'delete' Then
         Begin
                URI := SA[k]     //extract the uri of K^th ele-
ment to remove
                Delete the instance from Ont;
                Delete SA[k];
            End if
     Else //update
         Update the k^th instance of Ont;
   End for;
End.
```

The ontology evolution algorithm presented above takes as input, the database, the ontology and a set of operations, which can apply to both database and ontology. There are three principal operations: insert, delete and update. If any modification applied to the database (operation = any) then exit the program else execute the ontology evolution. The main idea in this algorithm is to use the special attribute SA for facilitating the reference of the instance to delete or to update [2]. In the insert operation, the first step consists to insert a new instance according to the new record. The second step allows giving the URI of this new instance and updates the SA value by this URI without changing of ontology. Therefore, the algorithm affects the ontology evolution according to the evolution of its relational databases.

4   Experimentation

In this section, we provide an experimentation of our approach. To do that, we have implemented the above processes into our prototype implementation of automatic ontology generation that based on Java as a programming language. The prototype provides an intuitive interface to represent the results of different processes of ontology learning. The following figure shows the execution of our prototype (Fig. 2).

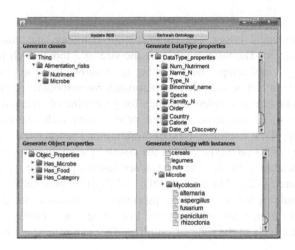

Fig. 2. A screenshot of generation ontology prototype

The interface of our prototype is split into four parts (01, 02, 03, 04) which represent results of four processes (generate classes, generate datatype properties, generate object properties and generate instances). Part 04 shows the generated ontology with its instances. The interface also offers the possibility to update relational databases using the button 'update RDB'. After updating the database, we can refresh the generated ontology by 'refresh ontology' button and show the ontology evolution in part 04.

Otherwise, in order to identify the advantage of our prototype compared to other similar tools, we compare it to three popular tools which are DataMaster [19], KAON2 [11] and RDBToOnto [12]. Hence, we use the same relational database among the five databases in alimentation risks field that we created for validating our approach. We are selected three comparative criteria: the functionality of a tool, which define the principal function of the tool, the features, which are a set of characteristics such as

Table 3. Comparison between our prototype and other similar tools

Tool	Functionality	Features	Size
Our prototype	- Ontology generation from many RDBs - Ontology evolution	- Complete ontology - Nothing errors - Quick building - Allows updating database - Evolutionary ontology	209 ko
DataMaster	- Ontology generation from one RDB	- Incomplete ontology (some relationships do not detected) - RDB constraints are not treated	775 ko
KAON2	- Manipulating OWL- DL ontologies - Extracting ontology instances from RDB	- Ontology manually generated which its instances have been extracted from RDB	2305 ko
RDBToOnto	- Ontology generation from RDB	- Less complete extraction rules	13119 ko

appearance of technical errors, the completeness ontology, the cost and time for ontology learning, etc. The third criterion is the size of a tool, which represent a disk space required by a tool setup. The following table shows the result (Table 3).

From the table, it can be seen that our approach for automatic ontology learning is more efficient and return a complete ontology for generating of essential components of an ontology. Thus, our prototype does not produce any technical errors during the ontology learning. So, it facilitates a quick building and reduces the cost and time for ontology learning. This prototype allows the ontology evolution as evolution evolves of its relational databases. In addition, we can see that our prototype having the smallest disk space compared to DataMaster, KAON2, and RDBToOnto. Unlike KAON2, is the framework for building ontology without instances; it contains a module for extracting ontology instances from relational databases. For future work, another criterion we will take into accounts such as the time of ontology generation, the performance, and the complexity.

5 Conclusion

We have presented a new approach for automatic ontology learning and evolution from relational databases RDB. This approach is based on five processes: Generate classes process which builds classes from relational tables, Generate datatype properties process which generates properties of the classes obtained from the first process, Generate object properties process which uses the relations between tables in order to create relationships between classes, Generate instances process which considers the data of RDB as instances of ontology, and Ontology evolution process which modifies the ontology as the modification of its RDB. We use in this approach a Wup's measure, which detects the synonymy relation between concepts used by Wordnet. Thus, the proposed approach has been validated by the development of the prototype to show the effectiveness and automation ontology learning. This prototype has been tested and evaluated by a comparison study with similar tools of ontology learning from databases and the result shows the novelty of our contribution.

Future work includes depth analysis of the performance of our prototype using big databases. The results of this work will help to solve several semantic problems such as semantic integration, semantic information retrieval, and semantic querying, etc.

References

1. Aggoune, A., Bouramoul, A., Kholladi, M.K.: Mediation system for dealing with semantic problems in databases. Int. J. Data Min. Model. Manag. **9**(2), 99–121 (2017)
2. Aggoune, A.: Traitement de l'Hétérogénéité Sémantique pour l'Exploration des Sources de Données Multimédias. Phd's thesis. Constantine 2 University, Algeria (2017)
3. Berners-Lee, T., Hendler, J., Lassila, O.: The semantic web. Sci. Am. **284**(5), 28–37 (2001)
4. Bumans, G.: Mapping between relational databases and OWL ontologies: an example. Sci. Pap. Univ. Latvia **756**, 99–117 (2010)
5. Dadjoo, M., Kheirkhah, E.: An approach for transforming of relational databases to OWL ontology. Int. J. Web Semant. Technol. (IJWesT) **6**(1), 19–28 (2015)
6. De Valk, H., Salvat, G.: Alimentation et risques infectieux: enjeux et stratégies pour limiter l'impact sur la santé. Les Tribunes de la santé **49**(4), 61–68 (2015)
7. Dogdu, E., Ozbayoglu, A.M., Benli, O., Akinc, H.E., Erol, E., Atasoy, T., Gurec, O., Ercin, O.: Ontology-centric data modelling and decision support in smart grid applications a distribution service operator perspective. In: International Conference on Intelligent Energy and Power Systems (IEPS), pp. 198–204. IEEE, Ukraine (2014)
8. Drame, K.: Contribution à la construction d'ontologies et à la recherche d'information: application au domaine médical. Phd's thesis. University of Bordeaux, French (2014)
9. Fahad, M.: ER2OWL: generating OWL ontology from ER diagram. In: Shi, Z., Mercier-Laurent, E., Leake, D. (eds.) IIP 2008. ITIFIP, vol. 288, pp. 28–37. Springer, Boston, MA (2008). https://doi.org/10.1007/978-0-387-87685-6_6
10. Ho, L.T.T., Tran, C.P.T., Hoang, Q.: An approach of transforming ontologies into relational databases. In: Nguyen, N.T., Trawiński, B., Kosala, R. (eds.) ACIIDS 2015. LNCS (LNAI), vol. 9011, pp. 149–158. Springer, Cham (2015). https://doi.org/10.1007/978-3-319-15702-3_15
11. http://kaon2.semanticweb.org/. Accessed 15 Nov 2017
12. http://www.taoproject.eu/researchanddevelopment/demosanddownloads/RDBToOnto.html
13. http://www.w3.org/TR/2004/REC-owl-guide-20040210/. Accessed 14 Nov 2017
14. Krivine, S., Nobécourt, J., Soualmia, L., Cerbah, F., Duclos, C.: Construction automatique d'ontologie à partir de bases de données relationnelles: application au médicament dans le domaine de la pharmacovigilance. In: Actes des 20es Journées Francophones d'Ingénierie des Connaissances, IC, pp. 1–12. Tunisia (2009)
15. Kumova, B.İ.: Generating ontologies from relational data with fuzzy-syllogistic reasoning. In: Kozielski, S., Mrozek, D., Kasprowski, P., Małysiak-Mrozek, B., Kostrzewa, D. (eds.) BDAS 2015. CCIS, vol. 521, pp. 21–32. Springer, Cham (2015). https://doi.org/10.1007/978-3-319-18422-7_2
16. Li, M., Du, X.Y., Wang, S.: Learning ontology from relational database. In: International Conference Machine Learning and Cybernetics, vol. 6, pp. 3410–3415. IEEE, China (2005)
17. Miller, G.A.: WordNet: a lexical database for English. Commun. ACM **38**(11), 39–41 (1995)

18. Nakhla, Z., Nouira, K.: Automatic approach to enrich databases using ontology: application in medical domain. Procedia Comput. Sci. **112**, 387–396 (2017)
19. Nyulas, C., O'Connor, M., Tu, S.: DataMaster - a plug-in for importing schemas and data from relational databases into Protégé. In: 10th International Protégé Conference, Hungary, pp. 1–3 (2007)
20. Tahat, S., Ahmad, K.: A method on lexical disambiguation in distributed heterogeneous autonomous database. In: International Conference on Research and Innovation in Information Systems, ICRIIS, pp. 330–335. IEEE, Malaysia (2013)
21. Upadhyaya, S., Kumar, P.: ERONTO: a tool for extracting ontologies from extended E/R diagrams. In: Proceedings of SAC 2005 ACM Symposium on Applied Computing, Santa Fe, USA, pp. 666–670 (2005)
22. Wu, Z., Palmer, M.: Verbs semantics and lexical selection. In: ACL 1994: Proceedings of the 32nd Annual Meeting on Association for Computational Linguistics, USA, pp. 133–138 (1994)
23. Cerbah, F.: Learning highly structured semantic repositories from relational databases. In: Bechhofer, S., Hauswirth, M., Hoffmann, J., Koubarakis, M. (eds.) ESWC 2008. LNCS, vol. 5021, pp. 777–781. Springer, Heidelberg (2008). https://doi.org/10.1007/978-3-540-68234-9_57

Towards the Prediction of Multiple Soft-Biometric Characteristics from Handwriting Analysis

Nesrine Bouadjenek, Hassiba Nemmour$^{(\boxtimes)}$ ⓘ, and Youcef Chibani

Laboratoire d'Ingénierie des Systèmes Intelligents et Communicants (LISIC), Faculty of Electronics and Computer Sciences, University of Sciences and Technology Houari Boumediene (USTHB), Algiers, Algeria
{nbouadjenek,hnemmour,ychibani}@usthb.dz

Abstract. Soft-biometrics prediction from handwriting analysis is gaining a wide interest in writer identification since it gives additional knowledge about the writer like its gender (man or woman), its handedness (left-handed or right-handed) and its age range. All research works developed in this context were focused on predicting a single soft-biometric trait. Nevertheless, it could be more interesting to develop a system that predicts several traits from a handwritten text. Presently, we investigate the feasibility of such multiple trait prediction. To reach this end, we propose two prediction schemes. The first combines individual prediction scores to aggregate a global prediction. The second scheme is based on a multi-class prediction. For both schemes, the prediction is based on SVM classifier associated with Gradient features. Experimental corpus is collected from IAM handwritten database. Conclusively, the second scheme proved to be more promising and evinced that the age characteristic is stable over time for a certain category of writers.

Keywords: Handwriting · Membership degree
Multiclass prediction · Soft-biometrics

1 Introduction

Soft-biometrics provides complementary information about the individual, without being able to fully authenticate him. It includes various traits such as gender, skin color, eyes color and ethnicity. These characteristics were recently used to reinforce biometric identification systems. Besides, in forensics applications, soft-biometrics allows the restriction of investigations to a limited category of persons or suspects [1]. First works in this field extracted soft-biometrics by analyzing individual face images which constitute the most practical identification tool [2,3]. Nevertheless, soft-biometrics can bring useful information to some forensics and handwriting recognition applications. For instance, when analyzing an anonymous threat letter soft-biometric information such as the writer's

© IFIP International Federation for Information Processing 2018
Published by Springer International Publishing AG 2018. All Rights Reserved
A. Amine et al. (Eds.): CIIA 2018, IFIP AICT 522, pp. 211–219, 2018.
https://doi.org/10.1007/978-3-319-89743-1_19

gender, handedness, age range, and educational level, has a precious contribution in investigations. Since 2001, researchers in the handwriting recognition field
started to predict soft-biometric traits from handwritten text. The first work proposed by Cha et al. [4] tried to classify US population into some demographic
sub-categories defined by gender, ethnicity and educational level. Then, some
other works have followed later by dealing with various traits such as gender,
handedness, age range and nationality [5–14]. However, in the state of the art,
soft-biometrics prediction systems were developed to deal with a single trait. This
is mainly due to two reasons. First, there is a lack of datasets providing several
soft-biometric characteristics and second, the prediction of one characteristic is a
challenging task. In fact, the results reported on several benchmark datasets vary
from 55% to 85% [5–14]. Thereby, the following question have come up: Could
we predict two or more characteristics from the same analysis and if so, how
much will be the prediction score? The present work attempts to answer these
questions by proposing two multi-class prediction schemes. In the first scheme,
we employ the same handwritten text to develop individual systems that predict writer's gender, handedness and age range. Then, the predictions obtained
are grouped to get a global prediction on the three characteristics. Whilst, the
second scheme adopts directly a multiclass prediction based on the one against
all implementation. In both schemes, the prediction process is based on SVM
classifier associated with gradient features. The rest of this paper is arranged as
follows: Sect. 2 introduces multi-trait prediction schemes. Section 3 presents the
experimental evaluation while the last section reports the main conclusions of
this work.

2   Multiclass Prediction of Soft-Biometrics

Gender is the social definition of a man and a woman, while handedness defines
the preference for use of a hand, known as the dominant hand (left or right). As
to age, it is perceived by ranges. Until now, predicting such characteristics from
handwriting is performed for only one characteristic at a time. In this work we
investigate the feasibility of a multiclass prediction from the analysis of the same
handwritten text. Whatever the adopted scheme, the prediction task is founded
on two main steps that are feature generation and prediction. As feature generation several texture, gradient, shape and geometric features were proposed
[10,14]. Also, for the prediction step, several classifiers were employed such as
artificial neural networks, Support Vector Machine (SVM) and decision tree algorithms [8,10]. Nevertheless, findings report that SVM is the best candidate for
solving the prediction task [10]. So, in this work we employ SVM associated with
the Gradient Local Binary Patterns (GLBP) which showed a high performance
for predicting a single soft-biometric trait [10].

2.1   Dataset Description

Up to now, the prediction of writer's soft-biometrics is not widely investigated
because of the lack of public datasets. Precisely, for the Latin script IAM is the

only public dataset which provides gender, handedness and age range of writers. IAM was developed by a research group on computer vision and artificial intelligence at Bern University in Switzerland[1]. It contains handwritten sentences of more than 200 writers grouped into two age categories that are "25–34 years" and "35–56 years". So, by considering the three available traits, we can define 8 classification categories as shown in Table 1. Presently, 534 samples are collected to perform multi-class soft-biometrics prediction. Specifically, we considered only one handwritten sentence per writer for right-handed writers. For classes including left-handed classes we considered more than one sample per writer since the IAM dataset contains only 20 left-handed writers.

Table 1. Data distribution for multiclass prediction.

Classes	# Training samples	# Test samples
Class 1: Female/Left/25–34	44	22
Class 2: Female/Left/35–56	44	22
Class 3: Female/Right/25–34	44	22
Class 4: Female/Right/35–56	44	22
Class 5: Male/Left/25–34	40	20
Class 6: Male/Left/35–56	40	20
Class 7: Male/Right/25–34	50	25
Class 8: Male/Right/35–56	50	25

2.2 Multiclass Soft-Biometrics Prediction Based on Individual Systems

Since soft-biometrics prediction is commonly evolved in systems predicting a single characteristic, the direct extension for a multi-class prediction consists of grouping individual decisions of such systems. The idea of this scheme is to predict each soft-biometric characteristic independently from the others, so that the global system will be composed of "j" individual binary systems, if we have "j" characteristics to predict. In this respect, three systems are developed to predict gender, handedness and age range by grouping the training data according to the considered characteristic. Hence, we use 176 female samples and 180 male samples for gender prediction, 168 left-handed samples and 188 right-handed samples for handedness prediction, and finally, 178 samples for age ranges prediction. For the test stage, each sample is simultaneously presented to the three systems as shown in Fig. 1. Then, predictions on gender, handedness and age range are grouped and compared to the ground truth of the considered sample. In experiments, classes in Table 1 were grouped to perform a binary classification. For gender prediction training samples of classes 1, 2, 3 and 4 were grouped to constitute the Female class while the remaining classes grouped to form the Male class.

---

[1] http://www.iam.unibe.ch/fki.

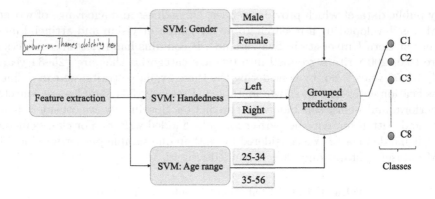

Fig. 1. Multiclass soft-biometrics prediction based on the individual systems.

2.3   Multiclass Soft-Biometrics Prediction Based on One-Against-All SVM

The One-Against-All (OAA) SVM builds "j" binary SVM to solve a j-class classification problem. Each SVM is dedicated to separate one class from all other classes. After the training stage, a test sample is presented to all SVM to produce 8 decisions according to the classes of interest. Then, the sample is assigned to the class with the highest decision as depicted in Fig. 2.

Note that test samples are common for the two schemes, in order to get a fair comparison of the prediction scores.

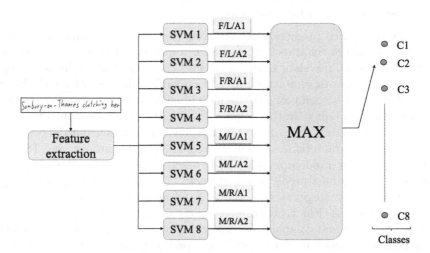

Fig. 2. OAA SVM for multiclass soft-biometrics prediction system; F: female, M: male, R: right, L: left, A1: 24–34 years, A2: 35–56 years.

3 Experimental Evaluation

The proposed multiclass prediction schemes are evaluated on the selected IAM sub-set. For performance evaluation the confusion matrix is used to highlight the precision per class and the global prediction accuracy. Recall that the prediction system is better the more the confusion matrix approaches a diagonal matrix.

3.1 Prediction Based on Individual Systems

Prediction results expressed through overall accuracies are exhibited in Fig. 3. The overall accuracy based on the combination of individual decisions that is 32,48% is much lower than those given by each individual predictions. This can be explained by the proliferation of prediction errors of each binary system, when aggregating the final decision. Specifically, the decrease in the prediction accuracy is mainly due to the age range prediction system that gives a medium prediction, which is about 52,86%. This finding leads us to move towards multi-class implementation to improve these results.

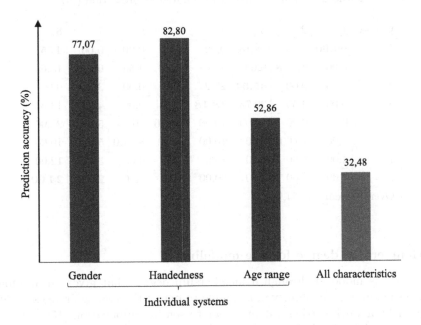

Fig. 3. Multiclass prediction results of combined individual systems.

3.2 Prediction Based on the OAA SVM

Compared to the first scheme, the OAA implementation improves the overall prediction accuracy to 54.09%. To understand this result, which remains low, we present the confusion matrix in Table 2. From a look at this table, we note that classes 3, 4, 7 and 8 that correspond to right-handed writers are poorly predicted. These classes are problematic, as the addition of the age range characteristic, especially for right-handed writers, increases the complexity of the prediction task. More precisely, for right-handed females the age range doesn't show any behavioral differences (classes 3 and 4). This means that writers of this category keep almost a stable and stationary age characteristic over time. Similar behavior is observed for right-handed males that are highly confused with the right-handed females by 36.36% in precision. To get a more precise interpretation of the soft-biometric behavior, we reproduced the OAA test by considering two soft-biometrics that are gender and handedness. This was done by merging classes age ranges according to gender and handedness, which yields a 4-classes prediction. Experiments report an overall prediction score of 66.17%. Based on this outcome as well as the results derived by the first scheme, we suggested that the age range is the most critical trait. So, to improve the description of age range information, we developed a fuzzy membership model that gives an additional knowledge about the writer's age.

Table 2. Confusion matrix for multiclass prediction (%).

Classes	1	2	3	4	5	6	7	8
1	**90,90**	4,55	18,18	0,00	9,09	0,00	0,00	4,55
2	0,00	**68,18**	0,00	4,55	0,00	4,55	0,00	13,64
3	4,55	0,00	**45,54**	27,27	0,00	0,00	36,36	9,09
4	0,00	4,55	22,73	**18,18**	0,00	0,00	4,55	13,64
5	0,00	0,00	10,00	10,00	**70,00**	10,00	5,00	25,00
6	0,00	10,00	0,00	10,00	10,00	**80,00**	5,00	10,00
7	4,00	4,00	4,00	8,00	0,00	0,00	**36,00**	12,00
8	0,00	8,00	0,00	20,00	8,00	4,00	20,00	**24,00**
Overall accuracy: 54,09%								

– **Membership degree for age modeling**

Age range is modeled through a membership degree that gives an automatic information about the affinity of a sample to one of the two age ranges. Indeed, inspired by a work carried out in a remote sensing application [15], we define a fuzzy membership degree to age categories based on Mahalanobis distance. Specifically, all training samples were grouped into two sets according to the age range. For each set, we calculated the mean and the covariance matrix. Then, for each sample, the fuzzy membership degree to the age range categories is calculated according to the steps presented in Algorithm 1.

The membership degree is generated for all samples and concatenated with GLBP features. Table 3 illustrates the confusion matrix obtained be adding the fuzzy meberships of age range.

Algorithm 1. Membership degree calculation for age modeling

1. Compute the Mahalanobis distance of each sample X as:

$$dist^2_{Xi} = (X - mean_i)^t cov_i^{-1}(X - mean_i)$$

$mean_i$: mean vector of the ith age category $i = \{1,2\}$
cov_i^{-1} : inverse of the covariance matrix of the ith age category
2. Compute the membership degree of the sample X as:

$$h_{Xi} = \frac{(\frac{1}{|dist^2_{Xi}|})^2}{\sum_{u=1}^{2}(\frac{1}{|dist^2_{Xui}|})^2}$$

Table 3. Confusion matrix for multiclass prediction with membership degree contribution (%).

Classes	1	2	3	4	5	6	7	8
1	**77,27**	0,00	13,64	4,55	9,09	0,00	0,00	4,55
2	0,00	**77,27**	0,00	9,09	0,00	4,55	0,00	13,64
3	27,27	4,55	**59,09**	4,55	0,00	0,00	36,36	9,09
4	4,55	9,09	0,00	**45,45**	0,00	0,00	4,55	13,64
5	15,00	10,00	5,00	0,00	**70,00**	10,00	5,00	25,00
6	0,00	5,00	5,00	5,00	10,00	**80,00**	5,00	10,00
7	4,00	0,00	24,00	8,00	0,00	0,00	**36,00**	12,00
8	4,00	8,00	4,00	8,00	8,00	4,00	20,00	**24,00**
Overall accuracy: 61,51%								

As can be seen, the membership degree allows a gain of 7,42% given an overall prediction about 61.51%. Indeed, an improvement of 13,64% and 27,27% are reached for right-handed females of both age ranges. Moreover, an improvement of 24% is noticed for right-handed males of the second age range. However, we observe a negative effect on left-handed writers. For instance, the precision of left-handed females aged between 25 and 34 years old drops from 90,91% to 77,27% which corresponds to 3 samples wrongly predicted.

In summary, predicting gender, handedness and age range simultaneously is very challenging as it is limited by the difficulty of separating sub-categories according to the age characteristic. For this reason, prediction accuracy drops from 66.17% to 54.09% without and with age characteristic, respectively. These, all outcomes reveal that the right-handed writers which represent the majority of writers in the database, are not distinguished according to age. This allows to say that the characteristic that is supposed to evolve over time, has generally stagnated for this category of writers.

4   Concluding Remarks

This work addresses the possibility of a simultaneous prediction of writer's gender, handedness and age range from one single analysis of the handwriting giving an 8-classes prediction problem. In this respect, using a set of data extracted from an English benchmark dataset, we investigate two prediction schemes. In the first scheme, three systems designed to predict a single characteristic are developed. Then, predictions are grouped to give an overall prediction of the three characteristics. The second scheme adopts a multiclass prediction based on the one against all implementation of SVM to solve the 8 classes prediction problem. Experimental findings reveal that the age characteristic is problematic as it seems to be stable and unchanged over time, especially for right-handed writers. This is perhaps due to the fact that all contributers are adult since the dataset doesn't contain young and very old writers. So, it seems necessary to perform other experiments with larger age range categories to get more concluding results. Nevertheless, the best multiclass prediction accuracy is about 61,51%. This remains a promising result and can be improved by finding a better modeling of the age characteristic, not necessarily through a classifier but through a model representation such as regression. Also, perhaps other features or classification methods can deal better with this characteristic.

References

1. Tome, P., Vera-Rodriguez, R., Fierrez, J., Ortega-Garcia, J.: Facial soft biometric features for forensic face recognition. Forensic Sci. Int. **257**, 271–284 (2015)
2. Jain, A.K., Park, U.: Facial marks: soft biometric for face recognition. In: International Conference on Image Processing, Cairo, Egypt, pp. 37–40, November 2009
3. Zhang, H., Beveridge, J.R., Draper, B.A., Phillips, P.J.: On the effectiveness of soft biometrics for increasing face verification rates. Comput. Vis. Image Underst. **137**, 50–62 (2015)
4. Cha, S.H., Srihari, S.N.: A priori algorithm for sub-category classification analysis of handwriting. In: International Conference on Document Analysis and Recognition, Seattle, USA, pp. 1022–1025, September 2001
5. Liwicki, M., Schlapbach, A., Loretan, P., Bunke, H.: Automatic detection of gender and handedness from on-line handwriting. In: Conference of the International Graphonomics Society, Melbourne, Australia, pp. 179–183, November 2007
6. Liwicki, M., Schlapbach, A., Bunke, H.: Automatic gender detection using on-line and off-line information. Pattern Anal. Appl. **14**, 87–92 (2011)
7. Al-Maadeed, S., Ferjani, F., Elloumi, S., Hassaine, A.: Automatic handedness detection from off-line handwriting. In: GCC Conference and Exhibition, Doha, Qatar, pp. 119–124, November 2013
8. Al-Maadeed, S., Hassaine, A.: Automatic prediction of age, gender, and nationality in offline handwriting. EURASIP J. Image Video Process. **2014**, 10 (2014)
9. Bouadjenek, N., Nemmour, H., Chibani, Y.: Age, gender and handedness prediction from handwriting using gradient features. In: International Conference on Document Analysis and Recognition, Tunisia, pp. 1116–1120, August 2015
10. Bouadjenek, N., Nemmour, H., Chibani, Y.: Robust soft-biometrics prediction from off-line handwriting analysis. Appl. Soft Comput. **46**, 980–990 (2016)

11. Bouadjenek, N., Nemmour, H., Chibani, Y.: Fuzzy integral for combining SVM-based handwritten soft-biometrics prediction. In: 12th IAPR Workshop on Document Analysis Systems, Santorini, Greece, pp. 311–316, April 2016
12. Tan, J., Bi, N., Suen, C.Y., Nobile, N.: Multi-feature selection of handwriting for gender identification using mutual information. In: International Conference on Frontiers in Handwriting Recognition, Shenzhen, China, pp. 578–583, October 2016
13. Mahreen, A., Rasool, A.G., Afzal, H., Siddiqi, I.: Improving handwriting based gender classification using ensemble classifiers. Expert Syst. Appl. **85**(C), 158–168 (2017)
14. Bouadjenek, N., Nemmour, H., Chibani, Y.: Fuzzy integrals for combining multiple SVM and histogram features for writer's gender prediction. IET Biom. **6**(6), 429–437 (2017)
15. Nemmour, H., Chibani, Y.: Fuzzy neural network architecture for change detection in remotely sensed imagery. Int. J. Remote Sens. **27**(4), 705–717 (2006)

Gearbox Fault Diagnosis Based on Mel-Frequency Cepstral Coefficients and Support Vector Machine

Tarak Benkedjouh$^{(\boxtimes)}$, Taha Chettibi$^{(\boxtimes)}$, Yassine Saadouni,
and Mohamed Afroun$^{(\boxtimes)}$

Ecole Militaire Polytechnique, Bordj El-Bahri, Alger, Algérie
bktarek@gmail.com, tahachettibi@yahoo.fr, yassine.saadouni@gmail.com,
mohamed.afroun@gmail.com

Abstract. The enhancement of the machine condition monitoring process is a key issue for reliability improvement. In fact, in order to produce quickly, economically, with high quality while decreasing the risk of production break due to a machine stop, it is necessary to maintain the equipment in a good operational condition. This requirement can be satisfied by implementing appropriate maintenance strategies such as Condition Based Maintenance (CBM) and using updated condition monitoring technologies for faults detection and classification. In this context, a new method for machinery condition monitoring based on Mel-Frequency Cepstral Coefficients (MFCCs) and Support Vector Machine (SVM) is proposed to automatically detect the mechanical faults by maximized the generalization ability. Hence, the purpose is to design an automatic detection system for mechanical components defects based on supervised classification by trained to maximize the margin. The proposed approach consists in a sequence of binary classifications after extracting a set of relevant features such as temporal indicators and MFCC coefficients. The diagnosis accuracy assessment is carried out by conducting various experiments on acceleration signals collected from a rotating machinery under different operating conditions.

Keywords: Diagnostics · Fault detection
Mel-Frequency Cepstral Coefficients · Support Vector Machines
Gearbox

1 Introduction

The ability to forecast machinery failure can help reducing maintenance costs, operation breakdowns and safety risks and is gaining importance in industry since it may limit the loss of production due to a machine stopping [1]. Fault diagnosis can be seen as a problem of pattern recognition for which several

© IFIP International Federation for Information Processing 2018
Published by Springer International Publishing AG 2018. All Rights Reserved
A. Amine et al. (Eds.): CIIA 2018, IFIP AICT 522, pp. 220–231, 2018.
https://doi.org/10.1007/978-3-319-89743-1_20

artificial intelligence methods like hidden Markov Models (HMM) [2]; artificial neural network (ANN) [3] and support vector machines [4] have been applied. A challenging problem in rotating machinery diagnostic is how to construct and evaluate an effective feature sub-space from available features that can accurately represent the fault. Implementation difficulties of rotating machinery diagnostic systems are inherent to the random nature of defect growth by crack propagation in mechanical components, because each feature is effective for a defect at certain stage [5]. Yan et al. [6] provided a review on utilizing wavelets as a powerful tool for signal analysis with the purpose of rotary machines faults diagnosis. Lei et al. [7] provide a review of applying EMD to fault diagnosis of rotating machinery. In the review, all reported applications of EMD in fault diagnosis are divided into a few main aspects based on the key components of rotating machinery, namely, rolling element bearings, gears and rotors. Liu et al. [8] propose a novel fault diagnose method based on short-time matching and SVM to overcome the limitations of traditional sparse representation and fault diagnosis methods.

Condition monitoring based classifier has existed for some time, by using a variety of features, and artificial intelligence-based approaches to distinguish between fault and normal condition. The other problem is mainly associated with selecting a features set to allow the classifier discriminate between the classes without confusion. Nyanteh et al. [9] discusses the faults in rotating machines and describes a fault detection technique using artificial neural network (ANN) which is an expert system to detect short-circuit fault currents in the stator windings of a permanent-magnet synchronous machine (PMSM).

In this paper, we analyze the use of the SVM classifier [10]. This technique used for enhancing mechanical components fault diagnosis has been developed by fusion of multiple feature extraction through support vector machine. Particularly, we investigate how best to select features from the available data in order to maximize the performance of the classifier. Another main challenge for condition monitoring performance prognostics is how to construct and evaluate an effective feature sub-space from available features extraction, which can always represent the degradation state and how the performance of dimensionality reduction (DR) techniques may be improved; various techniques for the data reduction have been proposed [11]. Several features extraction techniques are used in signal recognition systems such linear prediction coefficients (LPC), linear predictive cepstral coefficients (LPCC), perceptual linear predictive analysis (PLP), and Mel-Frequency Spectrum Coefficients (MFCC) which is currently the most popular and it is discussed in this paper.

The main contribution of this paper is to use the MFCC and SVM. This approach is divided in two phases: (i) a features extraction phase by calculating the Mel-frequency cepstral coefficients and (ii) applying the SVM for data classification and visualization phase. The Support vector machine technique has been successfully applied in different applications such as in communication [12], financial time series [13] and biomedicine [14].

This paper is organized as follows. Section 2 presents the description of the proposed method. Section 3 presents the feature extraction based on MFCCs technique. Section 4 describes the proposed method based on support vector machine for classification. Section 5 is dedicated to the experimental verification and results discussion and finally, Sect. 6 concludes the paper.

2   Description of the Proposed Method

Various conditions monitoring research works have been conducted for improving the performance classification. In Fig. 1, the three main steps of a generic condition based maintenance CBM process are indicated; namely: data acquisition, processing and maintenance decision making steps. Data acquisition step is intended to collect the data related to system health. Data processing phase is devoted to analyze the acquired data and finally, in the maintenance decision-making step, effective maintenance policies will be obtained based on information analysis.

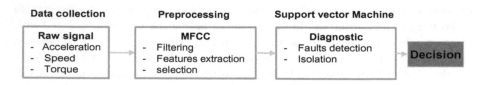

Fig. 1. Steps of a condition monitoring System.

3   Features Extraction Based on MFCC

In signal processing, The feature extraction is very important operation because the large data sets cause difficulties. Feature extraction using the MFCCs is widely known in speaker recognition. The MFCCs are commonly extracted from signals through cepstral analysis. Figure 2 shows the proposed steps of extraction of MFCCs from an raw signal. The input signal must first be broken up into small sections framed and windowed, these sections can be considered as stationary and exhibit stable characteristics. The Fourier transform is then taken and the magnitude of the resulting spectrum is warped by the Mel scale. The log of this spectrum is then taken and the DCT is applied [15].

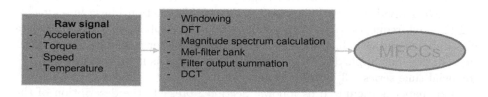

Fig. 2. Extraction of MFCCs from raw signals.

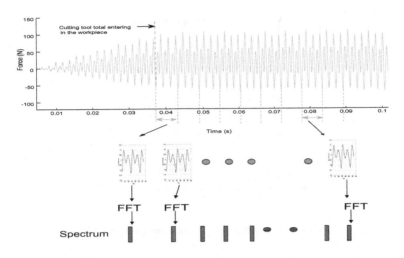

Fig. 3. Sequence of spectral vectors and time duration selection

The Input data is a raw signal in the time domain from different sensors (vibrations, force and acoustic emission) representation with duration in the order of 10 s (Fig. 3).

1. The first processing step is the computation of the frequency domain of (a windowed excerpt of) a signal. This is achieved by computing the Discrete Fourier Transform.
2. The second step is the computation of the mel-frequency spectrum. The powers of the spectrum obtained above onto the mel scale, using triangular overlapping windows.
3. The third step computes the logarithm of the signal; Take the logs of the powers at each of the mel frequencies.
4. The fourth step is to Take the discrete cosine transform of the list of mel log powers, as if it were a signal.
5. The fifth step tries to eliminate the information dependent characteristics by computing the cepstral coefficients. The MFCCs are the amplitudes of the resulting spectrum.

4   Data Classification by SVM

Support vector machine is a powerful technique for data classification [16]. SVM is developed from the optimal separation plane under linearly separable condition. Its basic principle can be illustrated in two-dimensional way as shown in Fig. 4.

Assume that a training set S is given by

$$S = \{x_i, y_i\}_{i=1}^{n},\tag{1}$$

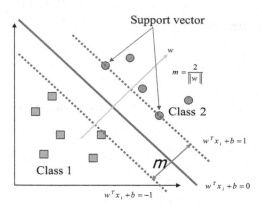

Fig. 4. Classification of data by using SVM

Where $x_i \in R^N$, and $y_i \in \{-1, +1\}$. The goal of SVM is to find an optimal hyperplane such that

$$\begin{cases} w^T x_i + b \geq 1 & for \ y_i = +1, \\ w^T x_i + b \leq 1 & for \ y_i = -1, \end{cases} \tag{2}$$

Where the weight vector $w \in R^N$, and the bias b is a scalar. If the inequality in Eq. 2 holds for all training data, it will be a linearity separable case. Therefore, in the linearly separable case, for finding the optimal hyperplane, one can solve the following constrained optimization problem:

Minimize

$$\Phi(w) = \frac{1}{2} w^T w \tag{3}$$

Subject to

$$y_i(w^T x_i + b) \geq 1 - \xi_i, \ \xi_i \geq 0, \quad i = 1, 2, ..., n. \tag{4}$$

By introducing a set of Lagrange multipliers α_i, β_i for constraints 4, the problem becomes the one of finding the saddle point of the lagrangian. Thus, the dual problem becomes

Minimize

$$Q(\alpha) = \sum_{i=1}^{n} \alpha_i - \frac{1}{2} \sum_{i=1}^{n} \sum_{j=1}^{n} \alpha_i \alpha_j y_i y_j x_i^T x_j \tag{5}$$

Subject to

$$\sum_{i=1}^{n} \alpha_i y_j = 0, \tag{6}$$

$$0 \leq \alpha_i \leq C, \qquad i = 1, 2, ..., n. \tag{7}$$

If $0 \leq \alpha_i \leq C$, the corresponding data points are called support vectors (SVs). SVMs map the input vector into a higher dimensional feature and thus can solve the nonlinear case. By choosing a nonlinear mapping function $\varphi(x) \in R^M$, where $M \succ N$, the SVM can construct an optimal hyperplane in this new feature space. $K(x, x_i)$ is the inner product kernel performing the nonlinear mapping into feature space $K(x, x_i) = K(x_i, x) = \varphi(x)^T \varphi(x_i)$.

$$0 \leq \alpha_i \leq C, \qquad\qquad i = 1, 2, ..., n. \tag{8}$$

Hence, the dual optimization problem becomes

Minimize

$$Q(\alpha) = \sum_{i=1}^{n} \alpha_i - \frac{1}{2} \sum_{i=1}^{n} \sum_{j=1}^{n} \alpha_i \alpha_j y_i y_j K(x_i x_j) \tag{9}$$

Subject to the same constraints as Eqs. 6 and 7, the only requirement on the kernel $K(x, x_i)$ is to satisfy the Mercer's theorem [16]. Using Kernel functions, without treating the high dimensional data explicitly, unseen data are classified as follows:

$$x \in \begin{cases} positive\ class, & if\ g(x) \succ 0, \\ negative\ class, & if\ g(x) \prec 0, \end{cases} \tag{10}$$

Where the decision function is

$$g(x_i) = y_i \left(\sum_{j=1}^{N} y_i \alpha_j K(x_i, x_j) + b \right), \tag{11}$$

The other different functions kernel used are:

Table 1. Different function kernel used

Function	Polynomial	Gaussian	Sigmoid
$k(z, z_i)$	$(z.z_i + 1)^d$	$\exp\left(\frac{z-z_i}{\sigma^2}\right)^2$	$S[v(z.z_i) + c]$

5   Results and Discussion

5.1   Experimental Setup

Figure 5 illustrates the test rig used to accomplish our experience and data collection. The shaft is driven by an electric motor and the rotation speed was variated between 0 and 6000 rpm. A radial load is added to the shaft and bearings. The bearings type MB Manufacturing ER-10K have 8 ball rollers in a single row, the pitch diameter is 33.5 mm, the roller element diameter is 7.93 mm and the contact angleis of 0°. The measured signals consist of two acceleration signals given by an Endevco 6259M31 Accelerometer (10 mv/g, +/−1% error, Resonance $\succ$45 KHz) which is installed in input and output position on the gearbox

Fig. 5. Experimental setup.

housing. The data sampling rate was 66666.67 Samples per Second (200 KHz/3). The gearbox contains three shafts, 4 gears (the number of teeth is 32, 96, 48 and 80) and 6 bearings. The overall objective of the data was to specify the condition of each of the mechanical components and to specify the particular fault if it was not in a healthy state. The detail of the gearbox inside is shown in Fig. 5. A *B&K* high frequency accelerometer was mounted vertically on the housing of the test roller bearing to pick up the vertical acceleration. A filter with a cutoff frequency of 24 KHz was used to filter out the unwanted signals. Signals were then sent to the *B&K* 3560C Signal Analyzer. Readings were directly taken from the digital readout on the analyzer and a graphical representation of the data was displayed on the screen and the data were analyzed.

5.2   Experimental Verification

The diagram of the SVM method proposed for conditions monitoring is given in Fig. 6. The method is decomposed into two main steps. The first step is done off-line and aims at MFCCs generating and classification. When the SVM classifier is trained, the kernel function must be determined by user. The second step, which is achieved on-line, utilizes the trained data to predict the faults.

Figures 7 show the sensor measurements of the healthy and degraded state of the system (Acceleration) respectively.

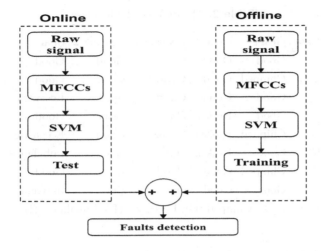

Fig. 6. Framework of the faults detection procedure

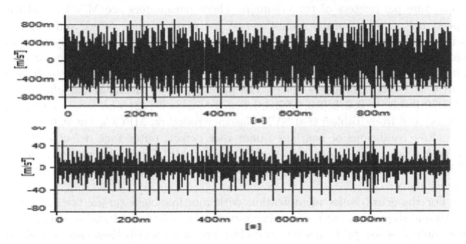

Fig. 7. Acceleration signals measurement of healthy system (top) and bearing defect (bottom) (speed 30 Hz)

We decompose the monitoring signals of each loading data above two conditions with MFCCs method for computing the feature extraction. It is noticed by signal analysis that the defect information of bearings and gears is mainly included in the first three MFCCs components. The above discussion deals with binary classification where the class labels can take only two values: +1 and −1. To find more than two classes in fault diagnosis of rotating machinery there are several fault classes such as bearing faults, gears broken, chipped, misalignment...etc. The different classes used in this paper are shown in Table 2.

Table 2. The different faults class

Case	Gear1	Gear 2	Bearing 1	Bearing 2	Shaft
1	Good	Good	Good	Good	Good
2	Chipped	Good	Good	Good	Good
3	Good	Good	Good	Inner	Eccentric
4	Good	Broken	Good	Inner	Good
5	Chipped	Broken	Outer	Good	Eccentric
6	Good	Broken	Outer	Good	Imbalance
7	Chipped	Good	Good	Inner	Good
8	Good	Good	Outer	Good	Imbalance

Broken (B); Chipped (C); Eccentric (E); Imbalance (I)

The total 13 features (16 signals for input and output) are calculated from 13 feature parameters of time domain. These parameters are MFCCs and the speed motor. The normal conditions of the system as $y = -1$ and the one with the defect as $y = +1$. The decision function $f(x)$ obtained by the linear kernel function and according to Eqs. (3) and (6) the parameters of classifier SVM, $\alpha = [0.0030, 0, 0.0056, 0, 0, 0, 0, 0, 0.0126, 0, 0, 0, 0]^T$, $\omega = 0.1628$ and $b = 2.4856$. For gears defect, the parameters of the SVM classifier, $\alpha = [0.0070, 0, 0.0028, 0, 0, 0, 0, 0, 0.0223, 0, 0, 0, 0]^T$, $\omega = 0.1421$ and $b = -3.4291$. It can be seen from Table 5 that SVM classifier based on MFCCs can still classify the three conditions of bearings (inner race defect, outer race defect and ball defect) which confirm fully that the SVM based MFCCs can be applied successfully to the faults recognition even in cases where only limited training samples are available.

For the gears faults identification with multiple-class (crack teeth, broken teeth and shipped ... etc.), generalizing method can be introduced to decompose the multiple-class problems into two-class problems which then can be trained with SVM.

In general, vibration signals of healthy bearings are Gaussian in distribution. The value of speed and load, therefore the value of the kurtosis is close to three for the vibration signals of a healthy system.

To select the optimal feature MFCCs that can well represent the condition of rotating machinery, a feature selection method based on the performance classification is shown in Tables 3 and 4.

Table 3. Motor speed influence for the classification

MFCCs+ speed			MFCCs		
Broken	Chipped	Eccentric	Broken	Chipped	Eccentric
100	99.285	100	100	97.857	100

Table 4. Window size influence for the classification

Window (ms)	MFCCs+ speed			Kurtosis+ speed		
	B	C	E	B	C	E
40	100	94.28	100	72.85	78.57	78.57
60	100	96.42	100	69.28	78.57	79.28
100	100	97.14	100	67.85	78.57	81.42
120	100	98.51	100	67.85	78.57	81.42
140	100	99.28	100	67.14	78.57	80.71

Table 5. Kernel used for classification

Kernel RBF $\sigma =$	Window (ms)	MFCCs+ speed			MFCCs+ RMS+ speed		
		B	C	E	B	C	E
(0.007)	100	100	100	99.285	100	100	99.285
(0.008)	120	100	100	99.285	100	100	99.285
(0.011)	140	100	100	99.285	100	100	99.285

The results shown in Table 3 compare the classification rate when including the motor speed as features with MFCCs. The classification ratio increases with the different kinds of faults. Note that the duration time of windowing equal to ($w = 140$ ms) and the kernel is RBF with ($\sigma = 0.002$).

In Table 4, classification process by SVM performed on the original feature (MFCCs) added the motor speed and compared with the fourth moment order (Kurtosis). The classification ratio of this process among 67.14% until 100%. The bad performance of this classification is due to the existence of irrelevant and useless features such as kurtosis.

Table 4 compares the classification rates for different windows size with different features used in this study by using the fourth moment order and the speed motor compared with MFCCs and speed. In this study, the RBF kernel are used as the basic kernel function of SVMs. The goal of this guideline is to identify optimal choice of the kernel parameter that the classifier can accurately classify the data input with a good classification rates.

In the specialized literature, no method is available for choosing the best kernel function. The most appropriate kernel function and the values of kernel function parameters (σ) for RBF. The selection of RBF kernel width is one of the major problems in SVMs for good performance of classification. For choosing the optimum values of the parameters (σ) of the RBF kernel, a large number of studies has been carried out by varying the values of parameters.

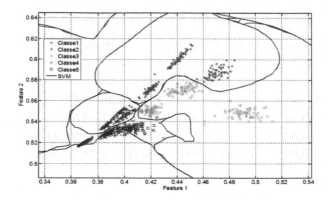

Fig. 8. Faults classification using RBF kernel

Table 5 compares the classification rates for different kernel function shown in Table 1. The Radial basis function (RBF) kernel gives a good classification results with a small number of the support vectors and learning time. The experiments are performed on three data sets with 60(%) training samples and 60(%) test samples (Fig. 8).

It is worth noting that the Gaussian kernel is the only kernel function used in our experiments. In fact, on each dataset we perform search for optimal combination of kernel width and the number of principal components for transformation. To speed up the search, we discard any eigenvector whose corresponding eigen value is smaller than 10^4. To achieve this, the SVM based on MFCCs is proposed; as it is a very powerful tool that can determine a good classification of the system.

6   Conclusion

In this paper, we applied the combination of MFCCs and SVMs for intelligent fault diagnosis of rotating machinery. MFCCs were successfully applied for feature extraction step. However, the training feature using SVM is better than the other features such as kurtosis and the root mean square of signals. The feature extraction is an important step in fault diagnosis process. The proposed method were developed based on the acceleration signals measurements. In this paper; the potential of MFCCs-SVM has been highlighted for classification. Particularly, the simulation results of SVM classifier have verified that the proposed method has good efficiency in classifying eight types of defect with different characteristics. SVMs based MFCCs for multi-class classification is applied to the faults classification. The results show that SVMs achieved high performance in using multi-class classification strategy for one-against-all.

Acknowledgement. This research work is partially supported by the ATRST "Agence Thématique de Recherche en Science et Technologie" (Algeria). (Project code: 129/2016/P8/LMS).

References

1. Zio, E.: An Introduction to the Basics of Reliability and Risk Analysis, vol. 13. World Scientific, Singapore (2007)
2. Geramifard, O., Xu, J.-X., Panda, S.K.: Fault detection and diagnosis in synchronous motors using hidden markov model-based semi-nonparametric approach. Eng. Appl. Artif. Intell. **26**(8), 1919–1929 (2013)
3. Janssens, O., Slavkovikj, V., Vervisch, B., Stockman, K., Loccufier, M., Verstockt, S., Van de Walle, R., Van Hoecke, S.: Convolutional neural network based fault detection for rotating machinery. J. Sound Vib. **377**, 331–345 (2016)
4. Jedliński, Ł., Jonak, J.: Early fault detection in gearboxes based on support vector machines and multilayer perceptron with a continuous wavelet transform. Appl. Soft Comput. **30**, 636–641 (2015)
5. Su, Z., Tang, B., Liu, Z., Qin, Y.: Multi-fault diagnosis for rotating machinery based on orthogonal supervised linear local tangent space alignment and least square support vector machine. Neurocomputing **157**, 208–222 (2015)
6. Yan, R., Gao, R.X., Chen, X.: Wavelets for fault diagnosis of rotary machines: a review with applications. Sig. Process. **96**, 1–15 (2014)
7. Lei, Y., Lin, J., He, Z., Zuo, M.J.: A review on empirical mode decomposition in fault diagnosis of rotating machinery. Mech. Syst. Signal Process. **35**(1), 108–126 (2013)
8. Liu, R., Yang, B., Zhang, X., Wang, S., Chen, X.: Time-frequency atoms-driven support vector machine method for bearings incipient fault diagnosis. Mech. Syst. Signal Process. **75**, 345–370 (2016)
9. Nyanteh, Y., Edrington, C., Srivastava, S., Cartes, D.: Application of artificial intelligence to real-time fault detection in permanent-magnet synchronous machines. IEEE Trans. Ind. Appl. **49**(3), 1205–1214 (2013)
10. Saimurugan, M., Ramachandran, K.: A comparative study of sound and vibration signals in detection of rotating machine faults using support vector machine and independent component analysis. Int. J. Data Anal. Techn. Strat. **6**(2), 188–204 (2014)
11. Van der Maaten, L., Postma, E., Van Den Herik, H.: Dimensionality reduction: a comparative review. J. Mach. Learn. Res. **10**, 1–41 (2009)
12. Qian, Z.-L., Juan, D.-C., Bogdan, P., Tsui, C.-Y., Marculescu, D., Marculescu, R.: A support vector regression (SVR)-based latency model for network-on-chip (NoC) architectures. IEEE Trans. Comput. Aided Des. Integr. Circuits Syst. **35**(3), 471–484 (2016)
13. Law, T., Shawe-Taylor, J.: Practical Bayesian support vector regression for financial time series prediction and market condition change detection. Quant. Financ. **17**(1), 1–14 (2017)
14. Du, W., Cheung, H., Johnson, C.A., Goldberg, I., Thambisetty, M., Becker, K.: A longitudinal support vector regression for prediction of ALS score. In: 2015 IEEE International Conference on Bioinformatics and Biomedicine (BIBM), pp. 1586–1590. IEEE (2015)
15. Kinnunen, T., Li, H.: An overview of text-independent speaker recognition: from features to supervectors. Speech Commun. **52**(1), 12–40 (2010)
16. Vapnik, V.: The Nature of Statistical Learning Theory. Springer, New York (1995)

A Modified Firefly Algorithm with Support Vector Machine for Medical Data Classification

Brahim Sahmadi[1]([⊠]), Dalila Boughaci[2], Rekia Rahmani[1],
and Noura Sissani[1]

[1] LMP2M Laboratory, University Yahia Fares of Medea, Medea, Algeria
sahmadi.brahim@univ-medea.dz, sh_brahim@yahoo.fr,
rekia_rah@yahoo.fr, infores287@gmail.com
[2] LRIA/Computer Sciences Department, University of Sciences and Technology
Houari Boumediene (USTHB), Algiers, Algeria
dalila_info@yahoo.fr, dboughaci@usthb.dz

Abstract. Clinical information systems store a large amount of data in medical databases. In the use of medical dataset for diagnosis, the patient's information is selectively collected and interpreted based on previous knowledge for detecting the existence of disorders. Feature selection is important and necessary data pre-processing step in medical data classification process. In this work, we propose a wrapper method for feature subset selection based on a binary version of the Firefly Algorithm combined with the SVM classifier, which tries to reduce the initial size of medical data and to select a set of relevant features for enhance the classification accuracy of SVM. The proposed method is evaluated on some medical dataset and compared with some well-known classifiers. The computational experiments show that the proposed method with optimized SVM parameters provides competitive results and finds high quality solutions.

Keywords: Medical data classification · Machine learning · Feature selection
Binary firefly algorithm · Support vector machine (SVM) · Cross-validation

1 Introduction

Clinical information systems store a large amount of information in medical databases. So, the manual classification of this information is becoming more and more difficult. Therefore, there is an increasing interest in developing automated evaluation methods to follow up the diseases. Classification is one of the techniques of data mining which involves extracting a general rule or classification procedure from a set of learning examples. Medical data classification refers to learning classification models from medical datasets and aims to improve the quality of health care [3].

As medical datasets are generally characterized as having high dimensionality, and many of the feature attributes in a typical medical dataset are collected for reasons other than data classification. Some of the features are redundant while others are irrelevant adding more noise to the dataset, although in medical diagnosis, it is desirable to select

© IFIP International Federation for Information Processing 2018
Published by Springer International Publishing AG 2018. All Rights Reserved
A. Amine et al. (Eds.): CIIA 2018, IFIPAICT 522, pp. 232–243, 2018.
https://doi.org/10.1007/978-3-319-89743-1_21

the clinical tests that have the least cost and risk and that are significantly important in determining the class of the disease [1].

Feature selection is important and necessary data pre-processing steps to increase the quality of the feature space. It aims to select a small subset of important (relevant) features from the original full feature set. It can potentially improve the performance of a learning algorithm significantly in terms of the accuracy; increase the learning speed, and simplifying the interpretation of the learnt models [2, 12]. Feature selection is used in different tasks of learning or data mining, in the fields of image processing, pattern recognition, data analysis in bioinformatics, categorization of texts, etc.

The methods used to evaluate a feature subset in the selection algorithms can be classified into three main approaches: filter methods, wrapper methods and embedded methods. Filter methods perform the evaluation independently of any classification algorithm; they are based on data and attributes [2]. Wrapper methods use the learning algorithm as an evaluation function. It therefore defines the relevance of the attributes through a prediction of the performance of the final system. Embedded methods combine the exploration process with a learning algorithm. The difference with wrapper methods is that the classifier not only serves to evaluate a candidate sub-set, but also to guide the selection mechanism.

Wrapper approaches conducts a search in the space of candidate subsets of features, and the quality of a candidate subset is evaluated by the performance of the classification algorithm trained on this subset [15]. Several wrapper methods have been proposed for feature selection, among them, the stochastic local search methods and the population based optimization metaheuristic methods, like the Genetic Algorithms (GA), the Memetic Algorithm (MA), Particle Swarm Optimization (PSO), and the Harmony Search Algorithm (HAS) [16].

In this work, we apply a wrapper method based on a binary version of Firefly Algorithm to the feature selection problem in medical data classification, in order to extract an ideally minimal subset of features with strong discriminative power. The proposed approach uses the SVM classifier for evaluating a feature subset.

This paper is organized as follows: first, we briefly outline the main idea of support vector machine methods (SVM) and the Binary Firefly Algorithm in Sect. 2. In Sect. 3, we describe the proposed approach for feature selection and classification of medical data. The experimental results are presented and discussed in Sect. 4. Finally, we conclude this study and discuss possible future work in Sect. 5.

2 Background

2.1 Support Vector Machines

Support Vector Machines (SVM) are a class of supervised learning algorithms introduced by Vladimir Vapnik [4]. The main principle of SVM is the construction of a function f called decision function that for an input vector x matches a value y, $y = f(x)$, where x is the example to classify and y is the class which corresponds to the example input. SVM are originally defined for binary classification problems, and their extension to nonlinear problems is offered introducing the kernel functions. SVM are widely used in statistical learning and has proved effective in many application areas such as image processing, speech processing, bioinformatics, natural language processing, and even data sets of very large dimensions [13].

SVM classifiers are based on two key ideas: the notion of maximum margin and the concept of kernel function. The first key idea is the concept of maximum margin. We seek the hyperplane that separates the positive examples of negative examples, ensuring that the distance between the separation boundary and the nearest samples (margin) is maximal, they are called support vector. And as it seeks to maximize the margin, we will talk about wide margin separators [14].

The second key idea in SVM is the concept of kernel function. This is transforming the data space entries in a space of larger dimension called feature space in which it is likely that there is a dividing line, in order to deal with cases where the data are not linearly separable. Some examples of kernel functions are:

- Linear kernel: $K(x_i, x_j) = x_i . x_j$
- Polynomial kernel: $K(x_i, x_j) = (\gamma x_i . x_j + r)^d$, $\gamma > 0$.
- RBF kernel: $K(x_i, x_j) = e^{-\frac{|x_i - x_j|^2}{2\gamma^2}}$, $\gamma > 0$.
- Sigmoid kernel: $K(x_i, x_j) = tanh(\gamma x_i . x_j + r)$.

Where γ, r and d are kernel parameters. In this study, we utilized the LIBSVM toolset and chose Radial Basis Function (RBF) as the kernel function, and its C and γ parameters are optimized using an iterative search method. Previous studies show that these two parameters play an important role on the success of SVMs [8].

2.2 Binary Firefly Algorithm

The Firefly algorithm is a recent bio-inspired metaheuristic developed by Xin She Yang in 2008 and it has become an important tool for solving the hardest optimization problems in almost all areas of optimization [6]. The algorithm is based on the principle of attraction between fireflies and simulates the behavior of a swarm of fireflies in nature, which gives it many similarities with other meta-heuristics based on the collective intelligence, such as the PSO (Particle Swarm Optimization) algorithm or the bee colony optimization algorithm. He uses the following three idealized rules:

- All fireflies are unisex, meaning that one firefly is attracted by another, regardless of sex.
- The attractiveness and brightness are proportional, so that for two flashing fireflies, the less bright will move towards the brighter. Attractiveness and brightness decrease with increasing distance. If there is not one firefly brighter than the other, they will place themselves randomly.
- The brightness of a firefly is determined by the point of view of the objective function to optimize. For a maximization problem, the luminosity is simply proportional to the value of the objective function.

Since the attractiveness is proportional to the luminosity of the adjacent fireflies, then the variation in the attractiveness β with the distance r is defined by:

$$\beta = \beta_0 \, e^{-\gamma r2} \tag{1}$$

Where β_0 is the attractiveness at $r = 0$ and γ is the absorption coefficient. The distance r_{ij} between 2 fireflies is determined by the formula (2).

$$r_{ij} = X_i - X_j = \sqrt{\sum_{k=1}^{d} \left(X_i^k - X_j^k \right)^2} \tag{2}$$

Where X_i^k is the kth component of the spatial coordinate of the ith firefly and d is the number of dimensions.

The movement of a firefly X_i to another firefly X_j more attractive is calculated by:

$$X_i^{t+1} = X_i^t + \beta_0 e^{-\gamma r_{ij}^2} \left(X_j^t - X_i^t \right) + \alpha \left(rand - \frac{1}{2} \right) \tag{3}$$

Where X_i^t and X_j^t are the current position of the fireflies X_i and X_j, and X_i^{t+1} is the ith firefly position of the next generation. The second term is due to attraction. The third term introduces randomization, with α being the randomization parameter.

The basic steps of the firefly algorithm can be formulated as the pseudo code shown in the Algorithm 1.

The original firefly algorithm is designed for optimization problems with continuous variables. Recently, several binary firefly algorithms were developed to solve discrete problems, such as scheduling, timetabling and combination. Compared with the original firefly algorithm, binary firefly algorithm obeyed similar fundamental principles while redefined distance, attractiveness, or movement of the firefly. In this study, we use a binary firefly algorithm for feature selection with new definitions of distance and movement of a firefly, similar to the approach used in [10].

3   The Proposed Method for Feature Selection and Classification

The feature selection task is a typical combination problem in essence, with the objective of selecting an optimal combination of features from a given feature space. Theoretically, for an n-dimensional feature space, there will be 2^n possible solutions (NP-hard problem). The use of metaheuristics methods as random selection algorithms, capable of effectively exploring large search spaces, which is usually required in case of feature selection. In this work, a binary firefly algorithm (BFA-SVM) is proposed; where the feature space is explored by a population of fireflies and the SVM classifier is used for evaluating a feature subset. The normalized Hamming distance was used to calculate attractiveness between a pair of fireflies and in order to increase the diversity

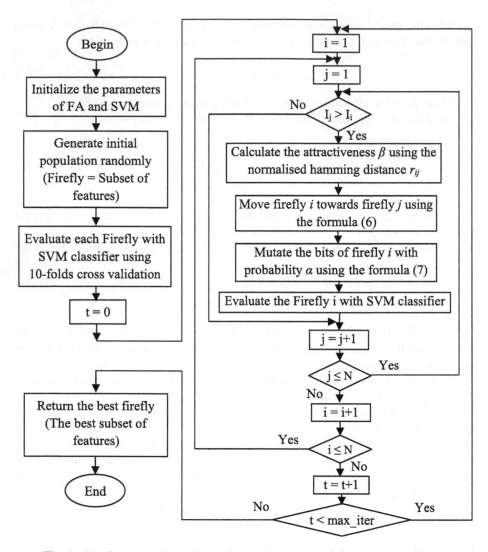

Fig. 1. The flowchart of the proposed binary firefly algorithm for feature selection

of fireflies a dynamic mutation operator was introduced. The flowchart of BFA-SVM method is shown in Fig. 1.

3.1 Fireflies Representation and Initialization

To represent the subset of selected features, we chose a binary representation of a solution in the multidimensional search space. Every firefly x_i in the binary firefly algorithm represents a subset of the feature space (i.e. a possible solution for feature selection problem) as an N-dimensional binary array: bit values 1 and 0 represent a selected and unselected feature, respectively. The Initial population of fireflies is generated randomly; the bit positions for each firefly are randomly assigned as 1 or 0.

Algorithm 1. Pseudo code of the Firefly algorithm

Inputs : *n (number of fireflies)*, **max_iter** *(maximum number of iterations)*, *α (randomness parameter)*, *γ (absorption coefficient)*.

1. Define objective function $f(x)$, $x = (x_1, ..., x_d)^T$; // *d is positions dimension*
2. Initialize parameters *n, max_iter, α* and *γ*;
3. Initialize a population of fireflies x_i *(i = 1, 2, ..., n)*;
4. Calculate the light intensity I_i (fitness) for each x_i by $f(x_i)$;
5. Set t ← 0;
6. **while** (*t <max_iter*) **do**
   6.1. **for** i = 1 **to** n // all n fireflies
      for j = 1 **to** n // all n fireflies
         If ($I_j > I_i$)
         - Move firefly *i* towards *j* in all *d* dimensions;
         - Attractiveness varies with distance r_{ij} via $e^{-\gamma r_{ij}^2}$;
         - Evaluate new solutions and update light intensity;
         end if;
      end for j;
   end for i;
   6.2. Rank the fireflies and find the current best;
   6.3. Set t ← t+1;
   end while;
7. Return the global best firefly (i.e. the best solution);

3.2   Objective Function

The objective function of the BFA-SVM algorithm when searching for the optimal features subset is to maximize the accuracy rate in classifying the testing dataset. This is equivalent to an optimization problem seeking for a maximum solution. The classification rate *ACC* is calculated using cross-validation with 10-Folds [11]. This measure is calculated by the formula (4):

$$ACC = \left(\frac{Totalcorrect}{L}\right) * 100 \qquad (4)$$

Where *Total correct* is the number of examples correctly classified by the SVM classifier, and *L* is the total number of examples. The classification rate indicates whether the candidate subset permits good class discrimination.

3.3   The Attractiveness of Fireflies

For two fireflies X_i and X_j, the distance r_{ij} is defined based on the similarity ratio of the two fireflies using the normalized Hamming distance of the two position vectors as follow:

$$r_{ij} = 1 - \left(\sum_{k=1}^{d} \left(X_i^k \oplus X_j^k \right) \right) / d \tag{5}$$

Where $\oplus$ denotes the XOR operation and d is the positions dimension. The attractiveness β between a pair of fireflies is calculated using the formula (1).

3.4   The Movement of Fireflies

The original firefly algorithm is designed for optimization problems with continuous variables. For the binarization of continuous metaheuristics, there are two main groups of binarization techniques. The first group of techniques allows working with the continuous metaheuristics without operator modifications and includes steps of binarization of the continuous solution after the original continuous iteration. The second group of techniques is called continuous-binary operator transformation; it redefines the algebra of the search space, thereby reformulating the operators [18]. In this work, we use a modification in the movement of a firefly by the reformulation of the formula (2). When a firefly X_i moves to another firefly X_j more attractive, every bit in its representation vector will make a decision to change its value or not. Changing a bit X_i^k in firefly X_i is done in two steps: the β-step (attraction) as indicated in the formula (6), which is regulated by the attractiveness β, and the α-step (mutation) were using the formula (7), which is controlled by a parameter α.

$$X_i^k = \begin{cases} X_j^k & \text{if } X_i^k \neq X_j^k \text{ and } rand(0.1) < \beta \\ X_i^k & \text{otherwise} \end{cases} \tag{6}$$

$$X_i^k = \begin{cases} 1 - X_j^k & \text{if } rand(0.1) < \alpha \\ X_i^k & \text{otherwise} \end{cases} \tag{7}$$

β is the probability of a hetero-bit in the moving firefly changes to the corresponding bit in the brighter firefly ($0 \rightarrow 1$ or $1 \rightarrow 0$). The parameter α regulates the random moving behavior (mutation) of a bit X_i^k, and it is calculated in each iteration of the BFA-SVM algorithm by the following formula:

$$\alpha = 1 - \frac{0.5 * iter}{max\_iter} \tag{8}$$

The mutation probability α is high in initial iterations, which makes BFA-SVM focus on exploration. As the number of iteration increases, the mutation probability will decrease, and BFA-SVM will accelerate its converging pace gradually.

4   Experiments

The proposed BFA-SVM algorithm was implemented on a PC with an Intel Core 2 Duo CPU 2.93 GHz, 4 GB of memory and the Windows 7 operating system. The programs are coded in Java language and we have used the LIBSVM package [5] as a library for the SVMs.

4.1   Dataset

To evaluate the performance of the proposed method, we have used 11 medical datasets obtained from the UCI Machine Learning Repository [17]. Table 1 describes the main characteristics of these datasets. The prediction process with the SVMs requires that the dataset must be normalized. The main advantages of such operation are to avoid attributes in greater numeric ranges dominating those in smaller numeric ranges, and to avoid numerical difficulties during the computation step. The range of each feature value is linearly scaled to the range $[-1, +1]$ using the WEKA tools [7].

Table 1. The dataset description

Dataset	Number of features	Number of instances	Number of classes
Arrhythmia	279	452	16
Breast cancer	10	683	2
Colon cancer	2000	62	2
Dermatology	34	366	6
Diabetes	8	768	2
Heart-c	13	303	5
Heart-stat	13	270	2
Hepatitis	19	155	2
Liver-disorders	6	345	2
Lung cancer	56	32	3
Lymphography	18	148	4

4.2   Parameter Settings

The parameter values of the proposed algorithm are fixed by an experimental study. After a series of experiments, the different parameters are fixed empirically. The values of each parameter for the proposed method are given in Table 2.

4.3   Numerical Results

Due to the non-deterministic nature of the proposed method, several executions (20) were considered for each dataset. The minimum value, the maximum value, and the average of the accuracy rate of the classification for each dataset are reported. The best results are in bold font.

Table 2. Parameters of BFA-SVM algorithm

Parameters	Values
Population size	30
Maximum number of generations	200
β_0	1.0
γ	1.0
Number of folds in cross-validation	10
Number of runs	20

Table 3 gives a comparison between the results (mean of accuracy rate) obtained by the use of SVM classifier with default parameters for RBF kernel function, the use of SVM with optimized parameters given with a grid-search method, and the results obtained with by BFA-SVM with optimized parameters. The best results are obtained with BFA-SVM algorithm for all datasets, confirming that the feature selection and optimization of SVM parameters improves significantly the classification accuracy.

Table 3. Comparison between SVM $_{default}$, SVM $_{Grid-search}$ and BFA-SVM $_{Optimized}$

Dataset	SVM$_{default}$ mean (%)	SVM$_{Optimized}$ mean (%)	BFA-SVM$_{Optimized}$ mean (%)
Arrhythmia	55.93	71.99	**75.84**
Breast cancer	97.01	97.12	**97.66**
Colon cancer	80.89	87.34	**91.94**
Dermatology	97.95	97.99	**99.36**
Diabetes	77.14	77.39	**79.23**
Heart-c	82.67	83.12	**85.07**
Heart-stat	82.44	83.94	**86.56**
Hepatitis	85.06	86.06	**92.46**
Liver-disorders	58.41	73.59	**75.70**
Lung cancer	73.91	77.50	**99.53**
Lymphography	83.04	84.56	**89.66**

– The SVM with optimized parameters is used as classifier with the algorithm BFA-SVM
– Method did not use the dataset for test.
– The best results are in bold.

In order to evaluate the effectiveness of the proposed algorithm BFA-SVM, a comparison of the experimental results obtained by the method with the results of the works cited in [8, 9] is presented in the Table 4, where gives the average (Mean), the best (Max), the worst (Min) values of the classification accuracy, and the standard deviation (Sd) obtained by different methods. In [9], a hybrid search method based on both harmony search algorithm and stochastic local search, combined with a support vector machine (HAS + SVM) is given for feature selection in data classification.

Table 4. Comparison between BFA-SVM Optimized, HAS + SVM, MA + SVM and GA + SVM

Dataset	BFA-SVM Optimized				HSA+SVM				MA+SVM				GA+SVM			
	Mean (%)	Max (%)	Min (%)	Sd (%)	Mean (%)	Max (%)	Min (%)	Sd (%)	Mean (%)	Max (%)	Min (%)	Sd (%)	Mean (%)	Max (%)	Min (%)	Sd (%)
Arrhythmia	**75.84**	**76.77**	**74.78**	0.57	69.15	71.24	66.59	1.28	–	–	–	–	–	–	–	–
Breast cancer	**97.66**	**97.80**	**97.51**	0.05	97.44	97.65	97.22	0.10	97.44	**97.80**	97.49	0.20	97.20	97.36	96.63	0.14
Colon cancer	**91.94**	**91.94**	**91.94**	0.00	89.94	90.32	88.71	0.69	–	–	–	–	–	–	–	–
Dermatology	99.36	**100.00**	98.91	0.41	**100.00**	**100.00**	**100.00**	0.00	99.86	**100.00**	98.90	0.02	98.69	98.90	98.63	0.08
Diabetes	**79.23**	**79.56**	**78.78**	0.17	77.79	78.13	77.47	0.17	77.71	78.26	77.34	0.21	77.53	77.99	76.82	0.27
Heart-c	**85.07**	**85.81**	**84.16**	0.48	59.80	61.05	58.41	0.49	57.88	63.38	56.44	1.11	59.30	60.73	58.08	0.63
Heart-stat	**86.56**	**87.04**	**86.30**	0.27	84.64	85.56	83.70	0.57	84.56	86.30	83.70	0.57	84.11	85.18	83.70	0.35
Hepatitis	92.46	93.55	90.97	0.71	**99.94**	**100.00**	**99.35**	0.19	99.66	100.00	98.06	0.45	88.83	90.00	86.45	0.89
Liver-disorders	75.70	76.23	75.36	0.22	72.60	73.33	71.30	0.42	71.75	73.04	70.43	0.58	63.49	64.35	62.60	0.55
Lung cancer	99.53	**100.00**	96.88	1.14	98.75	**100.00**	93.75	2.26	**99.96**	**100.00**	**99.86**	0.18	88.22	90.62	87.50	0.62
Lymphography	**89.66**	**91.22**	**88.51**	0.93	49.00	55.41	43.92	3.09	48.73	54.70	46.62	1.82	50.83	52.02	49.32	0.56
Average	88.46	**89.08**	**87.65**	**0.45**	81.73	82.97	80.04	0.84	81.95	83.72	80.98	0.57	78.69	79.68	77.75	0.47

- The SVM with optimized parameters is used as classifier with the algorithm BFA-SVM.
- Method did not use the dataset for test.
- The best results are in bold.

And the authors of [8] propose a genetic algorithm (GA) and memetic algorithm (MA) with SVM classifier for feature selection and classification.

As shown in Table 4, BFA-SVM algorithm succeeds in finding the best results for almost the checked datasets compared to HAS + SVM, MA + SVM and GA + SVM methods in term of classification accuracy point of view (in 8 datasets among 11, the BFA-SVM algorithm gives the best classification rate average and for the max value of the classification accuracy is reached in 10 datasets among 11). The small standard deviations of the classification accuracies presented show the consistency of the proposed algorithm. This proves the ability of the proposed algorithm as a good classifier in medical data diagnosis.

5 Conclusion

Health care systems generates vast amount of information and it is accumulated in medical databases, and the manual classification of this data becoming more and more difficult. Therefore, there is an increasing interest in developing automated methods for medical data analysis. In this work we proposed a wrapper method for feature selection and classification of medical dataset based on a Binary firefly algorithm combined with the SVM classifier. The results obtained from tests carried out on several public medical dataset indicate that the proposed BFA-SVM method is competitive with other meta-heuristics (Genetic algorithm, Memetic algorithm and Harmony search algorithm) for the feature selection, and experiments have shown us that the method greatly improves the learning quality and ensures the stability of the generated prediction model. It also reduces the size of the representation space by eliminating noise and redundancy.

As a continuation of this work, it would be desirable to work on the reduction of the computation time by proposing a parallel implementation of the proposed method. It is also possible to use the binary firefly algorithm with other classification algorithms such as neural networks and Naïve Bayes.

Acknowledgments. The authors would like to thank the developers of the Library for Support Vector Machines (LIBSVM) and the developers of Waikato Environment for Knowledge Analysis (WEKA) for the provision of the open source code.

References

1. Almuhaideb, S., El Bachir Menai, M.: Hybrid metaheuristics for medical data classification. In: El-Ghazali, T. (ed.) Hybrid Metaheuristics. Studies in Computational Intelligence, vol. 434, pp. 187–217. Springer, Heidelberg (2013). https://doi.org/10.1007/978-3-642-30671-6_7
2. Liu, H., Yu, L.: Toward integrating feature selection algorithms for classification and clustering. IEEE Trans. Knowl. Data Eng. **17**(4), 491–502 (2005)
3. Almuhaideb, S., El-Bachir Menai, M.: Impact of preprocessing on medical data classification. Front. Comput. Sci. **10**(6), 1082–1102 (2016). https://doi.org/10.1007/s11704-016-5203-5aydin

4. Vapnik, V.: The Natural of Statistical Learning Theory. Springer, New York (1995). https://doi.org/10.1007/978-1-4757-2440-0
5. Chang, C.C., Lin, C.J.: LIBSVM: a library for support vector machines. ACM Trans. Intell. Syst. Technol., **2**(3), Article 27 (2011). http://www.csie.ntu.edu.tw/~cjlin/libsvm/. Accessed 11 Oct 2017
6. Yang, X.S.: Nature-Inspired Metaheuristic Algorithms. Luniver Press, UK (2008)
7. Eibe, F., Mark, A.H., Witten, I.H.: The WEKA Workbench. In: Online Appendix for Data Mining: Practical Machine Learning Tools and Techniques, 4th edn. Morgan Kaufmann (2016)
8. Nekkaa, M., Boughaci, D.: A memetic algorithm with support vector machine for feature selection and classification. Memetic Comput. **7**, 59–73 (2015). https://doi.org/10.1007/s12293-015-0153-2
9. Nekkaa, M., Boughaci, D.: Hybrid harmony search combined with stochastic local search for feature selection. Neural Process. Lett. **44**, 199–220 (2015). https://doi.org/10.1007/s11063-015-9450-5
10. Zhang, J., Gao, B., Chai, H., Ma, Z., Yang, G.: Identification of DNA-binding proteins using multi-features fusion and binary firefly optimization algorithm. BMC Bioinf. **17**, 323 (2016). https://doi.org/10.1186/s12859-016-1201-8
11. Han, J., Kamber, M.: Data Mining Concepts and Techniques, 2nd edn. Morgan Kaufmann, San Francisco (2006)
12. Huerta, E.B., Duval, B., Hao, J.-K.: A hybrid GA/SVM approach for gene selection and classification of microarray data. In: Rothlauf, F., Branke, J., Cagnoni, S., Costa, E., Cotta, C., Drechsler, R., Lutton, E., Machado, P., Moore, Jason H., Romero, J., Smith, George D., Squillero, G., Takagi, H. (eds.) EvoWorkshops 2006. LNCS, vol. 3907, pp. 34–44. Springer, Heidelberg (2006). https://doi.org/10.1007/11732242_4
13. Kecman, V.: Learning and Soft Computing: Support Vector Machines, Neural Networks and Fuzzy Logic Models. The MIT press, London (2001)
14. Burgers, C.J.C.: A tutorial on support vector machines for pattern recognition. Data Min. Knowl. Discov. **2**, 121–167 (1998)
15. Duval, B., Hao, J.K.: Advances in metaheuristics for gene selection and classification of microarray data. Brief. Bioinf. **11**(1), 127–141 (2009)
16. El Aboudi, N., Benhlima, l.: Review on wrapper feature selection approaches. In: International Conference on Engineering and MIS (ICEMIS), pp. 1–5 (2016)
17. Lichman, M.: UCI machine learning repository. University of California, School of Information and Computer Science, Irvine (2013). http://archive.ics.uci.edu/ml
18. Crawford, B., Soto, R., Astorga, G., García, J., Castro, C., Paredes, F.: Putting continuous metaheuristics to work in binary search spaces. Complexity **2017**, 19 (2017). https://doi.org/10.1155/2017/8404231

Ensemble Learning for Large Scale
Virtual Screening on Apache Spark

Karima Sid$^{(\boxtimes)}$ and Mohamed Batouche

Computer Science Department,
University of Constantine 2 – Abdelhamid Mehri, Constantine, Algeria
sidk.karima@gmail.com,
mohamed.batouche@univ-constantine2.dz

Abstract. Virtual screening (VS) is an *in-silico* tool for drug discovery that aims to identify the candidate drugs through computational techniques by screening large libraries of small molecules. Various ligand and structure-based virtual screening approaches have been proposed in the last decades. Machine learning (ML) techniques have been widely applied in drug discovery and development process, predominantly in ligand based virtual screening approaches. Ensemble learning is a very common paradigm in ML field, where many models are trained on the same problem's data, to combine in the end the results in one improved prediction. Applying VS to massive molecular libraries (Big Data) is computationally intensive; so the split of these data to chunks to parallelize and distribute the task became necessary. For many years, MapReduce has been successfully applied on clusters to solve the problems with very large datasets, but with some limitations. Apache Spark is an open source framework for Big Data processing, which overcomes the shortcomings of MapReduce. In this paper, we propose a new approach based on ensemble learning paradigm in Apache Spark to improve in terms of execution time and precision the large-scale virtual screening. We generate a new training dataset to evaluate our approach. The experimental results show a good predictive performance up to 92% precision with an acceptable execution time.

Keywords: Virtual screening · Big data · Apache Spark · Machine learning
Ensemble learning

1 Introduction

The discovering of new drug is a very expensive and long process. High Throughput Screening (HTS) is a widely used experimental tool in the drug discovery process, where large molecular libraries are screened in fully automated environments [1]. However, with the very fast increase in the size of these libraries, HTS will be expensive and provides a small number of hits with a high false positive and false-negative rate [1, 2]. As an alternative, Virtual Screening (VS) is a pre-screening technique, cheaper and faster than HTS, successfully applied to decrease (filter) the number of compounds to be screened by generating new drug leads [2, 3]. There are two strategies for Virtual Screening: Ligand based (LBVS) and Structure based (SBVS) [3]. In LBVS, the existing information about the ligands is used to find compounds that

© IFIP International Federation for Information Processing 2018
Published by Springer International Publishing AG 2018. All Rights Reserved
A. Amine et al. (Eds.): CIIA 2018, IFIP AICT 522, pp. 244–256, 2018.
https://doi.org/10.1007/978-3-319-89743-1_22

best match a given query; this strategy can work in the absence of structural information of the target [4]. However, in SBVS strategy, the structural information of the target (generally proteins) is required [4].

Machine learning (ML) is a very active branch in artificial intelligence domain; it aims to build models that can predict the output value of input data. The application of ML in VS process is not recent; various methods have been developed in this context. Generally, there are two main applications of ML techniques in VS process. Firstly, in LBVS, where the common task is to distinguish between active and inactive compounds in a given dataset [3–5]; secondly, in SBVS, as scoring functions to improve structure-based binding affinity prediction [6, 7]. Artificial Neural Networks (ANN), Support Vector Machines (SVM), Decision Trees (DT) are the most popular used ML techniques. Ensemble learning is a recent paradigm in machine learning area, where the basic concept is to train a set of base learners on the same dataset, and combine their predictions into a single output prediction that should have better performance [8].

Recently, the number of compounds in the molecular libraries has increased considerably. Applying machine learning techniques on massive libraries (Big Data) in VS process is computationally expensive [9]. The need to sophisticated frameworks for efficient Big Data analytics is becoming more important [9]. Apache Hadoop [10] is one of the most popular used platforms for Big Data analytics; it includes Google's MapReduce model [11] as processing tool and Hadoop Distributed File System (HDFS) as storage system. Some limitations have been known with Google's MapReduce, such as acyclic data flow model, the absence of some features such as in-memory data caching (cache reusable data), and broadcast variables (reusable data), make it inappropriate for some applications [1]. Apache Spark [12, 13] is an open source framework for large-scale datasets processing on clusters, which overcomes the shortcomings of MapReduce, while providing similar scalability and fault tolerance properties [13]. It includes a set of libraries to support a variety of compute intensive tasks for instance *Spark MLlib* for Big Data machine learning [14, 15].

The rest of this paper is organized as follows. Section 2 discusses briefly some works carried out in the same context. Section 3 explains a set of methods and techniques that are used to develop our work. Section 4 details the proposed approach and the main sub-workflows. Section 5 presents and discusses the obtained results, while the paper is concluded in Sect. 6.

2 Related Works

Lately, research works concerning machine learning on Big Data in VS process is quite active. In [16] the authors used Spark and MapReduce programming model to implement SVM based virtual screening. The work showed how HDFS and Spark could be used in combination to distribute and process data in parallel, with a satisfactory scaling behavior. In the study [17], the author developed a general pipeline to perform machine learning on big datasets to derive predictive models using Apache Spark. These models are generated by learning from already tested chemical substances; the results showed the effectiveness of Spark to create pipelines based on machine learning techniques with a good scaling behavior in a distributed environment.

Dries Harnie and his teamwork [18] re-implemented the Chemogenomics pipeline using Apache Spark (S-CHEMO). The Chemogenomics project attempts to derive new candidate drugs from existing experiments through a set of machine learning predictor programs. The S-CHEMO aims to scale the existing pipeline to a multi-node cluster without changes. The authors benchmarked S-CHEMO pipeline against the original; the results showed almost linear speedup up to eight nodes. Where in the study [19], the author used large unbalanced dataset to train a homogeneous ensemble learning based on Support Vector Machine techniques. The evaluation was for two metrics, the predictive performance of the ensemble model, and the scalability (weak and strong scaling). The results proved that Apache Spark is a very powerful tool for Big Data machine learning.

This synthesis allowed us to see the effectiveness of Apache Spark, machine learning, and ensemble learning paradigm to improve the virtual screening process in terms of execution time and predictive performance. This paper aims to exploit this effectiveness to propose a novel approach for large scale VS, where the originality of the approach and the main contributions can be summarized in the following points:

- Propose a process to generate a new training dataset to be used for performance evaluation;
- Construct a heterogeneous ensemble learning model in Apache Spark using a different type of classifiers.

3   Methods and Materials

In this section, we will explain the methods and techniques used to develop the proposed approach, which is based principally on *MLlib* library for Big Data machine learning in Apache Spark, and heterogeneous ensemble learning using three different classifiers SVM, DT and MLP.

3.1   Machine Learning Algorithms

In VS context, the common task of machine learning algorithms is the assignment of molecular descriptors[1] vector (sample) to a class. Generally, ML algorithms are divided into two main categories:

- *Supervised learning*, which requires the vector of class-labels, both the input and the target (class) values for each sample are used in the training to derive classification/regression models [20]. The most common algorithms in this category are; Artificial Neural Networks (ANN), Support Vector Machines (SVMs), and Decision Trees (DT) [20];
- *Unsupervised learning*: which is used when the vector of class-labels is unknown [20].

---

[1] http://www.moleculardescriptors.eu/.

Support Vector Machines. Support vector machine (SVM) form a class of supervised machine learning algorithms, which train the classifier model using pre labeled data [21]. The SVM algorithm in LBVS intended to separate compounds that are represented by vectors of molecular descriptors in the training dataset into active and inactive compounds [20]. Each vector is represented as a *support vector* composed by its attributes (molecular descriptors), SVM search for one target known as the optimal hyperplane ρ that separates the support vectors. The optimal hyperplane ρ maximizes the margin of separation between the hyperplane and the closest data points (support vectors) on both sides of the hyperplane [21, 22].

Decision Trees. Decision Trees (DTs) is one of the most popular used supervised machine learning technique in LBVS. The general reason of using DT is to create a training model which can be used to predict classes of input compounds by learning decision rules inferred from prior data (training dataset) [23]. DT algorithm attempts to solve the problem by using tree representation. Each leaf node corresponds to a class label, whereas non-leaf node (root or internal node) corresponds to molecular descriptor, and branch represents a test on corresponding molecular descriptor [20, 23]. The class of unknown compound is a leaf node that it achieved over a series of questions (nodes) and answers (deciding which branches to take) from the root node [20]. Random Forests (RF) or Decision Forests (DF) is an ensemble classifier including many DTs based on bagging technique [8]. For each DT classifier the training set is constructed by random sampling with replacement from the original dataset [20].

Multi-Layer Perceptron. Multi-Layer Perceptron (MLP) is a feedforward Artificial Neural Network (feedforward ANN) consists of multiple layers (input layer, hidden layers and output layer) [20]. Each layer consists of a variable number of neurons, where the output layer represents the class-labels (two neurons for two class-labels: active -inactive). Each neuron has multiple inputs associated with weights and one output, and related to an activation function. During the training, MLP model seeks to solve an optimization problem by optimizing the weights of each neuron, through back propagation and gradient descent techniques. The optimization problem aims to minimize the mean-square error that represents the difference between the model outputs and the correct answers (correct outputs) [20].

Ensemble Learning. In the classification context, the main idea behind the ensemble learning paradigm is to weight several base classifiers (weak learners), and combine them to obtain a more efficient classifier (strong learner) [8, 24]. To derive an ensemble learning model, we need to follow two main steps. Firstly, many base classifiers are generated and trained in a parallel (e.g. bagging) or in a sequential (e.g. Boosting) manner [8]. Secondly, the results of base classifiers are combined, the most popular technique of combination for classification is majority voting [8], according to Eqs. (1) and (2) [24].

$$class\ (x) = \arg\ \max\nolimits_{L_i \in dom(c)} \left(\sum\nolimits_k g(c_k(x), L_i) \right) \tag{1}$$

Where k is the number of classifiers, $c_k(x)$ is the classification result of the $k'th$ classifier and g(c, L) defined as [24]:

$$\begin{cases} 1 \text{ if } c = L \\ 0 \text{ if } c \neq L \end{cases} \tag{2}$$

3.2   Apache Spark for Big Data Machine Learning

For many years, Apache Hadoop [10] was the de facto standard for Big Data analytics [25], because of its ecosystem structure that includes a set of modules appropriate to manage this type of data; principally, Google's MapReduce [11] for the processing and Hadoop Distributed File System (HDFS) for the storage [25]. The Hadoop MapReduce provides many benefits such as flexibility, scalability and fault-tolerance, but at the same time, it has some limitations that make it not suitable for some applications such as iterative jobs (e.g. Machine learning algorithms), and interactive analytics [1, 13, 22].

Apache Spark [12] is a highly scalable, fast and in-memory Big Data processing engine [9]; it overcomes the shortcomings of Hadoop MapReduce model, while retaining scalability and fault tolerance [13]; it offers an ability to develop distributed applications using Java, Python, Scala, and R programming languages [9]. It consists of a set of libraries for different compute intensive tasks, including Apache Spark Streaming, Apache Spark SQL, Apache Spark GraphX, and Apache Spark MLlib [9]. In cluster mode, Spark supports three cluster managers, standalone, YARN of Hadoop and Mesos. It can access diverse data sources including HDFS, Cassandra, HBase, and S3 [12, 22]. The main abstraction in Spark is the Resilient Distributed Dataset (RDD) [25], which is defined as an immutable collection of objects partitioned across the nodes in the cluster, it can be cached in memory (as reusable data) and rebuilt if a partition is lost through a notion of *lineage* [13, 22].

Spark MLlib is a distributed machine learning library; it consists of fast and scalable implementations of standard learning algorithms, including classification, regression, collaborative filtering, clustering, and dimensionality reduction [15]. It also provides a variety of underlying statistics, linear algebra, and optimization primitives. As part of Spark ecosystem, *MLlib* provides a high-level API to simplify the development of machine learning pipelines [15, 25].

3.3   The Chemistry Development Kit

The Chemistry Development Kit (CDK)[2] is an open source toolkit implemented in Java for Structural Chemo-and Bio- informatics. It provides methods for many common tasks in molecular informatics such as the calculation of molecular descriptors [26]. In this paper, we used CDK version 2.0 to generate a vector of molecular descriptors for each molecule for succeeding machine learning steps [17].

---

[2] http://sourceforge.net/projects/cdk/.

4   Proposed Approach

Our proposed approach is based on Spark's master-worker architecture as shown in the figure below (see Fig. 1), using Standalone cluster manager to acquire resources on the cluster, and HDFS as storage system; where the master node acts as NameNode of HDFS, and each worker node acts as DataNode of HDFS.

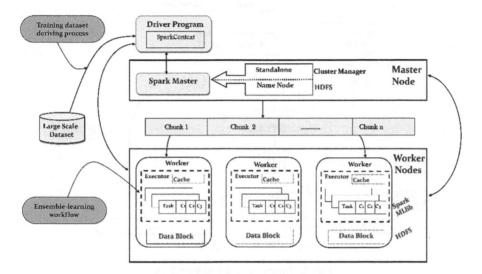

Fig. 1. Schematic representation of the proposed approach on Spark cluster.

Firstly, we create SparkContext via "`new SparkContext (conf)`" instance, which allows accessing the cluster through a resources manager (Standalone). "`conf`" represents SparkConf instance created via the instruction, "`new SparkConf().setAppName().set("parameters")`" that stores the configuration parameters to pass it to SparkContext such as the number of cores and memory size used by the executors on worker nodes.

When loading a dataset from HDFS to an RDD via "`sc.textFile("hdfs://...")`" method, Spark normally splits the input data into chunks or partitions. Next, the data partitions are distributed in the cluster (over the worker nodes) and conserved at DataNode; while NameNode only contains the metadata about the dataset kept at each DataNode. Then, in each worker the dataset is processed, a task for each partition is launched.

To transfer our application, we create a JAR file using Apache Maven[3], which includes all the dependencies. This JAR file is then submitted to a Spark cluster through the command line spark-submit as follows:

"`bin>spark-submit --master spark://master URL --class Main_class / path/to/JARfile.jar`".

---

[3] https://maven.apache.org/.

One JAR file includes all the dependencies ensure the availability of all these dependencies in each worker node, which reduce the overall processing time. The worker nodes process the dataset stored at DataNode (Data Block) using Ensemble-learning workflow (see Fig. 2). When all the worker nodes complete the processing, the master node gets back the final results.

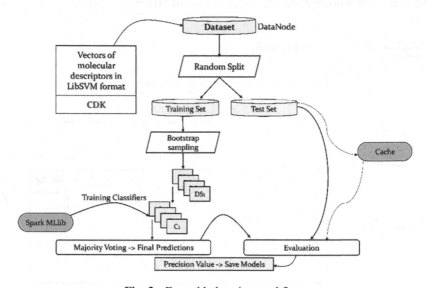

Fig. 2. Ensemble-learning workflow.

5   Experimental Results

This section describes the proposed process to generate our training dataset to be used to evaluate the predictive performance and the scalability, in the first section. Next, in the second section we will present and discuss the obtained results.

5.1   Dataset Generation

In this paper, we chose to study HIV/AIDS disease, which still does not have a practical drug. The main obstacle in the treatment of HIV is the ability of the virus to mutate rapidly into drug-resistant variants [27]. A major target in HIV disease is its protease (HIV protease receptor). Many studies have been developed to discover new HIV protease inhibitors that aims principally to prevent viral replication by selectively binding to viral proteases [28]. The following steps are used to derive our training dataset.

Step 1: HIV protease inhibitors preparation. Firstly, we extract a set of HIV protease inhibitors in SDF format from ChEBI[4] database (ID = CHEBI: 35660) that

---

[4] https://www.ebi.ac.uk/chebi/.

contains 23 inhibitors. Next, we generate the vectors of molecular descriptors of these inhibitors using CDK toolkit.

Step 2: Dataset preparation. The dataset used in this study was established from the open chemical database ChEMBL[5]. ChEMBL is a structured database, which is containing more than one million bioactive drug-like substances. From ChEMBL version 23, we extract randomly 50E+4 compounds. Next, we generate a vector of molecular descriptors for each compound using CDK toolkit.

Step 3: Training dataset generation. To derive our training dataset (see Fig. 3) we used Tanimoto index [29] to measure the similarity between two compounds (*i:* compound, *j: inhibitor*). Tanimoto coefficient *sim(i, j)* can be calculated as follows:

$$sim(i,j) = \frac{\sum_{k=1}^{l} x_{ki} x_{kj}}{\sum_{k=1}^{l} (x_{ki})^2 + \sum_{k=1}^{l} (x_{kj})^2 - \sum_{k=1}^{l} x_{ki} x_{kj}} \tag{3}$$

Where l represents the number of descriptors and x represents the molecular descriptors vector. The threshold values for similarity compounds are typically in the range of 0.8 to 0.9 [29]. We selected 0.9 as threshold to classify the compounds as inhibitors (actives) and non-inhibitors (inactive).

```
Inputs:
IN: a set of 23 inhibitors
CM: a set of compounds
s: dataset size
Output:
L: vector of class-labels
Begin
  for i =1 to s do
    for j=1 to 23 do
      Calculate sim(CM(i), IN(j)); according to eq. (3)
      if sim>=0,9 then
        L(i)=1
      else
        L(i)=0
      end
    end
  end
End
```

Fig. 3. Pseudo code of training dataset deriving process.

5.2 Results and Discussion

We implemented the proposed approach using Spark 1.6 version, Hadoop 2.6 version and Scala as programming language with Scala IDE for eclipse 4.3 version. The configuration of the computer used for experiments is a local machine Intel Core i7 with 2.10 GHz speed and 8 GB of RAM.

---

[5] https://www.ebi.ac.uk/chembl/.

We created Spark Standalone cluster locally (see Fig. 4), we launched firstly one spark master using the command line "bin>spark-class org.apache.spark. deploy.master.Master".

Next, we launched a set of workers using the command "bin>spark-class org.apache.spark.deploy.worker.Worker spark://master URL". Where master URL takes format @*IP:port* (ex. 192.168.223.1:7077).

Spark 1.6.0   **Spark Master at spark://192.168.223.1:7077**

URL: spark://192.168.223.1:7077
REST URL: spark://192.168.223.1:6066 *(cluster mode)*
Alive Workers: 3
Cores in use: 24 Total, 0 Used
Memory in use: 20.6 GB Total, 0.0 B Used
Applications: 0 Running, 0 Completed
Drivers: 0 Running, 0 Completed
Status: ALIVE

Workers

Worker Id	Address	State	Cores	Memory
worker-20171026225349-192.168.223.1-52605	192.168.223.1:52605	ALIVE	8 (0 Used)	6.9 GB (0.0 B Used)
worker-20171026225456-192.168.223.1-52647	192.168.223.1:52647	ALIVE	8 (0 Used)	6.9 GB (0.0 B Used)
worker-20171026225618-192.168.223.1-52687	192.168.223.1:52687	ALIVE	8 (0 Used)	6.9 GB (0.0 B Used)

Running Applications

Application ID	Name	Cores	Memory per Node	Submitted Time	User	State	Duration

Completed Applications

Application ID	Name	Cores	Memory per Node	Submitted Time	User	State	Duration

Fig. 4. Example of locally Standalone cluster with three workers.

Precision Evaluation. To evaluate the predictive performance of our approach, we used 10-fold cross validation technique, where the dataset is randomly divided into ten subsets, nine of them are used for training and the last one is used as a test set to validate the classifier [5]. We selected the precision as metric evaluation that represents the fraction of correctly identified positives over the total amount of instances classified as positive [19]. The results are illustrated in the following figure (see Fig. 5).

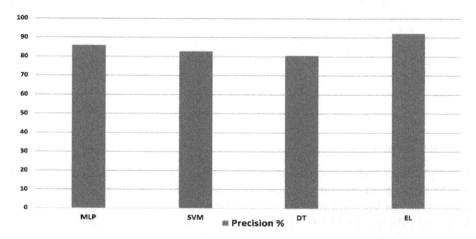

Fig. 5. Precision results for MLP, SVM and DT classifiers compared with ensemble learning model (EL).

The results show that Multi-Layer Perceptron (MLP) classifier gave higher precision on our dataset than the other classifiers SVM and DT, while the ensemble learning model (EL) made a good improvement in the results with precision up to 92%.

Scalability Evaluation. In order to evaluate the scalability of our approach, we carried out a scalability test on two aspects: scaling with training dataset size (weak scaling) and scaling with cluster size (strong scaling).

The figure below (see Fig. 6) shows the effect of training dataset size on the execution time. With more instances, ensemble learning model takes longer to train. These results are obtained for a number of worker nodes equal to 5.

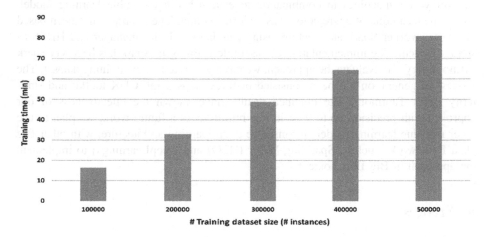

Fig. 6. Scaling with training dataset size (# instances).

The figure below (see Fig. 7) illustrate the obtained results where scaling the approach with cluster size. We started with 2 workers, and the number of workers increases in each test with 2 nodes. The results show that the time is decreasing

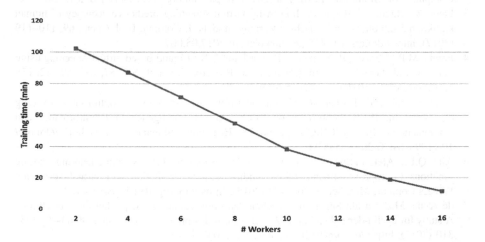

Fig. 7. Scaling with cluster size (# workers).

linearly, where with 2 workers the workflow takes average of 102 min to return the results and approximately 11 min with 16 workers. We can say that ensemble learning model is significantly faster when using more workers.

6 Conclusion and Future Work

In this paper, we have presented a new approach based on ensemble learning paradigm and Apache Spark to enhance the performance of large-scale virtual screening process. We have used three classifiers which are multi-layer perceptron, decision trees and support vector machines in combination to establish our ensemble learning model, where the technique of aggregation was majority voting. The approach has been based on master-worker Spark architecture, using standalone cluster manager and HDFS as storage system. The implementation of ensemble learning in Spark has been via Spark MLlib library. To evaluate the approach, we have generated a new training dataset. The process of generation has been consisted of three steps using CDK toolkit and similarity index Tanimoto. The obtained results have shown the effectiveness of our approach with precision up to 92% in few minutes. As a future work, we plan to use other machine learning models for instance deep learning architecture, with other Big Data frameworks such as Sparkling Water (H2O) and DeepLearning4j to implement the approach in Big Data context.

References

1. Capuccini, M., Ahmed, L., Schaal, W., Laure, E., Spjuth, O.: Large scale virtual screening on public cloud resources with Apache Spark. J. Cheminformatics **9**(15), 1–6 (2017). https://doi.org/10.1186/s13321-017-0204-4
2. Pradeep, P., Struble, C., Neumann, T., Sem, D.S., Merrill, S.J.: A novel scoring based distributed protein docking application to improve enrichment. J. IEEE/ACM Trans. Comput. Biol. Bioinform. **12**(6), 1–8 (2015). https://doi.org/10.1109/TCBB.2015.2401020
3. Fang, X., Bagui, S., Bagui, S.: Improving virtual screening predictive accuracy of human kallikrein 5 inhibitors using machine learning models. J. Comput. Biol. Chem. **69**, 110–119 (2017). https://doi.org/10.1016/j.compbiolchem.2017.05.007
4. Preeja, M.P., Hemant, P., Soman, K.P., Prashant, S.K.: Ligand-based virtual screening using random walk kernel and empirical filters. J. Procedia Comput. Sci. **57**, 418–427 (2015). https://doi.org/10.1016/j.procs.2015.07.508
5. Upul, S., Rahal, P., Roshan, R.: Machine learning based search space optimisation for drug discovery. In: IEEE Symposium on Computational Intelligence in Bioinformatics and Computational Biology (CIBCB), pp. 68–75. IEEE Press, Singapore (2013). https://doi.org/10.1109/CIBCB.2013.6595390
6. Ain, Q.U., Aleksandrova, A., Roessler, F.D., Ballester, P.J.: Machine-learning scoring functions to improve structure-based binding affinity prediction and virtual screening. WIREs Comput. Mol. Sci. **5**, 405–424 (2015). https://doi.org/10.1002/wcms.1225
7. de Ávila, M.B., et al.: Supervised machine learning techniques to predict binding affinity. A study for cyclin-dependent kinase 2. J. Biochem. Biophys. Res. Commun. **494**(1–2), 305–310 (2017). https://doi.org/10.1016/j.bbrc.2017.10.035

8. Yun, Y.: Temporal data mining via unsupervised ensemble learning, Chap. 4. In: Ensemble Learning, pp. 35–56. Elsevier (2017). https://doi.org/10.1016/B978-0-12-811654-8.00004-X
9. Mehdi, A., Ehsun, B., Liu, G., Ahmad, P.T.: Big data machine learning using Apache Spark MLlib. In: IEEE International Conference on Big Data (BIGDATA), pp. 3492–3498. IEEE Press, Boston (2017). https://doi.org/10.1109/BigData.2017.8258338
10. Apache Hadoop. http://hadoop.apache.org. Accessed 07 Feb 2018
11. Dean, J., Ghemawat, S.: MapReduce: simplified data processing on large clusters. In: The 6th Symposium on Operating Systems Design and Implementation, San Francisco, pp. 137–149 (2004)
12. Apache Spark™. http://spark.apache.org. Accessed 07 Feb 2018
13. Zaharia, M., et al.: Spark: cluster computing with working sets. In: The 2nd USENIX Conference on Hot Topics in Cloud Computing, HotCloud 2010, USA, pp. 1–7 (2010)
14. Wei, H., et al.: In-memory parallel processing of massive remotely sensed data using an Apache Spark on Hadoop YARN model. IEEE J. Sel. Top. Appl. Earth Obs. Remote Sens. 10(1), 3–19 (2017). https://doi.org/10.1109/JSTARS.2016.2547020
15. Meng, X., et al.: MLlib: machine learning in Apache Spark. J. Mach. Learn. Res. 17(34), 1–7 (2016)
16. Ahmed, L., Edlund, A., Laure, E., Spjuth, O.: Using iterative MapReduce for parallel virtual screening. In: IEEE International Conference on Cloud Computing Technology and Science, pp. 27–32. IEEE Press, Bristol (2013). https://doi.org/10.1109/CloudCom.2013.99
17. Staffan, A.: Automating model building in ligand-based predictive drug discovery using the Spark framework. Degree Project in Bioinformatics, Masters Programme in Molecular Biotechnology Engineering, Uppsala University School of Engineering (2015)
18. Harnie, D., et al.: Scaling machine learning for target prediction in drug discovery using Apache Spark. J. Future Gener. Comput. Syst. 67, 409–417 (2017). https://doi.org/10.1016/j.future.2016.04.023
19. Simon, L.: Distributed ensemble learning with Apache Spark. Degree Project in Bioinformatics, Masters Programme in Molecular Biotechnology Engineering, Uppsala University School of Engineering (2016)
20. Antonio, L.: Machine-learning approaches in drug discovery: methods and applications. J. Drug Discov. Today 20(3), 318–331 (2015). https://doi.org/10.1016/j.drudis.2014.10.012
21. Bissan, G., Joe, N.S.: High dimensional data classification and feature selection using support vector machines. J. Eur. J. Oper. Res. 265(3), 993–1004 (2018). https://doi.org/10.1016/j.ejor.2017.08.040
22. Karima, S., Mohamed, B.: Big data analytic techniques in virtual screening for drug discovery. In: The 2nd International Conference on Big Data, Cloud and Applications (BDCA), Article 9, 7 p. ACM, Morocco (2017). https://doi.org/10.1145/3090354.3090363
23. Introduction to Decision Tree Algorithm. http://dataaspirant.com/2017/01/30/how-decision-tree-algorithm-works/. Accessed 07 Feb 2018
24. Rokach, L.: Ensemble-based classifiers. Artif. Intell. Rev. 33(1–2), 1–39 (2010)
25. Bilal, A., Ying, Z., Uwe, R.: On the usability of Hadoop MapReduce, Apache Spark & Apache flink for data science. In: IEEE International Conference on Big Data (BIGDATA), pp. 303–310. IEEE Press, Boston (2017). https://doi.org/10.1109/BigData.2017.8257938
26. Christoph, S., et al.: The Chemistry Development Kit (CDK): an open-source Java library for chemo-and bioinformatics. J. Chem. Inf. Comput. Sci. 43(2), 493–500 (2003). https://doi.org/10.1021/ci025584y

27.
    Maris, L., et al.: Proteochemometric modeling of HIV protease susceptibility. J. BMC Bioinf. **9**, 181 (2008). https://doi.org/10.1186/1471-2105-9-181
28. Protease   inhibitor   (pharmacology).   https://en.wikipedia.org/wiki/Protease_inhibitor_ (pharmacology). Accessed 07 Feb 2018
29. Han, B., et al.: Development and experimental test of support vector machines virtual screening method for searching Src inhibitors from large compound libraries. J. Chem. Central J. **6**, 139 (2012). https://doi.org/10.1186/1752-153X-6-139

Trace Based System in TEL Systems: Theory and Practice

Tarek Djouad[1]([⊠]) and Alain Mille[2]

[1] Icosy Laboratory, Khenchela University, Khenchela, Algeria
tarek.djouad@gmail.com
[2] Liris, Lyon1 University, UMR CNRS 5205, Lyon, France
alain.mille@univ-lyon1.fr

Abstract. We present in this paper an easier way to manage activity traces and to compute human learning indicators activities in Technology Enhanced Learning TEL systems. We review our research work related to Trace based system TBS and we explain how we use TBS to develop new and generic model to represent the indicator life cycle from its creation to its reuse. This paper presents the underlying theory and how this theory is implemented to compute human learning indicators activities available for use with any other learning platform, provided the TBS can access the learning platform traces.

Keywords: Activity traces · Trace based system
Human learning indicator activity · Indicators engineering
Technology Enhanced Learning systems

1 Introduction

Most Technology Enhanced Learning (TEL) systems use activity traces to make diagnoses during the learning process in order to propose appropriate assistance for learners, tutors or teachers. Some TEL systems can be used by designers or teachers to set up these diagnoses and describe the information to be collected at the design step [1], but in most cases, both the data collection and diagnostic processes are embedded into the code and the IT developers are needed for designing and modifying them. In practice, therefore, it is very difficult to personalize the tracing and diagnostic processes, and, consequently, in most cases distance education systems offer very little in the way of personalization and remediation services during learning activities. In the field of online education research, researchers and analysts are also very interested in activity traces and are increasingly equipping themselves with tools to build interesting data from application logs, and then treat them as activity traces before carrying out treatments to understand current learning processes [2–4]. Learning analytics is now a big issue for researchers [5] and it is very important for the efficiency of the research to be able to get clear information on the semantics of the logs, which

© IFIP International Federation for Information Processing 2018
Published by Springer International Publishing AG 2018. All Rights Reserved
A. Amine et al. (Eds.): CIIA 2018, IFIP AICT 522, pp. 257–266, 2018.
https://doi.org/10.1007/978-3-319-89743-1_23

remains currently a big challenge. In order to overcome these difficulties, we have been working for several years to consider activity traces as knowledge objects available to designers, teachers, tutors, learners, researchers and, in general, to those involved in learning processes. Since activity traces are constructed to analyze, diagnose and understand expected or feared learning phenomena, we have also developed representations of learning indicators and their computation processes, based on traces of modelled activities. These new implementation versions of Trace based system we propose in this paper are intended to compute human learning indicators activities. In this paper, we present the state of our current research and its latest developments, illustrated with practical examples while recalling the theoretical foundations of the modelled activity trace. We describe on the one hand the formalization of the approach and on the other hand the developed and experienced software environments. We conclude by showing the strong potential of this approach for professional, individual and opportunistic learning activities, in particular with regard to the ethical aspects of e-learning.

2   Activity Traces in TEL Systems

Most of learning platforms produce activity traces. These traces are saved in different formats: Logs files, Databases, video and audio files, etc. Several works propose to use and to exploit these traces in TEL systems. For example, Betbeder [6], Guraud [7] and Dyke [8] define experiments to collect multimodal data in order to analyze traces. Mazza [9], France [10] and Cram [11] visualize traces in real time and provide a feedback to teachers and learners on their own activities. Ferraris [12] and Voisin [13] propose model driven engineering approaches to understand learning scenarios using trace. As activity traces represent very important knowledge containers, it is meaningful to provide specific environment for managing them as such. Several frameworks have been developed in order to manage activity traces [14,15] as for helping the designers to integrate them in the target application [16] or for helping researchers and analysts to understand the learning process in a batch way [17,18]. These approaches are very useful for managing and requesting traces as data, but do not consider traces as knowledge containers per se. In this paper, we will focus on one knowledge based approach for collecting, representing, managing, requesting and transforming activity traces. As far as we know, this approach is original and we will illustrate in which extent this approach allows new usages in the context of Technology Enhanced Learning systems.

3   Trace Based System

The Trace based system or TBS is proposed and implemented [19] by the TWEAK [20] research group to manage modelled activity traces or m-traces [21–23]. An m-trace in TBS is a structured object: the trace model and the corresponding trace instance in the form of a sequence of observed elements or obsels. Each instance obsel part of an m-trace is temporally situated by a time stamp and satisfies the trace model part of the m-trace. TBS proposes explicit transformation operators to be applied to a set of m-traces (transformation sources) in order to obtain other transformed m-traces (transformation targets). All m-trace obsels are represented by structured information resulting from a transformation operation using source m-trace obsels. Each m-trace is the result of some transformation of a lower level m-trace, except for the lowest level, directly built from an observation process constructing the primary m-trace. TBS proposes three steps for using m-traces (Fig. 1):

1. Users, as teachers/tutors or learners, use learning platforms. These platforms provide raw data as a source of observation. TBS connects to learning platforms, collects raw data and uses these data to build a primary m-trace (model and instance). This primary m-trace is then saved in an m-trace base,
2. TBS uses this primary m-trace and transforms it into other transformed m-traces according to the semantics of these transformations. The transformed m-trace is saved in the m-trace base. In turn, these transformed m-traces can be transformed again into other transformed m-traces. Starting from one primary m-trace, a transformation graph is progressively built and saved for providing explicit explanations of any transformation, i.e. providing the semantics of any m-trace in the m-trace base,
3. Moreover, the m-trace base can be used by any assistant to manage indicators, allow indicator computation, provide smart visualization, etc.

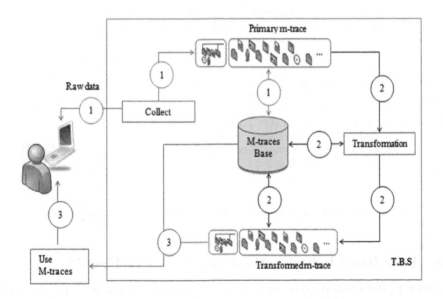

Fig. 1. Trace based system used by our research team to manage m-traces.

4   Case Studies: Indicator Computations Using TBS

We propose in this section some case studies of how to use TBS in real learning situations to compute human learning indicators activities. The new developed systems we propose here are based on TBS and their goal is to compute indicators activities using modelled traces. We will illustrate two systems we built: Trace based indicator management system TB-IMS, and multi-agents indicator computation system IC-MAS. But first we will illustrate the latest implemented KTBS: a kernel trace based system which is a concrete implementation of TBS.

4.1   Short Presentation of the Used Kernel Trace Based System (KTBS)

Kernel trace based system [24] is an implementation of TBS. KTBS considers an activity trace as a model and a set of obsels (m-trace). Is allows to create m-traces bases, m-traces' models, m-traces' obsels, transformed m-traces, etc. Each obsel has a set of attributes like type, value, time-begin, time-end, etc. KTBS stores modelled traces as RDF ontology and manipulates them using a set of operators like: filter, fusion, SPARQL query, etc. It proposes to use JSON [25], REST [26] and TURTLE [27] to describe m-traces. Figure 2 presents an example of m-trace model in KTBS using Turtle syntax.

```
@prefix : <http://liris.cnrs.fr/silex/2009/ktbs#> .
@prefix rdfs: <http://www.w3.org/2000/01/rdf-schema#>.

<.> :contains <model1> .

<model1> a :TraceModel ;
    :hasUnit :millisecond .

<#EnterChatRoom> a :ObselType .
<#SendMsg> a :ObselType .
<#MsgReceived> a :ObselType .
<#LeaveRoom> a :ObselType .
```

Fig. 2. Example of creating a model in KTBS using Turtle syntax.

4.2   Trace Based Indicator Management System TB-IMS

In order to demonstrate concretely how to compute indicators activities using TBS, we developed an effective method and its implementation [28] to carry out the entire life cycle of indicators activities. We recall that in TEL systems an indicator is a mathematical variable with a set of values [29]. In our work, we propose an indicator as composed of a model descriptor and a set of instances, and we propose four steps to describe and compute an indicator instance using a TB-IMS. Each indicator instance is computed using these four steps:

- **Step1:** To compute a new indicator, we propose to associate it with a set of empty m-traces. These m-traces will be instancied later for evaluating the indicator formula to compute indicator instances,

- **Step2:** We associate a transformation sequence to the empty indicator. The transformation sequence related to these m-traces describes how to move from a primary m-trace (described by its model) to a specific indicator (described by its formula),

- **Step3:** Collects data from the learning environment and builds the primary m-trace instance,

- **Step4:** Finally, we execute the transformation sequence, from the primary m-trace instantiated in step 3 to the indicator defined in step 1.

Indicator computation and m-trace transformation are managed by the prototype we developed: Trace based-indicator management system TB-IMS [30]. Figure 3 shows the TB-IMS collector module, the transformation module, the equation editor and the visualization result. In this example we compute the proportion between chat messages and private messages related to 'user15' according to a time interval T. Computation is possible without coding in machine. The system saves in its databases the indicator, its transformation, intermediate m-traces and the primary m-trace, which allows the indicator to be reused to compute new other indicators.

4.3   A Multi-agent System to Compute Human Learning Indicators Activities

We propose also to use Multi-agents technology to compute indicators activities based on TBS [31]. The indicator computation- multi agents system IC-MAS [32] we propose allows creating, updating, deleting, reusing and sharing indicator. These agents are important in order to:

- **Minimize the indicator's computation complexity.** If the indicator computation is difficult (Example: the case of a high level indicator defined by sub-indicators) then our system reuses existing indicators to compute new one. We use information like transformation sequence, computation rule and category to perform this computation,

- **The need to compute a same indicator in different contexts is very high.** Several users attempt to compute the same indicator with closest time intervals. The system creates for each new computation a list of agents that will minimize the indicator computation time.

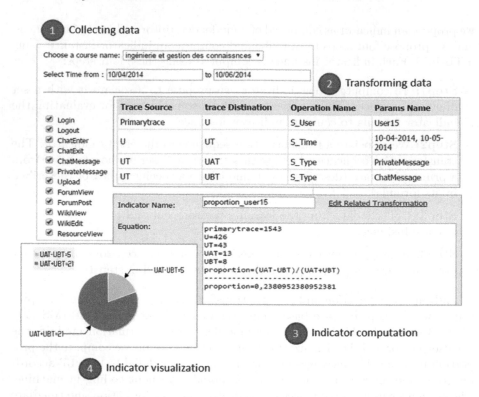

Fig. 3. Computing the proportion indicator using TB-IMS.

We propose these following agents to build our system:

- **The collecting agent** uses raw data issued from learning platforms (Example: database, Log files, etc.) to select useful data used to compute indicator. The collecting agent is guided by variables which are listed in the computation rule to optimize the collecting step. Once completed, it stores the collecting results in an m-traces base as a primary m-trace and keeps the collecting history to reuse it later. This agent aims to prepare things for having a better performance in a future collecting step,

- **The transformation agent** transforms a primary m-trace created by the collecting agent. Each transformed trace becomes a variable labelled according to the name of the transformed trace and takes its value from the number of instances present in the transformed m-trace. This agent uses transformation operators such as filtering, merging, matching, etc. to transform m-traces,

- **The computation agent** executes the computation rule using variables created by the transformation agent and saves it in the indicator base,

- **The interface agent** displays indicators using graphical views (Pie charts, Bar graphs, Histograms, etc.). It also retrieve information provided by teachers to define indicators,
- **The coordinator agent** coordinates and synchronizes the different agents to provide a consistent functioning of the system.

Figure 4 explains an example of indicator computation using IC-MAS.

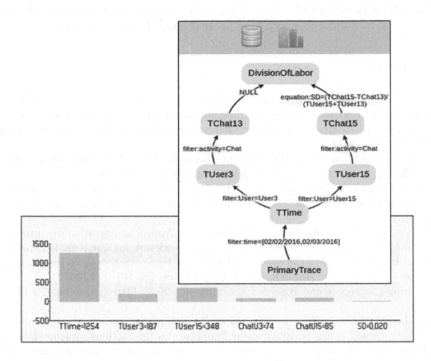

Fig. 4. Computing the proportion indicator using IC-MAS.

5   Conclusion

Our research was an opportunity to work on the consolidation of collaborative work around the theory of modelled traces. Indeed, we were able to experiment concretely with its potential to formalize and represent the notion of indicator as a structured computer object that provides the knowledge necessary for its elaboration. This knowledge structure integrates the modelled source trace as described in the learning environment, the set of logical and temporal transformations necessary to allow the calculation of indicators whose semantics are thus made completely explicit. This explicit representation is proposed in the form of an operational IT structure, adaptable by designers or teachers to new situations without requiring reprogramming, but requiring only a modification of the description of the mobilized knowledge. For researchers, the availability of a basic semantics for activity traces makes it possible to integrate this knowledge into the parameterization of Learning Analytics tools, but also to design specific and documented experiments to serve the object of their research. Finally, and these are the latest developments under way in our research, the ability to explain activity traces opens up very promising solutions in terms of ethics of use of personal data by involving learners and teachers in controlling the use of activity traces during learning. Learners and tutors can even, during the activity itself, suggest new ways of interpreting their learning behavior to make easier their task or to build new knowledge about human learning processes.

References

1. Ouali, M.A., Iksal, S., Laforcade, P.: The strategic organization of the observation in a TEL system: studies and first formalizations. In: Proceedings of the 6th International Conference on Computer Supported Education, Barcelona, Spain, pp. 317–324 (2014)
2. Georgeon, O.L., Mille, A., Bellet, T., Mathern, B., Ritter, F.E.: Supporting activity modelling from activity traces. Expert Syst. **29**, 261–275 (2012)
3. Ji, M.: Exploiting activity traces and learners' reports to support self-regulation in project based learning. Ph.D. thesis in Computer Science, INSA, Lyon, France (2015)
4. Lebis, A., Lefevre, M., Luengo, V., Guin, N.: Approche narrative des processus d'analyses de traces d'apprentissage: un framework ontologique pour la capitalisation. In: Proceedings of the 8ième conférence sur les Environnements Informatiques pour l'Apprentissage Humain, Strasbourg, France, pp. 101–112 (2017)
5. Learning Analytics and Knowledge Conference 2017. http://educ-lak17.educ.sfu.ca/
6. Betbeder, M.L., Reffay, C., Channier, T.: Environnement audiographique synchrone: recueil et transcription pour l'analyse des interactions multimodales. In: Proceedings of the premiéres journées communication et apprentissage instrumentés en réseau, Amiens, France, pp. 406–420 (2006)
7. Guéraud, V., Michelet, S., Adam, J.M.: Suivi de classe á distance: propositions génériques et expérimentation en électricité. In: Proceedings of the 3ième conférence sur les Environnements Informatiques pour l'Apprentissage Humain, Lausanne, Switzerland, pp. 167–178 (2007)
8. Dyke, G., Lund, K., Girardot, J.J.: Tatiana, un environnement d'aide à l'analyse de traces d'interactions humaines. Technique et Science Informatiques. **29**, 1179–1205 (2010)
9. Mazza, R., Botturi, L.: Monitoring an online course with the GISMO tool: a case study. J. Interact. Learn. Res. **18**, 251–265 (2007)
10. France, L., Heraud, J.M., Marty, J.C., Carron, T.: Visualisation et règulation de l'activitè des apprenants dans un EIAH tracè. In: Proceeding of the 3ième conférence sur les Environnements Informatiques pour l'Apprentissage Humain, Lausanne, Switzerland, pp. 179–184 (2007)
11. Cram, D., Mathern, B., Mille, A.: A complete chronicle discovery approach: application to activity analysis. Expert Syst. **29**, 321–346 (2012)

12. Ferraris, C., Vignollet, A.L., David, J.: Modèlisation de scènarios d'apprentissage collaboratif pour la classe. In: Proceedings of the 2ième confèrence sur les Environnements Informatiques pour l'Apprentissage Humain, Montpellier, France, pp. 285–296 (2005)
13. Voisin, J., Vidal, P.: Une approche conduite par les modèles pour le traçage des activités. Sciences et Technologies de l'Information et de la Communication pour l'Éducation et la Formation **14**, 1–18 (2007)
14. Santos, O.C., Rodríguez, A., Gaudioso, E., Boticario, J.G.: Helping the tutor to manage a collaborative task in a web-based learning environment. Artif. Intell. Educ. **4**, 153–162 (2003)
15. Avouris, N., Fiotakis, G., Kahrimanis, G., Margaritis, M., Komis, V.: Beyond logging of fingertip actions: analysis of collaborative learning using multiple sources of data. J. Interact. Learn. Res. **18**(2), 231–250 (2007)
16. Bouhineau, D., Luengo, V., Mandran, N.: Concevoir, Produire, Dècrire, Évaluer et Partager des Donnèes, Opérateurs, Processus d'analyse et Résultats d'études sur l'apprentissage Humain avec Ordinateur. In: Proceedings of the 8ième conférence sur les Environnements Informatiques pour l'Apprentissage Humain, Strasbourg, France, pp. 125–136 (2017)
17. Dyke, G.: A model for managing and capitalising on the analyses of traces of activity in collaborative interaction. Doctorate thesis, Ecole Nationale Supérieure des Mines, Saint-Etienne, France (2009)
18. Toussaint, B.-M., Luengo, V., Jambon, F.: Proposition d'un Framework de Traitement de Traces pour l'Analyse de Connaissances Perceptivo-Gestuelles. (Le cas de la chirurgie orthopédique percutanée). In: Proceeding of the 7th Conférence sur les Environnements Informatiques pour l'Apprentissage Humain, Agadir, Morocco, pp. 222–233 (2015)
19. Kernel of Trace Based System. https://kernel-for-trace-based-systems. readthedocs.io/en/latest/
20. TWEAK Research Team. https://liris.cnrs.fr/equipes/?id=75
21. Settouti, L.S., Marty, J.C., Mille, A., Prié, Y.: A trace-based system for technology-enhanced learning systems personalisation. In: Proceedings of the 9th International Conference on Advanced Learning Technologies, Riga, Latvia, Italy, pp. 93–97 (2009)
22. Zarka, R., Champin, P.A., Cordier, A., Egyed-Zsigmond, E., Lamontagne, L., Mille, A.: TStore: a trace-base management system using finite-state transducer approach for trace transformation. In: Proceeding of the 1st International Conference on Model-Driven Engineering and Software Development, Barcelona, Spain, pp. 117–122 (2013)
23. Cordier, A., Lefevre, M., Champin, P.A., Georgeon, O., Mille, A.: Trace-based reasoning-modeling interaction traces for reasoning on experiences. In: Proceedings of the 26th International Florida Artificial Intelligence Research Society Conference, Pete Beach, Florida, USA, pp. 363–368 (2013)
24. KTBS. http://tbs-platform.org/tbs/doku.php/tools:ktbs
25. JSON. https://www.json.org/json-fr.html
26. REST. https://www.w3.org/2001/sw/wiki/REST
27. TURTLE. https://www.w3.org/TR/turtle
28. Djouad, T., Mille, A.: Observing and understanding an on-line learning activity: a model-based approach for activity indicator engineering. Technol. Knowl. Learn. (2017). https://doi.org/10.1007/s10758-017-9337-9

29. Dimitracopoulou, A., Petrou, A., Martinez, A., Marcos, J.A., Kollias, V., Jermann, P., Harrer, A., Dimitriadis, Y., Bollen, L.: State of the art on interaction analysis for metacognitive support and diagnosis. Technical report, Deliverable D31.1.1 Kaleidoscope Project (2006). https://telearn.archives-ouvertes.fr/hal-00190146
30. Trace Based Indicator Management System. www.github.com/tb-ims/tb-ims/
31. Djouad, T., Mille, A.: A multi-agents system to compute human learning indicators activities based on model-driven engineering approach. Int. J. Technol. Enhanc. Learn. **10**(1/2), 91–110 (2018). https://doi.org/10.1504/IJTEL.2018.10008599
32. Indicator Computation-Multi Agent System. https://github.com/Activity-Traces/IC-MAS

Using a Social-Based Collaborative Filtering with Classification Techniques

Lamia Berkani$^{(\boxtimes)}$ ⓘ

Laboratory for Research in Artificial Intelligence (LRIA),
Department of Computer Science,
USTHB University, Bab Ezzouar, Algiers, Algeria
lberkani@usthb.dz

Abstract. In this paper, a social-based collaborative filtering model named SBCF is proposed to make personalized recommendations of friends in a social networking context. The social information is formalized and combined with the collaborative filtering algorithm. Furthermore, in order to optimize the performance of the recommendation process, two classification techniques are used: an unsupervised technique applied initially to all users using the Incremental K-means algorithm and a supervised technique applied to newly added users using the K-Nearest Neighbors algorithm (K-NN). Based on the proposed approach, a prototype of a recommender system is developed and a set of experiments has been conducted using the Yelp database.

Keywords: Recommendation of users · Collaborative filtering
Social filtering · Classification · K-means · Incremental K-means
K-nearest neighbors

1 Introduction

Due to the powerful ability of solving information overload, recommender systems have attracted a lot of attention in the last decade [1]. Recommender systems automatically predict the preferences of users for some given items, in order to provide them with the useful recommendations. There are three main methods to generate recommendations [2]: content-based filtering, collaborative filtering (CF) and a hybrid approaches combining in different way the two aforementioned algorithms [3]. Content-based filtering algorithms identify features that appear in item contents that users have appreciated before, thereafter suggest more items which contain these relevant features to users. CF is a widely-exploited technique in recommender systems to provide users with items that well suit their preferences [2]. The basic idea is that a prediction for a given item can be generated by aggregating the ratings of users with similar interest. However, one of the shortcomings of this algorithm is the sparsity and cold start problems due to insufficient rating [4].

Nowadays, interactions and sharing knowledge over the Social Web have become the main way of communication between people. The study of social-based recommender systems has emerged as users are unable to reach the relevant information due to the exponential growth of data.

© IFIP International Federation for Information Processing 2018
Published by Springer International Publishing AG 2018. All Rights Reserved
A. Amine et al. (Eds.): CIIA 2018, IFIP AICT 522, pp. 267–278, 2018.
https://doi.org/10.1007/978-3-319-89743-1_24

In this paper, we focus on the friends' recommendation in social networks. Our approach first enhanced the CF recommendations with social information. The social dimension is characterized by some social-behavior metrics such as friendship, commitment and trust degrees between users, to cope with problems of rating diversity and sparsity. Then, in order to optimize the performance of the recommendation process, we have used two classification techniques along with collaborative and social similarity measures: (1) an unsupervised technique using the Incremental K-means algorithm, applied initially to all users of the social network; and (2) a supervised technique using the K-Nearest Neighbors algorithm, applied to new users.

The remainder of the paper is organized as follows: Sect. 2 gives an overview on some related work about social recommendation. Section 3 presents our recommendation approach. In Sect. 4, we give an overview on the experimentation we carried out. Finally, we give the conclusion with some perspectives in Sect. 5.

2 Related Work

The review of the literature about social recommendation shows that much work proposes various approaches to provide personalized recommendations to users. For instance, Liu and Lee [5] developed a way to increase recommendation effectiveness by incorporating social network information into CF. Chang and Chu [6] proposed a recommendation approach which calculates similarity among users and users' trust-ability and information collected from the social networks. Banati et al. [7] explored the role of explicit social relationship by presenting two novel similarity metrics. The first metric is based on the social behavior that measures similarity between two users on the basis of "how similar they are in their social relationship". The second metric integrates the social similarity with the interest similarity between two users. Su et al. [8] proposed a music recommender system integrating social information (to cope with problems of rating diversity and sparsity) and collaborative information (to cope with problem of lack of rating information) to predict users' preferences.

In order to improve the performance of social recommendation, some studies have used classification techniques, for instance: Guo et al. [9] developed a multi-view clustering method through which users are iteratively clustered on the basis of rating patterns in one view and social trust relationships in the other. Najafabadi et al. [10] tried to improve the accuracy of CF recommendations using clustering and association rules mining on implicit massive data.

3 A Social-Based CF Approach

We propose in this section an enhanced-based CF algorithm by combining collaborative and social information in the recommendation process.

3.1   The User-User Based Collaborative Filtering (CF)

We chose a memory-based CF approach and used the user-user based recommendation. In this approach, the system offers the possibility of identifying the best neighbors for a given user, using the usage matrix. The matrix can be constructed using the ratings of users on items. We used the *Pearson Correlation function* to calculate the similarity between users. The collaborative similarities between users allow the identification of neighborhoods and therefore building communities of users who evaluate in the same way, according to a given threshold.

3.2   The Social Filtering (SocF)

The social information of the user's profile is based on two metrics: (1) *the friendship;* and (2) *the credibility degree.* Two parameters are considered to identify the credibility of an active user u: Commitment and trust degrees.

3.2.1   Friendship Metric

This metric computes the similarity weight between two users u_1 and u_2, based on their social relationships which is defined as the size of the intersection divide by the size of the union of friend sets:

$$Sim_{soc}(u_1, u_2) = \frac{F(u_1) \cap F(u_2)}{F(u_1) \cup F(u_2)} \qquad (1)$$

where:

 $F(u_1)$, respectively $F(u_2)$: represent the number of friends of u_1, respectively u_2.

3.2.2   Commitment Degree

Two parameters are considered: (1) *the participation degree* of an active user u, including the degree or rate of evaluations he carried out; and (2) *the sociability degree* represents his friendship rate in the social network.

$$Commitment\ (u) = \alpha_1.Participation(u) + \beta_1.Sociability(u) \qquad (2)$$

where: α_1 and β_1 are weights that express a priority level, with $\alpha_1 + \beta_1 = 1$

* **The participation degree of u:** concerns mainly the degree of performed evaluations by u. This degree is calculated based on the number of evaluations performed by u, *NbEval (u)*, according to all the evaluations carried out in the system, *NbTotalEval.*

$$Participation\ (u) = \frac{NbEval(u)}{NbTotalEval} \qquad (3)$$

* **The sociability degree of u:** this degree is calculated based on the number of friends of u according to all the registered users of the social network.

$$Sociability\,(u) = \frac{NbFriends(u)}{NbUsers - 1} \tag{4}$$

3.2.3   Trust Degree

This metric takes into account the seniority level of u and his degree of competence in the social network using the following formula:

$$Trust(u) = \alpha_2.Seniority(u) + \beta_2.Competency\,(u) \tag{5}$$

where: α_2, β_2 are weights that express a priority level, with $\alpha_2 + \beta_2 = 1$

- **The seniority level of u:** is calculated based on the date of his registration in the social network.
- **The Competency degree of u:** is calculated on two steps, based the following assumption [11]: *"A friend is competent if he has evaluated correctly the resources compared to their average evaluations in the social network"*:

– **Step 1:** Calculate the competency degree of a friend F regarding a given item R_j. We start by calculating the average of ratings for each item. Then, we compare the rating given by F for the same item with the average value.

$$Competency(F, R_j) = \begin{cases} \dfrac{avg(R_j)}{v_{i,j}} & \text{if } Avg(R_j) \leq v_{i,j} \\[2mm] \dfrac{v_{i,j}}{avg(R_j)} & \text{if } v_{i,j} \leq Avg(R_j) \end{cases} \tag{6}$$

where:

$avg\,(R_j)$: is the average evaluation of the item regarding all the users.

$v_{i,j}$: is the evaluation of the friend F for R_j

– **Step 2:** Calculate the global trust degree of the friend, using this formula:

$$Trust(F) = \frac{1}{n}\sum_{j=1}^{n} Degree_{comp}(F,\ R_j) \tag{7}$$

where: n represents the number of items evaluated by the friend.

4   Combining SBCF Approach and Classification Techniques

The main objective of the classification in our recommendation process is to group similar users according to the collaborative/social dimensions.

This will reduce the search space for the identification of neighbors and allow us to cluster the different groups. So, each user of the social network will have both a collaborative class and a social one. Moreover, if any user has friends but has not yet made enough evaluations, the system can recommend him other friends based only on the social dimension. Similarly, if a user has performed enough evaluations but has not yet added any friend, the system can recommend him new friends based on the collaborative dimension.

4.1 Unsupervised Classification Using Incremental K-means

We have first used the K-means classical algorithm, which is considered to be the most popular algorithm because of its simplicity and its ability to handle large data sets. However, the disadvantage of this algorithm is that each initialization corresponds to a different solution (local optimum) that can be far from the optimal solution (global optimum). A naive solution to this problem is to run the algorithm several times with different initializations and to retain the best grouping found. The use of this solution is limited because of its cost and that in some cases we cannot explain the partitioning of the clusters where we can find an optimal partition in a single execution [12].

Then, we applied the incremental K-means proposed in [12] which is a variant of the K-means algorithm. This algorithm eliminates the initialization problem of centroids. It is based on the principle of global K-means, which aims to achieve an optimal solution, i.e. instead of having a single center of the whole population (global K-means). The algorithm chooses two objects, where each being the center of a cluster so that the two latter are the more distant. The next step is to choose the next center. A simple function allows calculating the distance between the center of the cluster and its neighbors. The farthest element of the center is the elected candidate to be the new centroid. After this operation the clusters are reconstructed by affecting the set of objects where the distance between the object and the center is minimal. This action is repeated until *K* clusters are obtained.

4.1.1 Collaborative Incremental K-means (Col-Inc-K-means)

The set of users to be classified includes all the social network users. The application of col-Inc-K-means implies the application of the incremental K-means algorithm using the Pearson similarity functions for assigning objects to the corresponding clusters.

4.1.2 Social Incremental K-means (Soc-Inc-K-means)

The set of users to be classified includes all the social network users. The application of soc-Inc-K-means implies the application of the incremental K-means algorithm using the social similarity measure already presented in the previous section.

4.2 Supervised Classification Using K-NN Algorithm

We have chosen the k-nearest neighbors algorithm (K-NN), an instance-based classification method. The complexity of this algorithm is equal to $O(n)$, where n being the total number of users of the training set. This algorithm determines for each newly added user, the list of nearest neighbors among those already classified in the clustering step (unsupervised classification). The newly issued ratings cannot be quickly used for updating the classification obtained by Incremental K-Means as this operation is costly in terms of computation time. To overcome this obstacle, we classify newer users using both collaborative and social K-NN algorithms, adapted respectively to the collaborative and social classification. The collaborative-based K-NN algorithm (resp. social-based K-NN algorithm) uses the same collaborative distance measure already used with the Col-Inc-K-means (resp. Soc-Inc-K-means).

5 Recommendation Algorithm

We present in this section our algorithms for friends' recommendation. Collaborative and social clusters have already been calculated, before running this algorithm.

5.1 SBCF Algorithm

The following is the proposed algorithm of recommendation combining the SocF and the CF without using the classification techniques.

Algorithm 1: SBCF Recommendation Algorithm

Input: User's profile (*U'Profile*), Usage matrix
Output: Rec-list /* A list of recommended friends for a given user *u* */
Begin
  Rec-list: =∅;
  If u is a new user **Then**
      Rec-list: = Rec-list ∪ {list of leaders}
  Else
    If not enough ratings **Then**
        SocF
    Else
        SBCF/* combines the CF with the SocF*/
    EndIf
  EndIf
  Sort Rec-list (u);
  Display (u, Rec-list)
End.

Leaders are, for example, users who are very active in the social network, i.e. evaluate items correctly (based on the average item evaluations given by other users) and/or those with a significant number of friends, having strong/the most important social affinities in the social network.

Sort Rec-list (u): is the function that sorts in descending order the list of recommended friends for *u* (from the most similar friend to the less similar one).

Display (u, Rec-list): is the function that displays the list of recommended friends *Rec-list* for the user *u*.

The SBCF combination algorithm considers the recommendations made using the CF and then applies the SocF (i.e. the SoF is applied using the generated recommendations of users suggested by the CF).

5.2   Recommendation Algorithm Using SBCF and Classification

C-SBCF is the SBCF recommendation of users using the social and collaborative classification techniques.

Algorithm 2: C-SBCF Recommendation Algorithm

Input:

   UserTable /*contains the collaborative classes (ColClass) and social classes (SocClass) of users*/

Output:

   Rec-list /* A list of recommended friends for a given user u */

   For each user u from UserTable **Do**

   　　Rec-list: $=\varnothing$;

   　　$\alpha = 0, \beta = 0$ /* α and β represent a weighted levels of the collaborative and social classes*/

   　　If u has ColClass and SocClass **Then**

   　　　　$\alpha = 0.5, \beta = 0.5$

   　　Else IF u has ColClass and has no CollClass **Then**

   　　　　$\alpha = 1, \beta = 0,$

   　　Else if u has SocClass and has no SemClass **Then**

   　　　　$\alpha = 0, \beta = 1,$

   　　Else /* u has no ColClass and has no SocClass*/

   　　　　Rec-list: = Rec-list $\cup$ {list of leaders};

   　　　　Display (u, Rec-list)

   　　　　Exit

   　　End If.

   For the rest of users u' not friends with u and having same ColClass or SocClass as u **Do**

   　　Sim (u',u) = $(\alpha * Sim_{Col}$ (u', u)) + $(\beta * Sim_{Soc}$ (u', u))

   　　Credibility (u') = $(0.7* Trust$ $(u'))$ + 0.3* $Commitement$ (u')

   　　Rec-Val (u', u) = (0.8 * Sim (u', u)) + 0.2 * Credibility (u'))

   　　IF Rec-Val (u', u) >= $Recommendation\text{-}Threshold$ **Then**

   　　　　$Rec\text{-}list:$ = Rec-list $\cup$ {u'} /*Add u' to the recommended list of u*/.

   　　End If.

   Done.

   Sort Rec-list (u)

   Display (u, Rec-list)

   Done.

End.

---

6 Evaluation

In this section, we conduct experiments on a real dataset to validate the effectiveness of our approach. We used the Yelp social network (http://www.yelp.com/), which aims to connect people with local businesses and we chose the "Restaurant" category because of its frequency use in this social network. The experiments carried out have two main objectives:

1. Show the contribution of combining social information with the user-based CF recommendations;
2. Compare the use of K-means and incremental K-means in the recommendation process and show the added value of the K-NN algorithm.

6.1 Dataset

Before using the Yelp Database, we have performed some pretreatment operations for the inclusion of implicit data (i.e. interests, preferences, commitment and trust degrees, average rating and number of assessments per restaurant, etc.). The resulting database includes 4823 restaurants, 65 categories and 5436 users who have performed 118,709 assessments on these restaurants (only users who have evaluated more than 9 restaurants have been considered).

6.2 Evaluation Metrics

We considered the following evaluation metrics:

- Precision (P): measures the relevance of the recommendations:

$$P = \frac{TruePositive}{TruePositive + FalsePositive} \tag{8}$$

- Recall (R): measures the ability to make relevant recommendations:

$$R = \frac{TruePositive}{TruePositive + FalsePositive} \tag{9}$$

- F-measure (F): combines P and R metrics:

$$F = \frac{2 * P * R}{P + R} \tag{10}$$

- Accuracy (A): measures the performance of the system:

$$A = \frac{TruePositives + FalseNegatives}{TruePositives + TrueNegatives + FalsePositives + FalseNegatives} \tag{11}$$

where:

- True Positives (TP): represents the number of recommendations made to users who were originally friends,
- True Negatives (TN): represents the number of recommendations that are not made to users who were not initially friends,
- False Positives (FP): represents the number of recommendations made to users who were not initially friends,
- False Negatives (FN): represents the number of recommendations not made to users who were initially friends.

6.3   Evaluation Results

We started first by evaluating the CF using the Pearson similarity function and we have varied the values of similarity rate from 0.1 to 0.9 (Sim-Threshold).

Then we evaluated the SocF by varying the weights of the three parameters: the friendship as well as the commitment and trust degrees. The results of the tests carried out revealed that the combination $\alpha1 = 0.1$; $\beta1 = 0.6$; $\gamma1 = 0.3$ gives the best results in terms of precision and F-measure than the two other combinations of $\alpha1$; $\beta1$ and $\gamma1$ (see Fig. 1), where: $\alpha1$, $\beta1$ and $\gamma1$ represent respectively friendship, commitment and trust degrees parameters.

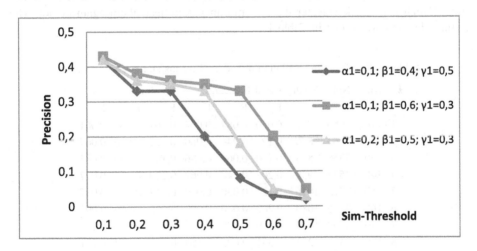

Fig. 1. Identification of the social parameters' weights

Finally, we have evaluated the SBCF using the best parameters for each algorithm. Figure 2 shows that the integration of social information enhanced the CF recommendation accuracy. We have obtained a better precision and F-measure using the SBCF compared to the CF.

In order to see the evolution of the two algorithms K-means and incremental K-means, we have considered in this experimentation only the social classification and

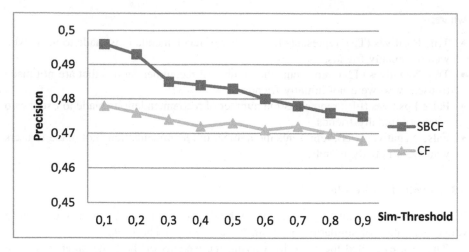

Fig. 2. Contribution of social information on CF recommendations

we have simulated the evolution of the social network using a partition of the database including: 200 users, 351 restaurants, 4852 ratings. We fixed the number of social classes (K = 3) and the recommendation threshold value to 0.3 and we have varied the number of evaluations (NbE), number of users (NbU) and number of deleted friendships (NbDF) to check whether the system can recommend them again or not. The results obtained are shown in Table 1.

Table 1. Results of the evolution of social K-means and Soc-Inc-K-means

I	NbU	NbE	NbDF	social K-means			Soc-Inc-K-means		
				P	R	F	P	R	F
1	110	2950	775	0,050	0,087	0,064	0,107	0,043	0,061
2	120	3163	826	0,123	0,040	0,060	0,121	0,046	0,067
3	130	3356	873	0,130	0,064	0,086	0,161	0,056	0,083
4	140	3563	934	0,207	0,040	0,067	0,220	0,062	0,097
5	150	3773	1036	0,127	0,030	0,049	0,212	0,055	0,087
6	160	3955	1080	0,203	0,060	0,093	0,228	0,06	0,095
7	170	4202	1149	0,329	0,079	0,127	0,350	0,073	0,121
8	180	4415	1237	0,331	0,073	0,120	0,339	0,074	0,121
9	190	4591	1277	0,332	0,086	0,137	0,395	0,081	0,134
10	200	4852	1358	0,354	0,055	0,095	0,472	0,032	0,060

We have presented the difference between the evolution of social K-means and social Incremental K-means in Fig. 3.

To simulate the evolution of the social network, we have sorted the evaluations and users (by registration dates), from the least recent to the most recent. Then, we gradually injected this data into our system.

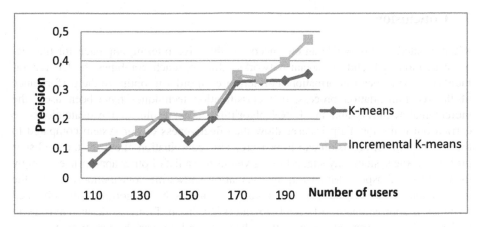

Fig. 3. Evolution of social K-means and incremental social K-means

We applied the Soc-K-Means for every 900 new users and col-K-means for every 20000 evaluations. The results presented in Fig. 4 were obtained with the following parameters for the different iterations (number of added users/new evaluations): col-K-NN = 6, social K-NN = 6, collaborative threshold = 20000, social threshold = 900, recommendation threshold value = 0.3. Between two applications of K-means, the K-NN algorithm is applied (Soc-K-NN is applied for each 300 new added users and Col-K-NN is applied for each 7000 evaluations). These results show the performance of the recommendation algorithm (accuracy value between 0.76 and 0.86) and confirm the contribution of the K-NN algorithm given that the system recommends friends between two applications of Incremental K-means.

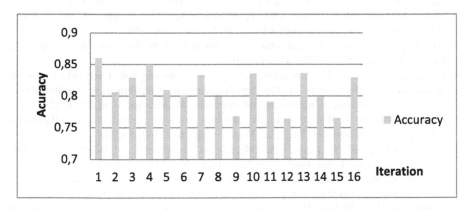

Fig. 4. The accuracy of our system with the evolution of the social network

7 Conclusion

We presented in this article an enhanced collaborative filtering approach for the recommendation of friends in social networks. Our approach combines the CF recommendation with social information. Furthermore, in order to optimize the performance of the recommendation process, two classification techniques have been used, the incremental K-means and the K-NN algorithms. The results of the evaluations we carried out using the Yelp Dataset show the effectiveness of our system compared to the CF in terms of precision and F-measure. This combination alleviates the cold start problem as the system may suggest to a given user u a list of other appropriate users by using the social aspect. Moreover, the evaluation shows the contribution of using the incremental K-means and the added value of the K-NN algorithm (the obtained accuracy is approximately between 0.76 and 0.86). In our future work, we envisage to make further experiments using other datasets and to enrich our approach with semantic information in order to take into account the user's preferences and to benefit from the advantage of the semantic representation of the user's profile.

References

1. Bobadilla, J., Ortega, F., Hernando, A., Gutiérrez, A.: Recommender systems survey. Knowl.-Based Syst. **46**, 109–132 (2013)
2. Adomavicius, G., Tuzhilin, A.: Toward the next generation of recommender systems: a survey of the state-of-the-art and possible extensions. IEEE Trans. Knowl. Data Eng. (TKDE) **17**, 734–749 (2005)
3. Burke, R.: Hybrid recommender systems: survey and experiments. User Model. User Adapted Interact. **12**(4), 331–370 (2002)
4. Pazzani, M.J., Billsus, D.: Content-based recommendation systems. In: Brusilovsky, P., Kobsa, A., Nejdl, W. (eds.) The Adaptive Web. LNCS, vol. 4321, pp. 325–341. Springer, Heidelberg (2007). https://doi.org/10.1007/978-3-540-72079-9_10
5. Liu, F., Lee, H.J.: Use of social network to enhance collaborative filtering performance. Expert Syst. Appl. **37**(7), 4772–4778 (2010)
6. Chang, C.-C., Chu, K.-H.: A recommender system combining social networks for tourist attractions. In: Proceedings of the 5th International Conference on Computational Intelligence, Communication Systems and Networks, Madrid, Spain, pp. 42–47 (2013)
7. Banati, H., Mehta, S., Bajaj, M.: Social behaviour based metrics to enhance collaborative filtering. Int. J. Comput. Inf. Syst. Ind. Manag. Appl. **6**, 217–226 (2014)
8. Su, J.-H., Chang, W.-Y., Tseng, V.S.: Effective social content-based collaborative filtering for music recommendation. Intell. Data Anal. **21**(1), 195–216 (2017)
9. Guo, G., Zhang, J., Yorke-Smith, N.: Leveraging multi-views of trust and similarity to enhance clustering-based recommender systems. KBS **74**, 14–27 (2015)
10. Najafabadi, M.K., Mahrin, M.N., Chuprat, S., Sarkan, H.M.: Improving the accuracy of collaborative filtering recommendations using clustering and association rules mining on implicit data. Comput. Hum. Behav. **67**, 113–128 (2017)
11. Berkani, L.: SSCF: a semantic and social-based collaborative filtering approach. In: Proceedings of the 12th ACS/IEEE International Conference AICCSA, pp. 1–4 (2015)
12. Guellil, Z., Zaoui, L.: Proposition d'une solution au problème d'initialisation cas du K-means. Technical report (2012)

Flexibility in Classification Process

Ismaïl Biskri[(⊠)]

Laboratoire de Mathématiques et Informatique Appliquées,
Université du Québec à Trois-Rivières, Trois-Rivières, Canada
Ismail.Biskri@uqtr.ca

Abstract. A whole classification process is the result of a discovery process that requires constant back and forth between theoretical description of the solution, software implementation, testing and refinement of the theoretical description in the light of the results of experimentation. This process is iterative. It should be, always, under the control of the user according to his subjectivity, his knowledge and the purpose of his analysis. In the last years, several platforms for digging data where classification is the main functionality have emerged. Some of these platforms allow a rapid prototyping and support a re-use of existing "computational modules" from existing "computational tool cases". However, they lack flexibility and sound formal foundations. We propose, in this paper, a formal model with strong logical foundations, based on typed applicative systems. In this model, "computational modules" are considered as operators followed by their operands. A specific processing chain becomes a specific arrangement of a set of modules according to the needs of the user. The model ensures a firm compositionality of this arrangement.

Keywords: Classification · Flexibility · Applicative systems

1 Introduction

Language processing, text processing, social Medias processing, knowledge extraction, applications in education, etc. are a broad field of research including information retrieval, indexation, classification, and information's analysis. As Web is a big source of information, this field can have many implications on several sectors of society. Compared to the quick expansion of data quantity, the evolution of their analysis is too slow and insufficient.

Classification is a major component of many processing chains for language processing, text processing, social Medias processing, knowledge extraction, applications in education, etc. Classification organizes documents, files, data, etc. in such a way that similar documents, files, data, etc. are grouped together. From a computational point of view, classification relies on statistical comparison of content-based descriptors identified by the user according to his needs and goals.

The whole classification process may be divided in three steps. The first one is the features extraction. Extracted features are used as content - based descriptors. The user decides the nature of these descriptors. He may decide to replace them by other descriptors. Indeed, the main difficulty is not extracting the features, over the years,

© IFIP International Federation for Information Processing 2018
Published by Springer International Publishing AG 2018. All Rights Reserved
A. Amine et al. (Eds.): CIIA 2018, IFIP AICT 522, pp. 279–289, 2018.
https://doi.org/10.1007/978-3-319-89743-1_25

many libraries have been developed and shared to facilitate the features extraction; but it is choosing the right features.

The second step is the classification process itself. Different classifiers such as neural network based classifier can be used to perform this step. One of the limitations of these classifiers is that they are opaque and it's not always easy to evaluate the contribution of a specific descriptor and a specific classifier in the classification result.

The third step is the interpretation by the user of the classification results. A user can use different tools to make an enlightened reading of the results. The choice of the tool strongly depends on the purpose of the analysis, or even the knowledge and subjectivity of the user.

Features extraction, classifiers and interpretation tools have an impact on the classification process. Different combinations of features, classifiers and interpretation tools should continue to be explored in order to improve the computer assisted classification process or in some cases better customize the classification process to specific application areas.

In the literature about classification, data-mining, text-mining, and Big-Data, many projects aim to allow the creation of complex processing chains. ALADIN [9], D2K/T2K [5], RapidMiner [8], Knime [11] and WEKA [12] use *processing chains* for language and data engineering, Gate [4] use it for linguistic analysis. The processing chains are widely used, but the solutions previously mentioned suffer from limitations. They are strongly bonded to their specific platforms and programming languages. To take the best advantage of them, the user needs to have knowledge about the developed software and sometimes about programming language. The user is not, always expert, in programming language. A big challenge in these fields is the multitude of disciplines needed to go further. So, experts of different domains need to work together.

In our paper, we propose a formal model based on typed applicative systems, in which the validation of the construction of a processing chain is performed by a logical calculation on types. Typed application systems are widely used in the field of automatic processing of natural language with the current of categorial grammars [1–3] and applicative grammars [10]. They allowed the construction of parsers for many languages like French, Arabic, English, Dutch, Korean, etc. They also allowed logical representations for linguistic or cognitive operators.

Before presenting the formal model itself, we will first introduce, in the next section, typed applicative systems and combinatory logic.

2   Combinatory Logic and Typed Applicative Systems

Combinatory logic was introduced by Moses Schönfinkel in 1924, and extended by Curry and Feys [6, 7]. This logic uses abstract operators called combinators in order to eliminate the need of the variables and, thus, to avoid variable telescoping. Combinators act as functions over argument within a typed operator-operand structure. Their action is expressed by a unique rule called β-reduction rule; which defines the equivalence between the logical expression without combinator and the one with combinator. Elementary combinators can be associated to others to create complex combinators. In our paper, we use only the four elementary combinators **B, C, S, W**, whose notations and β-reductions are shown in the table below.

Combinator	Role	β-Reduction
B	Composition	**B** x y z → x (y z)
C	Permutation	**C** x z y → x y z
S	Distributive composition	**S** x y u → x u (y u)
W	Duplication	**W** x y → x y y

The composition combinator **B** combines two operators x and y together and constructs the complex operator **B** x y that acts on an operand z, z being the operand of y and the result of the application of y to z being the operand of x.

The permutation combinator **C** uses an operator x in order to build the complex operator **C** x that acts on the same two operands as x but in reverse order.

The composition combinator **S** distributes an operand u to two operators x and y.

The duplication combinator **W** takes an operator x that acts on the same operand y twice and constructs the complex operator **W**x that acts on this operand only once.

We can combine elementary combinators together to construct more complex combinators. For example, we could have an expression such as "**B S C** x y z u v". Its global action is determined by the successive application of its elementary combinators (first **B** secondly **S** and finally **C**).

B S C x y z u v
S (**C** x) y z u v
(**C** x) z (y z) u v
C x z (y z) u v
x (y z) z u v

The resulting expression, without combinators, is called a normal form. This form, according to Church-Rosser theorem, is unique.

Two other forms of complex combinators exist: the power and the distance of a combinator. Let χ be a combinator.

The power of a combinator, noted by χ^n, represents the number n of times its action must be applied. It is defined recursively by $\chi^1 = \chi$ and $\chi^n = \mathbf{B} \, \chi \, \chi^{n-1}$. For example, the action of the expression $\mathbf{C}^2$ x y z would be:

$\mathbf{C}^2$ x y z
B C C x y z
C (**C** x) y z
(**C** x) z y
C x z y
x y z

The distance of a combinator, noted by χ_n, represent the number n of steps its action is postponed. It is defined by $\chi_0 = \chi$ and $\chi_n = \mathbf{B}^n \chi$. For example, the action of the expression $\mathbf{C_2}$ x y z u v will be:

$\mathbf{W_2}$ x y z u v
$\mathbf{B}^2 \mathbf{W}$ x y z u v
$\mathbf{B} \mathbf{B} \mathbf{B} \mathbf{W}$ x y z u v
$\mathbf{B} (\mathbf{B} \mathbf{W})$ x y z u v
$(\mathbf{B} \mathbf{W})$ (x y) z u v
$\mathbf{B} \mathbf{W}$ (x y) z u v
$\mathbf{W}$ ((x y) z) u v
((x y) z) u u v
x y z u u v

Applicative systems assign to each operator and to each operand an applicative type to express how they work. The set of applicative types is recursively defined as follows:

1. Basic types are types.
2. If x and y are types, Fxy is a type.

Fxy is the applicative type of an operator whose operand is of type x and the result of its application to its operand is of type y.

An operator function having two x typed operand and returning a y typed result will be of type FxFxy. This can also be read as: an operator taking an x type operand and giving back an operator taking an x type operand and returning an y type result.

3   Formal Model

In our model, we aim at explicitly defining the set of operations contained in programs. In the applicative modeling, these operations are translated into functional terms represented by typed modules. This translation allows a more formal definition of an operation in terms of its internal structure and relation with other operations. Also, this translation allows for a better specification of the processing chain design. Typed modules are organized in series and as such they form processing chains. A typed module acts then like a mathematical function that takes several arguments, process them and return an output. Here, we are not interested in the internal programming of the modules but only in their representation as functions and how they are organized to create processing chains.

A processing chain must be syntactically correct. Its semantic interpretation depends, mainly, on the user's point of view regarding the expected analysis.

Our model tends to answer two questions:

- Given a set of typed modules, what are the allowable arrangements that lead to coherent processing chains?
- Given a coherent processing chain, how can we automate as much as possible its assessment?

Fig. 1. Module schematisation

To do that, we must, first, assign to each module an applicative type. For example the type Fxy is assigned the module M1 in (Fig. 1) since its input is of type x and its output is of type y. We note the module M1 of type Fxy as follow: [M1 : Fxy].

As a general notation, $[M1 : Fx_1 \ldots Fx_n y]$ is a module M1 with n inputs of different types, input in place "i" is of type x_i, and an output of type y, $[M2 : (Fx)^n y]$ is a module M2 with n inputs of type x and an output of type y.

A processing chain is the representation of the order of application of several modules on their inputs. To be valid, the type of an input must be the same as the output linked to it (Fig. 2). It also can be seen as a module itself as it has inputs and output (Fig. 3).

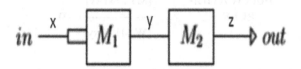

Fig. 2. Valid chain of two modules in series

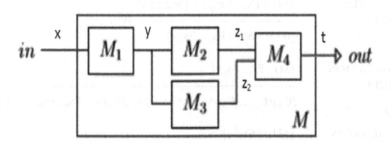

Fig. 3. Processing chain as a new module

Our model allows the reduction of a processing chain to this unique module representation. The combinatory logic keeps the execution order and the rules take type in account to check the syntactic correctness. To reduce a chain, we only need the modules list, their type, and their execution order.

Let us show these rules:

APPLICATIVE RULE	$[X : x] + [M1 : Fxy]$ ----------------------------- $[Y : y]$
COMPOSITION RULE	$[M1 : Fxy] + [M2 : Fyz]$ ---------------------------------**B** $[\mathbf{B}\ M2\ M1 : Fxz]$
DISTRIBUTIVE COMPOSITION RULE	$[M1 : Fxy] + [M2 : FxFyz]$ ---------------------------------**S** $[\mathbf{S}\ M2\ M1 : Fxy]$
PERMUTATION RULE	$[M1 : FxFyz]$ ---------------------**C** $[\mathbf{C}\ M1 : FyFxz]$
DUPLICATION RULE	$[M1 : FxFxy]$ ---------------------**W** $[\mathbf{W}\ M1 : Fxy]$

The above rules are only the core set of the model. Extended rules are provided so they can be applied to any number of inputs.

COMPOSITION RULE	$[M1 : Fx_1...Fx_ny] + [M2 : Fyz]$ ------------------------------------- $\mathbf{B^n}$ $[\mathbf{B^n}\ M2\ M1 : Fx_1...Fx_nz]$
PERMUTATION RULE	$[M1 : Fx_1...Fx_ny]$ -- $\mathbf{C^n}$ $[\mathbf{C}_{p-1}(\mathbf{C}_p(...(\mathbf{C}_{m-2}M1))) : Fx_1...Fx_{p-1}Fx_mFx_p...Fx_{m-1}Fx_{m+1}...Fx_ny]$
DUPLICATION RULE	$[M1 : (Fx)^ny]$ --------------------- $\mathbf{W^n}$ $[\mathbf{W^n}M1 : Fxy]$

The composition rule is used when two modules are in series (as in Fig. 2). If M1 has n inputs, the power of the B combinator is n. For these rules, the inputs number of M2 can be more than one. The duplication rule transforms a module with n identical inputs to a module with only one input. It can be applied only if the chain give the same value to each of its inputs (Fig. 4). The permutation rule allows to change the order of inputs. It takes the input at position m and moves it to the position p, with $p < m$. It's used to reorganize input to make the other rules applicable.

Fig. 4. Module getting a single value in its three inputs

4   Application of the Approach

In this section, we will show how the rules given in the previous section are applied and illustrate the reduction of a processing chain with an example.

Let us consider the linear connection of two modules (Fig. 2). The module [M1 : FxFxy] applies on two identical inputs of type x and yield an output of type y. The module [M2 : Fyz] applies on this output to yield an output of type z. This chain is expressed by the expression: [M1 : FxFxy] + [M2 : Fyz]. The composition rule can be applied and returns the complex module [$\mathbf{B}^2$ M2 M1 : FxFxy]. If the type of M1 output and M2 output where not the same, we could not have applied the composition rule. So, the application of the rules is a proof of syntactic correctness of the chain.

The module [$\mathbf{B}^2$ M2 M1 : FxFxy] can be reduced a second time with the duplication rule. It is reduced to the complex module [$\mathbf{W}$ ($\mathbf{B}^2$ M2 M1) : Fxy].

The permutation rule allows reorganising the inputs of a module to apply another rule. Let M be a module with four inputs of types x, y, z and x and an output of type t: [M : FxFyFzFxt]. Let X be the value given to the first and fourth inputs (Fig. 5a). If the fourth was in second position, we could apply the duplication rule to M. So, we want to move the fourth input to second position. The permutation rule returns the complex module [$\mathbf{C_1}$ ($\mathbf{C_2}$ M) : FxFxFyFzt] (Fig. 5b). On this new module, the duplication rule can be applied to get a complex module [$\mathbf{W}$ ($\mathbf{C_1}$ ($\mathbf{C_2}$ M)) : FxFyFzt] (Fig. 5c).

Let us now give the analysis of a somewhat complex processing chain (Fig. 6). This chain is a combination of five modules.

- M1 of type FxFyz
- M2 of type Fzx
- M3 of type Fzx
- M4 of type Fzy
- M5 of type FxFxFyt

To reduce this chain, we will start with the last module and process from left to right. So we start with [M5 : FxFxFyt]. His first input takes the output of [M2 : Fzx]. The composition rule gives a new complex module [$\mathbf{B}$ M5 M2 : FzFxFyt] (Fig. 7). This new module and [M3 : Fzx] can be reduced with the distributive composition rule to get the module [$\mathbf{S}$ ($\mathbf{B}$ M5 M2) M3 : FzFyt] (Fig. 8). The first input of this module can be reduced with the composition rule to get a new module [$\mathbf{B}^2$ ($\mathbf{S}$ ($\mathbf{B}$ M5 M2) M3) M1 : FxFyFyt] (Fig. 9).

To reduce this module with [M4 : Fzy] we want to use the composition rule. But to apply it, M4 output must be the first input of our module. We use the permutation rule to reorganise the inputs and got a new module [$\mathbf{C}$ ($\mathbf{C_2}$ ($\mathbf{B}^2$ ($\mathbf{S}$ ($\mathbf{B}$ M5 M2) M3) M1)) : FyFxFyt] (Fig. 10). Finally, we can apply the combination rule that returns the module

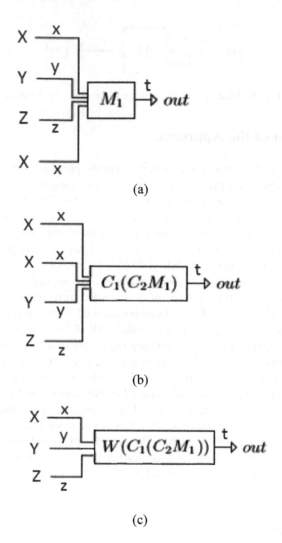

(a)

(b)

(c)

Fig. 5. Inputs reorganisation

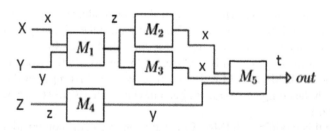

Fig. 6. A complex processing chain

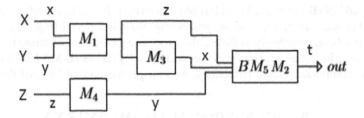

Fig. 7. Reduction step 1

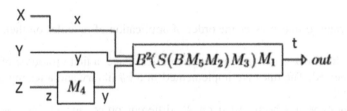

Fig. 8. Reduction step 2

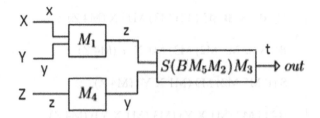

Fig. 9. Reduction step 3

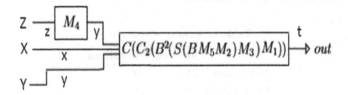

Fig. 10. Reduction step 4

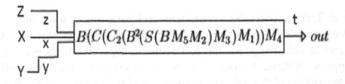

Fig. 11. Reduction last step

[**B** (**C** (**C$_2$** (**B^2** (**S** (**B** M5 M2) M3) M1))) M4 : FzFxFyt]. As we have only one module, and no other rule can be applied, the processing chain is reduced (Fig. 11).

As it has been completely reduced, the processing chain is considered as syntactically correct. Its combinatory expression is: **B** (**C** (**C$_2$** (**B^2** (**S** (**B** M5 M2) M3) M1))) M4. Using combinatory logic reductions, we can get the normal form of this expression.

$$\textbf{B} (\textbf{C} (\textbf{C}_2 (\textbf{B}^2 (\textbf{S} (\textbf{B}\ M5\ M2)\ M3)\ M1)))\ M4\ Z\ X\ Y$$

$$\textbf{C} (\textbf{C}_2 (\textbf{B}^2 (\textbf{S} (\textbf{B}\ M5\ M2)\ M3)\ M1))\ (M4\ Z)\ X\ Y$$

$$\textbf{C}_2 (\textbf{B}^2 (\textbf{S} (\textbf{B}\ M5\ M2)\ M3)\ M1)\ X\ (M4\ Z)\ Y$$

$$\textbf{B}^2 (\textbf{S} (\textbf{B}\ M5\ M2)\ M3)\ M1\ X\ Y\ (M4\ Z)$$

$$\textbf{S} (\textbf{B}\ M5\ M2)\ M3\ (M1\ X\ Y)\ (M4\ Z)$$

$$\textbf{B}\ M5\ M2\ (M1\ X\ Y)\ (M3\ (M1\ X\ Y))\ (M4\ Z)$$

$$M5\ (M2\ (M1\ X\ Y))\ (M3\ (M1\ X\ Y))\ (M4\ Z)$$

This normal form expresses the order of application of modules on their inputs (X, Y and Z).

Even if this work is currently at the theoretical stage, a first prototype of the model was implemented. The rules are implemented in a F# library and a testing software in C# language.

The prototype has been tested on 40 different processing chains containing 15 syntactically incorrect chains and 25 correct chains. The results are shown in Table 1. We are, currently, working on the implementation of modules with effective functionalities in the domain of classification.

Table 1. Results of reduction

	Reduced	Not reduced
Valid chain	25	0
Invalid chain	0	15

5   Conclusion

The need for flexible, adaptable, consistent and easy-to-use tools and platforms is essential. But, many challenges are yet to be solved. The user stays in center of its experience and he can change his mind. The flexibility of the tools is really important when it happens. Without it, user needs to, constantly, go back and forth between theoretical description of the solution, software implementation, testing and refinement of the theoretical description in light of experimentation results. The model that we propose allows rapid prototyping and support a maximal re-use and composition of existing modules. It also ensures a firm compositionality of the different modules in the different processing chains. Moreover, our approach provides a general framework in which users would be able to build multiple language and text analysis processes according to their own objectives.

References

1. Biskri, I., Desclés, J.P.: Applicative and combinatory categorial grammar (from syntax to functional semantics). In: Recent Advances in Natural Language Processing. John Benjamins Publishing Company, Amsterdam (1997)
2. Biskri, I., Anastacio, M., Joly, A., Amar Bensaber, B.: A typed applicative system for a language and text processing engineering. In: International Journal of Innovation in Digital Ecosystems. Elsevier, Amsterdam (2015)
3. Biskri, I., Anastacio, M., Joly, A., Amar Bensaber, B.: Integration of sequence of computational modules dedicated to text analysis: a combinatory typed approach. In: Proceedings of AAAI. FLAIRS 2013, St. Pete Beach (2013)
4. Cunningham, H., Maynard, D., Bontcheva, K., Tablan, V.: GATE: a framework and graphical development environment for robust NLP tools and applications. In: Proceedings of the 40th Anniversary Meeting of the Association for Computational Linguistics (ACL 2002), Philadelphia (2002)
5. Downie, J.S., Unsworth, J., Yu, B., Tcheng, D., Rockwell, G., Ramsay, S.J.: A revolutionary approach to humanities computing: tools development and the D2K datamining framework. In: Proceedings of the 17th Joint International Conference of ACH/ALLC (2005)
6. Curry, B.H., Feys, R.: Combinatory Logic, vol. I. North Holland, Amsterdam (1958)
7. Hindley, J.R., Seldin, J.P.: Lambda-Calculus and Combinators, an Introduction. Cambridge University Press, Cambridge (2008)
8. Mierswa, I., Wurst, M., Klinkemberg, R., Scholz, M., Euler, T.: YALE: rapid prototyping for complex data mining tasks. In: Proceedings of the 12th ACM SIGKDD International Conference on Knowledge Discovery and Data Mining (KDD 2006). ACM Press (2006)
9. Seffah, A., Meunier, J.G.: ALADIN: Un atelier orienté objet pour l'analyse et la lecture de Textes assistée par ordinaleur. In: International Conferencence on Statistics and Texts, Rome (1995)
10. Shaumyan, S.K.: Two paradigms of linguistics: the semiotic versus non-semiotic paradigm. Web J. Formal Comput. Cogn. Linguist. 2, 1–72 (1998)
11. Warr, A.W.: Integration, Analysis and Collaboration. An Update on Workflow and Pipelining in Cheminformatics. Strand Life Sciences (2007)
12. Witten, I., Frank, E., Hall, M.: Data Mining: Practical Machine Learning Tools and Techniques. Morgan Kaufmann Publishers, Burlington (2011)

An Evolutionary Scheme for Improving Recommender System Using Clustering

ChemsEddine Berbague[1]([✉]), Nour El Islem Karabadji[1,2], and Hassina Seridi[1]

[1] Electronic Document Management Laboratory (LabGED),
Badji Mokhtar-Annaba University, BP 12, Annaba, Algeria
{berbague,Karabadji,Seridi}@labged.net
[2] High School of Industrial Technologies, P.O. Box 218, 23000 Annaba, Algeria

Abstract. In user memory based collaborative filtering algorithm, recommendation quality depends strongly on the neighbors selection which is a high computation complexity task in large scale datasets. A common approach to overpass this limitation consists of clustering users into groups of similar profiles and restrict neighbors computation to the cluster that includes the target user. K-means is a popular clustering algorithms used widely for recommendation but initial seeds selection is still a hard complex step. In this paper a new genetic algorithm encoding is proposed as an alternative of k-means clustering. The initialization issue in the classical k-means is targeted by proposing a new formulation of the problem, to reduce the search space complexity affect as well as improving clustering quality. We have evaluated our results using different quality measures. The employed metrics include rating prediction evaluation computed using mean absolute error. Additionally, we employed both of precision and recall measures using different parameters. The obtained results have been compared against baseline techniques which proved a significant enhancement.

1 Introduction

A recommender system RS is an extension of the conventional information retrieval techniques. It plays a major role in e-commerce websites as well as social networks such as Google, Facebook, Netflix and others whereas RS aims to estimate user ratings or to personalize a bag of recommendation for every user based on his preferences made explicitly by bias of his ratings or implicitly during the interaction with the system (browsing sequences). Besides, that contextual information (time, location), as well as social and demographic information, are used to make high-quality recommendations. There are different recommendation algorithms each of which fits a specific type of information and deals with a well-defined limitation; two well-known algorithms are listed in the items below.

One, Content based filtering algorithms denoted CBF [1] uses available information about items (features, attributes) to make recommendations for a target use & here they look for a set of nearest neighbors for the previously seen items; which will be used to estimate user's preferences. CBF is a very beneficial filtering

© IFIP International Federation for Information Processing 2018
Published by Springer International Publishing AG 2018. All Rights Reserved
A. Amine et al. (Eds.): CIIA 2018, IFIP AICT 522, pp. 290–301, 2018.
https://doi.org/10.1007/978-3-319-89743-1_26

algorithm since it can deal with low users overlap by focusing only on user's profile; however it involves a hard content preprocessing step which raises a complicated task especially for media content (images, videos). Besides, for a given user, CBF algorithm tends to recommend a set of very similar items to those already seen in the past, by consequence limiting the diversity of recommendations, this effect is denoted in the literature by the overspecialization problem.

Second, memory based collaborative filtering algorithms denoted CF [2,3] have given good results in terms of accuracy and used widely in many researches [4]; User based CF bases its recommendation on the concept of the collective users trends toward products and items whereas the algorithm predicts the unrated items for a specific user by considering similarities with the rest of users around him, So close users would influence strongly the predicted rating value while far users would have a limited effect; Similarities distance between users may be computed in many ways such as cosine similarity, Pearson correlation, etc. K nearest neighbors (KNN) [2] is the most used algorithm for collaborative filtering, KNN takes on consideration only a limited number of users; where it determines for a given user k nearest neighbors then calculates predictions for the unrated items by aggregating nearest users ratings. Some contributions [3,5] have restricted the neighbors set size to include only users who (1) satisfy a minimum similarity threshold; (2) simultaneously allow making both accurate and diverse recommendations.

For a given user $u \in U$ and item $i \in I$ the estimated rating is calculated using the Eq. 1:

$$Pr(u,i) = \bar{r_u} + \frac{\sum_{j \in N(u)} sim(u,j) \times R(j,i)}{\sum_{j \in N(u)} sim(u,j)} \tag{1}$$

Where U denote the set of all users in the dataset, $|.|$ the size of the set while i_u is the set of users who have rated the item i.

An ideal collaborative filtering engine should cope with scalability problem, and recommendation quality since rating prediction in whole datasets might be a very complicated task and involves long time and high computation complexity which are proportional to the number of users and items in the rating matrix. A common approach to deal with scalability issues is to make a preprocessing step consists of clustering users into groups of similar users and restrict collaborative computing on the scale of the cluster to which a target user belongs. K-Means is a partitioning algorithm used widely in recommendation system to address the scalability issue; the algorithm is known by its suitability for large scale datasets and fast convergence, while it does not guarantee the best results since it usually gets stuck at local optimality solutions. The initialization step plays a crucial role in clustering quality; random based initialization of K-means in the classical version influences strongly the quality of obtained results. K-means clustering involves specifying initially a fixed number of clusters, the algorithm stops after getting a lower error rate according to a predefined threshold or after reaching a number of iteration.

In more details, the algorithm assigns each data vector to the closest center and iteratively looks for the best center inside each newly formed cluster. Clustering results depend on the initially chosen seeds which raises the worst drawback.

Many papers have addressed the initial seeds selection dilemma in K-means algorithm; it exists two considerable classified solutions one based on statistic and another one based on the evolutionary algorithms. The first type scans data samples to look for their statistical features, density and variance are some measurements used to select the appropriate initial seeds; By revenge mainly evolutionary approaches perform a research in the search space of possible solutions by adopting an adequate formulation of the clustering problem and optimizing a clustering quality measure, density, mean square error have been used to guide the optimization.

In this paper, we present a partitioning based genetic algorithm to enhance user based collaborative filtering. This latter' objective is achieved by pulling up an optimal partitioning of users over k groups. This may allow us to deal with scalability problem in recommender systems. The proposed optimization algorithm guided by a clustering quality measure explores possible solutions over the whole ways allowing for partitioning a set of n users into k non-empty subsets. Mainly, our contribution in this work consists in (1) proposing a new genetic individuals encoding that represents possible solutions where each solution is represented as a set of centers and the number of the most similar users around them. (2) designing a multi-objective optimization fitness function to ensure similarity intergroup and diversity between centers. The experimental results on the benchmark Movielens dataset using different parameters and compared with baseline approaches proved a significant enhancement.

The remainder of this paper includes in Sect. 2 related work explaining clustering in recommendation systems and discussing the initial seed selection problem in K-means algorithm, Then in Sect. 3, we cite minutely our proposed clustering approach based completely on a genetic algorithm. In Sect. 4, we present the experimental results on MovieLens dataset compared to K nearest neighbors algorithm and k-means.

2   Related Works

Collaborative filtering (memory based) is a widely used recommendation algorithm that bases its recommendation on selecting a set of neighbor users by measuring similarity between each pair of users, this step is a time consuming and a high computation complexity task, these problems have been treated by proposing distributed collaborative filtering [6], other approaches consist of clustering users into similar groups on which computation is restricted.

Many statistical and bio-inspired clustering algorithms have been proposed in the literature for recommender system. However, K-means is still the most used algorithm due to its simplicity and fast convergence. In contrary this algorithm suffers the initial seeds selection dilemma. A comparative study [7] has taken the different existing approaches of k-means initialization in terms of complexity.

In [8] Zahra et al. have proposed many variants of K-means algorithm, it employs mainly similarity measures as well as density, variance, average to select best initial seeds, in some other extensions authors have made a filtering on users based on their statistical measures to improve their proposed K-means algorithms, the filtering has extracted users with high similarity distance from the remaining data samples or users who have larger rating size, the proposed algorithms in the paper have been applied to many recommendation system data sets and their results have been compared in terms of mean square error and precision measure, however, the proposed algorithms have not dealt with sparsity problem, measuring similarity between two users profiles might reflect inaccurate similarity.

In [9] Cao et al. have proposed an iterative progressive initial seed selection for k means algorithm, their method consists of a random selection of the first seed, however, the remainder $k-1$ seeds are selected in regardless to the actually inserted seeds in the seed set, two selection criteria are employed to ensure the coherence of the seeds: coupling combining indicates the probability of belonging of two samples to the same cluster, cohesion give an index about the breadth of neighborhood of a given sample, the efficiency of the algorithm is highly influenced by the selection of the first seed, selecting inappropriate first seed would affect the rest of seeds.

In [10] Li has proposed a center clustering initialization based on euclidean distance measurement and neighbors selection, the algorithm constitutes a set of pairs each of which contains a data sample and its nearest neighbors found by scanning the dataset and calculating the euclidean distance, the authors have put some initial seed selection rules based on two main assumptions: A data sample and its NN are in the same cluster or the overlap of two clusters; The more two pair of user and his NN are dissimilar the more they belong to different clusters, these algorithms are still can deal only with a limited number of cluster equal 2.

Besides statistically based clustering methods, there are evolutionary approaches [11] which have been used widely for clustering in recommendation systems since the conventional clustering techniques are subject to failing stuck at local optimality solutions, evolutionary clustering techniques have been classified into two categories: Algorithms that require specifying the number of clusters and algorithms that do not require such information.

In [12] Kuo et al. proposed a k-means algorithm based on ant colony, the execution starts by initializing a number of clusters and its correspondents randomly selected centroids, equal pheromone values are laid on each possible path, ants are assigned to random samples, at every iteration pheromone values are updated so ants move from a source to a center respecting a transmission probability based on a statistical based variance value.

In [13] Kalyani and Swarup have proposed a supervised k-means algorithm whereas a PSO is integrated into the clustering process; a set of randomly generated particles represent initial seeds which are fed to a classical k-means algorithm, the quality measure computed by k-means results is used to supervise

PSO behavior. However, supervision process is complex since the quality computation involves executing the k-means algorithm to obtain the clustering quality value.

Globally most proposed evolutionary (EA) based clustering algorithms are combined with a baseline clustering technique, such as k-means. The objective of EA is enhancing the clustering by exploring the search space to find the best initial seeds. Genetic algorithms are an evolutionary approach proposed early in 1989 and used mainly for optimization problems. The algorithm is inspired from the natural selection and survive for the fittest principals. It has been used for recommendation systems in many aspects such as similarity computation [14], clustering [15] and recommendation algorithms hybridization [16]. Most presented works using genetic algorithms have discussed encoding scheme problem, fitness function and genetic operators as well as the population initialization choice. We present in our paper a completely genetic based clustering algorithm that finds the appropriate number of clusters without requiring k-means post clustering.

3   Partitioning Based Genetic Algorithm

On the contrary of traditional algorithms that fail stuck at local optimality, genetic algorithms (GA) are a powerful optimization techniques since it looks for a global optimum solution. GAs apply genetic operations at every iteration to produce completely new different solutions, these operations are applied initially on a first generation generated randomly. In particular, we fixed the initial number of individuals to 20, using a single point crossover operator with a probability of 0.5 as well as employing bit flip mutation operator with a probability of 0.1.

3.1   Encoding and Decoding Phase

This phase is an essential step of our proposed genetic algorithm. Formally, this phase consists of designing a bijective function that allows us representing each solution of whole ways allowing for partitioning a set of n user into k non-empty groups.

Encoding. The encoding step consists in representing solutions as arrays of 0s and 1s (i.e., binary strings). Therefore, we propose a new binary string encoding, each one must store the full information about its corresponding partition solution. Our designed encoding is presented in Fig. 1, where a chromosome is composed of k genes and each gene stores two informations: (a) an integer C_i representing a center; (b) an integer α_i representing the number of must nearest users over the center C_i. This encoding allows representing the whole possible partition configurations. Where k is the number of partition for which we take as parameters two thresholds minimum Min_C and maximum Max_C number of clusters and look for the best users partition over k groups. Each group (i.e.,

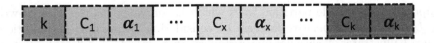

Fig. 1. Chromosome encoding scheme

010	0000000000010	0100	0000000000111	1110
k	C_1	α_1	C_2	α_2

(a)

100	01010	0100	00101	0100	01111	0100	00001	0100
k	C_1	α_1	C_2	α_2	C_3	α_3	C_4	α_4

(b)

011	00000001	0100	00000101	1000	00001111	0110
k	C_1	α_1	C_2	α_2	C_3	α_3

(c)

100	00001	0100	00100	0100	00101	0100	01000	0100
k	C_1	α_1	C_2	α_2	C_3	α_3	C_4	α_4

(d)

Fig. 2. Chromosome instances

cluster) of the k ones is represented by its center C_i and the number of its members α_i. A C_i storage space is represented by X_C bits which encode a user identifier $[1..|U|]$. Even here we can note that the C_i identifiers must be different. While α_i storage space is represented by X_α bits which encode members number in each group and their sum must be equal to $(|U| - k)$. The chromosome size $|ch|$ is number of bits need to represent in binary: (1) Max_C; (2) $|U|$; (3) $|U| - k$. Thus $|ch| = X_{Max_C} + (Max_C * (X_\alpha + X_\alpha))$. To put it simply we propose to see Example 1.

Example 1. Assume a data with $|U| = 200$. Thus, the minimum Min_C and the maximum Max_C number of clusters are 2 and 10 respectively if the size of the cluster including the center is at least 20. Therefore, to encode all possible configurations, we consider the largest case which requires the greatest storage space. For this instance case we need to encode information over $X_k + Max_C * (X_C + X_\alpha)$ bits= 164 bits. First, to encode the clusters number, 4 bits are required to store a binary representation of integers between 2 and 10. Then, a maximum size value of a group including a center may be 180 for two clusters partition. According to this maximum size value (i.e., 180), the storage space X_α required is 8 bits. While the most value of center identifier may be 200, then the storage space X_C required is 8 bits.

Decoding. The decoding steps allow converting the binary gene codes encoded in a chromosome ch into the appropriate k groups. This conversion consists in determining centers' identifier (i.e., users considered as centers) and the groups' size over these centers. Mainly this decoding phase follows two steps: (1) Binary subsequence represented the group's number is converted to the appropriate integer (i.e., k). (2) According to this groups number, couples of binary subsequences represented centers and groups size are converted to the appropriate user identifier and groups member size (i.e., an integer).

Example 2. Assume a data with $|U| = 20$, a minimum Min_C and a maximum Max_C number of clusters equal to 2 and 4 respectively. The search space is composed of the whole partition ways which is equivalent to the ways of writing $|U|$ as a sum of positive integers by considering all possible permutations. Figure 2 illustrates 4 chromosome instances that encode 4 different solutions which could be decoded into: (a) $k = 2$, $(C_1 = 2, \alpha_1 = 4)$, $(C_2 = 5, \alpha_2 = 14)$; (b) $k = 4$, $(C_1 = 10, \alpha_1 = 4)$, $(C_2 = 5, \alpha_2 = 4)$, $(C_3 = 15, \alpha_3 = 4)$, $(C_4 = 1, \alpha_4 = 4)$; (c) $k = 3$, $(C_1 = 1, \alpha_1 = 4)$, $(C_2 = 5, \alpha_2 = 8)$, $(C_3 = 15, \alpha_3 = 6)$; (d) $k = 4$, $(C_1 = 1, \alpha_1 = 4)$, $(C_2 = 4, \alpha_2 = 4)$, $(C_3 = 5, \alpha_3 = 4)$, $(C_4 = 8, \alpha_4 = 4)$.

All chromosomes are 0's and 1's strings of 39 length. Clusters number is encoded on 3 bits. Cluster centers C_i are encoded at least on 5 bits and at most 9 bits (i.e., 5 bits for chromosomes (b) and (d), 6 bits for chromosome (c) and 9 bits for chromosome (a)). Groups size are encoded on 4 bits. According to the decoding phase: first the k clusters number is designed where the three first bits of chromosomes are converted to integer (i.e., 010⇔2, 100⇔4, 011⇔3, 100⇔4 for chromosomes (a), (b), (c) and (d) respectively). Next, using k we can determined the couples of centers identifier and size of the groups over them. The number of bits required to encode centers is $log_2(|U|)$. However, if k is less than Max_C, extra 0's bits are added to X_c. According to this example, nine 0's and three 0's bits are added to centers identifier bits on chromosomes (a) and (b) respectively. Converting these centers identifier bits results users identifier considered as centers. Then, bits representing groups size are converted to the appropriate integer. Finally, each chromosome is converted to its corespondent groups, where each center C_i and its α_i most nearest neighborhoods compose a cluster.

3.2   Fitness Function

A good clustering quality in recommendation context implies maximizing accuracy on the level of each cluster as well as keeping a meaningful distance between centers, our fitness function combines these two measure whereas we have computed chromosome fitness as the next:

$$group\_precision(ch) = (max(r) - min(r)) - (\frac{1}{k} \times (\sum_{i=1}^{k} MAE(G_i)) \qquad (2)$$

$$center\_diversity(ch) = \frac{1}{k \times (k-1)} \times (\sum_{i=1}^{k} \sum_{j=i+1}^{k-1} (1 - sim(C_i, C_{j+1}))) \qquad (3)$$

$$fitness(ch) = group\_precision(ch) + center\_diversity(ch) \qquad (4)$$

Where max(r) and min(r) denote the biggest and lowest rating values equivalent to 5 and 1 respectively, G_i is the ith cluster and C_i is the center of the cluster i. In fact, we have chosen to combine two statistical terms: the first one has the goal of giving a significant indicator about centers positions. Logically, those two centers should not be very close or belonging to the same cluster. However targeting only maximizing distances between pair of centers will lead to searching for the farthest users in search space. For that reason, we added a second term to control the internal prediction error in each cluster.

4 Experiments

In this section, we validate the proposed clustering algorithm experimentally, the obtained results are compared to both KNN algorithm applied directly on the whole dataset, by applying kmeans data reduction and PCA-$GAKM$ a well known collaborative filtering.

4.1 Dataset Description

Experiments are performed on movielens data that contains 100.000 ratings assigned by 943 users on 1682 movies, every user gives a rating in a scale of 1 to 5, the main objective of a recommender is to predict the unrated movies which represent about 93% missing values of all possible ratings. We have kept 90% of ratings for training while carrying out randomly 10% of ratings for the test.

4.2 Evaluation Measure

There are several evaluation metrics in the literature. The available evaluators include rating prediction error and recommendation set evaluators. We have evaluated our algorithm from both of these aspects. We cite in next items the employed evaluation metrics which we used in the experimental section.

Rating Prediction Evaluation. Mean Absolute Error (MAE) is a statistical measure used widely over different studies to evaluate the recommendation accuracy. Whereas MAE computes the proportion of the sum absolute difference between every real rating in the test set and its equivalent predicted by the recommender, as shown in the Eq. 5:

$$MAE = \frac{1}{|U|} \sum_{u \in U} \frac{\sum_{i \in I_u} |p_u^i - r_u^i|}{|I_u|} \tag{5}$$

Where $|U|$ is the number of user, P_u^i is the predicted value, $|I_u|$ is the number of rated items by the user u.

Recommendation Set Evaluation. In contrary of rating prediction evaluation, recommendation set evaluation gain an increasing importance. We used both of precision and recall for validation our results. Whereas for a u's recommendation set I_u, correct u's recommendation set I_u^c and a minimum relevance threshold β we define precision and recall in the next items.

A. Precision: this metric computes the portion of relevant items in the recommendation set among the total list.

$$Precision = \frac{1}{|U|} \sum_{u \in U} \frac{|\{i \in I_u | R_{ui} > \beta\}|}{|I_u|} \qquad (6)$$

B. Recall: this metric computes the portion of relevant items in the recommendation set among the total relevant items.

$$recall = \frac{1}{|U|} \sum_{u \in U} \frac{|\{i \in I_u | R_{ui} > \beta\}|}{|\{i \in I_u^c | R_{ui} > \beta\}|} \qquad (7)$$

4.3   Experimental Design

Our main objective is to validate the efficiency of the proposed algorithm against two of state of the art algorithms, the proposed GA configurations consist at initializing the number of chromosomes at 20, the maximal iteration number at 200. While for both K-means and *PCA-GAKM* we chose the best clustering configuration that allows minimizing their MAE.

4.4   Analyzing Neighbor Size Parameter

Collaborative filtering algorithm involves two stages: a neighbor selection process followed by a rating prediction step. In particular, neighbors selection deals with two important parameters. Expressly the distance measure choice and the size of the neighborhood. For the first parameter we have employed euclidean distance however for the second one we moved the number of neighbors from 5 to 60 by an increment of 5 each time.

KNN algorithm has been applied one time on the whole dataset another one separately on each cluster gotten using the different clustering algorithms: GA based clustering, K-means, PCA-GAKM. In detail, We have applied different evaluation metrics on KNN considering both rating accuracy and recommendation set evaluators, Whereas we adopted MAE metric for validating rating accuracy. While for recommendation set evaluation, We have used precision and recall by fixing the number of recommendation at 5.

Prediction Rating Accuracy. We have used MAE measure in wide neighbor size range to evaluate our results against baseline recommender. Notably that The results shown in the Figs. 3 and 4 show the superiority of our algorithm against both of classical KNN and the clustering techniques. Whereas we observe that in general MAE trace descends by increasing the size of neighbors. In fact K-means clustering is almost stable in MAE range of 0.81 to 0.82 while our algorithm gives widely better results than KNN as well as it overpasses PCA-GAKM after a neighbors size of 10.

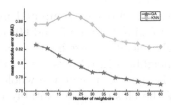

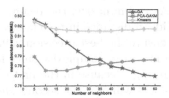

Fig. 3. Proposed algorithm against KNN

Fig. 4. Proposed algorithm against Kmeans and PCA-GAKM

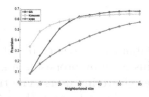

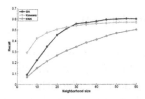

Fig. 5. Precision comparison

Fig. 6. Recall comparison

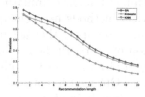

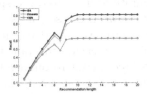

Fig. 7. Comparison of precision with different recommendation length

Fig. 8. Comparison of recall with different recommendation length

Recommendation Set Accuracy. We examine our proposed algorithm in terms of precision and recall against KNN and K-means algorithm. We observe in Figs. 5 and 6 that both of clustering algorithms are better than KNN. Notably that K-means starts better than our proposed algorithm however after a neighborhood size of 30 it becomes worse.

4.5   Analyzing Recommendation Length Parameter

We analyze in this section the number of recommendation parameter in terms of accuracy and recall. The length of recommendation has been increased incrementally from 1 to 20. The results in Figs. 7 and 8 show precision and recall variation while increasing recommendation length. We observe that precision corresponds inversely and recall corresponds directly the recommendation length. Additionally, both of clustering algorithms keep giving better results than KNN with a clear superiority of our algorithm against Kmeans.

As a summarize, the results show that the users closeness computed using euclidean distance in the classical KNN for a k ranging between 5 and 60 could not ensure the best precision values. In fact, higher neighbor size increases accuracy in cost of complexity since looking for 60 neighbors in KNN is harder than clustering. Because in the first case the search is open to include the whole set members while in the second one is limited by the cluster's size. Notably that the clustering approach has achieved a better balance between results accuracy and the scalability. Whereas, the proposed genetic algorithm overpassed KNN as well as both of Kmeans and PCA-GAKM clustering.

5   Conclusion

Scalability problem is a major drawback in collaborative memory based filtering algorithm. We have targeted in this paper this issue by proposing a genetic based partitioning algorithm in the aim of better clustering users. The treated problem have been encoded to reduce search space by indicating the number of possible clusters as well as specifying their maximum and minimum size. Besides that, we have maximized the quality of clustering in a way to achieve more accurate results. However, sparsity problem might influence similarity measurement between the set of users. Many statistical methods have emerged to make users profiles more dense, in addition, it starts to appear that accuracy is an insufficient measure to evaluate the satisfaction of users. Pushed by a large amount of redundant recommendations, diversity and novelty of recommendation are becoming more interesting quality measures. Our next intent is orienting clustering to diversity enrichment purpose passing first by addressing sparsity problem faced during our current research.

References

1. Balabanović, M., Shoham, Y.: Fab: content-based, collaborative recommendation. Commun. ACM **40**(3), 66–72 (1997)
2. Candillier, L., Meyer, F., Boullé, M.: Comparing state-of-the-art collaborative filtering systems. In: Perner, P. (ed.) MLDM 2007. LNCS (LNAI), vol. 4571, pp. 548–562. Springer, Heidelberg (2007). https://doi.org/10.1007/978-3-540-73499-4_41
3. Karabadji, N.E.I., Beldjoudi, S., Seridi, H., Aridhi, S., Dhifli, W.: Improving memory-based user collaborative filtering with evolutionary multi-objective optimization. Expert Syst. Appl. **98**, 153–165 (2018)
4. Mustafa, N., Ibrahim, A.O., Ahmed, A., Abdullah, A.: Collaborative filtering: techniques and applications. In: 2017 International Conference on Communication, Control, Computing and Electronics Engineering (ICCCCEE), pp. 1–6, January 2017
5. Liu, H., Hu, Z., Mian, A., Tian, H., Zhu, X.: A new user similarity model to improve the accuracy of collaborative filtering. Knowl. Based Syst. **56**, 156–166 (2014)

6. Narang, A., Srivastava, A., Katta, N.P.K.: Distributed scalable collaborative filtering algorithm. In: Jeannot, E., Namyst, R., Roman, J. (eds.) Euro-Par 2011, Part I. LNCS, vol. 6852, pp. 353–365. Springer, Heidelberg (2011). https://doi.org/10.1007/978-3-642-23400-2_33
7. Celebi, M.E., Kingravi, H.A., Vela, P.A.: A comparative study of efficient initialization methods for the k-means clustering algorithm. Expert Syst. Appl. **40**(1), 200–210 (2013)
8. Zahra, S., Ghazanfar, M.A., Khalid, A., Azam, M.A., Naeem, U., Prugel-Bennett, A.: Novel centroid selection approaches for KMeans-clustering based recommender systems. Inf. Sci. **320**, 156–189 (2015)
9. Cao, F., Liang, J., Jiang, G.: An initialization method for the K-means algorithm using neighborhood model. Comput. Math. Appl. **58**(3), 474–483 (2009)
10. Li, C.S.: Cluster center initialization method for K-means algorithm over data sets with two clusters. Procedia Eng. **24**(2011), 324–328 (2011). International Conference on Advances in Engineering
11. Hruschka, E.R., Campello, R.J.G.B., Freitas, A.A., de Carvalho, A.C.P.L.F.: A survey of evolutionary algorithms for clustering. IEEE Trans. Syst. Man Cybern. Part C (Appl. Rev.) **39**(2), 133–155 (2009)
12. Kuo, R., Wang, H., Hu, T.L., Chou, S.: Application of ant K-means on clustering analysis. Comput. Math. Appl. **50**(10), 1709–1724 (2005)
13. Kalyani, S., Swarup, K.: Particle swarm optimization based K-means clustering approach for security assessment in power systems. Expert Syst. Appl. **38**(9), 10839–10846 (2011)
14. Alhijawi, B., Kilani, Y.: Using genetic algorithms for measuring the similarity values between users in collaborative filtering recommender systems. In: 2016 IEEE/ACIS 15th International Conference on Computer and Information Science (ICIS), pp. 1–6, June 2016
15. Kim, K.J., Ahn, H.: A recommender system using GA K-means clustering in an online shopping market. Expert Syst. Appl. **34**(2), 1200–1209 (2008)
16. Fong, S., Ho, Y., Hang, Y.: Using genetic algorithm for hybrid modes of collaborative filtering in online recommenders. In: 2008 Eighth International Conference on Hybrid Intelligent Systems, pp. 174–179, September 2008

Drug-Target Interaction Prediction in Drug Repositioning Based on Deep Semi-Supervised Learning

Meriem Bahi$^{(\boxtimes)}$ and Mohamed Batouche

Computer Science Department, Faculty of NTIC,
University Constantine 2 - Abdelhamid Mehri Biotechnology Research Center
(CRBt) & CERIST, Constantine, Algeria
{meriem.bahi,mohamed.batouche}@univ-constantine2.dz

Abstract. Drug repositioning or repurposing refers to identifying new indications for existing drugs and clinical candidates. Predicting new drug-target interactions (DTIs) is of great challenge in drug repositioning. This tricky task depends on two aspects. The volume of data available on drugs and proteins is growing in an exponential manner. The known interacting drug-target pairs are very scarce. Besides, it is hard to select the negative samples because there are not experimentally verified negative drug-target interactions. Many computational methods have been proposed to address these problems. However, they suffer from the high rate of false positive predictions leading to biologically interpretable errors. To cope with these limitations, we propose in this paper an efficient computational method based on deep semi-supervised learning (DeepSS-DTIs) which is a combination of a stacked autoencoders and a supervised deep neural network. The objective of this approach is to predict potential drug targets and new drug indications by using a large scale chemogenomics data while improving the performance of DTIs prediction. Experimental results have shown that our approach outperforms state-of-the-art techniques. Indeed, the proposed method has been compared to five machine learning algorithms applied all on the same reference datasets of DrugBank. The overall accuracy performance is more than 98%. In addition, the DeepSS-DTIs has been able to predict new DTIs between approved drugs and targets. The highly ranked candidate DTIs obtained from DeepSS-DTIs are also verified in the DrugBank database and in literature.

Keywords: Drug repositioning · Drug-target interactions
Deep learning · Semi-supervised learning · Stacked autoencoders
Deep neural network

1 Introduction

Over the past decades, de novo drug discovery has become increasingly difficult and risky. This process has grown to be time consuming and expensive. It can

© IFIP International Federation for Information Processing 2018
Published by Springer International Publishing AG 2018. All Rights Reserved
A. Amine et al. (Eds.): CIIA 2018, IFIP AICT 522, pp. 302–313, 2018.
https://doi.org/10.1007/978-3-319-89743-1_27

take about 17 years and costs at least one billion dollars. In 2015, Pharmaceutical Research and Manufacturers of America (PhRMA) members had invested more than half a trillion dollars in research and development of a new drug [1], while the number of newly approved drugs and clinical compounds known as New Molecular Entities (NMEs) is steadily declining annually. Therefore, it is beneficial to develop strategies to reduce this time frame, decrease costs and improve success rates [2]. Discovering potential uses for existing drugs, also known as drug repositioning [1], is one strategy which has attracted increasing interests from both the pharmaceutical industry and the research community.

Discovering new indications for existing drugs can be attained through identification of new interactions between drugs and target proteins. The in silico prediction of drug target interaction (DTI) is a challenging task in drug repositioning which lies on two main aspects. First, the volume of chemogenomic data available on drugs and proteins is growing in an exponential manner. Second, the known drug-target interactions pairs are rare. Besides, it is hard to select the negative samples because there are not experimentally verified negative drug-target interactions. To date, a variety of computational methods have been proposed to solve these problems and to accurately predict new interactions between known drugs and targets. They fall into two categories (i) Network-based and (ii) learning based. However, they suffer from the high rate of false positive predictions leading to biologically interpretable errors.

To overcome these limitations, we propose in this work a novel computational method, namely DeepSS-DTIs, based on deep semi-supervised learning to accurately predict potential new drug-target interactions using large-scale chemical-protein data. This method nicely combines the advantages of the two different methods of feature-based and semi-supervised learning. The rest of the paper is organized as follows. Drug repositioning field is described in Sect. 2. The different computational methods using for drug repurposing are briefly reviewed in Sect. 3. Section 4 is dedicated to the description of the proposed approach based on the hybrid deep learning architecture. In Sect. 5, the performance of the proposed approach is assessed. In Sect. 6, the list of new predicted interactions is presented. Finally, conclusions and future work are drawn.

2   Drug Repositioning

Drug repositioning or repurposing, rescue or reprofiling (the terms are sometimes used interchangeably) refers to studying drugs that are already approved to treat one disease or condition to see if they are effective for treating other diseases [3]. Finding a new indication of existing drugs is an accelerated route for drug discovery. The process of drug repurposing is generally approved in shorter time frames (3 years). It can reduce about 70% of development cost and decrease the drug safety risk. Because the information about safety, efficacy, and toxicity of an existing drug have been extensively studied and therefore data have already been accumulated toward gaining approval by the U.S. Food and Drug Administration (FDA) for a specific indication.

Most drugs are small compounds that target and interact with therapeutic proteins implicated in a disease of interest to induce perturbation in the protein network [4]. However, approximately 90% of drugs interact not only with the therapeutic target proteins but also with additional proteins resulting in unexpected side effects. The drug side effect may be beneficial for identifying new therapeutic indications [5]. For example, thalidomide is a drug that was developed as a sleeping pill, but it was also found to be useful for easing morning sickness in pregnant women. Unfortunately, it damaged the development of unborn babies. The drug led to the arms or legs of the babies being very short or incompletely formed. More than 10,000 babies were affected around the world. As a result of this disaster, thalidomide was banned [6]. But, thalidomide was redeveloped and repurposing and now it is used as a treatment for leprosy and bone cancer. Many drugs have enormous potential for new therapeutic indications in terms of polypharmacology.

3   In Silico Methods for Drug Reprofiling

Identifying drug-target interactions to find new uses of existing and abandoned drugs is a crucial prerequisite and is a major challenge in drug repositioning. Currently, experimental methods of identifying new interactions between drugs and targets are cumbersome. In silico approaches can provide a promising and efficient tool to alleviate this problem, and thus significantly reduce both experimental time and cost of identifying potential DTI. Therefore, so far, there is a strong incentive to seek and develop computational methods to better predict new drug-target interactions. Traditional in silico approaches can be categorized into the ligand-based approach, structure-based approach and text mining approach [7]. The ligand-based approach is based on the concept that similar ligands (or molecules) tend to have similar biological properties. One of these methods is Quantitative Structure-Activity Relationship (QSAR) that predict the bioactivity of a ligand on a target. Given a certain amount of targets, each target builds a predictive model using its known active ligands. Then these built models are used to screen all the drugs to predict the DTIs between drugs and targets. Unfortunately, the problem with this category of a method is that many target proteins have little or no ligand information available. Structure-based methods or molecular docking represent the second category of approaches for drug repositioning. They have been successfully used for predicting drug-target interactions [8]. These methods are based on the same principle of similarity observed for ligands. Proteins with similar structures are likely to have similar functions and to recognize similar ligands. They use the crystallographic structure of target to screen the small molecules and to identify secondary targets of an approved drug. The limitation of these methods is that they require the three-dimensional (3D) structure of a target which is a problem because not all proteins have their 3D structures available [3]. Indeed, for most membrane proteins, like GPCRs, their 3D structure information is still unavailable, as determining their structures is a challenging task. Another approach is the text mining techniques which are based on keyword searching in the huge number of literature [9], but it suffers from the problem of redundancy in the compound/protein names in the literature.

To overcome challenges of traditional methods, chemogenomic approaches have recently attracted increasing attention in drug discovery and repositioning to find new Drug-Protein interactions on a large scale. They simultaneously utilize both the drug and target features (e.g., drug-induced gene expression, chemical structures, side effects, target protein sequences, and biological pathways) and also disease information (e.g., symptomatic state and phenotype) to perform better predictions [10]. Chemogenomic methods can be divided broadly into network-based techniques and learning-based approach. Network-based methods aim at organizing the relationships among drugs and targets in the form of networks to infer unknown drug-target interactions. The drug-target network can be depicted as a connected graph, where each node represents either a drug or a target and the known interactions between drugs and targets corresponding to the lines that link the nodes. These methods have been widely used for computational drug repositioning. For example, Yamanishi et al. [10] integrated the relationship between pharmacological, chemical, and topology spaces of drug-target interaction networks to predict new associations between drugs and targets. Also, Chen et al. [11] developed an effective model of a heterogeneous network, named NRWRH, to predict potential drug-target interactions on a large scale. Liu et al. [12] have developed a network-based inference model for the prediction of potential DTI. A common limitation of these network-based methods is that they mainly look for novel targets which are close to known targets in the network. Learning-based techniques have been extensively used to cope with the drawbacks of the previous methods, under the assumption that similar drugs are likely to interact with similar proteins. The learning-based methods can be divided into supervised and semi-supervised. The supervised-learning approach has been used in two ways including the similarity based-methods and feature-based methods [13]. Similarity-based methods have been developed to predict potential drug-target interactions through the constructed similarity matrices of drug and protein. Nascimento et al. [14] incorporated multiple heterogeneous information sources using multiple kernel learning method for the identification of new DTIs. Furthermore, a key disadvantage of the similarity-based methods is that they cannot be used on large-scale datasets due to the significant computational complexity of measuring similarity matrices. In contrast, the feature-based methods are regarded as more advantageous strategies where drugs and targets are represented by sets of descriptors (i.e., feature vectors). These methods provide meaningful solutions for discovering interest drug-target interactions by identifying features that are highly more discriminative [13]. They can easily be applied to such a dataset and their computational complexity is moderate. The commonly used learning method is to build a supervised binary classification model where the positive class consists of interacting drug-target pairs and the negative class consists of non-interacting drug-target pairs. It takes drug target pairs (DTPs) as input, and the output is whether there is an interaction between the drug target pair (DTP). However, these models exhibit complicated issues. Since the known DTIs are rare and negative DTIs are difficult or even impossible to achieve because experimentally validated negative samples are not reported and unavailable [15], these methods consider the unknown drug-target interactions as negative samples. This would largely influence the prediction accuracy. Accordingly, the semi-supervised learning approach has been applied to address this problem of imbalanced datasets in drug-target interaction prediction by using the small number of labeled data in conjunction with the numerous unlabeled data. There are only a few studies that have published on semi-supervised learning. That is why researchers are investigating more efforts to develop semi-supervised methods to improve the prediction performance of drug-target interactions. With the increasing of experimental data and increasing complexity of the machine learning algorithms that perform poorly, deep learning methods have been widely applied in many fields of bioinformatics, biology and chemistry [16]. Deep learning methods attract a lot of attention for its better performance and ability to learn representations of data with multiple levels of abstraction. In the drug-repositioning, Wen et al. [17] developed a deep learning method based on deep belief network algorithm to predict new DTIs. They found that deep learning outperforms other state-of-the-art machine learning methods. Wang et al. [18] proposed a stacked autoencoders incorporated with the random forest as the final classifier for predicting interactions between drugs and targets.

4   Materials and Methods

4.1   Data Preparation

The drugs and targets data used in this study were collected from a recent publication [15] which are extracted from DrugBank database. The latter is a unique bioinformatics and cheminformatics resource that combines detailed drug data with comprehensive drug target information. The interactions of drugs and targets were downloaded from drug Target Identifiers category of Protein Identifiers in DrugBank. The Drug target space (DTS) is defined as all possible drug-target pairs (DTPs). In total, there are 5877 drugs and 3348 targets. DTS has 19676196 (that is, $5877 * 3348$) DTPs. Among them, 12674 pairs are positive DTIs (Drug-Target interactions marked as Yes or $+1$) which have known interaction, and the others are not known (unlabeled data). Because the number of no interaction pairs is much more than the number of interaction pairs, the negative dataset can be randomly selected from the DTS. In this work, we randomly select 12674 drug-target pairs from the DTS as a negative dataset (marked as -1). Therefore, the whole labeled dataset contains 25348 samples, as depicted in Fig. 1.

4.2   Drug-Target Representation

Drugs and targets are represented by sets of descriptors (i.e. feature vectors). These features are classified into two categories: chemical structure of drugs (or

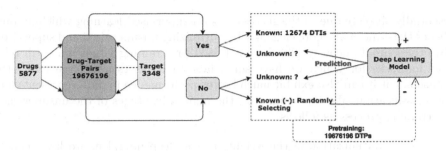

Fig. 1. The flowchart of data flow.

molecular fingerprints) and protein sequence (molecular descriptors). We collected the best features from the recent publication of Ezzat et al. [15]. The latter used the Rcpi package to calculate drug features. Examples of drug features include constitutional, topological and geometrical descriptors among other molecular properties. The target features were obtained using the PROFEAT web server. The features that have been used to represent targets are descriptors related to amino acid composition, dipeptide composition, autocorrelation, composition, transition, and distribution, quasi-sequence-order, amphiphilic pseudo-amino acid composition and total amino acid properties. Thus, we obtained 193 and 1290 features for drugs and targets, respectively.

After collecting the features, each drug-target pair is represented by feature vectors that are formed by concatenating the feature vectors of the corresponding drug and target involved. For example, a drug-target pair is represented by the feature vector:

$$[d_1, d_2, d_3, , d_{193}, t_1, t_2, t_3, , t_{1290}]$$

where $[d_1, d_2, ..., d_{193}]$ is the feature vector corresponding to drug d, and $[t_1, t_2, ... , t_{1290}]$ is the feature vector corresponding to target t. We refer to these drug-target pairs as instances, and we associate a label ($+1$ or -1) to each sample.

4.3   DeepSS-DTIs: The Proposed Method for Drug Repositioning

The number of known interactions between drugs and targets is limited (less than 0.2% among the DTS) and no negative sample of drug-target interaction is verified experimentally [19]. Thus, it is hard to use only the small part of DTIs to represent the whole sample space and applicability of the model may be biased. In this case, it is necessary to use a semi-supervised learning approach for addressing this problem in drug-target interaction prediction with the small number of labeled data and numerous unlabeled data. In addition, with the sheer size of drug-target pairs available (over twenty million DTPs), it is imperative to use the deep learning method.

The unsupervised pre-training followed by supervised fine-tuning is a way of applying with success the semi-supervised deep learning method. Pre-training is essentially obsolete due to the success of semi-supervised learning which accomplishes the same goals more elegantly by optimizing unsupervised and supervised objectives simultaneously [20]. Unsupervised pre-training is not only still relevant for tasks for which we have small labeled datasets and large unlabeled datasets, but it can also exhibit much better performance in data representation and classification. We can summarize the main advantages of the unsupervised pre-training process as follows:

– A better initialization of the weights in the deep neural network instead of randomly initialized weights which may lead to better convergence and better performing classifiers.
– It acts as some special kind of regularization process which yields a better generalization power.

In this study, the training procedure of our deep learning model DeepSS-DTIs can be divided into two consecutive processes: the layer-wise unsupervised pre-training process using stacked autoencoders, and the supervised fine-tuning process of the deep neural network.

Stacked Autoencoders

Stacked Autoencoders (SAE) is one of popular deep learning model, built with multiple layers of autoencoders, in which the output of each layer is connected to the input of the next layer [21], as depicted in Fig. 2.

An autoencoder (AE) can be considered as a special neural network with one hidden layer. It tries to reconstruct the same features at the output layer using its hidden activations. The AE takes the input and puts it through an encoding function to get the encoding of the input, and then it decodes the encodings through a decoding function to recover (an approximation of) the original input [22]. More formally, let $x \in R^d$ be the input:

$$h = f_e(x) = s_e(W_e x + b_e) \tag{1}$$

$$x_r = f_d(x) = s_d(W_d h + b_d) \tag{2}$$

where $f_e : R^d \rightarrow R^h$ and $f_d : R^h \rightarrow R^d$ are encoding and decoding functions respectively, W_e and W_d are the weights of the encoding and decoding layers, and b_e and b_d are the biases for the two layers. s_e and s_d are elementwise non-linear functions in general, and common choices are sigmoidal functions like tanh or logistic [21].

In general, N-layer stacked autoencoders with parameters $P = \{P^i \mid i \in \{1, 2, ...N\}\}$, where $P^i = \{W_e^i, W_d^i, b_e^i, b_d^i\}$ can be formulated as follows:

$$h^i = f_e^i(h^{i-1}) = s_e^i(W_e^i h^{i-1} + b_e^i) \tag{3}$$

$$h_r^i = f_d^i(h_r^{i+1}) = s_d^i(W_d^i h_r^{i+1} + b_d^i) \tag{4}$$

$$h^0 = x \tag{5}$$

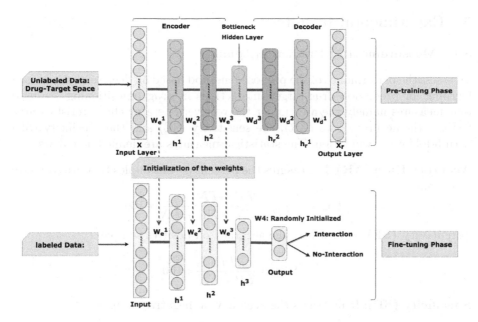

Fig. 2. The proposed drug semi-supervised model. Top: the stacked autoencoders after training. Down: the pre-trained deep neural network initialized with stacked autoencoders' weights. For the simplicity, all biases are excluded from the figure.

SAE plays a fundamental role in unsupervised learning. It is based on a greedy layer-wise training [23]. It can better learn the features of the input information and reduce the original data dimension [24] where the raw data are transformed from layer to layer up to the top layer.

The layer-wise unsupervised pre-training of stacked autoencoders process is as follows:

1. Train the bottom most autoencoder using the unlabeled data.
2. After training, we remove the decoder layer, we construct a new auto-encoder by taking the latent representation of the previous auto-encoder as input.
3. Train the new autoencoder. Note the parameters (weights and bias) of the encoder from the previously trained autoencoder are fixed when training the newly constructed autoencoder.
4. Repeat step 2 and 3 until all encoding layers are trained. The activation function is usually the sigmoid function or tanh function.

The supervised fine-tuning process is as follows:

1. After training, we use the weights of the unsupervised stacked autoencoders model to initialize the weights of the supervised deep neural networks model (DNN).
2. Initialize randomly the output layer parameters of deep neural networks.
3. Fine-tune all the parameters of all deep neural networks with stochastic gradient descent using back-propagation. As shown in Fig. 2.

5   Experimental Results

5.1   Measurement of Prediction Quality

To assess the performance of the proposed method based on deep semi-supervised learning for prediction drug-target interactions in drug repositioning, we used four measures namely the area under the receiver operator characteristic curve (AUC), the accuracy rate (AR), the sensitivity (SE) and the specificity (SP) with 5-fold cross-validation. The statistical measures are defined as follows:

Accuracy Rate (AR): It measures the percentage of samples that are correctly classified.

$$(AR) = \frac{TP + TN}{TP + TN + FP + FN} * 100$$

Sensitivity (SE): It measures the accuracy on positive samples.

$$(SE) = \frac{TP}{TP + FN} * 100$$

Specificity (SP): It measures the accuracy on negative samples.

$$(SP) = \frac{TN}{TN + FP} * 100$$

With TP, FP, TN and FN the numbers of true-positive, false-positive, true-negative and false-negative predictions, respectively. In a two-class prediction problem, the outcomes are labeled either as positive (p) or negative (n). If the prediction and actual value are all p, it is called a TP; if the prediction value is p while the actual value is n, it is called a FP. Conversely, if the prediction and actual value are all (n), it is called a TN; if the prediction value is n while the actual value is p, it is called a FN.

5.2   Cross-Validation Results

We compared our approach to five state-of-the-art machine learning algorithms reported in the literature [15] which are Random Forest, SVM, Decision Tree, Nearest Neighbor and ensemble learning. The obtained results are summarized in Table 1 and show that our method outperforms other methods in all measurements.

As shown in Table 1, the results obtained by our method DeepSS-DTIs using H2O platform are more than 0.98 (98%) in almost all measurements. The AUC, accuracy, sensitivity, and specificity of test set are 0.9980, 0.9853, 1 (100%) and 1 (100%) respectively. Because the number of positive DTIs is much fewer than that of negative in Drug-Target space and the purpose of the model is to predict the true positive DTI, the sensitivity (SE) is a more important evaluation metric among the four evaluation metrics. The obtained results by our approach are clearly better than the ones reported in [15]. Our method achieved an AUC of 0.998, which is 9.8% higher than the ensemble classifier learning

Table 1. Performance assessment of the proposed method

Prediction model	AUC (%)	AR (%)	SE (%)	SP (%)
Our method (DeepSS-DTIs)	99.80	98.53	100	100
SVM	80.40	81.08	76.31	85.85
Nearest neighbor	81.40	82.26	78.50	85.37
Ensemble classifier	90.00	89.65	88.80	88.54
Random forest	85.5	81.84	79.58	84.11
Decision tree	76.00	75.05	51.67	66.80

(or class imbalance method) with an AUC of 0.900. This method is well suited for the prediction of new drug-target interactions. The other methods such as Decision Trees, SVM, Nearest Neighbor and Random Forest, yield to heterogeneous results. This supports our claim that using both a semi-supervised and deep learning technique is important for improving the prediction performance. Overall, the cross-validation between the results of our approach (DeepSS-DTIs) and those of five different machine learning algorithms applied all on the same datasets, clearly demonstrates that the DeepSS-DTIs method gained the best performance in *AUC*, *AR*, *SP* and *SE*. This indicates that the built DeepSS-DTIs model is reliable and can be further applied for novel DTIs prediction.

6   Predicting New Drug-Target Interactions

After confirming the performance of our method (DeepSS-DTIs) in comparison against other state-of-the-art methods, we tested the ability of our built model to correctly predict interactions on the remaining of the drug-target space (DTS) and ranked them by their probability. The Table 2 shows the list of the top 10 probability predicted DTIs by DeepSS-DTIs with H2O platform.

Table 2. Top10 probability scoring DTIs predicted by our model.

Rank	Drug-ID	Target-ID	Description
1	DB00246	P08909	Ziprasidone, 5-hydroxytryptamine receptor
2	DB00126	P40238	Vitamin C, Thrombopoietin receptor
3	DB00363	P08909	Clozapine, 5-hydroxytryptamine receptor
4	DB00463	Q16478	Metharbital, Glutamate receptor ionotropic
5	DB00898	P15587	Ethanol, Xylose isomerase
6	DB00312	Q13002	Pentobarbital, kainate 2
7	DB01045	Q16850	Rifampicin, Lanosterol 14-alpha demethylase
8	DB00933	P21728	Mesoridazine, D(1A) dopamine receptor
9	DB00114	P19235	Pyridoxal Phosphate, Erythropoietin receptor
10	DB00128	Q13332	L-Aspartic Acid, Receptor-type tyrosine-protein phosphatase S

In order to evaluate the reliability of new predicted interactions, we consulted the literature and the DrugBank database with the predicted relationships between drugs and targets. We found that some of the drug predicted by our method are validated by relevant literature and show potentiality for further study.

7   Conclusion and Future Work

Identifying drug-target interactions (DTIs) is a key area in drug repositioning. In this paper, we have presented an effective method for predicting both new drugs and detecting new targets for drug repositioning based on deep semi-supervised learning dealing with unbalanced data using a small number of known interactions in conjunction with the many unknown interactions. The cross-validation experiments demonstrated that the proposed approach (DeepSS-DTIs) outperforms the previous methods for drug-target interaction prediction. As future work, we expect to scale up the proposed approach by using sparkling water (Spark+H2O) to handle big data and improve performances.

References

1. Barratt, M.J., Frail, D.E.: Drug Repositioning: Bringing New Life to Shelved Assets and Existing Drugs. Wiley, Hoboken (2012)
2. Zhang, R.: An ensemble learning approach for improving drug–target interactions prediction. In: Wong, W.E. (ed.) Proceedings of the 4th International Conference on Computer Engineering and Networks. LNEE, pp. 433–442. Springer, Cham (2015). https://doi.org/10.1007/978-3-319-11104-9_51
3. Li, J., Zheng, S., Chen, B., Butte, A.J., Swamidass, S.J., Lu, Z.: A survey of current trends in computational drug repositioning. Briefings Bioinform. **17**(1), 2–12 (2015)
4. Li, Y.Y., Jones, S.J.: Drug repositioning for personalized medicine. Genome Med. **4**(3), 27 (2012)
5. Napolitano, F., Zhao, Y., Moreira, V.M., Tagliaferri, R., Kere, J., DAmato, M., Greco, D.: Drug repositioning: a machine-learning approach through data integration. J. Cheminformatics **5**(1), 30 (2013)
6. Shim, J.S., Liu, J.O.: Recent advances in drug repositioning for the discovery of new anticancer drugs. Int. J. Biol. Sci. **10**(7), 654 (2014)
7. March-Vila, E., Pinzi, L., Sturm, N., Tinivella, A., Engkvist, O., Chen, H., Rastelli, G.: On the integration of in silico drug design methods for drug repurposing. Front. Pharmacol. 8 (2017)
8. Phatak, S.S., Zhang, S.: A novel multi-modal drug repurposing approach for identification of potent ACK1 inhibitors. In: Pacific Symposium on Biocomputing, p. 29. NIH Public Access (2013)
9. Peng, L., Liao, B., Zhu, W., Li, K.: Predicting drug-target interactions with multi-information fusion. IEEE J. Biomed. Health Inf. **21**(2), 561–572 (2015)
10. Yamanishi, Y.: Chemogenomic approaches to infer drug-target interaction networks. In: Mamitsuka, H., DeLisi, C., Kanehisa, M. (eds.) Data Mining for Systems Biology: Methods and Protocols, vol. 939, pp. 97–113. Humana Press, Totowa (2013). https://doi.org/10.1007/978-1-62703-107-3_9

11. Chen, X., Liu, M.X., Yan, G.Y.: Drug-target interaction prediction by random walk on the heterogeneous network. Mol. BioSyst. **8**(7), 1970–1978 (2012)

12. Cheng, F., Liu, C., Jiang, J., Lu, W., Li, W., Liu, G., Zhou, W., Huang, J., Tang, Y.: Prediction of drug-target interactions and drug repositioning via network-based inference. PLoS Comput. Biol. **8**(5), e1002503 (2012)

13. Mousavian, Z., Masoudi-Nejad, A.: Drug-target interaction prediction via chemogenomic space: learning-based methods. Expert Opinion Drug Metabol. Toxicol. **10**(9), 1273–1287 (2014)

14. Nascimento, A.C., Prudêncio, R.B., Costa, I.G.: A multiple kernel learning algorithm for drug-target interaction prediction. BMC Bioinformatics **17**(1), 46 (2016)

15. Ezzat, A., Wu, M., Li, X.L., Kwoh, C.K.: Drug-target interaction prediction via class imbalance-aware ensemble learning. BMC Bioinformatics **17**(19), 509 (2016)

16. Chen, H., Zhang, Z.: A semi-supervised method for drug-target interaction prediction with consistency in networks. PLoS One **8**(5), e62975 (2013)

17. Wen, M., Zhang, Z., Niu, S., Sha, H., Yang, R., Yun, Y., Lu, H.: Deep-learning-based drug-target interaction prediction. J. Proteome Res. **16**(4), 1401–1409 (2017)

18. Wang, L., You, Z.H., Chen, X., Xia, S.X., Liu, F., Yan, X., Zhou, Y., Song, K.J.: A computational-based method for predicting drug-target interactions by using stacked autoencoder deep neural network. J. Comput. Biol. **24**, 1–13 (2017)

19. Zhao, J., Cao, Z.: A label extended semi-supervised learning method for drug-target interaction prediction. Studies **13**, 21 (2015)

20. Erhan, D., Bengio, Y., Courville, A., Manzagol, P.A., Vincent, P., Bengio, S.: Why does unsupervised pre-training help deep learning. J. Mach. Learn. Res. **11**, 625–660 (2010)

21. Zhou, Y., Arpit, D., Nwogu, I., Govindaraju, V.: Is joint training better for deep auto-encoders? (2014). arXiv preprint arXiv:1405.1380

22. Oliveira, T.P., Barbar, J.S., Soares, A.S.: Multilayer perceptron and stacked autoencoder for internet traffic prediction. In: Hsu, C.-H., Shi, X., Salapura, V. (eds.) NPC 2014. LNCS, vol. 8707, pp. 61–71. Springer, Heidelberg (2014). https://doi.org/10.1007/978-3-662-44917-2_6

23. Candel, A., Parmar, V., LeDell, E., Arora, A.: Deep Learning with H2O. H2O. ai Inc., Mountain View (2016)

24. Zabalza, J., Ren, J., Zheng, J., Zhao, H., Qing, C., Yang, Z., Du, P., Marshall, S.: Novel segmented stacked autoencoder for effective dimensionality reduction and feature extraction in hyperspectral imaging. Neurocomputing **185**, 1–10 (2016)

Optimization

FA-SETPOWER-MRTA: A Solution for Solving the Multi-Robot Task Allocation Problem

Farouq Zitouni[1,2]([✉]) and Ramdane Maamri[1,2]

[1] Kasdi Merbah University, Ouargla, Algeria
[2] Constantine 2 — Abdelhamid Mehri University, Constantine, Algeria
{farouq.zitouni,ramdane.maamri}@univ-constantine2.dz

Abstract. The Multi-Robot Task Allocation problem (MRTA) is the situation where we have a set of tasks and robots; then we must decide the assignments between robots and tasks in order to optimize a certain measure (e.g. allocate the maximum number of tasks...). We present an effective solution to resolve this problem by implementing a two-stage methodology: first a global allocation that uses the Firefly Algorithm (FA), next a local allocation that uses the set theory properties (Power Set algorithm). Finally, results of the different simulations show that our solution is efficient in terms of the rate of allocated tasks and the calculated allocations are locally optimal.

Keywords: Multi-Robot Systems · Task allocation
Firefly algorithm · Set theory properties · Power set algorithm

1 Introduction

Nowadays, Multi-Robot Systems (MRS) are receiving a great attention and become omnipresent in our daily life. Consequently, intensive researches have been conducted on them to validate their applicability and adequacy with different real-life issues. These systems have several advantages that allow them solving easily various problems, such as: industrial applications, surveillance, target tracking and rescue missions [1].

It should be noted that the design of MRS must respect the different interaction and coordination aspects between their components (e.g. robots), otherwise it may produce non-deterministic systems with reduced performances [2]. Thus, one of the coordination issues that researchers must consider is the Task Allocation (TA) problem, which is intuitively the process of assigning a set of tasks to some robots [3]. In MRS, the TA problem can be expressed by the following expression: given a set of robots and tasks, how to select the appropriate robots to perform the desired tasks? so as to achieve in a cooperative manner the overall objective of the considered system. Therefore, the goal here is to coordinate the robot behaviors and find an optimal way for the allocation of different tasks [4].

© IFIP International Federation for Information Processing 2018
Published by Springer International Publishing AG 2018. All Rights Reserved
A. Amine et al. (Eds.): CIIA 2018, IFIP AICT 522, pp. 317–328, 2018.
https://doi.org/10.1007/978-3-319-89743-1_28

Mainly, we can find two taxonomies for the categorization of MRTA problems. In fact, the main objective of these taxonomies is to propose a classification of different TA configurations in categories, so that the different problems encountered in the real-life can be inserted into one of them (i.a. a mathematical formulation is generally given).

Authors of [5] have categorized the TA problem by proposing a classification which is articulated around three axes. The first axis distinguishes robots according to their abilities to perform a single task or probably several tasks at a time {single-task robots (ST), multi-task robots (MT)}. The second axis distinguishes tasks according to their needs to be executed by a single robot or several robots at the same time {single-robot tasks (SR), multi-robot tasks (MR)}. The third axis distinguishes between tasks that require an instantaneous assignment without considering future allocations and tasks that require assignments that consider both current and future allocations {instantaneous assignment (IA), time-extended assignment (TA)}. Finally, despite its good coverage of the majority of encountered MRTA problems; however, the authors of this taxonomy state that it does not capture problems with interrelated utilities and constraints on tasks [5], hence the need to propose a taxonomy that addresses these limitations.

To remedy the limitations of the previous taxonomy, the authors of [6] have taken and modified it by adding a higher level. In reality, this level expresses interdependence degrees of utilities between robots and tasks. Thus, this taxonomy is a hierarchy that has two levels, where in the second one we find the different classes proposed in the previous taxonomy. On the other hand, in the first level we find four interdependence degrees, i.e. No Dependencies (ND), In-Schedule Dependencies (ID), Cross-Schedule Dependencies (XD), Complex Dependencies (CD).

The rest of the paper is organized as follows. First, we give, in the Sect. 2, an overview of the related works already done to address the MRTA problem. Then, we present, in the Sect. 3, the solution that we have proposed to address this problem and explain its different algorithms. After, we simulate, in the Sect. 4, our solution and discuss the obtained results. Finally, we give, in the Sect. 5, a conclusion and some perspectives.

2   Related Works

In this section, we will give some solutions that have been proposed in the literature to address the MRTA problem. More specifically, these solutions will be categorized according to the first level of the taxonomy 2, that is to say these approaches will be divided into four categories. Moreover, the list of approaches presented bellow is not exhaustive, but shows the most frequently encountered in the literature [4].

In the first family, we can clearly see that several approaches have been identified in the literature [5] to address more specifically ND[ST-SR-IA] problems. For example, we can cite the paper [7] that uses potential fields and the papers [8,9] that use auction-based methods.

In the second family, we can find several works devoted to the coordination in MRS. Generally, these approaches focus on ID[ST-SR-TA] problems. For example, we can cite the work exposed in [10] that uses the Traveling Salesman Problem (TSP) and multiple TSP to model the TA problem. Also, the paper [11] provides a good solution for the routing problem in MRS using mixed linear programming. Finally, auction-based approaches [12–15] have been widely used to resolve this problem, because of their distributed nature which is well suited to MRS.

In the third family, we can easily identify lot of developed works. For example, we find the system M+ [16] which addresses the problem of instantaneous allocation of tasks using a market system with supporting precedence constraints. Also, we cite the work presented in [17] that proposes a solution manipulating constraints between tasks and adopting a market-economy approach. In addition, the work exposed in [18] addresses the problem of routing, where some robot teams must accomplish a set of scientific missions. Besides, we can cite the work presented in the paper [19], which exposes a solution for the MRTA problem by using quantum genetic algorithms and reinforcement learning. Also, in the paper [20], the authors manipulate the case of a simple task scheduling using the heuristic: "the task x must be executed n second(s) before the beginning of the task y". Another way to consider the TA problem is to bring it to a coalition formation problem [21, 22]. Accordingly, the authors of the paper [23] have taken the works presented in the last two papers and improved the proposed algorithms by minimizing communications and imposing constraints on agents' capabilities. Likewise, the authors of the papers [24, 25] use auctions and coalitions to address the TA problem. On the other hand, a Framework has been proposed in [26] that imposes the constraint of shared resources, e.g. communication mediums and processors. Finally, the papers presented in [27, 28] address the TA problem, where robots must consider current tasks and future allocations.

In the fourth family, we can unfortunately make a short list of done works. In fact, this is due to the high difficulty of these problems and the lack of mathematical formulations [6]. However, we can cite the work exposed in [29] which addresses the TA problem, with time-extended assignment, by managing an environment hit by a natural disaster. In addition, we can also mention the work presented in [30] which exposes a task allocation approach using coalitions (this approach has two versions: centralized version or ASyMTRe and distributed version or ASyMTRe-D). Finally, we mention the work given in [31], which is an improved version of [12], that manipulates the time-extended assignments.

3   Proposed Solution

In this section, an auction-based solution (named FA-POWERSET-MRTA for Firefly Algorithm-Power Set algorithm-Multi Robot Task Allocation) is presented to address the MRTA problem. In fact, auction-based approaches have been widely used because of the multiple advantages they offer [32]. In this kind of approaches, we consider an auctioneer, a set of bidders and a set of goods. In a MRTA problems, a particular robot is the auctioneer, the rest of robots are the bidders and tasks are the goods. In our solution, we use the Contract Net Protocol (CNP) [33] for exchanging messages between the auctioneer and bidders.

Algorithm 1. Behavior of the auctioneer a

Input   : the set of bidders B and the set of unallocated tasks T.
Output: allocations between bidders and tasks.

```
1  while (a is active) do
2  │   PeriodicBehaviour {n₁ seconds}
3  │   │   if (T ≠ ∅) then
4  │   │   │   foreach bᵢ ∈ B do
5  │   │   │   │   sendMessageTo(bᵢ,CHOOSE-TASK,T);
6  │   │   │   end
7  │   │   end
8  │   end
9  │   CyclicBehaviour
10 │   │   message ← receiveMessageFrom(bᵢ);
11 │   │   if (message.subject = CHOSEN-TASK) then
12 │   │   │   foreach t ∈ T do
13 │   │   │   │   if (t is feasible) then
14 │   │   │   │   │   compute the allocation;
15 │   │   │   │   │   T ← T/{t};
16 │   │   │   │   │   inform concerned bidders;
17 │   │   │   │   end
18 │   │   │   end
19 │   │   end
20 │   end
21 end
```

3.1   Problem Configuration and Assumptions

In our methodology (i.e. solution), we use a particular robot called "auctioneer" and several robots called "bidders". The auctioneer must first communicate the tasks (i.e. goods) to be allocated to the different bidders and receive their offers; then it must decide the best assignments between robots and tasks; finally, it notifies bidders of its results. Although the use of an auctioneer can be seen as a bottleneck for the system (e.g. if it breaks down then the system will be too); in fact, this limitation can be omitted, because it decreases considerably the communication rates and maintains an overall view of different allocations. Also, the auction-based methods have the advantage of sharing computations on the different bidders of the system (scalability), which would maintain the assignment process even if a bidder breaks down (robustness) [32].

Furthermore, the bidders calculate their utilities for the considered tasks according to the different information they have. So, the overall cost function (i.e. fitness) is divided into sub-functions, which are estimated in a decentralized and independent way by bidders. In our solution, the allocation of a given task means the presence of the resources it requires. In other words, a given task is allocated if and only if the resources it requires are all offered. On the other hand, these resources are offered by bidders, i.a. generally these resources are sensors and actuators. In our solution, we use confusedly the term "skill" to represent these sensors and actuators.

Algorithm 2. Behavior of a bidder b_i

Input : S_i, C_i, v_i, e_i, a_i and p_i.

Output: the chosen task and its costs.

1 **while** *(b_i is active)* **do**
2 | **PeriodicBehaviour** $\{n_2$ seconds$\}$
3 | | **if** *($T \neq \emptyset$)* **then**
4 | | | **if** *(state = AVAILABLE)* **then**
5 | | | | **for** $k \leftarrow 1$ **to** 1000 **do**
6 | | | | | **foreach** $t \in T$ **do**
7 | | | | | | move the bidder b_i towards the task t;
8 | | | | | **end**
9 | | | | **end**
10 | | | | $t \leftarrow$ choose the closest task to the bidder b_i;
11 | | | | $U \leftarrow$ estimate the costs of the bidder b_i for the task t;
12 | | | | sendMessageTo(a,CHOSEN-TASK,t,U);
13 | | | **else**
14 | | | | sendMessageTo(a,CHOSEN-TASK,$\perp$,$\perp$);
15 | | | **end**
16 | | | $T \leftarrow \emptyset$;
17 | | **end**
18 | **end**
19 | **CyclicBehaviour**
20 | | message $\leftarrow$ receiveMessageFrom(a);
21 | | **if** *(message.subject = CHOOSE-TASK)* **then**
22 | | | $T \leftarrow$ message.content;
23 | | **end**
24 | **end**
25 **end**

Now we consider the following assumptions, we assume the class of ST-MR-IA problems, that is to say that each robot can execute only one task at the same time, the tasks might be executed by several robots at the same time and the allocation of tasks to robots is instantaneous. Moreover, we assume two non-empty sets of bidders B and tasks T. Also, we consider the both cases where all the tasks are previously known or gradually inserted into the system.

Accordingly, each bidder $b_i \in B$ is defined by six variables $\langle S_i, C_i, v_i, e_i, a_i, p_i \rangle$, where S_i is a vector representing the skills that has the bidder b_i, C_i is a vector representing the skill costs of the bidder b_i and the rest of variables represent respectively the speed, energy level, aging factor and spatial position of the bidder b_i. In a similar way, each task $t_j \in T$ is defined by three variables $\langle S_j, d_j, p_j \rangle$, where S_j is a vector representing the skills required by the task t_j and the rest of variables represent respectively the execution duration and spatial position of the task t_j. Now, we present and explain the proposed algorithms to address the MRTA problem.

3.2   Proposed Algorithms

The Algorithms 1 and 2 show respectively the behaviors adopted by the auctioneer and bidders to address the TA problem. In fact, we can clearly see that these algorithms have two sub-behavior types that we have named "PeriodicBehavior" and "CyclicBehavior", where in the first one we have put the instructions that are executed in a periodic manner (i.e. every n seconds) and in the second one we have put the instructions that are executed immediately after every message receiving.

Now, we will explain the behaviors (i.e. instructions) of previous algorithms. To do this, we can resume their working principle in the following three sections.

A. Task Announcement

Firstly, the task allocation process begins when the auctioneer announces that there are tasks that must be allocated (i.e. "PeriodicBehavior" of the Algorithm 1). To do this, the auctioneer broadcasts a message, i.e. to all bidders, which contains the list of tasks to be allocated, their required skills and spatial locations.

B. Global Allocation

Secondly, when the bidders receive a message from the auctioneer (i.e. "CyclicBehavior" of the Algorithm 2), each one must choose a task to perform, calculate its costs and answer the auctioneer (i.e. "PeriodicBehavior" of the Algorithm 2).

To choose a task, one bidder executes the firefly algorithm [34, 35] on all the received tasks and converges gradually towards one of them. For the convergence of a bidder to a given task, we have used the equation adopted by the firefly algorithm, which is:

$$p_r = p_r + \beta_0 e^{-\gamma(d_{rt})^2}(p_t - p_r) + \alpha(rand - 0.5) \tag{1}$$

Where p_r and p_t are the respective locations of the bidder r and a task t, β_0 is a regularization parameter and its value is always 1, the term $\beta_0 e^{-\gamma(d_{rt})^2}(p_t - p_r)$

is called the attraction degree of the robot r by the task t, the parameter γ characterizes the attraction variation and its value is very important for the convergence speed of the algorithm and the behavior of the robot r ($\gamma \in [0.01, 100]$), the parameter α represents physically the environment noise and its value affects the visibility of a task by a robot ($\alpha \in [0, 1]$), $rand \in [0, 1]$ is a random variable and the d_{rt} represents the weighted distance between the robot r and the task t and it is given by the following equation:

$$d_{rt} = \delta \|p_t - p_r\| + (1 - \delta)H(S_t, S_r) \tag{2}$$

Where the term $\|p_t - p_r\|$ represents the Euclidean distance between the locations of the robot r and the task t, the term $H(S_t, S_r)$ represents the Hamming distance between the skill vectors of the robot r and the task t and the parameter δ makes the balance between the two distances.

Once a task has been selected, one bidder must now estimate its costs for each offered skill to the chosen task. To do this, we have proposed the following equation:

$$u_{i \in \Delta_{rt}}^{rt} = [\delta c_i^r + (1 - \delta)(\frac{v_r}{\|p_t - p_r\|})^{\frac{e_r}{a_r}}] \times |\Delta_{rt}| \tag{3}$$

Where the term Δ_{rt} represents the common skills between the robot r and the task t (i.e. the skills that the robot r offers to the task t), again the parameter δ make the balance between the two costs and the other variables are explained above.

Finally, the bidder sends its result to the auctioneer. It should be noted that the symbol $\perp$ means that the bidder has sent an empty result to the auctioneer (i.e. it is not able to perform any task: busy or do not have the required skills).

C. Local Allocation

Thirdly, when all the responses of bidders are received, now the auctioneer should estimate an allocation for each selected task ("CyclicBehavior" of the Algorithm 1). To do this, first the auctioneer browses the list of the chosen tasks and considers for each one the set of all skills offered by the bidders, then it calculates all the useful subparts (the power set algorithm) of this set of skills (i.e. the concept of a useful subpart will be explained in the following example), finally the fitness of each useful subpart is estimated and the one with the smallest fitness value will be considered as an allocation for the considered task. The fitness value of a given subpart is the sum of cost values of its bidder skills. At the end of this step, the bidders concerned by the calculated allocations are notified.

4   Simulation and Result Discussion

In this section, we simulate the proposed algorithms on some numerical data and we evaluate the found results. Moreover, we assume that all the tasks are known in advance; however, our algorithms can also handle the case where tasks are inserted gradually in the system.

4.1   Generation of Simulation Data

In fact, the numerical data that we have used for our simulations are generated randomly, on a 2D grid of 100×100 cells, and the number of used skills is 10. To simplify things, we suppose that the speed, energy level and aging factor values are constant for all bidders.

Besides, these simulation data are divided into three different data sets named "dataset 1", "dataset 2" and "dataset 3". First, in the "dataset 1" there is from 10 to 50 bidders (bidders are incremented by 10) and 100 tasks (this dataset is created to see the effect of the number of used bidders on allocation performances). Second, in the "dataset 2" there is from 10 to 100 tasks (tasks are incremented by 10) and 10 bidders (this dataset is created to see the effect of the number of used tasks on allocation performances). Thirdly, in the "dataset 3" there is also from 10 to 100 tasks (the tasks are incremented by 10) and 50 bidders (similarly this dataset is created to see the effect of the number of used tasks on allocation performances).

4.2   Results and Analysis

In the Fig. 1, we present in the left sub-figure the allocation times needed to allocate all tasks and in the right sub-figure the fitness values found for the "dataset 1". For this experimentation, we remember that we have used 100 tasks to be allocated and the number of bidders varies from 10 to 50.

As a first observation, we can clearly notice that the number of used bidders does not greatly improve the allocation time of all tasks (approximately we have an allocation time of 20,199 s, whatever the number of used bidders); therefore, the increase in the number of bidders for the allocation of all tasks is not imperative in our proposed solution, hence its power to allocate the tasks with a small number of bidders.

In the Fig. 2, we present in the left sub-figure the allocation times needed to allocate all tasks and in the right sub-figure the fitness values found for the "dataset 2" and "dataset 3". For this experimentation, we remember that we have used respectively 10 bidders ("dataset 2") and 50 bidders ("dataset 3") and the number of tasks to be allocated varies from 10 to 100 in both datasets.

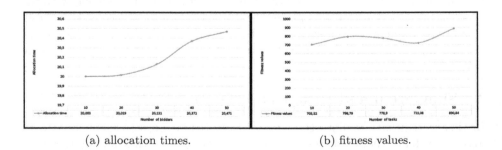

(a) allocation times.   (b) fitness values.

Fig. 1. Measures for the "dataset 1".

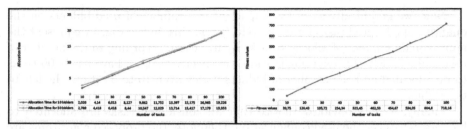

(a) allocation times for "dataset 2" and "dataset 3".

(b) fitness values the "dataset 2".

Fig. 2. Measures for the "dataset 2" and "dataset 3".

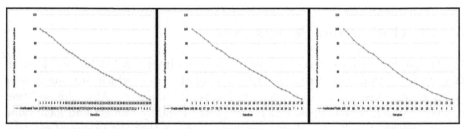

(a) number of bidders = 10. (b) number of bidders = 20. (c) number of bidders = 30.

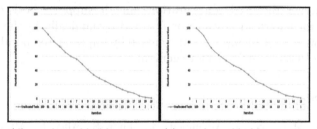

(d) number of bidders = 40. (e) number of bidders = 50.

Fig. 3. Number of allocated tasks for each iteration according to the number of used bidders

In fact, we can clearly notice again that the number of bidders does not greatly improve the allocation time, whatever the number of tasks to be allocated (with 10 or 50 bidders we have approximately the same times). However, the allocation times increase almost linearly with the increase in the number of tasks, which is quite natural.

It should be noted that the allocation times shown in the Figs. 1 and 2 encompass all the possible actions for the allocation of tasks, i.e. sending and receiving messages between the auctioneer and bidders, selection of tasks by bidders (global allocation) and allocation of tasks by the auctioneer (local allocation) and the periodicity of the different behaviors.

Now, we graphically show the impact of the number of used bidders on the quantity of allocated tasks for each cycle. To do this, in the Fig. 3 we present the approximate number of allocated tasks by iteration, according to the number of used bidders. For this experimentation, we specify that we have used 100 tasks to be allocated and the number of bidders varies from 10 to 50 (from the left to the right).

Finally, we can obviously notice that all considered tasks are allocated to bidders, whatever the used configuration; therefore, the proposed algorithms are efficient in terms of the rate of allocated tasks. On the other hand, we can also say and affirm that the increase in the number of used bidders only enhances the number of allocated tasks by iteration but does not minimize the allocation times.

5   Conclusion and Perspectives

In this paper, we have exposed a solution to address the MRTA problem. In fact, this solution uses auctions with two robot types, which are: auctioneer (with a single instance) and bidder (with several instances). Also, this solution exploits two allocation types, which are: global done by bidders and local done the auctioneer. Finally, the discussion section shows that the proposed solution is effective in terms of the rate of allocated tasks (the rate of allocated tasks is 100%, whatever the used configuration) and it supports both cases where the tasks are all known in advance or inserted dynamically in the system. As perspectives, we expect to improve our solution by imposing temporal and spatial constraints on tasks to make it more general.

References

1. Ahmed, H., Mohamed, A., Mohamed, B., Omar, S., Alaa, K.: Multi-robot task allocation for search and rescue missions. J. Phys. Conf. Ser. **570**(5), 052006 (2014)
2. Ahmed, H., Alaa, K.: Market-based approach to multi-robot task allocation. In: International Conference on Individual and Collective Behaviors in Robotics (ICBR), pp. 69–74 (2013)
3. Yanyan, H., Deshi, L., Jian, C., Xiangguo, Y., Yuxi, H., Guangmin, Z.: A multi-robots task allocation algorithm based on relevance and ability with group collaboration. Int. J. Intell. Eng. Syst. **3**(2), 33–41 (2010)
4. Khamis, A., Hussein, A., Elmogy, A.: Multi-robot task allocation: a review of the state-of-the-art. In: Koubâa, A., Martínez-de Dios, J.R. (eds.) Cooperative Robots and Sensor Networks 2015. SCI, vol. 604, pp. 31–51. Springer, Cham (2015). https://doi.org/10.1007/978-3-319-18299-5_2
5. Gerkey, B.P., Mataric, M.J.: A formal analysis and taxonomy of task allocation in multi-robot systems. Int. J. Robot. Res. **23**(9), 939–954 (2004)
6. Korsah, A.G., Stentz, A., Dias, B.M.: A comprehensive yaxonomy for multi-robot task allocation. Int. J. Robot. Res. **32**(12), 1495–1512 (2013)
7. Vail, D., Veloso, M.: Multi-robot dynamic role assignment and coordination through shared potential fields. In: Schultz, A., Parker, L., Schneider, F. (eds.) Multi-Robot Systems. Kluwer, Dordrecht (2003)

8. Gerkey, B.P., Mataric, M.J.: Sold!: auction methods for multirobot coordination. IEEE Trans. Robot. Autom. **18**(5), 758–768 (2002)

9. Simmons, R.G., Apfelbaum, D., Burgard, W., Fox, D., Moors, M., Thrun, S., Younes, H.L.S.: Coordination for multi-robot exploration and mapping. In: Proceedings of the Seventeenth National Conference on Artificial Intelligence and Twelfth Conference on Innovative Applications of Artificial Intelligence, pp. 852–858 (2000)

10. Barnhart, C., Johnson, E.L., Nemhauser, G.L., Savelsbergh, M.W.P., Vance, P.H.: Branch-and-price: column generation for solving huge integer programs. Oper. Res. **46**(3), 316–329 (1998)

11. Melvin, J., Keskinocak, J., Koenig, S., Tovey, C.A., Ozkaya, B.Y.: Multi-robot routing with rewards and disjoint time windows. In: IEEE/RSJ International Conference on Intelligent Robots and Systems, pp. 2332–2337 (2007)

12. Dias, B.M.: Traderbots: a new paradigm for robust and efficient multirobot coordination in dynamic environments. Carnegie Mellon University (2004)

13. Berhault, M., Huang, H., Keskinocak, P., Koenig, S., Elmaghraby, W., Griffin, P., Kleywegt, A.: Robot exploration with combinatorial auctions. In: IEEE/RSJ International Conference on Intelligent Robots and Systems, vol. 2, pp. 1957–1962 (2003)

14. Koenig, S., Tovey, C.A., Zheng, X., Sungur, I.: Sequential bundle-bid single-sale auction algorithms for decentralized control. In: Proceedings of the 20th International Joint Conference on Artificial Intelligence, pp. 1359–1365 (2007)

15. Lagoudakis, M.G., Markakis, E., Kempe, D., Keskinocak, P., Kleywegt, A., Koenig, S., Tovey, C., Meyerson, A., Jain, S.: Auction-based multi-robot routing. In: Proceedings of Robotics: Science and Systems (2005)

16. Silva, S., Botelho, C., Alami, R.: M+: a scheme for multi-robot cooperation through negotiated task allocation and achievement. In: IEEE International Conference on Robotics and Automation, pp. 1234–1239 (1999)

17. Mackenzie, D.C.: Collaborative tasking of tightly constrained multi-robot missions. In: Multi-Robot Systems: From Swarms to Intelligent Automata: Proceedings of the 2003 International Workshop on Multi-Robot Systems, vol. 2, pp. 39–50 (2003)

18. Chien, S., Barrett, A., Estlin, T., Rabideau, G.: A comparison of coordinated planning methods for cooperating rovers. In: Proceedings of the Fourth International Conference on Autonomous Agents, pp. 100–101 (2000)

19. Zitouni, F., Maamri, R.: Cooperative learning-agents for task allocation problem. In: Auer, M.E., Tsiatsos, T. (eds.) IMCL 2017. AISC, vol. 725, pp. 952–968. Springer, Cham (2018). https://doi.org/10.1007/978-3-319-75175-7_93

20. Lemaire, T., Alami, R., Lacroix, S.: A distributed tasks allocation scheme in multi-UAV context. In: IEEE International Conference on Robotics and Automation, vol. 4, pp. 3622–3627 (2004)

21. Shehory, O., Kraus, S.: Task allocation via coalition formation among autonomous agents. In: Proceedings of the 14th International Joint Conference on Artificial Intelligence, vol. 1, pp. 655–661 (1995)

22. Shehory, O., Kraus, S.: Methods for task allocation via agent coalition formation. Artif. Intell. **101**(1), 165–200 (1998)

23. Vig, L., Adams, J.A.: Multi-robot coalition formation. IEEE Trans. Rob. **22**(4), 637–649 (2006)

24. Jose, G., Gabriel, O.: Multi-robot task allocation strategies using auction-like mechanisms. In: Artificial Research and Development in Frontiers in Artificial Intelligence and Application, vol. 100, pp. 111–122 (2003)

25. Lin, L., Zheng, Z.: Combinatorial bids based multi-robot task allocation method. In: Proceedings of the 2005 IEEE International Conference on Robotics and Automation, pp. 1145–1150 (2005)
26. Shiroma, P.M., Campos, M.F.M.: CoMutaR: a framework for multi-robot coordination and task allocation. In: IEEE/RSJ International Conference on Intelligent Robots and Systems, pp. 4817–4824 (2009)
27. Koes, M., Nourbakhsh, I., Sycara, K.: Constraint optimization coordination architecture for search and rescue robotics. In: Proceedings 2006 IEEE International Conference on Robotics and Automation, pp. 3977–3982 (2006)
28. Korsah, A.G.: Exploring bounded optimal coordination for heterogeneous teams with cross-schedule dependencies. Carnegie Mellon University (2011)
29. Jones, G.E., Dias, B.M., Stentz, A.: Time-extended multi-robot coordination for domains with intra-path constraints. Auton. Robot. **30**(1), 41–56 (2011)
30. Parker, L.E., Tang, F.: Building multirobot coalitions through automated task solution synthesis. Proc. IEEE **94**(7), 1289–1305 (2006)
31. Zlot, R.M.: An auction-based approach to complex task allocation for multirobot teams. Carnegie Mellon University (2006)
32. Tang, F., Parker, L.E.: A complete methodology for generating multi-robot task solutions using ASyMTRe-D and market-based task allocation. In: IEEE International Conference on Robotics and Automation (2007)
33. Smith, R.: Communication and control in problem solver. IEEE Trans. Comput. (1980)
34. Yang, X.-S.: Firefly Algorithm: Nature-Inspired Metaheuristic Algorithms. Luniver Press, Bristol (2008)
35. Yang, X.-S.: Firefly algorithms for multimodal optimization. In: Watanabe, O., Zeugmann, T. (eds.) SAGA 2009. LNCS, vol. 5792, pp. 169–178. Springer, Heidelberg (2009). https://doi.org/10.1007/978-3-642-04944-6_14

Effective Streaming Evolutionary Feature Selection Using Dynamic Optimization

Abdennour Boulesnane[1,2]([⊠]) [iD] and Souham Meshoul[1]

[1] Department of Computer Science, College of NTIC,
Abdelhamid Mehri-Constantine 2 University, 25000 Constantine, Algeria
{abdennour.boulesnane,souham.meshoul}@univ-constantine2.dz
[2] MISC Laboratory, Abdelhamid Mehri-Constantine 2 University,
Constantine, Algeria

Abstract. Feature selection is a key issue in machine learning and data mining. A great deal of effort has been devoted to static feature selection. However, with the assumption that features occur over time, methods developed so far are difficult to use if not applicable. Therefore, there is a need to design new methods to deal with streaming feature selection (SFS). In this paper, we propose the use of dynamic optimization to handle the dynamic nature of SFS with the ultimate goal to improve the quality of the evolving subset of selected features. A hybrid model is developed to fish out relevant features set as unnecessary by an online feature selection process. Experimental results show the effectiveness of the proposed framework compared to some state of the art methods.

Keywords: Dynamic optimization problems · OSFS
Feature selection · Streaming features · Classification

1 Introduction

Feature selection is a crucial task in data mining and machine learning especially when high dimensional datasets need to be processed. The purpose behind feature selection is to select a subset of the most relevant features for building powerful predictive models [1]. All traditional feature selection methods are time consuming and require all input features to be available at the beginning of the learning process. However, with the advent of new information technologies and in the big data era, many real world applications are forced to work with attributes occurring over time or in streaming. Therefore, a new challenge has emerged namely streaming feature selection (SFS) where new features are integrated on their arrival and calculations are carried out at the same time.

In this new challenge, the number of learning samples is fixed whereas the number of features increases with time as new attributes arrive. A critical challenge for SFS is the unavailability of the entire space of features at the beginning of the learning phase. Compared to traditional feature selection problems, there

© IFIP International Federation for Information Processing 2018
Published by Springer International Publishing AG 2018. All Rights Reserved
A. Amine et al. (Eds.): CIIA 2018, IFIP AICT 522, pp. 329–340, 2018.
https://doi.org/10.1007/978-3-319-89743-1_29

are two properties of SFS [2]. One is that the number of features could grow infinitely over time. Another is that features can be read one by one and each of them is processed online upon its arrival. However, up today and compared to traditional feature selection methods, few approaches have been proposed in the literature to tackle this new challenge [2–5]. This can be explained by the fact that the process of SFS requires new, fast and inexpensive methods.

In this context and as a first initiative of this kind, we present in this paper a new approach that deals with the problem of streaming feature selection by introducing dynamic optimization during the selection of the best attributes. Motivated by the fact that the problem of online feature selection is a dynamic problem whose dimension (feature) changes over time, we propose a hybridization between the WD2O dynamic optimization algorithm proposed in [6] and the Online Streaming Feature Selection algorithm (OSFS) [2], whose objective is to find a subset of optimal attributes to ensure better classification of unclassified data. Therefore, the main contribution of this work consists of a new hybrid approach called Dynamic Online Streaming Feature Selection (DOSFS) that exhibits the following features:

- This new feature selection system exhibits two properties: on one hand, the speed of OSFS, helps in providing quality attributes at any time; on the other hand, WD2O's self-adaptivity helps in exploring efficiently the space of redundant features taking into account the importance of the interaction between these features.
- This hybridization helps in fishing out relevant information previously treated as unnecessary data, which in turn helps to improve decision making in the future.
- DOSFS helps in strengthening the exploration capability of the OSFS algorithm and fill its gaps.

The rest of the paper is organized as follows. Section 2 presents some background material and related works. The proposed hybridization DOSFS is described in Sect. 3. Section 4 describes the conducted experimental study and obtained results. Finally, conclusions and perspectives are given in Sect. 5.

2 Background and Related Works

2.1 Feature Selection for Classification

Classification is the most common task of data mining and machine learning. It consists in identifying to which of a given set of classes a new incoming instance belongs. The purpose of classification is to obtain a model that can be used to classify unclassified data [7].

Many real world classification problems require supervised learning where the underlying class probabilities and class-conditional probabilities are unknown and each instance is related with a class label, i.e., relevant features are often unknown a priori [8]. Therefore, many candidate features are used in order to improve the representation of the domain, resulting in the existence of irrelevant and redundant attributes for the studied domain. This makes the decision algorithms complex, inefficient, less generalizable and difficult to interpret. For the majority of classification problems, it is hard to learn good classifiers before deleting these unwanted features because of the huge dimensionality of the data. Decreasing the number of irrelevant/redundant features can dramatically reduce the operating time of the learning algorithms and hence provide a more efficient classifier [7].

Therefore, feature selection techniques is considered one of the best methods to reduce the dimension in the feature space. It consists in finding the best subset among 2^n candidate features according to some evaluation functions. Generally, the feature selection methods used to evaluate a subset of features in the learning algorithms can be classified into three main categories, according on how they introduce the feature selection search with the construction of the classification model: filter methods, wrapper methods and embedded methods [9].

2.2 Streaming Feature Selection

In classical feature selection methods (filter, wrapper and embedded methods), all candidate attributes are assumed to be known a priori. These features are iteratively examined in order to select the best attribute. However, nowadays this does not extend to many real-world applications where one needs to deal with dynamic data streams and feature streams. For example, Twitter generates more than 320 millions of tweets daily and a large amount of words (features) are continually being produced. These new words quickly attract user's attention and become popular in a short period of time [9]. Therefore, because of the ineffectuality of traditional feature selection methods to applications involving streaming features, it should be preferable to use SFS to quickly adapt to the changes. Streaming features involve an attribute vector whose elements flow one by one over time while the number of instances in the training set remains fixed. The particularity of the SFS compared to the traditional feature selection, is as follows [2]:

- **The dynamic and uncertain nature:** Feature space's dimension may grow dynamically over time and may even extend to an infinite size.
- **The streaming nature:** Features flow one by one where each feature must be processed online upon its arrival.

2.2.1 Streaming Feature Selection Approaches

For the problem of SFS, the number of instances is considered as constant whereas the features arrive one at a time (in streaming). The task is to select in a timely manner a most relevant subset features from a huge number of available features. Compared to traditional methods and instead of searching in the entire attribute space that is very expensive, the SFS techniques process a new feature on its arrival [10].

In the literature, few approaches have been proposed to tackle the problem of SFS [2–5]. These approaches have different implementations especially in the way newly arrived features are checked. In the following section, we briefly present one of the most successful techniques widely used to resolve the problem of SFS.

2.2.2   Online Streaming Feature Selection Algorithm (OSFS)

Unlike the existing studies on traditional feature selection, online streaming feature selection aims to deal with feature streams in an online manner. For that purpose, a new algorithm is proposed in [2] called Online Streaming Feature Selection (OSFS). The authors in [2] study the SFS problem from an information theoretic perspective based on the criterion of Markov blanket. In OSFS, features are characterized into four types: redundant feature, irrelevant features, weakly relevant but non redundant features and strongly relevant features. An optimal feature selection approach should have strongly relevant and non-redundant features. OSFS finds an optimal subset of features based on online feature relevance and feature redundancy analysis [10]. A general framework of OSFS is presented as follows:

- Initialize the list of Best Candidate Features (BCF) in the model, $BCF = \emptyset$;
- **Step 1:** Generate a new feature x;
- **Step 2:** Online Relevance Analysis;
  - If x is relevant for the class label: $BCF = BCF \cup x$;
  - Otherwise, reject attribute x;
- **Step 3:** Online Redundancy Analysis;
- **Step 4:** Alternate Step 1 to Step 3 until some stop criteria are met.

In step 2 and in the relevance analysis phase, OSFS consists of discovering strongly and weakly relevant features, in order to add them into the best candidate features (BCF). If a new coming feature is relevant to the class label, it is added to BCF, otherwise it is discarded. In the redundancy analysis (step 3), OSFS dynamically removes the redundant features in the BCF subset. For each feature x in BCF, if there exists a subset within BCF making x and the class label conditionally independent, x is eliminated from BCF [2].

2.3   Optimization in Dynamic Environments

In every day life, each type of optimization problem has specific characteristics that make it distinct from others. However, almost all of them have a common feature which is their dynamic nature. Static optimization has known its limitations in solving such problems and therefore sophisticated methods are needed. More specifically, the field of research that addresses this kind of problems is commonly known as: *Optimization in Dynamic Environments* or simply: *Dynamic Optimization* [11].

Solving a dynamic optimization problem, requires not only finding the optimal global solution in a specific environment, but also following the trajectory of the evolution of this optimum in dynamic landscapes. Therefore, the main challenge for optimization algorithms in dynamic environments is how to increase or maintain the search diversity in such environments.

2.3.1 Dynamic Optimization Problems

We can formally describe a dynamic optimization problem (DOP) as the task that aims to find the sequence $(x_1^*, x_2^*, \ldots, x_n^*)$ that:

$$
\begin{aligned}
&\text{Optimize } f(x, t) \\
&\text{subject to. } h_j(x, t) = 0 \text{ for } j = 1, 2, ..., u \\
&\qquad\qquad g_k(x, t) \leq 0 \text{ for } k = 1, 2, ..., v. \\
&\text{with } x \in \mathbb{R}^n
\end{aligned}
\tag{1}
$$

Where $f(x, t)$ is a time dependent objective function, $(x_1^*, x_2^*, ..., x_n^*)$ is the sequence of n optima found as the fitness landscape changes. In other ways, it depicts optima tracking, $h_j(x, t)$ denotes the jth equality constraint and $g_k(x, t)$ denotes the kth inequality constraint. All these functions may change over time, as indicated by the dependence on the time variable t. A comprehensive review on dynamic optimization can be found in [12].

2.3.2 Wind Driven Dynamic Optimization Algorithm (WD2O)

In the literature, several algorithms have been proposed to deal with DOPs [12]. Each one has features that make it appropriate for solving specific problems than others. In other words and according to the No Free Lunch theorem, there is no universal algorithm which solves in the best way all optimization problems. Therefore, each optimization problem requires a thorough study that allows to find the best algorithm.

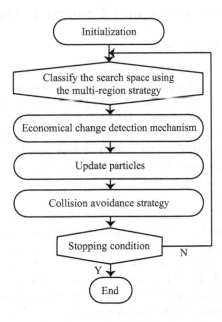

Fig. 1. Flow chart of WD2O algorithm.

Recently, a new dynamic optimization algorithm has been proposed in [6], called "Wind Driven Dynamic Optimization Algorithm (WD2O)". The characteristic property of this metaheuristic is the classification, it suggests to set regions of the search space as promising and non-promising regions with accordance to low and high pressure regions in the natural model. This new framework has been inspired from the meteorology. Compared to other dynamic optimization algorithms, the powerful feature of WD2O essentially resides in its efficiency and scalability against high-dimensional dynamic problems due to the new multi-region classification of search space. It is a multi-population metaheuristic.

Formally and as shown in Fig. 1, the WD2O algorithm can be described as follow: in the initialization phase, the particle's positions and velocities are randomly set for each dimension within the corresponding bounds. WD2O proceeds iteratively as described in Fig. 1. Firstly, the whole search space is divided into promising and non-promising areas using a multi-region strategy. This classification was beneficial in helping to find and track the global optimum as quickly as possible. Next to the objective function value, the pressure value is calculated for each particle in the population. These values have been exploited to achieve such a classification. A change detection mechanism is a significant step as it allows the algorithm to rapidly react to the possible environmental changes. In the next step, particle's positions are updated. Since this metaheuristic uses multiples populations, collision avoidance strategy is used in order to maintain several sub-populations on several peaks. This process continues in this manner till a stopping criterion is satisfied.

3   Research Problematic and Proposed Approach

3.1   Research Problematic

Generally, the evaluation criterion and the search strategy are two key elements in feature selection. According to the evaluation criterion, most feature selection algorithms (offline) are based either on filter methods or on wrapper methods. Furthermore, since the size of the search space for n features is 2^n, it is impossible to carry out an exhaustive search for the feature selection [13]. Therefore, the search strategy can drastically influence the results of a feature selection algorithm. Evolutionary techniques including particle swarm optimization, evolutionary algorithms, etc. have been widely applied to the traditional feature selection problems [14].

However, in online feature selection, these two problems need to be revisited. For the evaluation criterion, there is no general classification of the related methods, where most of them are based on information theory. On the other hand, since the feature space grows continuously over time, the size of the search space will be unknown or infinite. Therefore, it is impossible to adopt a global search technique because of the unavailability of the entire feature space.

Therefore, the questions that arise are, how to introduce computational intelligence to carry out streaming feature selection? And, to what extent will it be possible to solve the aforementioned issues? In a first initiative of its kind and motivated by the fact that the problem of the streaming feature selection is a dynamic problem whose dimension (feature) changes over time, we propose to treat this problem by introducing dynamic optimization during the selection of the best attributes.

3.2 Proposed Approach

As explained previously, the OSFS algorithm has been applied successfully to the problem of online feature selection. The idea of this simple technique is mainly focused on the exclusive selection of highly relevant features and weakly relevant non-redundant features. Whereas, irrelevant and redundant features are eliminated during the selection process. Although the results of this approach have been very encouraging in terms of accuracy, the interaction between the eliminated attributes and the future attributes is completely ignored in this model. Therefore, the logical question we have thus to ask is to what extent it would have been interesting to take these interactions into account?

To address the issues raised above, we propose a new model for selecting features. More precisely, we propose a new hybridization between the OSFS algorithm and the dynamic optimization algorithm WD2O.

The proposed approach can be summarized as shown in Fig. 2. We adopt the acronym DOSFS (Dynamic Online Streaming Feature Selection) to refer to the proposed approach. In this model, we propose a new system structure of SFS. The gist of this approach can be described as follows. In the first phase, which can be considered as an initialization phase, all parameters and components of either WD2O or OSFS are initialized. It should be mentioned that the selection process is automatically triggered by the arriving of new features, which carries the information relating to each Xi instance in the training data.

As soon as a new feature arrived (feature F44 in Fig. 2), the second phase is launched, in which the OSFS algorithm will run. As explained above, this algorithm retains only the relevant features for the classification task in question (class C in Fig. 2), which will be added to the set of best candidate feature (BCF) found so far. But before being selected, it must be proved that there is no subset S within BCF for which the features in BCF will be redundant.

Otherwise, irrelevant features are simply eliminated. However, eliminating redundant features is the key task for an optimal feature selection process, because according to the definition in [15], a redundant feature is a weakly relevant feature. Therefore, a redundant feature deleted at time t may become highly relevant in the future, through interaction with other newly arrived attributes.

In order to avoid such a scenario, or at least alleviate the problem, we propose to create a new feature space named Best Redundant Candidate Feature (BRCF). This set includes only the redundant attributes that are dependent on each other. Once the BRCF set is created, the third phase is launched, in which the dynamic optimization algorithm WD2O should find the best sequence of features independently of that found by OSFS.

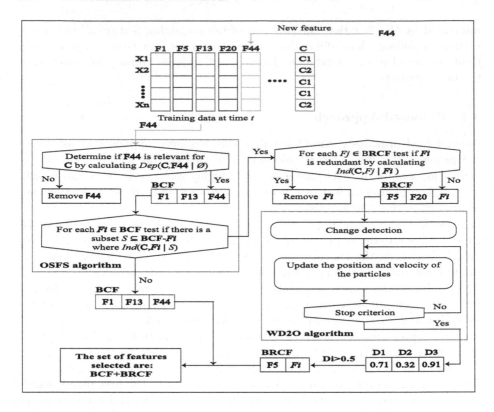

Fig. 2. Functional diagram of the proposed model DOSFS.

3.2.1   Objective Function and Solution Encoding

First of all, let us recall that the SFS problem is a dynamic problem whose dimension (D) is the only element that changes over time. In this type of problem, the objective function is always static, it consists, on the one hand, in maximizing the relevance between the features (F1, F2, ..., Fn) and the class labels (C1, C2), on the other hand, to minimize the redundancy between the selected features. The objective function used in this problem is inspired by the one proposed in [16] which can be defined as follows:

$$Max \left(Fitness = \sum_{x \in X} Z\left(x, C\right) - \sum_{x_i, x_j \in X} Z\left(x_i, x_j\right) \right) \qquad (2)$$

Where X is the set of selected attributes (BRCF) and C is the class label. In our case, since the data used are continuous, we will use the same objective function defined by Eq. (2), but with Fisher's Z test instead of mutual information as used in [16].

Otherwise, in WD2O, each particle's position is a vector of D dimension representing the number of features available in the subset BRCF. Each dimension corresponds to a decision variable, we adopt a real coding (between 0 and 1) because we are in a context of continuous optimization. A value greater than or equal to (respectively less than) 0.5 indicates that the attribute is selected (respectively not selected).

3.2.2 Change Detection

In this step, WD2O detects new environmental changes in the problem's dimension (D). Therefore, a simple technique has been adopted in WD2O in order to check the size of the dimension each time a new feature has arrived. If a change is detected, WD2O exploits the best solution found so far (stocked in the memory) (*gbest*) to quickly follow the new optimum. The idea is to change the *gbest*'s dimension to be compatible with the new problem, but without selecting the newly arrived feature (i.e., generating a value between 0 and 0.5 in this dimension). Furthermore, all particles in the population will be reinitialized to increase the diversity level in the search space.

3.2.3 Global Search

Once a change is detected (i.e., a new feature has been added to the BRCF set), the WD2O algorithm proceeds iteratively until a stopping criterion is met. At the end of the optimization process, the best features selected that maximize the objective function will be incorporated into the BCF set and considered as important features retained by the dynamic optimization algorithm.

4   Experimental Study

4.1   Experimental Setup

In order to assess the performance of the proposed approach and to verify its usefulness in practice, seven large-scale biological data sets were used (Feature Selection Datasets[1]; Kent ridge biomedical data set repository[2]) as shown in the Table 1. These biological datasets were provided for the purpose of selecting relevant features to a classification problem.

For the datasets: Breast-Cancer, Lung-Cancer, Leukemia-ALLAML, we use the training and validation datasets provided in Kent ridge biomedical data set repository. For the other three datasets, we adopt 2/3 of the instances for the training and 1/3 of the remaining instances for the test. Our comparative study compares the proposed model DOSFS algorithm with α-investing [4] and standard OSFS algorithm [2]. In order to evaluate a selected feature subset in the experiments, the SVM classifier is used. Two performances measures to evaluate our algorithm with the standard OSFS and α-investing are compactness (the number of selected features) and the prediction accuracy (the percentage of correctly classified test instances). All experiments were conducted on a computer with Intel i5-2450, 2.50 GHz CPU and 6 GB memory.

---

[1] http://featureselection.asu.edu/datasets.php.
[2] http://datam.i2r.a-star.edu.sg/datasets/krbd/.

Table 1. Summary of biological datasets used for evaluation.

Dataset	# Features	# Instances	Training data	Test data
Breast-Cancer	24481	97	78	19
Colon-Tumor	2000	62	41	21
Central Nervous System	7129	60	40	20
Lung-Cancer	12533	181	32	149
Leukemia-ALLAML	7129	72	38	34
Prostate-GE	5966	102	68	34
Arcene	10000	200	134	66

4.2 Experimental Results and Comparisons

4.2.1 Usefulness of WD2O for Streaming Feature Selection

To analyze closely how the dynamic optimization algorithm WD2O react to the arrival of a new streaming feature, we recorded the evolution of WD2O over time during the streaming processing of the "Leukemia-ALLAML" dataset, as shown in Fig. 3. Similar behaviors have been noticed for the other datasets.

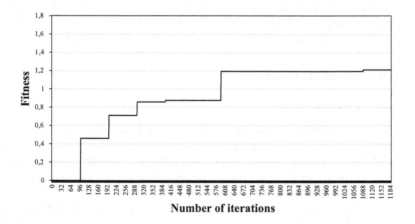

Fig. 3. Evolution of the best fitness value (corresponding to gbest) during a run of the proposed model using the Leukemia-ALLAML dataset.

From Fig. 3, one can observe the gradual improvement of the best fitness value corresponding to *gbest*, which implies both the high exploration capacity of WD2O in the search space and its efficiency to handle dynamic optimization problems. Furthermore, the performance stability recorded over certain periods of time can be explained in two ways: either the new arrival feature is redundant in the set BRCF, which implies the deletion of this attribute, or this new streaming feature is added in BRCF but its interaction with the other features does not significantly improve the classification task.

4.2.2 Comparison and Analysis

In this section we compare our proposed hybrid approach with the standard OSFS and α-investing algorithms in terms of prediction accuracy and the number of selected features, on the seven high-dimensional datasets previously presented in Table 1.

Using the SVM classifier, Table 2 presents the prediction accuracy values of the proposed DOSFS vs. OSFS and α-investing algorithms. The highest predictive accuracy values are shown in bold.

Table 2. Prediction accuracy of SVM (Acc) and the number of selected features (#).

Dataset	Algorithm					
	α-investing		OSFS		DOSFS	
	Acc	#	Acc	#	Acc	#
Breast-Cancer	0.63158	1	0.78947	2	**0.84211**	4
Colon-Tumor	0.57143	2	**0.71429**	2	0.57143	5
Central Nervous System	**0.8**	1	0	2	0.05	4
Lung-Cancer	0.92617	2	0.95302	2	**0.96644**	6
Leukemia-ALLAML	0.52941	1	0.91176	2	**0.94118**	6
Prostate-GE	0.88235	2	0.67647	2	**0.91176**	6
Arcene	0.56061	6	0.56061	3	**0.59091**	7

From the obtained results in Table 2, it can be seen that the proposed hybrid approach DOSFS has succeeded in improving the performance of OSFS on several cases, due to the features recovered by the dynamic optimization algorithm as shown in Table 2. According to the SVM classifier, the results indicate that the proposed DOSFS is very competitive and promising approach compared to OSFS and α-investing on most datasets.

5 Conclusion and Future Work

The major contribution of this paper is that we proposed a new model to solve streaming feature selection, the goal of which is to develop an online classifier involving only a small number of features. More precisely, we propose for the first time in the literature, a new hybridization between the streaming feature selection algorithm OSFS and the WD2O dynamic optimization algorithm.

Thanks to the proposed hybrid model, the exploration capability of the OSFS algorithm has been enhanced and many of the relevant features previously treated by the OSFS algorithm as unnecessary data has been fished out, which in turn helps to improve decision making. To analyze the performance of the proposed approach, high-dimensional biological data were used. The results obtained are encouraging and confirm that this new model is a promising way for a more stable and precise selection of streaming features.

As for future work, we intensify our work on studying the relationship between the dynamic problem of data streaming and dynamic optimization. This new convergence could bring many new lines of research in the field of data mining.

Acknowledgments. This work has been supported by the National Research Project CNEPRU under grant N: B*07120140037.

References

1. Wang, J., Zhao, P., Hoi, S., Jin, R.: Online feature selection and its applications. IEEE Trans. Knowl. Data Eng. **26**(3), 698–710 (2014)
2. Wu, X., Yu, K., Ding, W., Wang, H., Zhu, X.: Online feature selection with streaming features. IEEE Trans. Pattern Anal. Mach. Intell. **35**(5), 1178–1192 (2013)
3. Perkins, S., Theiler, J.: Online feature selection using grafting. In: Proceedings of the 20th International Conference on Machine Learning (ICML), pp. 592–599 (2003)
4. Zhou, J., Foster, D., Stine, R., Ungar, L.: Streamwise feature selection. J. Mach. Learn. Res. **7**, 1861–1885 (2006)
5. Yu, K., Wu, X., Ding, W., Pei, J.: Towards scalable and accurate online feature selection for big data. In: 2014 IEEE International Conference on Data Mining, pp. 660–669. IEEE (2014)
6. Boulesnane, A., Meshoul, S.: WD2O: a novel wind driven dynamic optimization approach with effective change detection. Appl. Intell. **47**(2), 488–504 (2017)
7. Tang, J., Alelyani, S., Liu, H.: Feature selection for classification: a review. In: Aggarwal, C.C. (ed.) Data Classification: Algorithms and Applications, pp. 37–64. CRC Press, Boca Raton (2014)
8. Dash, M., Liu, H.: Feature selection for classification. Intell. Data Anal. **1**(1–4), 131–156 (1997)
9. Li, J., Hu, X., Tang, J., Liu, H.: Unsupervised streaming feature selection in social media. In: Proceedings of the 24th ACM International on Conference on Information and Knowledge Management, pp. 1041–1050. ACM (2015)
10. Li, J., Cheng, K., Wang, S., Morstatter, F., Trevino, R., Tang, J., Liu, H.: Feature selection: a data perspective. arXiv preprint arXiv:1601.07996 (2016)
11. Cruz, C., González, J., Pelta, D.: Optimization in dynamic environments: a survey on problems, methods and measures. Soft Comput. **15**(7), 1427–1448 (2011)
12. Yang, S., Yao, X. (eds.): Evolutionary Computation for Dynamic Optimization Problems. SCI, vol. 490. Springer, Heidelberg (2013). https://doi.org/10.1007/978-3-642-38416-5
13. Kohavi, R., John, G.H.: Wrappers for feature subset selection. Artif. Intell. **97**(1), 273–324 (1997)
14. Xue, B., Zhang, M., Browne, W., Yao, X.: A survey on evolutionary computation approaches to feature selection. IEEE Trans. Evol. **20**(4), 606–626 (2016)
15. Yu, L., Liu, H.: Efficient feature selection via analysis of relevance and redundancy. J. Mach. Learn. Res. **5**, 1205–1224 (2004)
16. Cervante, L., Xue, B., Zhang, M., Shang, L.: Binary particle swarm optimisation for feature selection: a filter based approach. In: IEEE Congress on Evolutionary Computation (CEC 2012), pp. 881–888 (2012)

Evolutionary Multi-objective Optimization of Business Process Designs with MA-NSGAII

Nadir Mahammed[1($\boxtimes$)], Sidi Mohamed Benslimane[1($\boxtimes$)],
and Nesrine Hamdani[2($\boxtimes$)]

[1] LabRI-SBA, Ecole Supérieure en Informatique 08 Mai 1945,
BP 73 Bureau de poste EL WIAM, 22016 Sidi Bel Abbes, Algeria
{n.mahammed, s.benslimane}@esi-sba.dz
[2] LIO Laboratory, Ahmed Ben Bella University of Oran 1,
BP 1524 EL MNaouer, 31000 Oran, Algeria
hamdani.nesrine@edu.univ-oranl.dz

Abstract. Optimization is known as the process of finding the best possible solution to a problem given a set of constraints. The problem becomes challenging when dealing with conflicting objectives, which leads to a multiplicity of solutions. Evolutionary algorithms, which use a population approach in their search procedures, are advised to suitably solve the problem. In this article, we present an approach for an evolutionary combinatorial multi-objective optimization of business process designs using a variation of NSGAII, baptized MA-NSGAII. The variants of NSGAII are numerous. In fact, the vast majority deals either with the crossover operator or with the crowding distance. We discuss an optimization Framework that uses (i) a proposal of effective Fitness function, (ii) 02 contradictory criteria to optimize and (iii) an original selection technique. We test the proposed Framework with a real life case of multi-objective optimization of business process designs. The obtained results clearly indicate that an effectual Fitness function combined with the appropriate selection operator affects undeniably quality and quantity of solutions.

Keywords: Multi-objective optimization · Evolutionary computing
Genetic algorithm · Selection operator · Business process

1 Introduction

The optimizing of business processes (BP) is considered the problem of building feasible BPs while optimizing criteria such as reducing execution time and minimizing the resource cost. To achieve that goal, evolutionary algorithms (EA), and more specifically genetic algorithms (GA) have been much talked about. Generally, these optimization techniques are indoctrinated by 02 important questions: Exploration and Exploitation. If exploitation refers to the tendency of the algorithm to direct its search using information obtained in the past and to determine promising regions, for later research [1], exploration, for its part, explores new and unknown areas in the research space to find promising areas. In search of optimization, exploitation is carried out by selection operator, and exploration of the search space is carried out by other -search- operators in EAs [2].

© IFIP International Federation for Information Processing 2018
Published by Springer International Publishing AG 2018. All Rights Reserved
A. Amine et al. (Eds.): CIIA 2018, IFIP AICT 522, pp. 341–351, 2018.
https://doi.org/10.1007/978-3-319-89743-1_30

Fogel [3] describes selection operator as the action of selecting more fitting individuals, by analogy to Darwin's theory of evolution (survival of the strongest). All individuals have a chance of being selected in the population, but there is a chance that an individual can be selected more than once depending on its Fitness [4]. In an optimization matter, these characteristic determine the convergence of GA [5].

This article proposes a Framework that deals with a MOO of BP designs (called BPMOO), by reducing the cost and minimizing the duration. The Framework uses an original EA: MA-NSGAII, for Mass selection based NSGAII. MA-NSGAII tests and experiments the influence of the proposed new selection operator; inspired by viticulture, Mass selection. Section 2 presents a state of the art on BPMOO exclusively with NSGAII. Section 3 presents the proposed Framework with its Fitness function aiming the optimization issue, and introduces MA-NSGAII. Section 4 presents the experimental phase of the Framework applied on a test scenario based on BP designs optimization, then evaluates and discusses the obtained results. Finally, Sect. 5 summarizes the proposed research and provides future work directions.

2   Related Work

To overcome the question of BPMOO, NSGAII is one of the most widely used EAs [6, 7]. The first work to quote is [8]. They focus on how to appropriately allocate resources to activities, in BP designs to ensure its high performance, using NSGAII.

A series of work on BPMOO with EAs, in general, and NSGAII, in particular are introduced in [9]. The proposed approach uses a formal definition for a BP (based on [8]). Multi-objectivity is expressed in terms of cost and duration of BP designs. They propose and test a Framework using NSGAII, for generating new BP designs. Thereafter, [9, 10] present the most important work in this field. They have improved [9] work by adding (i) the ability to review or reconfigure any unfeasible BP design resulting from NSGAII iterations, (ii) to compare the efficiency of NSGAII with other EAs. Finally, to reach to the work of [11, 12], they propose a Framework for optimizing BP designs, where each task composing a BP can be seen a Web service. Wibig [13] proposes a Framework for BPMOO using Petri networks for modeling. He uses dynamic programming to reduce the computation time required. Farsani et al. [14] modifies the mutation and crossover operators used by NSGAII. Mahammed et al. [6] are interested in a BPMOO (up to 03 criteria). They proposed a Framework that combines an original Fitness function with NSGAII with a modified crossover operator.

The work on a multi criteria optimization with NSGAII is legion, and few to have proposed to review or modify NSGAII's selection operator. Ishibuchi et al. [15] propose a new selection technique with NSGAII, in MOO. The proposed technique is a two-stage selection mechanism (i) a standard selection based on individual Fitness is applied, then, (ii) tournament selection is used. Emmerich et al. [16] use NSGAII to arrive at an evolutionary steady-state algorithm (it produces only an individual at each iteration). They combine NSGAII with a selection operator based on hyper-volume measurement. Trivedi et al. [17] propose to review all genetic operators of NSGAII. The binary tournament selection technique limited by constraints without parameters is used to efficiently manage constraints. Phan et al. [18] study a method to aggregate existing indicator-based selection operators. They show that a boosted selection operator outperforms exiting ones in optimality, diversity and convergence velocity. Zhong et al. [19] present an interesting study on the reduction of solutions diversity following a MOO (02 criteria) using NSGAII with truncation selection. Mahammed et al. [20] propose an evolutionary multi-criteria approach based on a modified EA for generating optimized business processes. They replace the binary tournament selection with uniform selection, roulette wheel selection and ranking selection, combining the whole with a proposed crossover operator. The proposed framework improves the results obtained by [6]. In the present work, the authors present an approach for a BPMOO using NSGAII with an unusual selection operator, Mass selection.

3   Proposed Approach

The current study gives rise to a Framework capable of combinatorial MOO of BP designs, using a modified NSGAII. It aims to generate a series of optimized designs, with reduced cost and minimized duration. Thereafter, the main steps of the proposed Framework are depicted. Then, MA-NSGAII with Mass selection is explained.

3.1   Overall Architecture of the Proposed Framework

Following a number of steps, the proposed Framework is able to generate a series of optimized designs from an initial BP. The authors define a BP as a set of activities that takes one or more kinds of input and creates an output that has value to the customer [20]. Throughout the Framework course (Fig. 1), each design must fulfill a certain amount of constraints. MA-NSGAII is used to generate BP designs, each one has (i) a feasible graphical representation and (ii) optimized' attributes values.

1. Create an initial population. The first step of the proposed optimization Framework is to generate a random population of BP designs. It takes place only once in the Framework's progress, then the population evolves for a defined number of generations. The steps 2–5 are repeated for a predefined number of iterations.
2. Create designs representation. For each design, a mathematical representation is generated. 02 distinguished matrixes are used (i) the first for representing the relationship between resources and tasks composing a potential BP design. (ii) The second matrix portrays the tasks attributes (i.e. optimization criteria) of each design.
3. Verify and apply the restraints. The proposed Framework checks constraints prior the evaluation of each individual, because its design might be modified, thereafter. As restraints, we quote:

   - Each design must have a bounded size.
   - A task must appear once, in each design.
   - Each task must take birth form one or more resources, or BP inputs resources.
   - Each task must be linked to another task by one or more resources, or BP outputs resources.

- Replace or delete any task or resource useless in a design.
- Verify inputs and outputs of each design.

4. Assess designs. It involves calculating each BP design's Fitness value. It is based on each solution attributes values in its design. Knowing that the Fitness function is at the heart of an evolutionary computing application, the proposed Framework uses a Fitness function dealing with 02 criteria, (i) minimizing the delivery price of the service (i.e. cost) and (ii) minimizing the delivery duration of the service (i.e. duration). The solutions are evaluated after the restraints verification because only tasks that really participate in a BP design are taken into account

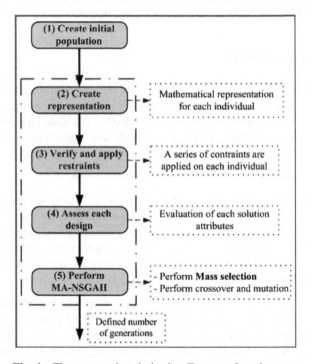

Fig. 1. The proposed optimization Framework main steps.

To correctly represent the Fitness function F(x), the authors propose to use the Pythagorean Theorem[1]. F(x) can be symbolized geometrically as the hypotenuse of the right-angled triangle formed by C and D. This choice was decided by the absolute necessity of not favoring any particular objective, in an MOO [20]. The Fitness function may be formulated:

$$F(x) = \sqrt{C^2 + D^2}$$

With $C = \sum_1^{n_d} c_i$, $D = \sum_1^{n_d} d_i$ Where C and D are normalized

---

[1] Maor, E. (2007). *The Pythagorean Theorem: a 4,000-year history*. Princeton Univ. Press.

C: Cost of the BP design.
D: Duration of the BP design.
n_d: Number of tasks in the BP design.
c_i: Cost of a task i being a part of the BP design.
d_i: Duration of a task i from the BP design

5. Perform MA-NSGAII. After the evaluation, MA-NSGAII is applied. First, a sim- ulated binary crossover is performed. The process does not check whether the solution is feasible, it is the concern of step 3. Then, mutation operator is performed. Finally, Mass selection is performed regarding to all new solutions. Subsection III-B introduces MA-NSGAII with Mass selection.

3.2 MA-NSGAII

MA-NSGAII is inspired from NSGAII proposed by [21], one of its main features is to ensure the diversity of the population throughout iterations. To do so, the authors propose to replace the simulated binary tournament selection operator recommended by [21] by Mass selection operator (see Fig. 2). This technique has never been used with NSGAII, by the past. The authors propose to add experiments comparison with other techniques of selection: Truncation selection [22] and roulette wheel selection [23], to demonstrate the effectiveness of MA-NSGAII.

Mass selection is native to traditional agriculture, mainly viticulture [24]. The use of such a technique goes back quite a long time [25–27] and never in the context of BPMOO. According to [24], Mass selection consists in choosing breeders according to their individual aspect performance(s). They explain that it is easy to achieve, and advised in the situation where no followed-up of individuals is available or required. To the knowledge of the authors, Mass selection has never been used with a GA, in a MOO context. This article proposes to replace the selection technique usually used within NSGAII by Mass selection, and compare the results using traditional tournament selection, truncation selection and roulette wheel selection. Knowing that Mass selection is purely used in biology research, the authors propose to implement it by taking inspiration from roulette wheel selection (see [28]).

Mass selection algorithm can be summarized as follows:

1. Evaluate Fitness f_i of each individual in the population.
2. Compute the probability of selecting each individual $p_i = f_i / \sum_{j=1}^{N} f_j$.
3. Calculate the sum of all individuals Fitnesses in the population $S = \sum_{i=1}^{N} f_i$.
4. Generate a uniform random number from interval $r \in \,]0, S]$.
5. Calculate the cumulative probability $q_i = \sum_{j=1}^{i} p_j + \dfrac{\sum_{j=1}^{m} c_j}{m} + m_{norm}$.
6. If $r \le q_i$ then select individual i.
7. Repeat 4 to 6 N times.

Mass selection operator considers, in addition to Fitness function value of each individual, 02 parameters (i) the mean between the chosen criteria MOO of these

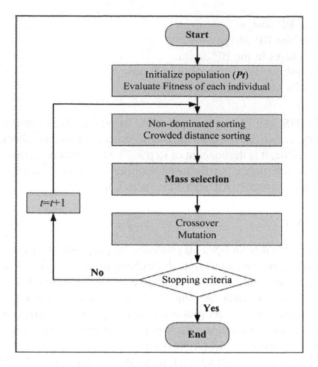

Fig. 2. MA-NSGAII's operating.

individuals and (ii) a value attributed to each solution for the appearance (e.g. aspect[2]). This implementation was chosen by the nature of Mass selection to take into account more external appearances of individuals. Which, compared to the others selection techniques adds more credibility, in its unfolding.

4 Experimentation and Results

The test scenario describes "Sales forecasting" from which new designs must be generated and optimized, proposed by [29]. The scenario takes as inputs (a) the company name and (b) the market update request, and produces as output (c) a report containing the forecast results of the contract.

The work required by the proposed Framework is to produce new optimized BP designs from the sales forecast scenario. The generation of these solutions is done using a tasks library (20 tasks), and a set of resources known and limited (09 resources). An evolutionary combinatorial MOO with 02 optimization criteria is used (i) the mini-mization of BP designs cost, and (ii) minimizing their duration.

---

[2] Genes' number of a solution.

Table 1. The proposed Framework parameters (inspired form [20]).

Parameter	Value
Tasks library size	20
Number of available resources	9
Number of attributes per task	2
Minimum BP design size	4
Maximum BP design size	6
Initial BP input resources	{a, b}
Initial BP output resources	{c}
Cost values' interval	[200 230]
Duration values' interval	[300 390]
Selection operator	Mass selection operator
Crossover operator	Simulated binary crossover
Mutation operator	One point mutation

Table 2. Fitness values according to selection techniques.

Population size	Selection operator	Solution size (tasks)		
		4	5	6
		Minimal Fitness		
500	Tournament	963	1192	1419
	Roulette wheel	963	1190	1424
	Truncation	958	1190	1421
	Mass	955	1185	1416
100	Tournament	963	1192	1419
	Roulette wheel	955	1188	1418
	Truncation	955	1183	1418
	Mass	954	1183	1411

Table 1 summarizes the parameters used by the proposed optimization Framework.

Table 2 shows the different Fitness values obtained by the proposed Framework according to the selection operator applied, for the BPMOO. The results vary according to the initial population size chosen (500 then 100 individuals) and solutions size obtained (depending on BP design's size: 4, 5 then 6 tasks).

A number of experiments have been performed to assess the capabilities of the proposed Framework using a MSI GT70 laptop with NetBeans 8.1 IDE and Java 8. The authors chose the minimum Fitness value of each solution as a parameter to evaluate each solution. The obtained results by [11] are resumed with the tournament selection. As explained in Subsection II-B, Mass selection is the only technique that gives importance to solution's appearance during the evaluation process. This feature makes it possible to add further criteria, for solutions estimation during BPMOO. Another interesting point is the proximity of resulting Fitness values. It can be explained by the

Table 3. Optimization criteria values according to selection techniques.

Selection	Solution size (task)	Cost	Duration
Tournament [11]	4	448	853
	5	553	1056
	6	662	1257
Roulette wheel	4	448	847
	5	550	1053
	6	659	1256
Truncation	4	447	844
	5	554	1050
	6	662	1252
Mass	4	446	843
		447	844
	5	555	1045
		558	1048
	6	658	1248
		658	1252

discrete nature of the values used in the BPMOO studied, even if the optimization Framework obtained the most interesting solutions.

Table 3 shows the optimization criteria values obtained by the proposed Framework, i.e. minimize BP design cost and duration. The results are obtained with an initial population of 500 then 100 individuals (see Table 2), a number of iterations up to 20 iterations. The design size ranges from 04 tasks to 06 tasks according to [11]. Table 3 shows -clearly- that BPMOO Framework provides the finest results with MA-NSGAII, in comparison with [11]. Figure 3 compares selection techniques used by the proposed

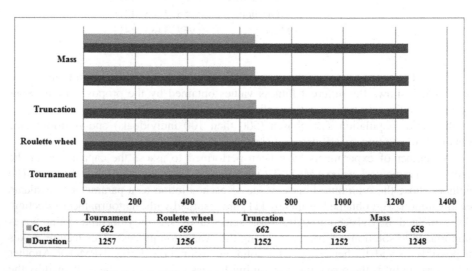

Fig. 3. Solutions comparison (Design size = 6 tasks).

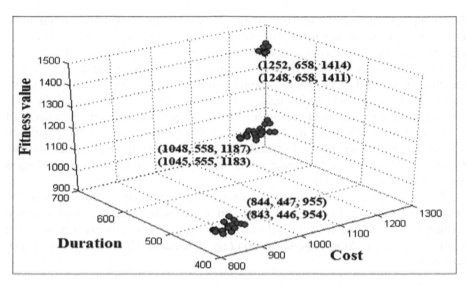

Fig. 4. Solutions and their Fitness (with MA-NSGAII).

Framework, with BP designs size equal to 06 tasks. The results show that using MA-NSGAII offers more solutions that meet optimization requirements (02 solutions per design size, against 01 solution with other selection techniques). The difference between the different solutions according to the selection technique used is certainly minimal. It may be explained by intervals' values (cost and duration) used by the Framework (Table 1). The results obtained by the proposed Framework with MA-NSGAII are shown in Fig. 4. Three (03) clusters of solutions are generated.

Each cluster corresponds to the average size of a solution, i.e. each BP design size should be between 04 and 06 tasks, making a cluster per value. A triplet as (843, 446, 951) represents: duration, cost and Fitness values of a BP design, respectively.

Although the results seem better than [11], it seems obvious that improvements can be made to the presented work. Adding other optimization criteria with a form of dependency between them could help to evaluate with confidence the quality of each potential solution. Another factor had a significant impact on the presented study is the tasks library size and the amount of resources. It turns out that a larger library (with more resources) could improve the obtained solutions (at least quantitatively). It appears to the authors that the selection operator as much as the crossover operator (see [20]) has a moderate effect on the generated solutions quality.

5 Conclusion

The presented work aims is to deal with a BPMOO using an EA, *i.e.* to generate feasible BP designs, in a combinatorial MOO environment, using an implemented Framework. The proposed Framework, apart from a mathematical representation of BP designs, uses MA-NSGAII, a modified version of NSGAII, with an original Fitness function proposal. One of MA-NSGAII specifications consists on replacing the traditional selection technique by a different one: Mass selection. To demonstrate the efficiency and effectiveness of the optimization Framework many selection techniques have been identified and used. It has an experimental case explaining an evolutionary combinatorial BPMOO problem. A rather confusing fact emerges. Mass selection, which has not been used in this field of research by the past, has yielded the most convincing results. The obtained outcomes are better either through the quality of solutions (Fitness value), the quantity (number of solutions) and even the population size used for experimentations. The results have demonstrated that the Framework with the help of MA-NSGAII, has enhanced its capability to generate diverse BP designs with optimal objective values.

Several issues can be addressed. For example, it is desirable to do more work on how an evolutionary bio inspired algorithm reacts within the proposed selection technique. Also, by incorporating information about the MOO context (e.g. servicing business processes). More experimentation (e.g. different real life scenarios) would be welcomed.

References

1. Thierens, D., Goldberg, D.: Convergence models of genetic algorithm selection schemes. In: Davidor, Y., Schwefel, H.-P., Männer, R. (eds.) PPSN 1994. LNCS, vol. 866, pp. 119–129. Springer, Heidelberg (1994). https://doi.org/10.1007/3-540-58484-6_256
2. Holland, J.H.: Adaptation in Natural and Artificial Systems. University of Michigan Press, Ann Harbor (1975)
3. Fogel, D.B.: Evolutionary Computation: Toward a New Philosophy of Machine Intelligence. IEEE, Piscataway (1995)
4. Goldberg, D.E., Deb, K.: A comparative analysis of selection schemes used in genetic algorithms. Found. Genet. Algorithms 1, 69–93 (1991)
5. Mitchell, M.: An Introduction to Genetic Algorithms. MIT Press, Cambridge (1998)
6. Mahammed, N., Benslimane, S.M.: Toward Multi Criteria Optimization of Business Processes Design. In: Bellatreche, L., Pastor, Ó., Almendros Jiménez, J.M., Aït-Ameur, Y. (eds.) MEDI 2016. LNCS, vol. 9893, pp. 98–107. Springer, Cham (2016). https://doi.org/10.1007/978-3-319-45547-1_8
7. Zhou, Y., Chen, Y.: The methodology for business process optimized design. In: IECON Proceedings (Industrial Electronics Conference), vol. 2, pp. 1819–1824 (2003)
8. Hofacker, I., Vetschera, R.: Algorithmical approaches to business process design. Comput. Oper. Res. 28(13), 1253–1275 (2001)
9. Vergidis, K., Tiwari, A., Majeed, B., Roy, R.: Optimisation of business process designs: an algorithmic approach with multiple objectives. Int. J. Prod. Econ. 109(1), 105–121 (2007)
10. Vergidis, K., Saxena, D., Tiwari, A.: An evolutionary multi-objective framework for business process optimisation. Appl. Soft Comput. 12(8), 2638–2653 (2012)
11. Vergidis, K., Turner, C., Alechnovic, A., Tiwari, A.: An automated optimisation framework for the development of re-configurable business processes: a web services approach. Int. J. Comput. Integr. Manufact. 28(1), 41–58 (2015)
12. Georgoulakos, K., Vergidis, K., Tsakalidis, G., Samaras, N.: Evolutionary multi-objective optimization of business process designs with pre-processing. In: 2017 IEEE Congress on Evolutionary Computation (CEC), pp. 897–904. IEEE (2017)

13. Wibig, M.: Dynamic programming and genetic algorithm for business processes optimisation. Int. J. Intell. Syst. Appl. **5**(1), 44 (2012)
14. Farsani, S.T., Aboutalebi, M., Motameni, H.: Customizing NSGAII to optimize business processes designs. Res. J. Recent Sci. **2**, 74–79 (2013)
15. Ishibuchi, H., Shibata, Y.: A similarity-based mating scheme for evolutionary multiobjective optimization. In: Cantú-Paz, E., et al. (eds.) GECCO 2003. LNCS, vol. 2723, pp. 1065–1076. Springer, Heidelberg (2003). https://doi.org/10.1007/3-540-45105-6_116
16. Emmerich, M., Beume, N., Naujoks, B.: An EMO algorithm using the hypervolume measure as selection criterion. In: Coello Coello, C.A., Hernández Aguirre, A., Zitzler, E. (eds.) EMO 2005. LNCS, vol. 3410, pp. 62–76. Springer, Heidelberg (2005). https://doi.org/10.1007/978-3-540-31880-4_5
17. Trivedi, A., Pindoriya, N.M., Srinivasan, D.: Modified NSGA-II for day-ahead multi-objective thermal generation scheduling. In: IPEC 2010, pp. 752–757. IEEE (2010)
18. Phan, D.H., Suzuki, J.: Boosting indicator-based selection operators for evolutionary multi objective optimization algorithms. In: 23rd IEEE International Conference on Tools with Artificial Intelligence (ICTAI), pp. 276–281. IEEE, November 2011
19. Zhong, X., Zhao, Y., Han, Q.: An improved non dominated sorting multi objective genetic algorithm and its application (2015)
20. Mahammed, N., Benslimane, S.M.: An evolutionary algorithm based approach for business process multi-criteria optimization. Int. J. Organ. Collect. Intell. (IJOCI) **7**(2), 34–53 (2017)
21. Deb, K., Agrawal, S., Pratap, A., Meyarivan, T.: A fast elitist non-dominated sorting genetic algorithm for multi-objective optimization: NSGA-II. In: Schoenauer, M., Deb, K., Rudolph, G., Yao, X., Lutton, E., Merelo, J.J., Schwefel, H.-P. (eds.) PPSN 2000. LNCS, vol. 1917, pp. 849–858. Springer, Heidelberg (2000). https://doi.org/10.1007/3-540-45356-3_83
22. Crow, J.F., Kimura, M.: Efficiency of truncation selection. Proc. Natl. Acad. Sci. **76**(1), 396–399 (1979)
23. Mühlenbein, H., Voigt, H.M.: Gene pool recombination in genetic algorithms. In: Osman, I. H., Kelly, J.P. (eds.) Meta-Heuristics, pp. 53–62. Springer, Boston (1996). https://doi.org/10.1007/978-1-4613-1361-8_4
24. Leroy, G., Verrier, E., Meriaux, J.C., Rognon, X.: Genetic diversity of dog breeds: within-breed diversity comparing genealogical and molecular data. Anim. Genet. **40**(3), 323–332 (2009)
25. Liyanage, D.V.: Identification of genotypes of coconut palms suitable for breeding. Exp. Agric. **3**(03), 205–210 (1967)
26. Magnussen, S., Yeatman, C.W.: Predictions of genetic gain from various selection methods in open pollinated Pinus banksiana progeny trials. Silvae Genetica **39**(3–4), 140–153 (1990)
27. Berthouly, C., Leroy, G., Van, T.N., Thanh, H.H., Bed'Hom, B., Nguyen, B.T., Maillard, J.C., et al.: Genetic analysis of local Vietnamese chickens provides evidence of gene flow from wild to domestic populations. BMC Genet. **10**(1), 1 (2009)
28. Mahammed, N., Benslimane, S.M., Hamdani N.: An approach to addressing a business process multi-objective optimization issue with MA-NSGAII. In: EDiS 2017, University of Oran 1, Oran, Algeria, 17–18 December. IEEE (2017). (in press)
29. Grigori, D.: Business process intelligence. Comput. Ind. **53**, 321–343 (2004)

Sink Mobility Based on Bacterial Foraging Optimization Algorithm

Ranida Hamidouche[1]([✉])[iD], Manel Khentout[1], Zibouda Aliouat[1],
Abdelhak Mourad Gueroui[2], and Ado Adamou Abba Ari[2,3]

[1] LRSD Laboratory, University Ferhat Abbes Setif 1, El Bez, Setif, Algeria
{ranida.hamidouche,zaliouat}@univ-setif.dz, manelkhentoutmk@gmail.com
[2] LI-PaRAD Laboratory, Universit Paris Saclay,
University of Versailles Saint-Quentin-en-Yvelines, Versailles, France
{mourad.gueroui,ado-adamou.abba-ari}@uvsq.fr
[3] Department of Computer Science, University of Maroua, Maroua, Cameroon

Abstract. Increasingly, the adoption of mobile sensors becomes imperative in the context of target tracking applications, especially for reliable data collection purpose. However, the design of a strategy that allows mobile sensors suitably to move in an autonomous, distributed and self-organized way is not evident to achieve by a deterministic polynomial algorithm. Solutions that are biologically inspired by the collective behaviour of individual social communities provide alternative tools and efficient algorithms that emerge from many interesting properties applicable to sensor technology. These solutions implement highly efficient systems that are structurally simple, powerful, highly distributed and fault-tolerant. Some biological societies, like colonies of the Escherichia coli bacteria, offer prospects to certain mobile sensors to acquire an artificial intelligence allowing them to move autonomously through the network. In this paper, we proposed a bio-inspired protocol named SMB-FOA (Sink Mobility based on Bacterial Foraging Optimization Algorithm). The main idea of this protocol was inspired by the autonomous movement of the Escherichia coli bacterium. Based on the simulation results, we concluded that our proposed SMBFOA protocol increases the throughput data rate and prolongs the network lifetime duration for 30% and 5% respectively compared to Clustering Duty Cycle Mobility aware Protocol (CDCMP).

Keywords: Sink · Mobile wireless sensor networks · Bio-inspired
BFOA · Throughput

1 Introduction

Nowadays, a number of sensor-based applications require the presence of sensors on mobile elements. The multiple advantages of using mobility in the Wireless Sensor Network (WSN) made the sensors powerful to be integrated in large networks. It can extend the coverage of a network or obtain more precise results.

© IFIP International Federation for Information Processing 2018
Published by Springer International Publishing AG 2018. All Rights Reserved
A. Amine et al. (Eds.): CIIA 2018, IFIP AICT 522, pp. 352–363, 2018.
https://doi.org/10.1007/978-3-319-89743-1_31

For example, in a surveillance application, sensors can be placed on animals that will evolve freely in a park, where also fixed nodes will be deployed. Moreover, it is very likely that, like IP network equipments, mobile nodes will no longer be confined within a single network but will evolve within multiple networks. This mobility is one of the key elements of the Internet of Things (IoT). Unfortunately, given to their small size, sensors devices are extremely constrained. Despite the easy and interesting solutions they offer and their various applications, sensors suffer from constraints such as: rapid depletion of the energy and routing information. Moreover, the mobility of the sensor brings new challenges such as network coverage, packets destined for a mobile node would not be able to be delivered to it when it moves out of its network, etc. Therefore, one of the proposed solutions for Mobile WSN, the Clustering Duty Cycle Mobility aware Protocol (CDCMP) [1]. It is a hierarchical routing protocol, combining three very important aspects in the WSN: a duty cycle mechanism with sink mobility in the context of nodes clustering organization. Although, the algorithm uses a predefined mobility. The Sink moves in a fixed path and its coordinates are random. After a while, the Sink may be away from some CHs that want to send data and privileges some CHs over others. Authors of [2–4] review data collection method with mobile sinks in WSN. Article [5] is also a survey of routing protocols of WSN with mobile sinks.

A new promising bio-inspired protocols, that can lead to quick convergence to quality solutions [6], and bridge the gap between microbiology and engineering, have emerged. Bacterial activities are complex and organized systems that give a new approach to solve complex optimization problems. Bacterial Foraging Optimization Algorithm (BFOA) was introduced by Passino in 2002 [7]. Indeed, the BFOA design is inspired by the social and cooperative intelligent behaviors of Escherichia coli bacteria, which are the most common kind of bacteria living in a human gut. According to paper [7], BFOA is better than PSO in terms of convergence, robustness and precision. In the area of WSNs research, Kulkarni et al. [6] in 2009 was the first authors that successfully used the BFOA for node localization. They compared the performances of PSO and BFOA in terms of the number of nodes localized, localization accuracy and computation time. They conclude that PSO determines the node coordinates more quickly and BFOA does it more accurately. Authors in [8] proposed a new routing algorithm to achieve low mean delay and energy efficiency, respecting the shortest path algorithm and based on the intelligent behavior of searching method of E.Coli. This approach could scale since no flooding of messages is required. However, the network is randomly deployed in a two dimension circle with the sink at the center. This could be applicable only in some application fields where the data collector is in the center of the sensing area. In the paper [9], BFOA is used for cluster head (CH) selection to provide improved energy efficiency in routing. Several simulations were carried out and results showed that BFOA had better performance compared to other clustering protocols. Rajagopal et al. [10] also proposed an improved CH selection for efficient data aggregation in sensor networks based on BFOA. This approach improved the average End-to-End delay (sec), the average packet drop ratio and lifetime computation. Another mechanism is proposed by Lalwani and Das [11]. Authors used BFOA method to propose a new manner to select CHs and to route sensed data. Homogeneous sensors are deployed randomly. Then the algorithm is applied centrally, by the base station (BS), to form several clusters. After that, the routing path is computed from each CH to the BS. Kavitha and Wahidabanu [12,13] combined both BFOA and GSA (Gravitational Search Algorithm) and incorporated the hybrid algorithm in LEACH to optimize its CH selection method. In [14], a new low energy intelligent clustering protocol (LEICP) is proposed for balancing the energy consumption in every cluster and prolonging the network lifetime. According to the residual energy and positions of nodes, in order to balance to balance the energy consumption in every cluster, a fitness function is defined. Authors in [15] wanted to deploy the nodes only in the terrains of interest identified by segmentation of the images captured by a camera on board the unmanned aerial vehicle (UAV). They used PSO and BFOA for image segmentation for autonomous deployment of WSN nodes and for localization of the deployed nodes in a distributed and iterative fashion. In [16], BFOA based Traveling Salesman Problem (TSP) is used for data collection in mobile sinks. Recently, Ari et al. [17] used the BFOA to introduce a mobile sensing scheme in WSNs. This approach guarantees a better network coverage and a high quality of service, especially in term of throughput.

This paper presents a new pattern of Sink mobility in WSN, to deal with the problems encountered by the CDCMP protocol [1], with better throughput and high availability. The proposed protocol takes into account new characteristics inspired by Escherichia coli bacteria. This solution introduces an autonomous and intelligent mobility. The entire network operational time is divided into some rounds. At the first, the clusters are constructed, and in the remainder of the round, the data are gathered, aggregated and transmitted to the sink. During transmission phase, Cluster Heads (CHs) can be awake only during the closest members transmissions. Clusters-members can switch to the sleep state except in their time slots.

This paper is organized as follows: An overview of Bacterial Foraging Optimization Algorithm is given in Sect. 2. Sink Mobility based on Bacterial Foraging Optimization Algorithm is discussed in Sect. 3. The experimental results are presented in Sect. 4 followed by the conclusion.

2   Bacterial Foraging Optimization Algorithm

This section presents the social foraging behavior of the Escherichia coli bacteria. To model the behavior of this bacterium, Passino [7] proposed four main operations in the bacterial feeding system: "Chemotaxis", "Swarming", "Reproduction" and "Elimination and Dispersal". In our proposed mobility design for WSNs, we are interest by only the two first operations.

1- Chemotaxis: This process aims to simulate the basic operations that allow the Escherichia bacteria to move via the flagella [18]. The chemotactic step corresponds to a swim, movement without changing the direction [6], followed by swim or tumble, random modification of movement direction [6].

Let j, k and l be the indexes for the chemotaxis step, the reproduction and the event of the elimination-dispersion respectively, such as: $\theta^i(j, k, l) \in \mathbb{R}, i = 1, 2, ..., S$ defines the position of each member in the population of S bacteria in the jth chemotactic step, k^{th} stage of reproduction and at the lth event the elimination-dispersion. Let $C(i) > 0$ be the size of the step taken in the random direction specified by the tumble. To represent a tumble, a random unit length of the direction, say $\phi(i)$, is generated, see Eq. 1.

$$\phi(i) = (\Delta^T(i)\Delta(i))^{-1/2} * \Delta(i) \tag{1}$$

Where $\Delta(i) \in \mathbb{R}^p$ is a randomly generated vector, where each element of $\Delta(i)$, $i = 1, 2, .., p$ is a number taken in $[-1, 1]$. Liu and Passino [19] have modeled the chemotaxis, i.e. the movement of the bacterium, by Eq. 2.

$$\theta^i(j + 1, k, l) = \theta^i(j, k, l) + C(i) * \phi(i) \tag{2}$$

2- swarming: This process consists of the formation of a group of bacteria in a self-organized manner. In Eq. 3, the function $J_{cc}(\theta, \theta^i(j, k, l))$, which models the cell-to-cell signaling via an attractant and a repellent, is given.

$$J_{cc}(\theta, \theta^i(j, k, l)) = - d_{attractant} * \exp(-w_{attractant} * \textstyle\sum_{m=1}^{p} (\theta_m - \theta^i{}_m)^2)$$
$$+ h_{repellant} * \exp(-w_{repellant} * \textstyle\sum_{m=1}^{p} (\theta_m - \theta^i{}_m)^2) \tag{3}$$

Where $d_{attractant}$ is the depth of the attractor released by the cell, $w_{attractant}$ is the attractive signal width, $h_{repellant} = d_{attractant}$ is the height of the repellent effect (magnitude), and $w_{repellant}$ is the width of the repellent.

The objective function representing the cell-to-cell signaling in the Escherichia coli swarm is defined by Eq. 4.

$$J_{cc}(\theta, P(j, k, l)) = \sum_{i=1}^{S} J_{cc}(\theta, \theta^i(j, k, l)) \tag{4}$$

Where $P(j, k, l) = \theta^i(j, k, l)$ with $i = 1, 2, ..., S$ is the position of each member in the population of S bacteria in the jth chemotactic step, k^{th} reproduction step and the lth elimination-dispersion event.

Equation 5 represents the cost, i.e., the *fitness* function, at the location $\theta^i(j, k, l)$ of the ith bacterium.

$$J(i, j, k, l) = J(i, j, k, l) + J_{cc}(\theta^i(j, k, l), P(j, k, l)) \tag{5}$$

3   SMBFOA Protocol

A flowchart of our proposal is represented in Fig. 1. It takes place in "rounds" which have approximately the same predetermined time interval. Each round

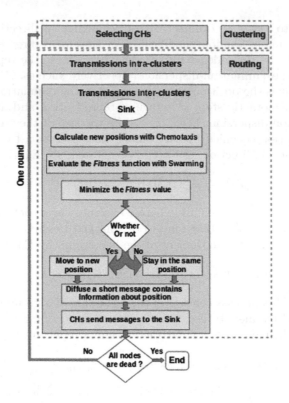

Fig. 1. Flow chart of the proposed SMBFOA protocol.

begins with a clustering phase, where the optimum location of CHs is found. Followed by a routing phase, where the intra-clusters sensed data are transferred to CHs as in CDCMP protocol [1]. After that, the Sink moves intelligently inspired by the BFOA to find the better position to receive more data. Like that, the inter-clusters transmissions are done.

In the proposed SMBFOA protocol (Fig. 1), the following assumptions about network model are fixed:

- The Sink is a resource-rich device and is not limited in terms of memory and computing power.
- The Sink is located inside the WSN.
- All sensor nodes are stationary after deployment.
- All sensors have GPS or other location determination devices.
- The WSN includes homogeneous sensor nodes.
- Initially, all nodes sensors have the same amount of energy.

3.1   Clustering Phase

As mentioned in [1], this phase consists of dividing the network into four regions, each containing a CH, in order to ensure an equitable distribution of CHs in the network. The selection of CHs is done according to the nodes topological density (D_t) during the round 0 and as a function of the residual energy of the sensors during the other rounds. The density of each node S is computed with:

$$D_t(S) = (D + L)/D$$

Or D is the degree of the node S and L is the neighbouring links of the node S. Each CH acts as a local coordinator to coordinate the data transmission within its group. It divides the cluster into two sub-levels (Level1 and Level2) to create a TDMA Schedule starting with the most remote nodes, and at the same time assigns each member node a time slot during which it can transmit its data. CH also realizes the notion of the Duty-Cycle to synchronize the wake-up and sleep time in the network.

3.2   Routing Phase

The data collection in our proposition is done on two steps. Firstly, the sensed data are transferred to CHs (intra-clusters transmissions). Then, each CH aggregates data received from its members to send them to the Sink (inter-clusters transmissions). Meantime, the Sink choose its best position according to BFOA.

3.2.1   Transmissions Intra-clusters

In [1], authors decomposed this stage into several time frames. Each frame contains a number of slots equal to the number of cluster-members set. A slot is dedicated to a single member node of a cluster, during which time it transmits its data to the Level concerned. These operations are repeated at each frame until the end of the round.

3.2.2   Transmissions Inter-clusters

At the end of each frame, the CH gathers the data of its members and sends them to the Sink.

a - The Sink moves according to the "Chemotaxis": The Sink moves like a bacterial, from the position $\theta(t)$ to $\theta(t+1)$ as follows:

$$\theta(t+1) = \theta(t) + C * \phi$$
$$\phi = (\Delta^T * \Delta)^{-1/2} * \Delta * Degre$$

ϕ: is a unit length random direction to represent a tumble.
C: let $C > 0$ be the size of the step taken in the random direction specified by the tumble.
Δ: is a vector of two elements Δ_1 and Δ_2, which are chosen randomly in the interval $[-1, 1]$ as mentioned in [7].
$Degre$: is the largest number of members of our clusters. The Sink moves in order to collect as much data as possible. It moves to the densest cluster.

$$\Delta * \Delta^T = \sum_{i=1}^{n} \Delta_i^2$$

$\Delta * \Delta^T$: is the product of a vector by its transpose. It is a mathematical law called the Scalar Square.

b - The Sink evaluates its *Fitness* according to "Swarming": The Sink evaluates the value of the *Fitness* function as follows:

$$J(\theta^k(t+1)) = -\alpha_1 * g(\beta_1) + \alpha_2 * g(\beta_2)$$

Where α_1, α_2, β_1 and β_2 are coefficients that should be chosen properly. Let $g(x)$ be the swarming function that governs the movement of a mobile sink.

$$g(x) = \exp\left(-x * \Gamma(\theta^k(t+1))\right)$$
$$\Gamma(\theta^k(t+1)) = (\theta_1(t+1) - \theta_1(t))^2 + (\theta_2(t+1) - \theta_2(t))^2$$

To minimize the *Fitness* function, we must check the following equation:

$$J(\theta^K(t)) > J(\theta^K(t) + \Delta\theta^k)$$

3.3 Illustrative Case

We suppose that the coordinates of our Sink are $(50; 50)$. The Sink will move to new position according to BFOA.

a- The Sink moves according to the "Chemotaxis": We take two random numbers between $[-1; 1]$. For example: $(-1; 1), (0.5; 0.5), (0.9; -0.5), (0.75; -0.75)$.

Firstly, we calculate the Scalar Square:

$$\Delta * \Delta^T = [-1, 1] * [-1, 1]^T = 2$$

Then, we compute:

$$\phi = (\Delta * \Delta^T)^{-1/2} * \Delta * Degre = 2^{-1/2} * [-1, 1] * 8 = [-5.6568, 5.6568]$$

Then, the coordinates of the new position (X, Y) are calculated with $C = 0.1$:

$$\theta_x(t+1) = 50 + (-5.6568) * 0.1 = 49.4343 \text{ and}$$
$$\theta_y(t+1) = 50 + (5.6568) * 0.1 = 50.5656$$

For each new position, the *Fitness* function must be evaluated.

b- The Sink evaluates its fitness according to the "Swarming": We must carefully choose the value of the following: α_1, α_2, β_1 and β_2.

$$J(\theta^k(t+1)) = -0.5 * g(0.1) + 1.5 * g(0.1)$$

If $J(\theta^K(t+1)) < J(\theta^K(t))$, then the Sink will move to the new position because it will minimize the *Fitness* function. So, there will be less moves. The Sink will ensure that its movement will collect as much data as possible without unnecessary move.

Figure 2 shows the movement of the Sink using BFOA.

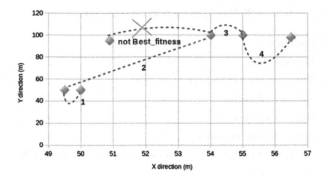

Fig. 2. Sink positions with BFOA mobility.

4   Simulation Results and Analysis

In this section, we perform simulations using NS2 to analyse and evaluate the performance of the proposed protocol. The number of nodes in the simulated network is 100, scattered randomly within a $100 \times 100\,\mathrm{m}^2$ sensor field as shown in Fig. 3. The simulation parameters are presented in Table 1.

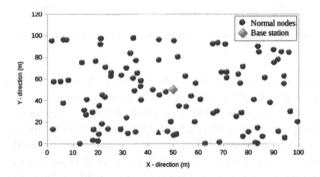

Fig. 3. Nodes deployment map.

In Fig. 4 the performance of the CDCMP and SMBFOA protocol was evaluated in terms of the number of data packets received by the Sink and the energy consumed by nodes. In our simulation, the number of nodes is 100. As it is evident from Fig. 4, inter-cluster communication minimizes the energy consumption in the data transmission to reach the Sink. Simulation results revealed that in a network which uses CDCMP protocol, it delivers 70475 data packets using 160 J at 460 s. By contrast, a network employing the SMBFOA protocol, Sink receives 99141 data packets using only 155 J. In other words, when 3/4 of the energy of a network is consumed, CDCMP transmits only 2/3 data packets comparing to SMBFOA protocol. We note an energy saving in the case where the intelligent

Table 1. Simulation parameters

Parameters	Value
Network size	$100\,m \times 100\,m$
The location of the Sink	(50, 50)
Number of nodes	100
Number of clusters	4
Initial energy of nodes	2 J
Position of nodes	Between (0, 0) and (100, 100)
Round time	20 s
Packet size	2000 bits
Simulation time	3600 s

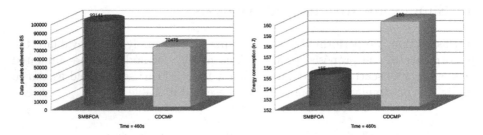

Fig. 4. Packets number received by the Sink and energy consumed by nodes.

mobility is integrated in the SMBFOA. If the Sink stay immobile for a long period of time, CHs standing far from it, will consume more energy and their lifetimes decrease rapidly, as in the case of the CDCMP protocol.

In Fig. 5 the performance of the CDCMP and SMBFOA protocols was evaluated in terms of the number of data packets received by the Sink. The amount of data packets received within a certain time period is an important index for measuring the quality of network service. As it is evident from Fig. 5, the simulation results clearly show that the SMBFOA protocol is better than CDCMP protocol in terms of the number of data packets received by the Sink. We see that practically the number of data packets received by the Sink using the SMB-FOA protocol is doubled ($16 * 10^4$ data packets received by the Sink) compared to the CDCMP protocol ($10 * 10^4$ data packets received by the Sink). Following the ideal use of time synchronization and *Fitness* functions, we have got a significant improvement in the amount of data transmitted, virtually no loss. The reason is simple, an intelligent mobile Sink moves closer to the CHs. Thus, the distance will be minimised. CHs consume less energy and transmit more data.

In Fig. 6 the performance of the CDCMP and SMBFOA protocols was evaluated in terms of the number of dead nodes. From this Fig. 6, it is noticed that the first node died after 470 s and all nodes died after 680 s in CDCMP protocol. However in SMBFOA protocol, the first node died also after 470 s and all nodes

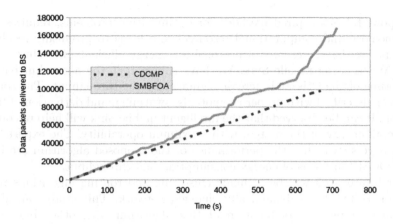

Fig. 5. Number of data packets received by the Sink.

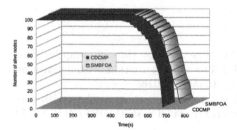

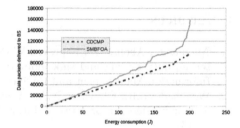

Fig. 6. Alive nodes number function of round time = 20 s.

Fig. 7. Energy consumed by nodes.

died after 710 s. It is clearly that the proposed protocol do not differ too much from CDCMP protocol. This is related to the fact that the SMBFOA protocol do not change the clustering and intra-cluster routing method from CDCMP protocol. Moreover, the energy consumption is distributed between the CHs due to the use of an intelligent mobile Sink. As a result, no node will be exploited more than its capability which in turn leads to optimization of the energy consumption in the whole network.

In Fig. 7, the performance of CDCMP and SMBFOA protocols was evaluated by the ratio between energy consumption and throughput. It can be seen that after 150 J consumed, CDCMP protocol transmits only 65000 data packets and SMBFOA protocol transmits 90000 data packets.

5   Conclusion

Researchers have used bio-inspired stochastic optimization methods instead of analytical methods due to their efficiency in computations. The latter methods require enormous computational efforts to solve optimization problems. Computations grow exponentially as the problem size increases. BFOA is one of these powerful bio-inspired methods for optimization problems, requires moderate memory and computational resources and leads to promising results. In this paper we have presented a protocol for energy efficient inter-cluster routing in WSNs using mobile Sink. We have shown that the algorithm balances the lifetime of the CHs and increase significantly the throughput. The solution have been described with proper chemotaxis, swarming, and derivation of fitness function. It can be observed that the intelligent mobile Sink will help to conserve the overall energy of the system with maximum operability. The experimental results have shown that the performance of the proposed algorithm are better than CDCMP in terms of energy consumption and throughput.

Our future work will be to made clustering and routing algorithms totally based on BFOA for a dynamically changed network. This study can also be supplemented by the combination of this algorithm with other bio-inspired approaches, which will yield an hybrid algorithm.

References

1. Aliouat, L., Aliouat, Z.: Improved WSN life time duration through adaptive clustering, duty cycling and sink mobility. In: The 8th International Conference on Information Management and Engineering (ICIME) (2016)
2. Wankhade, S.R., Chavhan, N.A.: A review on data collection method with sink node in wireless sensor network. Int. J. Distrib. Parallel Syst. (IJDPS) **4**(1), 67–74 (2013)
3. Anisi, M.H., Khan, A.W., Abdullah, A.H., Bangash, J.I.: A comprehensive study of data collection schemes using mobile sinks in wireless sensor networks. Sensors J. **14**(2), 2510–2548 (2014)
4. Nair, N.M., Arokya Jose, A.F.: Survey on data collection methods in wireless sensor networks. Int. J. Eng. Res. Technol. (IJERT) **12**(2), 2220–2225 (2013)
5. Keerthika, A., Berlin Hency, V.: A survey of routing protocols of wireless sensor network with mobile sinks. ARPN J. Eng. Appl. Sci. **11**(11), 6951–6963 (2016)
6. Kulkarni, R.V., et al.: Bio-inspired node localization in wireless sensor networks. In: IEEE International Conference on Systems, Man, and Cybernetics (SMC 2009) (2009)
7. Passino, K.M.: Biomimicry of bacterial foraging for distributed optimization and control. IEEE Control Syst. **22**(3), 52–67 (2002)
8. Hoa, T.D., Kim, D.S.: Bio-inspired and biased random walk routing in dense and lossy wireless sensor networks. In: The 2012 International Conference on Advanced Technologies for Communications (ATC 2012) (2013)
9. Sharma, N., Behal, S.: A systematic way of soft-computing implementation for wireless sensor network optimization using bacteria foraging optimization algorithm: a review. Int. J. Appl. Innov. Eng. Manag. (IJAIEM) **2**(2), 150–154 (2013)
10. Rajagopal, A., et al.: Soft computing based cluster head selection in wireless sensor network using bacterial foraging optimization algorithm. Int. J. Electron. Commun. Eng. **9**(3), 379–384 (2015)
11. Lalwani, P., Das, S.: Bacterial foraging optimization algorithm for CH selection and routing in wireless sensor networks. In: The 3rd International Conference on Recent Advances in Information Technology (RAIT) (2016)

12. Kavitha, G., Wahidabanu, R.S.D.: Improved cluster head selection for efficient data aggregation in sensor networks. Res. J. Appl. Sci. Eng. Technol. **7**(24), 5135–5142 (2014)
13. Kavitha, G., Wahidabanu, R.S.D.: Foraging optimization for cluster head selection. J. Theor. Appl. Inf. Technol. **61**(3), 571–579 (2014)
14. Li, Q., et al.: A low energy intelligent clustering protocol for wireless sensor network. In: IEEE International Conference on Industrial Technology (ICIT) (2010)
15. Kulkarni, R.V., Venayagamoorthy, G.K.: Bio-inspired algorithms for autonomous deployment and localization of sensor nodes. IEEE Trans. Syst. Man Cybern. **40**(6), 663–675 (2010)
16. Kavitha, G.: Bacterial Foraging Optimization (BFO) based Traveling Salesman Problem (TSP) for data collection in mobile sinks. Int. J. Sci. Comput. Intell. **2**(1), 76–94 (2015)
17. Ari, A.A.A., et al.: Bacterial foraging optimization scheme for mobile sensing in wireless sensor networks. Int. J. Wirel. Inf. Netw. **24**(3), 254–267 (2017)
18. Gu, H., et al.: How Escherichia coli lands and forms cell clusters on a surface: new role of surface topography. Technical report (2016)
19. Liu, Y., Passino, K.: Biomimicry of social foraging bacteria for distributed optimization: models, principles and emergent behaviors. J. Optim. Theory Appl. **115**(3), 603–628 (2002)

Social-Spider Optimization Neural Networks for Microwave Filters Modeling

Erredir Chahrazad$^{(\boxtimes)}$, Emir Bouarroudj, and Mohamed Lahdi Riabi

Laboratory of Electromagnetic and Telecommunication,
University Brothers Mentouri Constantine1, Constantine, Algeria
cerredir@yahoo.fr

Abstract. In this paper, Social-Spider optimization (SSO) algorithm is proposed for training artificial neural networks (ANN). Further, the trained networks are tested on two microwave filters modeling (Broad-band E-plane filters
with improved stop-band and H-plane waveguide filters considering rounded
corners). To validate the effectiveness of this proposed strategy, we compared
the results of convergence and modeling obtained with the results obtained by
NN used a population based algorithm namely Particle Swarm Optimization
(PSO-NN). The results prove that the proposed SSO-NN method has given
better results.

Keywords: Neural networks · Social-Spider optimization
Microwave structures · Modeling

1 Introduction

Microwave passive filters are essential components in the implementation of
telecommunications systems. Their main purpose is to pass selected signal and atten-
uate unwanted signals. Thus, it can clean the communication network by letting only
the system band signals to be transmitted or received.

The most popular method for modeling these kinds of filters is the mode matching
method (MMM). The mode matching method is often used to solve boundary-value
problems involving waveguides. The method consists of decomposing a complex
geometrical structure into many regions of simple geometrical form. Hence, in each
region, we can find a set of modal functions (or, modes) that satisfy the Maxwell's
equations except at the junctions. The problem is that the expanding all the modes in
each region with unknown modal coefficients and solving for these coefficients by
applying the boundary conditions at the junction of each region, thus making complex
mode matching method.

In the past several years, Artificial Neural Network (ANN) provides fast and
accurate models for the modeling, simulation, and optimization of microwave component [1–3]. In this paper, we propose a waveguide Filters (Broad-band E-plane filters
with improved stop-band and H-plane waveguide filters considering rounded corners)
modeling using a multilayer perceptron neural network (MLP-NN) to three layers
trained by a recently proposed algorithm called Social-Spiders Optimization (SSO) [4].
Each sub-net in the NN architecture shown in Fig. 1, possesses Ne inputs neurons in

© IFIP International Federation for Information Processing 2018
Published by Springer International Publishing AG 2018. All Rights Reserved
A. Amine et al. (Eds.): CIIA 2018, IFIP AICT 522, pp. 364–372, 2018.
https://doi.org/10.1007/978-3-319-89743-1_32

accordance with the number of the system geometry parameter of structure, Nc neurons in the hidden layer and one output associated with the value of reflection coefficient $S_{11}(f_k)$. The entire network consists of k distinct NNs corresponding to a particular frequency with k determined by the number of points in the frequency interval. Frequency responses of S_{11} parameters obtained in simulations compose the network database. The connection weight from the neurons of input layer to the neurons of hidden layer is *WE* and the connection weight from the neurons of hidden layer to the neurons of output layer is *WS*.

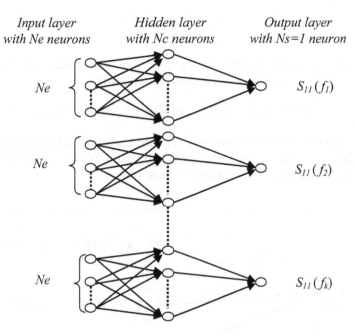

Input layer
with Ne neurons

Hidden layer
with Nc neurons

Output layer
with Ns=1 neuron

$S_{11}(f_1)$

$S_{11}(f_2)$

$S_{11}(f_k)$

Fig. 1. Neural networks architecture.

The paper is organized as follows. Section 2 describes the social-spiders optimization. Training of neural networks using social-spiders optimization is developed in Sect. 3. Application examples and results are presented in Sect. 4 and finally, Sect. 5 makes conclusions.

2   Social Spiders Optimization

Erik Cuevas et al. [4] proposed a new swarm optimization algorithm called Social-Spiders Optimization (SSO). The (SSO) algorithm is based on the simulation of the cooperative behavior of social-spiders. A (SSO) process consists of two parts: members and communal web. The communal web is used as a mechanism to transmit information among the colony members. This information is encoded as small vibrations depend on the weight and distance of the spider which has generated them, it is calculated according to:

$$Vib_{ij} = W_j * exp\left(-\left(d_{ij}\right)^2\right)$$ (1)

Where d_{ij} is the Euclidean distance between i^{th} and j^{th} spiders and W_j indicates the weight of the jth spider dependent on the fitness value; is calculated by the following equation

$$W_i = \frac{Worst - f(x_i)}{Worst - Best}$$ (2)

Where $f(x_i)$ is the fitness value of the spider (x_i). The values *Best* and *Worst* are defined as follows (considering a minimization problem):

$$Best = \min_{i=1...N} f(x_i) \text{ and } Worst = \max_{i=1...N} f(x_i)$$ (3)

Every spider (x_i) is able to consider three vibrations from other spiders as follows Fig. 2, Where Vib_{ci} the nearest spider subject to having higher fitness, Vib_{bi} the best spider in the swarm and Vib_{fi} the nearest female spider.

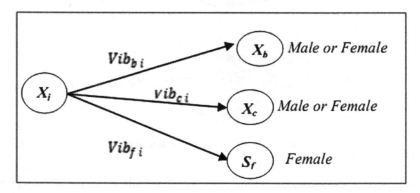

Fig. 2. Configuration of each special relation

The spiders members are divided by gender into two types: female spiders represent 65–90% of the entire population and the rest is the male spiders. They are calculated as follows:

$$Nf = floor[(0.9 - rand * 0.25) * N]$$ (4)

$$Nm = N - Nf$$ (5)

With N, N_f, Nm are the population, the number of females and the number of male respectively. (floor) rounds each element to the nearest integer, and *rand* is a random number in the unitary range [0,1]. A female spiders are updates its position as follows:

$$X_i = X_i + \alpha\, Vib_{ci}(X_c - X_i) + \beta\, Vib_{bi}(X_b - X_i) + \delta(rand - 0.5)\, \text{with probability}\, PF \quad (6)$$

$$X_i = X_i - \alpha\, Vib_{ci}(X_c - X_i) - \beta\, Vib_{bi}(X_b - X_i) \\ + \delta(rand - 0.5) \text{ with probability } 1 - PF \quad (7)$$

Where α, β, δ and *rand* are random numbers between [0, 1] whereas PF represents the threshold parameter. The individual X_c and X_b represent the nearest member to X_i that holds a higher weight and the best individual of the entire population, respectively.

The male population divided into two groups: Dominant members D with a weight value above the median value and non-dominant members ND with a weight value under the median value. The change of positions for the male spider can be modeled as follows:

$$X_i = X_i + \alpha\left(\frac{\sum_{h=1}^{N_m} X_h W_{N_f+h}}{\sum_{h=1}^{N_m} W_{N_f+h}} - X_i\right),\, \text{Male D} \quad (8)$$

$$Xi = Xi + \alpha Vibfi(Xf - Xi) + \delta.(rand - 0.5),\quad \text{Male ND} \quad (9)$$

Where the individual (X_f) represent the nearest female individual to the male member.

After all males and females spiders are update, the last operator is representing the mating behavior where only dominant males will participate with females who are within a certain radius called mating radius given by

$$R = \frac{\sum_{d=1...n}(X_d^h - X_d^l)}{2D} \quad (10)$$

Where X^h and X^l are respectively the upper and lower bound for a given dimension and n is the problem dimension.

Males and females which are under the mating radius generate new candidate spiders according to the roulette method. Each candidate spider is evaluated in the objective function and the result is tested against all the actual population members. If any member is worse than a new candidate, the new candidate will take the actual individual position assuming actual individual's gender.

3   Training Neural Networks Using Social Spiders Optimization

Training of neural networks is a complex task of great importance in the learning, where it is depends on adaptation of free network parameters, that is, on the proper selection of the neural weight values. The training of neural networks is to find an algorithm for optimized weights of networks to minimize the mean square error (MSE) described as follows

$$MSE = \frac{1}{PT} * \sum_{PT}\sum_{Ns}(Ys - Y)^2 \tag{11}$$

Where PT the total number of training samples, Ys is the output of the network and Y is the desired output.

$$Ys = f_2\left(\sum_{Nc} Ws * f_1\left(\sum_{Ne} WE * X\right)\right) \tag{12}$$

With f_2 and f_1 are the activation functions (typically: sigmoid, tanh …), X is the input vector of NN.

Specialized learning algorithms are used for adaptation of these weight values. Among those algorithms, the most popular algorithm is a back-propagation method (BP) [5] based on a gradient descending. Lately, many populations based algorithms were proposed for training a neural network such as Particle Swarm Optimization (PSO) [6], Genetic Algorithms [7].

In this paper, the neural networks are trained by a SSO algorithm. The step-wise procedure for the implementation of SSO-NN can be summarized as follows:

Step 1. Define the neural network architecture (number of layers, number of neurons in each layer and the activation functions) and define the parameters of the SSO algorithm (number of population N, the number of female N_f, the number of male N_m, maximum iteration and the threshold PF)

Step 2. Define the optimization parameters and the optimization problem:

– Design variables of the optimization problem: WE and WS the matrixes of input connection weights and output connection weights respectively. WE matrix of Nc rows and Ne columns and WS matrix of Ns rows and Nc columns.

– Define the optimization problem (fitness function): find the optimal WE and WS which minimizes the mean square error (MSE) Eq. (11).

Step 3. Initialize the population and evaluate the corresponding objective function value. For simplification, the population is decomposed into two groups, one represents the inputs weights population WEp and the second one represents the output weights population WSp. Where, each group generates randomly, according to the population female, the population male and the number of neurons as follows:

$$WEp = rand(N * Nc, Ne), \quad WSp = rand(Ns, N * Nc) \tag{13}$$

Step 4. Locate the best and the worst fitness, then calculate the weight of every spider in terms of its fitness Eq. (2)

Step 5. Move female spiders according to the female cooperative operator Eqs. (6) and (7)

Step 6. Move male spiders according to the male cooperative operator Eqs. (8) and (9)

Step 7. Perform the mating operation.

Step 8. If the stopping criteria are reached, the process is finished; otherwise, go back to Step 4.

4   Application Examples and Application

In this section, the ability of neural networks trained with SSO algorithm is assessed by applying it for the modeling of two microwave filters modeling test (Broad-band E-plane filters with improved stop-band [8] and H-plane waveguide filters considering rounded corners [9]) Fig. 3. The dimensions of the first filter and second filters are listed in Table 1.

The common parameters of each algorithms are chosen identical (population size = 30 and number of iterations equal 10000), neural network parameters (Number of hidden neurons, input parameter and their limit, and frequency interval) of each structure are presented in Table 2. The activation functions are hyperbolic tangent function (Tansig), and linear function (Purelin) respectively. The other specific parameters of algorithms are given below.

PSO Settings: c_1 and c_2 are constant coefficients $c_1 = c_2 = 2$, w is the inertia weight decreased linearly from 0.9 to 0.2.

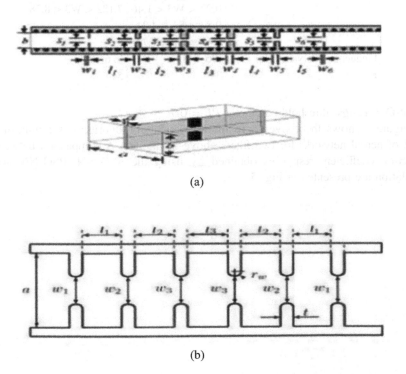

(a)

(b)

Fig. 3. (a) Broad-band e-plane filters with improved stop-band, (b) H-plane waveguide filters considering rounded corners

Table 1. The dimensions of the structures

Structure	Parameters and dimensions (mm)			
(a)	a	b	w1=w6	w2
	7.112	3.556	0.254	0.889
	w3=w4	s1=s6	s2=s5	s3=s4
	1.219	1.092	0.610	0.508
	l1=l5	l2	l3	d
	7.697	6.452	6.706	0.25
(b)	a	b	l1	l2
	34.85	15.79	22.88	25.57
	l3	w1	w2	w3
	25.57	15.10	9.03	7.98

Table 2. Neural network parameters

Structure	(a)	(b)
Nc	8	11
Input neurons & interval (mm)	0.20 < W1 < 0.30	13.95 < W1 < 16.61
	0.71 < W2 < 1.07	8.13 < W2 < 9.93
	0.97 < W3 < 1.46	7.182 < W3 < 8.78
	0.87 < S1 < 1.31	
	0.48 < S2 < 0.74	
Frequency interval (GHz)	[26–36]	[6.6–7.4]
K	34	54

SSO Settings: threshold parameter $PF = 0.7$

Figure 4 shows the convergence of SSO and PSO algorithms for minimize the MSE of neural networks for two filters above-mentioned. A comparison between the reflection coefficient responses obtained by using the SSO-NN, PSO-NN and the simulation are presented in Fig. 5.

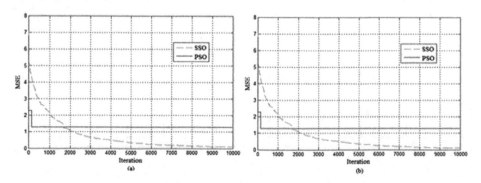

Fig. 4. Convergences of algorithms for minimize the MSE; (a) Broad-band e-plane filters with improved stop-band, (b) H-plane waveguide filters considering rounded corners.

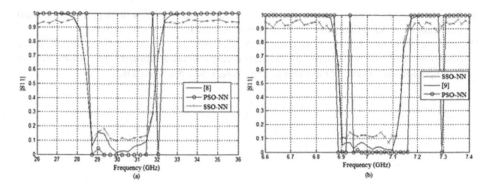

Fig. 5. The approximate reflection coefficient; (a) Broad-band e-plane filters with improved stop-band, (b) H-plane waveguide filters considering rounded corners.

It is observed from Figs. 4 and 5 that, the SSO algorithms perform better in terms of convergence and approximation than the PSO algorithm.

5 Conclusion

In the present work, we propose microwave filters (Broadband E-plane filters with improved stop-band and H-plane waveguide filters considering rounded corners) modeling using neural networks trained by a social spider optimization (SSO) algorithm. For validated this neural network (SSO-NN) we compared them with those obtained using neural networks trained by PSO (PSO-NN). Excellent results of convergence and modeling are obtained.

References

1. Wang, Y., Yu, M., Kabir, H., Zhang, Q.J.: Application of neural networks in space mapping optimization of microwave filters. Int. J. RF Microwave Comput. Aided Eng. **22**, 159–166 (2012)
2. Deshmukh, A.A., Kulkarni, S.D., Venkata, A.P.C., Phatak, N.V.: Artificial neural network model for suspended rectangular microstrip antennas. Procedia Comput. Sci. **49**, 332–339 (2015)
3. Sivia, J.S., Pharwaha, A.P.S., Kamal, T.S.: Analysis and design of circular fractal antenna using artificial neural networks. Prog. Electromagnet. Res. B **56**, 251–267 (2013)
4. Cuevas, E., Cienfuegos, M.: A new algorithm inspired in the behavior of the social-spider for constrained optimization. Expert Syst. Appl. **41**, 412–425 (2014)
5. Jwo, D.J., Chin, K.P.: Applying back-propagation neural networks to GDOP approximation. J. Navig. **55**, 97–108 (2002)
6. Gyanesh, D., Prasant, K.P., Sasmita, K.P.: Artificial neural network trained by particle swarm optimization for non-linear channel equalization. Expert Syst. Appl. **41**, 3491–3496 (2014)
7. Ding, S., Zhang, Y., Chen, J., Weikuan, J.: Research on using genetic algorithms to optimize elman neural networks. Neural Comput. Appl. **23**, 293–297 (2013)

8. Xu, Z., Guo, J., Qian, C., Dou, W.: Broad-band E-plane filters with improved stop-band performance. IEEE Microwave Wirel. Compon. Lett. **21**(7), 350–752 (2011)
9. Diaz Caballero, E., Morro, J., Belenguer, A., Esteban, H., Boria, V.: CAD technique for designing H-Plane waveguide filters considering rounded corners. In: IEEE MTT-S International Microwave Symposium Digest, WA, USA, pp. 1–3 (2013)

Planning and Scheduling

Scheduling in Real-Time Systems Using Hybrid Bees Strategy

Yahyaoui Khadidja$^{(\boxtimes)}$ and Bouri Abdenour

Department of Computer Science, Exact Sciences of Faculty,
University Mustapha Stambouli, Mascara, Algeria
yahyaouikhadidja@yahoo.fr, bouriabdenour17@yahoo.fr

Abstract. In the last decade, stochastic and meta-heuristic algorithms have been extensively used as intelligent strategies to resolve different combinatorial optimization problems. Honey Bee Mating Optimization is one of these most recent algorithms, which simulate the mating process of the queen of the hive. The scheduling algorithm is of paramount importance in a real-time system to ensure desired and predictable behavior of the system. Within computer science real-time systems are an important while often less known branch. Real-time systems are used in so many ways today that most of us use them more than PCs, yet we do not know or think about it when we use the devices in which they reside. Finding feasible schedules for tasks running in hard, real-time computing systems is generally NP-hard. In this work, we are interested in hybridizing this HBMO algorithm with other metaheuristics: Genetic Algorithms (GA), Greedy Random Adaptive Search Procedure (GRASP), Tabu Search (TS) and Simulated Annealing (SA) to resolve a real-time scheduling problem and obtain the optimal tasks schedule with respecting all temporal constraints. This is a complex problem which is currently the object of research and applications. In this scheduling problem, each task is characterized by temporal, preemptive and static periodicity constraints. The quality of the proposed procedure is tested on a set of instances and yields solutions which remain among the best.

Keywords: Real-time systems · Scheduling · HBMO · Optimization
Metaheuristics

1 Introduction

Finding feasible schedules for tasks running in hard, real-time computing systems is generally NP-hard [1]. A vast amount of work has been done in the area of real-time scheduling by both operations research and computer science communities.

The proposed approach is based on an optimization algorithm inspired from the Honey Bees Mating Optimization (HBMO). It combines an HBMO algorithm, Genetic Algorithms (GA), Simulated Annealing (SA), Tabu Search (TS) and the Greedy Randomized Adaptive Search Procedure (GRASP). The scheduling algorithms can be rated based on different parameters like makespan, flowtime, communication cost, reliability cost, and makespan [2]. The task duplication scheduling scheme has been utilized to minimize the processor idle time and lessen the mean task response time in [3]. In this

A. Amine et al. (Eds.): CIIA 2018, IFIP AICT 522, pp. 375–386, 2018.
https://doi.org/10.1007/978-3-319-89743-1_33

paper, we aim to obtain a schedule called feasible if both of the following conditions are met (1) No precedence constraints are violated and (2) all temporal constraints are satisfied. In order to realize this goal, we have used a new approach based on the hybridization of several algorithms. Our strategy uses the HBMO as a base algorithm which combines a number of different procedures. Each of them corresponds to a different phase of the mating process of the honey bees.

In the next section, we see also that the honey bee mating algorithm has never been applied to resolve real-time tasks scheduling problems. This prompted us to use this strategy and hybridizing it with other metaheuristics in order to see its performance to resolve this kind of problem. Our contribution consists in combining the global search represented by HBMO with some local search metaheuristics for the resolution of real-time task scheduling problems in order to intensify the search in promising zones detected by the HBMO exploration process. This hybridization is realized with methods which have been applied in isolation in the resolution of real-time task scheduling problems: GA, TS, SA and GRASP [4, 5] First, we have to select the population of honey bees that will make up the initial hive. In the proposed approach, this population is created by using GA alone, or GRASP alone, or random selection selected. Then a percentage is selected from these three populations using a hybridization of these processes. The best member of each initial population obtained is selected as the queen of the hive. All the other members of each population are the drones.

The two local search strategies SA and TS are used as workers for improving the broods.

The paper is organized in the following way. The related works are presented in the Sect 2. The presentation and formulation of real-time tasks scheduling problems is described in Sect. 3. Section 4 presents the proposed Scheduling Approach. The resolution process of the scheduling real-time tasks problem is detailed in Sect. 5. Section 6 presents the simulation and experimental results arrived at. The paper finishes with the conclusion and recommendations for future research.

2 Related Works

The task scheduling problem is defined as one of the many popular academic NP-hard problems. In [6], different aspects in scheduling and issues in various levels of real time systems are described. The use of metaheuristic methods shows their efficiency and effectiveness to solve these categories of complex problems [7]. In the literature, many metaheuristics have been proposed based on methods and approaches to task scheduling. In [8], Talbi et al. have hybridized Simulated Annealing (SA) with Genetic Algorithms (GA). The authors in [9] resolve the scheduling task problem through hybridization of Particle Swarm Optimization (PSO) with GA.

The comparison between various scheduling algorithms and simulations results is given in [10]. An scheduling algorithm has multiobjective to minimize the total tardiness and total number of processors is used in [11]. The authors combine Adaptive Weight Approach with genetic algorithm and simulated annealing method. In order to explore the computing power of quantum computation, an approach based on the hybrid Quantum genetic algorithm is used to resolve the real-time scheduling problem in multiprocessor environment in [12]. In [13], the authors applied Bee Colony Optimization (BCO) as a heuristic algorithm to solve the problem of static scheduling of independent tasks on identical machines. This strategy is based on the intelligent behavior of honey bees in the foraging process. Koudil et al. adapted the MBO algorithm to solve integrated partitioning/scheduling problems in co-design in [14]. This algorithm gives good results in terms of solution quality and execution time. The Fuzzy logic is also used to solve the real time tasks scheduling problem. In [15], the authors propose a model witch in the input stage consists of three linguistic variables i.e. CPU time, deadline and communication overhead.

3   Partitioning and Scheduling

A real-time application is normally composed of multiple tasks with different levels of criticality: Missing deadlines is not desirable. In this context, three categories of real-time applications are defined [2, 16]:

- Soft real-time applications: The system could still work correctly Although the tasks could miss some deadlines. However, missing some deadlines for soft real-time tasks will lead to paying penalties.
- Hard real-time applications: The tasks can not miss any deadline, otherwise, undesirable or fatal results will be produced in the system.
- Firm real-time applications: The tasks, which are such that the sooner they finish their computations before their deadlines, the more rewards they gain.

In this paper, a real-time task t_{ij} graph is defined by a set of temporal constraints such as $\left\{ st(t_{ij}^k), c(t_{ij}^k), ft1(t_{ij}^k), dl(t_{ij}^k) \right\}$ as well as preemption and static periodicity constraints. The Table 1 reports the temporal constraints or parameters description of real-time system to study.

Table 1. Description of the real-time tasks adopted

Temporal parameters	Temporal parameters description	Temporal parameters definition
$st(tij)$	Start time	Time at which a j^{ieme} task of i^{eme} graph starts its execution
$c(tij)$	Computing time	Time necessary for completion of a j^{eme} task of i^{eme} graph i instance
$ft(tij)$	Completion time	Time at which a j^{ieme} task of i^{eme} graph instance finishes its execution
$dl(tij)$	Deadline time	Time by which execution of the j^{ieme} task of i^{eme} graph instance should be completed, after the task is released
τij	The period of τ_{ij}	The periodic behavior of task, the study be restricted to the hyperperiod H which is the least common multiple of the periods. τ_i

The task model real-time system is usually described by a set of task graphs. The related concepts are defined as follows:

System: A real-time system is modeled as a set $G = \{Gi, i = 1, ..g\}$ of periodic task graph.

Task graph: A task graph Gi is a directed acyclic graph (DAG) of period τi and defined by a 2-tuple $Gi = \{Ti, Ei, \tau i\}$ where:

- $Ti = \{tij, j = 1, ..ni\}$ represents the set of tasks,
- $Ei = \{eijl, j = 1, ..ni\}$ represents the set of communications.

Task: A task *tij* is a node of task graph and defined by t_{ij}^k denote the processor implementation (instance of task).

The problem consists of three parts:

Part 1: Partitioning P, which determines the assignment of each task to p_u, $u \in \{1, ...p\}$, processors:

$$\forall t_{ij}, P(t_{ij}) = p_u, u \in \{1, ...p\} \tag{1}$$

Part 2: Scheduling on processors $SC1_u$, which associates each task instance assigned on processors $\{1, ...p\}$ a start time st.

$$\forall P(t_{ij}) = p_u, t_{ij} \rightarrow SCh_{1_u}\left(t_{ij}^k\right) = st\left(t_{ij}^k\right), \\ u \in \{1, ...p\} \tag{2}$$

Part 3: Scheduling on interconnexion system SCh_2_I, Which associates each inter-processors communication a start time st. The communication defined by e_{ijl} is a directed edge, which means task t_{ij} precedes t_{il}. Each communication is associated with a data transmission time on a specific communication channel. It is commonly assumed, in multiprocessor systems research, that communication between tasks assign to the same processor is effectively instantaneous, compared with inter processors communication.

$$e_{ijl}^k \rightarrow SCh\_2I\left(e_{ijl}^k\right) = st\left(e_{ijl}^k\right), \left\{e_{ijl}^k \in E_i^k \,/\, P(t_{ij}) \neq P(t_{il})\right\} \tag{3}$$

Scheduling in real-time systems requires that the order of execution be stipulated. A schedule is called feasible if both of the following conditions are met:

(1) *All* precedence constraints are respected and
(2) All timing constraints are satisfied such as the deadline defined by $dl(t_{ij}^k)$ which means the task must be finished before the deadline, otherwise the system will fail. In the multiprocessor system. The task scheduling is a mapping process of all tasks to p processors. The goal of the scheduler is to schedule the system's tasks on these p processors, so that every task is completed before the expiration of the deadline $dl(t_{ij}^k)$.

Here we use the similar cost function as [17]. The cost function formulated by Eq. (4) serves to calculate the tardiness scheduling of the real time system.

$$Tardiness = \sum_k \sum_i \sum_j fcost\left(t_{ij}^k\right)$$
$$= \begin{cases} \left| dl\left(t_{ij}^k\right) - ft\left(t_{ij}^k\right)\right| & if\ ft(t_{ij}^k) > dl(t_{ij}^k) \\ 0 & if\ ft(t_{ij}^k) < dl(t_{ij}^k) \end{cases} \tag{4}$$

With:

$$\begin{cases} ft(t_{ij}^k) = st(t_{ij}^k) + c(t_{ij}^k) \\ st(t_{ij}^{k-1}) \le st(t_{ij}^k) \le dl(t_{ij}^k) - c(t_{ij}^k) \end{cases} \tag{5}$$

The Real-time tasks complete their computing time before their deadlines as early as possible. Different comportments are showed in Fig. 1.

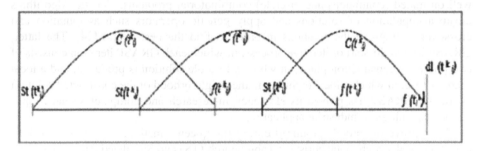

Fig. 1. Comportments of the real time instances of the task

4 Hybrid HBMO for Real Time Scheduling Resolution Problem

The HBMO algorithm was proposed by Abbass in [17]. Since then it has been used on a number of different applications [18–22]. HBMO was found to outperform some better known algorithms. However, it has not been applied to real-time scheduling problems. A honey-bee colony consists of the queen(s), drones, worker(s) and broods. The Honey-bee Mating Optimization algorithm mimics the natural mating behavior of the queen bee when she leaves the hive to mate with drones in the air. After each mating, the genetic pool of the queen is enhanced by adding sperm to her spermatheca. Before the mating flight begins, the queen is initialized with a certain amount of energy and only ends her mating flight when the energy level drops below a threshold level (which is close to zero) or when her spermatheca is full. The probability of a drone mating with a queen obeys the following annealing function:

$$Prob(D) = e^{[-\Delta(f\Delta)/speed(t)]} \qquad (6)$$

where $\Delta(f)$ is the absolute difference between the fitness of D and the fitness of the queen and $Speed(t)$ is the speed of the queen at a given time t. After each flight, the queen's speed and energy evolve according to the following equations:

$$Speed(t+1) = \propto \times Speed(t) \qquad (7)$$

$$Energy(t+1) = Energy(t) \qquad (8)$$

where factor $\alpha \in (0,1)$ is the amount of speed and energy reduction after each flight. The workers are presented as heuristics whose functionality is to improve the broods produced during the mating process.

To diversify the initial HBMO population and select the best solution as the queen, Genetic Algorithms (GA) and GRASP method are used. The former, GA, works very well on mixed (continuous and discrete) combinatorial problems. Genetic Algorithms create a population of solutions and apply genetic operators such as mutation and crossover to improve the solutions in order to find the best one(s) [24]. The latter, GRASP, is a multi-start or iterative process, in which each GRASP iteration consists of two phases, a construction phase, in which a feasible solution is produced, and a local search phase, in which a local optimum in the neighborhood of the constructed solution is sought. GRASP has proved its efficiency in research and computer science applications as well as in industrial applications.

To improve the broods produced during the queen's mating, we use two workers based on local search metaheuristics: Tabu Search (TS) and Simulated Annealing (SA).

5   Presentation of the Resolution Process

In this section, we present the approach adapted to resolve the issues addressed. To start with, a set of parameters must be defined:

- The bee population represents the set of scheduling plans of real-time tasks with temporal constraints,
- The queen represents the tasks best scheduling plan in the population generated by GA, by GRASP, randomly or hybridization of a percentage from each of the three populations obtained using the three aforementioned procedures and gradually improved by the implementation of the neighborhood generations iterative procedures (crossover in the HBMO) and two workers SA and TS.
- The drones make up the remaining task scheduling plans.

Algorithm 2 describes The HBMO approach for scheduling

Algorithm 2. Hybrid HBMO scheduling approach

---

1. Initialization
2. **Generate the initial population of the bees using separately GA, GRASP, Random or through Hybridization of percentages selected from these procedure;**
3. Evaluate the fitness of the plan;
4. Select the best plan of the population of plans-scheduling which represents the queen;
5. Selection of the best bee as the queen;
6. Sperm: Size of the spermatheca;
7. **E (t) et S (t)**: Energy and Speed between respectively [0.5, 1];
8. α : factor of reduction of energy and speed between [0,1];
9. **M**: maximum number of mating flights;
10. **For i=0 to M** (mating flights)
11. **do while** E(t) > 0 and Sperm is not full
12. Select a drone
13. **if** the drone passes the probabilistic condition $Prob(D) = e^{[-\Delta(f\Delta)/S(t)]}$
14. **Add** sperm (plan) of the drone in the spermatheca
15. **endif**
16. S(t+1) = α × S(t)
17. E(t+1) = α × E(t)
18. **enddo**
19. **do** j = 1 Size of Spermatheca
20. Select a sperm from the spermatheca;
21. Generate a brood by crossover the queen's genotype with the selected sperm;
22. **Improve the brood's fitness by applying the workers (Tabu Search /Simulated Annealing)**
23. **if** the brood's fitness is better than the queen's fitness
24. **Replace** the queen with the brood
25. **else Add** the brood to the population of drones
26. **Endif**
27. **Replace the drones by the broods**
28. **Enddo (for)**
29. **enddo**
30. **return** The Queen (Best Solution Found)

A summary of differences and similarities between the HBMO approach applied for the resolution of the real time scheduling problem and the original algorithm are showed in Table 2.

Table 2. Differences and similarities between our HBMO-real time scheduling and the original HBMO

Parameters	HBMO-basic algorithm [17]	HBMO-real time scheduling adopted
Number of queen	01	01
Initial population generation processs	Randomly	Randomly, Genetic algorithm, Grasp, Hybridization
Local search	Greedy SAT (GSAT)	Stochastic process of the drone selection
Brood generation	Crossover haploid & mutation	Genetic crossover operator
Improvement process (local search)	Mutation FLIP	Simulated annealing, Tabu search
Resultant broods	All broods are killed	All broods will be used in the next mating flight
Fitness function	Fitness function	Objective function $lateness(G)$ of the real time system

6   Simulation and Experimental Results

This part is devoted to the implementation and testing of the algorithmic methods developed above. In order to assess the effectiveness of the approaches under study, we constructed a testbench consisting three task systems, which are randomly generated. The systems were composed of 40, 60 and 80 task graphs. Each task graph had its own temporal characteristics. The parameters of the algorithm have been selected after thorough testing. A number of different values were tested and the ones selected are those that yielded the best results in both solution quality and computational time. The best values for these parameters appear in the Tables 3 and 4.

In the first time, we generate the initial population of the proposed algorithm randomly and with GA. After 10 runs, The Fig. 2 illustrate the graphic results obtained through the application of HBMO, GA and GRASP approaches. The best real-time task scheduling approach is to be selected taking into consideration the Cost function criteria. The graphics results show that the HBMO approach converge to the optimal results comparing with the GA and GRASP methods.

Table 3. Parameters and their values for the HBMO approach adopted

Parameters	Values
Size of population	50
Mating flight	100
Spermatheca	6
Speed	0.80
Energy	0.70
Alpha (α)	0.20

Table 4. Parameters and their values for the used approaches

Algorithms	Parameters	Values
GA	Size of population	50
	Probability of crossing	0.60
	Probability of mutation	0.20
	Size of population	70
GRASP	Size of the population	50
	Algorithm of amelioration	Local search
SA	Initial temperature	Fitness of brood
	Final temperature	0
	Number of iterations	Until the amelioration or temperature = 0
TS	Dimension of the list	5
	Number of iterations	5
	Type of neighborhood	With mutation and recoil

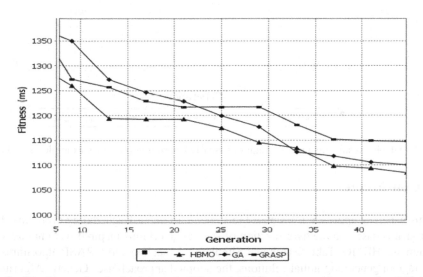

Fig. 2. Graphic results obtained through different hybridization processes example of 60 tasks

In order to ameliorate scheduling real-time problem results obtained through the application of basic HBMO (HBMO_Random), we have hybridized the HBMO algorithm with other metaheuristics: Genetic algorithms GA, GRASP and hybridization (three sub-populations made up of one percentage from random, one percentage from AG, one percentage from GRASP) in the stage of initial population generation. The Fig. 3 presents all graphic results obtained through different hybridization processes. The graphics results show that the results given by HBMO-hybridization approach were best than the results given by the HBMO with GA and HBMO with GRASP algorithms. We notice also that the results given by the HBMO-GRASP and HBMO-hybridization were nearly the same. (in the iterations number 34, 24, 16 and 9) the values obtained by HBMO_hybridization and the HBMO_GRASP are the same).

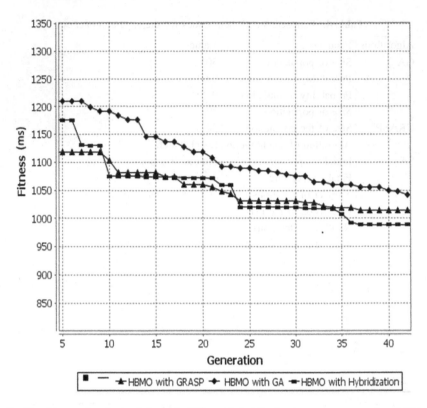

Fig. 3. Graphic results obtained through different hybridization processes example of 80 tasks.

7 Conclusion

In this paper, an approach is proposed to solve real-time task scheduling problems. This approach is based on the hybridization of several algorithms inspired from nature such as Genetic, HBMO, Tabu Search and Simulated Annealing and GRASP algorithms. In the stage of generating initial solutions, the adopted approach uses Genetic Algorithms alone, random selection alone, GRASP alone, and hybridization of the three processes selecting a percentage of the population from each. In the stage of solution improvement by the workers, HBMO uses Tabu Search and Simulated Annealing algorithms. The proposed approach was tested on three real-time task scheduling instances with 40 tasks, 60 tasks and 80 tasks where each task has its own temporal constraints. The gain coefficient from the optimal solution is used as the standard to evaluate the quality of the solutions obtained. Results confirm the positive impact of using a hybrid strategy with regard to function objective quality and computing time in comparison with the hybridization algorithms object of this paper.

The results arrived at show that for real-time scheduling problems (with a number of tasks 80), the best approach is HBMO-Hybridization. In the future works, we will focus our searches to resolve the problem of minimizing the energy consumption in the real-time embedded systems by using the proposed algorithms.

References

1. Azerou, M., Akl, S.G.: Scheduling algorithms for real-time systems. Technical report No. 2005-499, School of Computing, Queen's University Kingston, Ontario, Canada K7L 3N6, 15 July 2005
2. Rathna Devi, M., Anju, A.: Multiprocessor scheduling of dependent tasks to minimize makespan and reliability cost using NSGA-II. Int. J. Found. Comput. Sci. Technol. (IJFCST) **4**(2), 27–39 (2014)
3. Singh, R.: An optimized task duplication based scheduling in parallel system. Int. J. Intell. Syst. Appl. (IJISA) **8**(8), 26–37 (2016)
4. Fleischer, M.: Simulated annealing: past, present, and future. In: Proceedings of the Winter Simulation Conference, Department of Engineering Management, Old Dominion University, Norfolk, VA (1995)
5. Casey, S., Thompson, J.: GRASPing the examination scheduling problem. In: Burke, E., De Causmaeckerk, P. (eds.) Practice and Theory of Automated Timetabling IV. LNCS, vol. 2740, pp. 232–244. Springer, Heidelberg (2002)
6. George, D.I., Amalarethinam, A., Josphin, M.: Dynamic task scheduling methods in heterogeneous systems - a survey. Int. J. Comput. Appl. **110**(6), 12–18 (2015)
7. Talbi, E.G.: Metaheuristics: from Design to Implementation. Wiley, Hoboken (2009)
8. Davidovic, T., Selmic, M., Teodorovic, D.: Scheduling independent tasks: Bee colony optimization approach. In: 17th Mediterranean Conference on Control & Automation Makedonia Palace, Thessaloniki, Greece, June 2009
9. Pradhan, S.R., Sharma, S., Konar, D.: A comparative study on dynamic scheduling of real-time tasks in multiprocessor system using genetic algorithms. Int. J. Comput. Appl. **120** (20), 3 (2015)
10. Yoo, M., Yokoyama, T.: Multiobjective GA for real time task scheduling. In: Proceedings of the International MultiConference of Engineers and Computer Scientists (IMECS 2016), vol. 1, Hong Kong (2016)
11. Konar, D., Bhattacharyya, S., Sharma, K., Pradhan, S.R.: An improved hybrid quantum-inspired genetic algorithm (HQIGA) for scheduling of real-time task in multiprocessor system. Appl. Soft Comput. **53**, 296–307 (2016)
12. Davidovic, T., Ramljak, D., Selmic, M., Teodorovic, D.: Bee colony optimization for the p-center problem. Comput. Oper. Res. **38**, 1367–1376 (2011)
13. Koudil, M., Benatchba, K., Tarabet, A., Sahraoui, E.B.: Using artificial bees to solve partitioning and scheduling problems in codesign. Appl. Math. Comput. **186**(2), 1710–1722 (2007)
14. Nirmala, H., Girijamma, H.A.: Fuzzy scheduling algorithm for real – time multiprocessor system. Int. J. Sci. Eng. Res. (IJSER). **5**(7) (2014)
15. Monnier, Y., Beauvais, J.-P., Deplanche, A.-M.: A genetic algorithm for scheduling tasks in a real-time distributed system. In: Proceedings of 24th EUROMICRO Conference, pp. 708–714 (1998)
16. Greenwood, G.W., Lang Hurley, C.L.S.: Scheduling tasks in real-time systems using evolutionary strategies. In: Proceedings of the 3rd Workshop on Parallel and Distributed Real-Time Systems. IEEE (1995)
17. Abbass, H.A.: A single queen single worker honey–bees approach to 3-SAT. In: The Genetic and Evolutionary Computation Conference, GECCO, San Francisco, USA (2001)
18. Sahoo, R.R., Rakshit, P., Haidar, T.Md.: Navigational path planning of multi-robot using honey bee mating optimization algorithm (HBMO). Int. J. Comput. Appl. **27**(11), 0975–8887 (2011)

19. Marinakis, Y., Marinaki, M., Dounias, G.: Honey Bees Mating Optimization algorithm for large scale vehicle routing problems. Nat. Comput. **9**, 52–57 (2010)
20. Sabar, N., Ayob, M., Kendall, G.: Solving examination timetabling problems using Honey-Bee Mating Optimization (ETP-HBMO). In: Multidisciplinary International Conference on Scheduling: Theory and Applications (MISTA), Dublin, Ireland (2009)
21. Yousefi, N., Ebrahimian, H.: Optimal design of multi-machine power system stabilizers using interactive Honey Bee Mating Optimization. Trends Life Sci. (TLS) Dama Int. J. **4**(1) (2015)
22. Abedinia, O., Naderi, M., Ghasemi, A.: Robust LFC in deregulated environment: fuzzy PID using HBMO. In: 10th International Conference Environment and Electronical Engineering (EEEIC) (2011)
23. Holland, J.H.: Adaptation in Natural and Artificial Systems. University of Michigan Press, Ann Arbor (1975)

Solving an Integration Process Planning and Scheduling in a Flexible Job Shop Using a Hybrid Approach

Nassima Keddari[1]([⊠]) [iD], Nasser Mebarki[2] [iD], Atif Shahzad[3],
and Zaki Sari[1] [iD]

[1] Manufacturing Engineering Laboratory of Tlemcen MELT,
University of Tlemcen, Tlemcen, Algeria
keddarinassima@yahoo.fr, z_sari@mail.univ-tlemcen.dz
[2] LUNAM, Université de Nantes, LS2N, Laboratoire des Sciences
du Numérique de Nantes, UMR CNRS 6004, Nantes, France
nasser.mebarki@univ-nantes.fr
[3] Department of Industrial Engineering,
King Abdulaziz University, Jeddah, Saudi Arabia
atifshahzad@gmail.com

Abstract. Traditionally, process planning and scheduling functions are performed sequentially, where scheduling is implemented after process plans has been generated. Recent research works have shown that the integration of these two manufacturing system functions can significantly improve scheduling objectives. In this paper, we present a new hybrid method that integrates the two functions in order to minimize the makespan. This method is made up of a Shifting Bottleneck Heuristic as a starting solution, Tabu Search (TS) and the Kangaroo Algorithm metaheuristics as a global search. The performance of this newly hybrid method has been evaluated and compared with an integrated approach based on a Genetic Algorithm. Thereby, the characteristics and merits of the proposed method are highlighted.

Keywords: Integration · Process planning · Scheduling · Metaheuristics
Shifting bottleneck heuristic · Tabu search · Kangaroo algorithm

1 Introduction

The concept of Industry 4.0 promotes the integration of all aspects of production for greater efficiency. The factory of the future will be a smart manufacturing, holistic and flexible, where Internet of Things, augmented reality, automation and artificial intelligence will enable to adapt quickly the production system to a constantly changing environment. It necessitates a high level of enterprise integration. Yet, in most of manufacturing systems, two distinct functions are still handled independently to manage the production: the process planning function and the scheduling function.

Process planning determines how a product will be manufactured from its initial to a finished product. In other words, which sequence to use and which resource to select [1]. The scheduling is another manufacturing function that finds a mapping between

© IFIP International Federation for Information Processing 2018
Published by Springer International Publishing AG 2018. All Rights Reserved
A. Amine et al. (Eds.): CIIA 2018, IFIP AICT 522, pp. 387–398, 2018.
https://doi.org/10.1007/978-3-319-89743-1_34

jobs and resources to achieve some relevant criteria. The output of process planning is an input of scheduling. Therefore, scheduling is based on a fixed process planning. Moreover, process planning doesn't consider the current capacity of the resources while it is a strict constraint for the scheduling function. This sequential organization doesn't permit to take fully advantage of the flexibility provided by modern manufacturing systems. Research on scheduling has focused primarily in the construction of efficient algorithms to solve different types of scheduling problems: flow shop, job shop, open shop, and so on. However, research works show that the integration of the process planning function and the scheduling function permits to gain valuable insights [2–4].

The Flexible Job Shop Scheduling Problem (FJSSP) is an extension of the classical job shop scheduling problem (JSSP), where each operation Oj can be processed by a set of alternative machines, subset of the set of machines M. In that case, pjk denotes the processing time of operation Oj on machine Mk. Machines are always available and while operations are being processed, preemption is not allowed. The aim is to find a schedule for processing these n jobs on the m machines. The quality of a schedule is given by a performance measure objective function f, based on the jobs completion times denoted Ci. One of the most classical objective functions is to minimize the makespan, which measures the total length of the schedule [5]. The makespan, denoted Cmax is computed as $Cmax = max_{i=1..n}Ci$. Following the Lawler notation [6], this problem is noted J‖Cmax.

The integration of process planning and scheduling in a flexible job shop system (FJSSP-PPF) considers alternative machines for the operations, called operation flexibility (OF), and alternative operations' sequence, called sequencing flexibility (SF).

- In this paper, we address the problem of integrating process planning and scheduling on a flexible job shop. The main contribution of this paper is to propose a very effective hybrid method to evaluate and explore the search space intelligently in a reasonable time frame in order to solve the FJSSP-PPF problem. This method has been evaluated on a manufacturing model from the literature, with objective to find a schedule which minimizes the makespan. This hybrid method, described more precisely in Sect. 3, is based on three stages: Generation of an initial solution using an innovative local search procedure called Shifting Bottleneck Heuristic (SBH).

- The second stage is based on an exploration of the solutions' space using a Tabu Search method. The solution found by the SBH procedure becomes the initial solution of the Tabu Search. An efficient initialization of the solution is an essential aspect of a metaheuristic's performance in terms of solution quality and computing time. As the SBH is one of the most successful heuristics for the J‖Cmax problem, it is of the greatest interest to use it as initial solution for the Tabu Search method.

- Tabu Search gives very effective results to solve the JSSSP, but it is a neighborhood method that may be trapped in local optimum. To avoid this, the solution obtained from the Tabu Search becomes the initial solution of an iterated solution improvement metaheuristic called Kangaroo Algorithm. The Kangaroo method is a combination of local and global search. The algorithm tries to improve the current solution by exploring its neighborhood using an iterative stochastic descent procedure. When a new improvement is no longer possible a "jump" procedure is performed in order to escape from the attraction of a local minimum [7].

Within this hybrid approach, the SBH to provide an efficient solution, tabu search is used to define an effective neighborhood around the initial solution as a local search method and the kangaroo algorithm is used to perform the global search among neighborhood. The main positive effect of this hybridization is the convergence speed to local optimum and to intensify the local search and the intensification ability of the Kangaroo algorithm (KA) the of global optimum.

The remainder of this paper is organized as follows: the second section proposes a literature review of hybrid methods for flexible scheduling job shop problems, section three describes the case study and the framework of our hybrid approach. In section four experimental results are reported and then discussed in section five. The last section concludes this work and proposes different research perspectives.

2   Literature Review

To solve the FJSSP-PPF problems, which are harder than FJSSP, hybrid methods give promising results. [2] Has developed a linear mixed-integer programming model (LMIPM) for integration problem that relates a Tabu Search (TS) heuristic with branch-and-bound method. The TS algorithm is employed as a fast heuristic to find an initial solution for a branch-and-bound procedure in an LMIPM environment. The feasible initial solution is generated using a specially designed dispatching rule earliest completion time first. [8] Have proposed a symbiotic Genetic Algorithm, which uses an artificial intelligent search technique, to handle, at the same time, the process planning and the scheduling functions. [1] Have analyzed the effect of changing flexible process plan of a part-type in a production order. They use a simulation-based Genetic Algorithm (GA) in order to select the key part-type so that a performance measure can be further improved. This approach generates near-optimal performance for the makespan. [9] Have proposed a new hybrid algorithm to solve the FJSSP-PPF problem with stochastic processing time. This approach combines simulated annealing and Tabu Search heuristic. More recently, [10] have compared the performance of two different heuristics based on genetic algorithms and simulated annealing algorithm, for the FJSSP-PPF problem, in order to minimize the total completion time.

One of the key factors of hybrid methods is the synergetic between the different methods used. Another key factor is the quality of the initial solution. Due to its great efficiency to solve the J||Cmax problem, the Shifting Bottleneck Heuristic has been largely used in hybrid methods. [11] Have planned a hybrid method which combines a Shifting Bottleneck Heuristic and a Tabu Search algorithm. In their method, the re-optimization step in the Shifting Bottleneck algorithm is replaced by the Tabu search. [12] Have combined a Shifting Bottleneck Heuristic with an iterated local search method for re-optimizing already scheduled machines. Computational experiments show that this hybrid method improve existing results for benchmark instances. Moreover this combination of SBH with an iterated local search method is applicable to large instances of job shop (more than 100 jobs and 20 machines). [13] Have developed a hybrid method, combining the genetic algorithm with the SBH for the flexible job shop scheduling problem. They generate the initial population randomly in order to maintain the diversity of individuals.

Most of works related to application of Kangaroo Algorithm use it as a level of a hybridization method. [7] Have proposed a hybrid method formed by an Ant Colony System (ACS) and a Kangaroo Algorithm (KA) to solve the single machine scheduling problem. This work is based on the collaborative power of the ACS and the intensification ability of KA.

3   Framework of the Approach

3.1   Case Study

In order to evaluate our method and compare its performances, we have chosen a manufacturing model already used by [14] (see Table 1). [14] Have used a genetic algorithm to solve it and their results are presented in Sect. 4.

Table 1. Representation of the integration problem

Jobi	Oj	Set of alternatives machines (pj, k)	Process plan number (operation sequence)
J1	1	{M1(6), M2(6)}	1: (1-2-3)
	2	{M2(5), M1(6), M3(6)}	2: (1-3-2)
	3	{M3(4)}	
J2	4	{M1(3), M3(4)}	1: (4-5-6)
	5	{M2(7)}	2: (4-6-5)
	6	{M3(6), M1(5), M2(7)}	3: (6-4-5)
J3	7	{M1(7), M3(8)}	1: (7-8-9)
	8	{M3(5), M1(5), M2(6)}	2: (7-9-8)
	9	{M2(4)}	

This model presents two difficulties. The first one is to choose the order, in which the operations of each job are sequenced (sequencing flexibility), and the second is to assign each operation Oj to a machine Mk selected from the set of alternative machines (operation flexibility). For instance, job J2 has three process plans, each one with a fixed sequence; furthermore, operation O8 may be processed by M3, M1 or M2, with processing time 5, 5 and 6, respectively, but it must be processed after operation O7.

3.2   General Framework

As the FJSSP-PPF problem is a highly combinatorial optimization problem we propose a hybrid procedure where the first stage (the Shifting Bottleneck heuristic) is dedicated to solve a part of the problem as shown in Fig. 1.

In our hybrid procedure, as the SBH doesn't take into account alternative machines and alternative sequences as well, it is used only once, with an instance of a JSSP

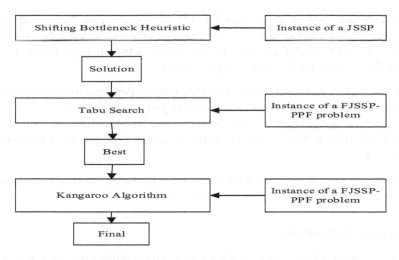

Fig. 1. General framework of our approach

derived from the FSP-PPF problem. Thus, at this stage, for each job a process plan is chosen arbitrarily and for each operation a machine is assigned arbitrarily. The Tabu Search (TS) and the Kangaroo Algorithm take into account both alternative sequences and alternative machines. The idea is that, at each stage, very good quality solutions are obtained permitting to improve this solution at the next stage.

3.3 Shifting Bottleneck Heuristic (SBH)

The Shifting Bottleneck Heuristic (SBH) was proposed by [15] to solve the minimum makespan problem for the job shop problem. The principle of this heuristic is to iteratively determine a machine considered as bottleneck and to optimally schedule this machine only. For each machine, a single machine scheduling problem 1 |rj| Lmax is solved using the branch and bound technique proposed by [16]. The bottleneck machine is the one with the highest Lmax. Then, using a disjunctive graph, the machines already scheduled are re-sequenced to include the optimal sequence of the current bottleneck machine. For each operation Oj, a release date rj and a due date dj are computed iteratively: rj is the earliest beginning date of operation Oj, computed from its predecessors already scheduled; dj is the latest completion time of operation Oj. For a more complete presentation of the SBH, the reader may refer to [17].

3.4 Tabu Search

Tabu Search (TS) algorithms are among the most effective approaches for solving JSSP [18]. They use a memory function to avoid being trapped in a local optimum [19]. Neighborhood structures and move evaluation strategies play the central role in the effectiveness and efficiency of the Tabu Search for the JSSP [20]. Most of the TS algorithms related to job shop scheduling are based on the pair-exchange method for generating neighborhood solutions. It depends on a permutation of operations placed on position i and $i + 1$ on a given machine [2]. During the Tabu Search exploratory, two types of neighborhood structures are added to take into account the two distinct decisions of the FJSSP-PPF problem: process plans' selection and machine's affectation. Finally, it gives the following movement operators:

- For a machine, permute the order of two sequential operations (operation exchange) when these two operations are on the critical path of the solution.
- For a critical operation, assign an alternative machine (machine exchange).
- For a job with at least a critical operation, change the process plan (sequence exchange).

These three movement operators will also be used during the stochastic descent of the Kangaroo Algorithm, described in the next section.

3.5   Kangaroo Algorithm

As most heuristics, there is no theoretical results ensuring the convergence of a Tabu Search procedure to a global optimum. Further progress is achieved by using a jump procedure to potentially move away from the previous local optimum. A very promising heuristic that includes a jump mechanism is the Kangaroo Algorithm proposed by [21]. It is described in Fig. 4. The principle of the Kangaroo method is similar to the simulated annealing algorithm but with a different research strategy. It generates a solution by using an iterative procedure that contains two parts: the stochastic descent procedure and the jump procedure [7].

During the stochastic descent, Kangaroo Algorithm seeks a solution that minimizes a function $f(S)$ in a neighborhood $N(S)$ of the current solution S using a local uniform mutation $\eta 1$. If the new solution S' is better than the previous solution, it is stored and a new solution is explored in the same neighborhood $N(S')$. The algorithm tries to improve the current solution A times, A being the maximum number of iterations in improving the current solution before a jump. If it is not possible to obtain a further improvement, the algorithm moves to another neighborhood with a jump procedure using a global uniform mutation $\eta 2$.

3.6   Solution Representation

To solve FJSSP-PPF using TS and KA, we first need to represent the solution of our problem as a chromosome. Each individual in FJSSP-PPF consists of three vectors with the same length, as shown in Fig. 2.

7	4	6	1	9	5	2	8	3	→	Operation number vector
2	3	3	1	2	3	1	2	1	→	Process plan number vector
3	3	1	2	2	2	1	1	3	→	Machine assignment vector

Fig. 2. Solution structure

With this representation, by selecting a chromosome, the FJSSP-PPF problem is transformed into a JSSP problem formulation. Depending on the selection of each job's sequence and each operation's machine it produces different instances of a JSSP. For example, the chromosome presented in Fig. 2 gives the following schedule (Fig. 3):

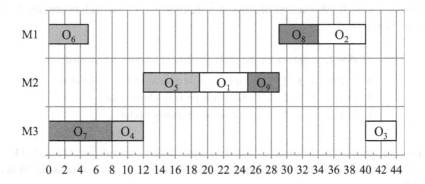

Fig. 3. Gantt chart of the solution derived from the chromosome selected in Fig. 2.

It can be noticed that the chromosome representation gives a semi-active schedule and not necessarily an active schedule. Active schedules are dominant for the make-span. For example, the example in Fig. 3 gives a semi-active schedule but not an active schedule.

4   Results

Our hybrid approach was coded in JAVA software Intel(R), Core (TM) i7 with a 2.2 GHz CPU. First, we present the solution obtained by the genetic algorithm developed by [14] in Fig. 4.

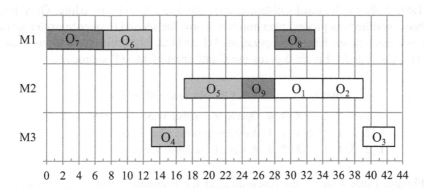

Fig. 4. Gantt chart of the solution proposed by [14]

The solution proposed by [14] is not an active schedule. For example, operations O1, O2, O3, O9 and O8 could begin earlier, on the same machines, without delaying any other operation, giving an active schedule. However, their solution gives a makespan of 43. In the following, we present the different solutions, on the same case study, obtained at each stage of our method. The SBH runs with the selections shown in Table 2: for each job the first process plan is chosen and for each operation the first machine is assigned.

Table 2. Solution for the SBH (critical operations are shaded)

4	5	7	8	6	1	2	3	9	Operation number vector	S(SBH)
1	1	1	1	1	1	1	1	1	Assigned process plan vector	
1	2	1	3	3	1	2	3	2	Assigned machines vector	

The outcome of the SBH gives the following critical operations: O4, O7, O1, O2 and O9. Gantt chart of the SBH solution is shown in Fig. 5. This solution is an active schedule.

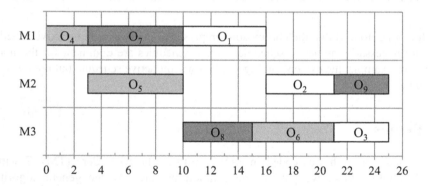

Fig. 5. Gantt chart of the SBH solution

Table 3 shows the results obtained after the Tabu Search procedure. Only job J1 has been modified: its process plan number is now the second (sequence exchange) and there is a machine exchange on operations O1 and O2. The critical operations become: O4, O7, O8 and O6. Figure 6 gives the Gantt chart of the TS solution.

Table 3. Presentation of the obtained SBH+TS solution

4	1	5	7	2	3	8	9	6	Operation number vector	S(SBH+TS)
1	2	1	1	2	2	1	1	1	Assigned process plan vector	
1	2	2	1	1	3	3	2	3	Assigned machines vector	

Finally, Table 4 shows the outcome after the Kangaroo Algorithm. This time, the modifications are on job J2 (sequence exchange and machine exchange for operations O4 and O6). The critical operations become O7, O6, O5. Figure 7 gives the Gantt chart obtained after the KA procedure.

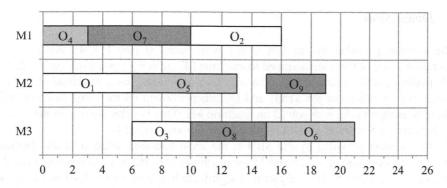

Fig. 6. Gantt chart of the SBH+TS solution

Table 4. Presentation of the obtained SBH+TS+KA solution

4	1	7	6	3	2	9	8	5	Operation number vector	S(SBH+TS+KA)
2	2	2	2	2	2	2	2	2	Assigned process plan vector	
3	2	1	1	3	1	2	3	2	Assigned machines vector	

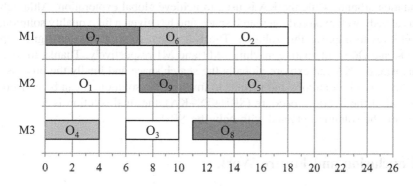

Fig. 7. Gantt chart of the SBH+TS+KA solution

Table 5 gives the number of generations, computational time and Cmax value obtained at each stage of our hybrid approach. The lowest Cmax value is found in 20 s by the hybrid method (SBH+TS+KA).

Table 5. Computational results

Method	Generations number	Comp. Time (seconds)	C_{max}	Improvement of the solution
GA [14]	Not provided	Not provided	44	
SBH	1	<1	25	43% (w.r.t. to GA)
SBH+TS	100	11	21	21% (w.r.t. to SBH)
SBH+TS +KA	100	20	19	5% (w.r.t. to SBH+TS)

5 Discussion

The solution provided by the GA is not well balanced (see Fig. 4). Among nine operations, four of them are assigned to machine M2, which gives a workload of 22 for this machine, while machine M3, with only two operations, has a workload of 8. Two successive operations of job J1:O1 and O2 are performed on the same machine M2. There is another cycle, with job J3 on machine M1. These cycles are one of the causes of the poor performance of the GA procedure proposed by [14].

The solution provided by the SBH is the most balanced solution of all obtained solutions (a workload of 16, 16 and 15 for machines M1, M2 and M3 respectively). As we can see in Table 5, the makespan is significantly improved by SBH w.r.t. to the solution obtained by [14]. This shows the importance of the initial solution when using metaheuristic. Moreover, as the SBH deals only with one JSSP instance, the solution is obtained immediately. Thanks to the high quality of the initial solution of the TS, the TS exploratory permits to improve again, and largely, the makespan (21%). With regards to the SBH solution, the new solution has permuted the machines of operations O1 and O2 (M1 and M2) and has changed the process plan of J1. Thus, during the TS search, only the operation and sequence flexibility of J1 has been updated.

Inside the hybrid system, SBH and TS methods are used to achieve local exploitation around, while the KA is used to achieve global exploration. Although the TS has already well explored the search space and has given a high quality solution, the KA achieves to improve this solution. The KA's final solution has reassigned operations O4 and O6 to alternative machines M3 and M1 respectively. Thanks to its jump mechanism, the KA can explore globally the search space and finally the process plan flexibility is used for all jobs and permits to improve the solution. It can be noticed that at the end of the hybrid procedure (SBH+TS+KA), the final solution is improved by 24% w.r.t. the solution obtained with only the SBH.

6 Conclusion and Future Work

In this paper, we developed a new approach that simultaneously integrates process planning and scheduling functions in a flexible job shop in order to minimize the makespan. This approach consists of hybridizing three heuristics: Shifting Bottleneck Heuristic (SBH), Tabu Search (TS) and Kangaroo Algorithm (KA). This approach has been compared with an integrated approach based on a Genetic Algorithm.

Our strategy is to start from a feasible initial solution created by the SBH. This heuristic gives a very good performance for the makespan in less than one second. This solution becomes a starting point for the Tabu Search. TS improves the solution obtained from the SBH by integrating operation flexibility and sequence flexibility. Again, the obtained solution becomes the initial solution for the next phase. In this phase, the search area is expanded by a global search algorithm called Kangaroo Algorithm, which improves the current solution by performing a jump procedure. The best makespan is produced by the hybridation SBH+TS+KA and outperforms the Genetic Algorithm by a factor of 2.3.

Although the scheduling problem studied is in this paper has only three machines and three jobs, due to its high flexibility, it is hard to get an optimal solution. However, it will be of the greatest interest to evaluate the efficiency (in terms of quality and speed) of our approach with a more complex problem. Comparisons of this approach with various optimization algorithms, such as exact methods or other metaheuristics, should also be very interesting. Another future work is to propose a multi-objective framework for solving the FSSP-PPF problem in order to minimize simultaneously the makespan and the maximum lateness.

References

1. Kumar, M., Rajotia, S.: Integration of process planning and scheduling in a job shop environment. Int. J. Adv. Manuf. Technol. **28**(1–2), 109–116 (2006)
2. Tan, W.: Integration of process planning and scheduling. A mathematical programming approach. Ph.D. Dissertation, University of Southern California (1998)
3. Tan, W., Khoshnevis, B.: Integration of process planning and scheduling – a review. J. Intell. Manuf. **11**(1), 51–63 (2000)
4. Li, W.D., McMahon, C.A.: A simulated annealing-based optimization approach for integrated process planning and scheduling. Int. J. Comput. Integr. Manuf. **20**(1), 80–95 (2007)
5. Samur, S., Bulkan, S.: An evolutionary solution to a multi-objective scheduling problem. In: Proceedings of the World Congress on Engineering 2010, WCE, vol. III (2010)
6. Lawler, E.L., Lenstra, J.K., Rinnooy Kan, A.H.G., Shmoys, D.: Sequencing and scheduling: algorithms and complexity. Handb. Oper. Res. Manag. Sci. **4**, 445–522 (1993)
7. Serbencu, A., Minzu, V., Serbencu, A.: An ant colony system based metaheuristic for solving single machine scheduling problem. Ann. Dunarea De Jos Univ. Galati Fascicle III **3**(1), 19–24 (2007). ISSN 1221-454X
8. Kim, Y.K., Park, K., Ko, J.: A symbiotic evolutionary algorithm for the integration of process planning and job shop scheduling. Comput. Oper. Res. **30**, 1151–1171 (2003)
9. Haddadzade, M., Razfar, M.R., Zarandi, M.H.F.: Integration of process planning and job shop scheduling with stochastic processing time. Int. J. Adv. Manuf. Technol. **71**, 241–252 (2014)
10. Botsali, A.R.: Comparison of simulated annealing and genetic algorithm approaches on integrated process routing and scheduling problem. Int. J. Intell. Syst. Appl. Eng. Adv. Technol. Sci. **4**, 101–104 (2016). ISSN 2147-67992
11. Bulbul, K.: A hybrid shifting bottleneck-tabu search heuristic for the job shop total weighted tardiness problem. Comput. Oper. Res. **38**(6), 967–983 (2011)
12. Braune, R., Zapfe, G.: Shifting Bottleneck scheduling for total weighted tardiness minimization: a computational evaluation of subproblem and re-optimization heuristics. Comput. Oper. Res. **66**, 130–140 (2016)
13. Gen, M., Gao, J., Lin, L.: Multistage-based genetic algorithm for flexible job-shop scheduling problem. In: Gen, M., et al. (eds.) Intelligent and Evolutionary Systems. Studies in Computational Intelligence, vol. 187, pp. 183–196. Springer, Heidelberg (2009). https://doi.org/10.1007/978-3-540-95978-6_13
14. Park, B.J., Choi, H.R.: A genetic algorithm for integration of process planning and scheduling in a job shop. In: Sattar, A., Kang, B.-H. (eds.) AI 2006. LNCS (LNAI), vol. 4304, pp. 647–657. Springer, Heidelberg (2006). https://doi.org/10.1007/11941439_69

15. Adams, J., Balas, E., Zawac, D.: The shifting bottleneck procedure for job shop scheduling. Manag. Sci. **34**(3), 391–401 (1988)
16. Carlier, J.: The one-machine sequencing problem. Eur. J. Oper. Res. **11**(1), 42–47 (1982)
17. Pinedo, M.: Scheduling Theory, Algorithms, and Systems. Springer, New York (2012). https://doi.org/10.1007/978-1-4614-2361-4
18. Jain, A.S., Meeran, S.: Deterministic job-shop scheduling: Past, present and future. Eur. J. Oper. Res. **113**, 390–434 (1998)
19. Zhang, C.Y., Li, P., Rao, Y., Guan, Z.: A very fast TS/SA algorithm for the job shop scheduling problem. Comput. Oper. Res. **35**(1), 282–294 (2008)
20. Jain, A.S., Rangaswamy, B., Meeran, S.: New and stronger job-shop neighborhoods: a focus on the method of Nowicki and Smutnicki. J. Heuristics **6**(4), 457–480 (2000)
21. Fleury, G.: Application des méthodes stochastiques inspirées du recuit simulé à des problèmes d'ordonnancement. Automatique Productique Informatique Industrielle **29**(4–5), 445–470 (1995)

A New Hybrid Genetic Algorithm to Deal with the Flow Shop Scheduling Problem for Makespan Minimization

Fatima Zohra Boumediene$^{(\boxtimes)}$ (iD), Yamina Houbad, Ahmed Hassam,
and Latéfa Ghomri

University of Tlemcen, B.P. 119, Tlemcen, Algeria
fatimazohra.boumediene@yahoo.fr, houbadyamina@yahoo.fr,
hassam.ahmed@yahoo.fr, ghomri@gmail.com

Abstract. In the last years, many hybrid metaheuristics and heuristics combine one or more algorithmic ideas from different metaheuristics or even other techniques. This paper addresses the hybridization of a primitive ant colony algorithm inspired from the Pachycondyla apicalis behavior to search prey with the Genetic Algorithm to find near optimal solutions to solve the Flow Shop Scheduling Problem with makespan minimization. The developed algorithm is applied on different flow shop examples with diverse number of jobs. A sensitivity analysis was performed to define a good parameter choice for both the hybrid metaheuristic and the classical Genetic Algorithm. Computational results are given and show that the developed metaheuristic yields to a good quality solutions.

Keywords: Scheduling · Flow shop · Pachycondyla apicalis algorithm
Hybridization · Optimization · Metaheuristic

1 Introduction

Flow Shop Scheduling Problem (FSSP) can be found in many real world problems because of the multitude of its application and its strong industrial background. A flow shop is a processing system characterized by a unidirectional flow of jobs being processed sequentially. Only small scale problems of Flow Shop Scheduling Problem can be solved by exact techniques, while others are classified NP-Hard problems and their resolution require dedicated methods to solve them in a reasonable computational time. Metaheuristics are among search techniques that offer near optimal solutions in generally low computational time. A metaheuristic can be seen as a search technique which can be applied to a wide set of different optimization problems with relatively few modifications to make them adapted to a specific problem. Genetic Algorithm [1], Ant Colony Optimization (ACO) [2, 3], Particle Swarm Optimization (PSO) [4], Artificial Bee Colony (ABC) [5, 6], Firefly Algorithm (FA) [7–9], Cuckoo-Search (CS) [10], just to name a few, are nature inspired metaheuristics and were largely used to solve many combinatorial optimization problems such as scheduling problems [11–13].

© IFIP International Federation for Information Processing 2018
Published by Springer International Publishing AG 2018. All Rights Reserved
A. Amine et al. (Eds.): CIIA 2018, IFIP AICT 522, pp. 399–410, 2018.
https://doi.org/10.1007/978-3-319-89743-1_35

Genetic Algorithm is among the well known metaheuristics and has been widely used to solve combinatorial optimization problems such as Flow Shop Scheduling Problems and performed well. The pioneering research on ants inspired metaheuristics was the Ant Colony Optimization (ACO) algorithm done by Dorigo [14], and an introduction to the ACO algorithm has been provided by Dorigo, Maniezzo and Colorni (1996) [15]. The ACO has been used to solve many combinatorial optimization problems [16–19].

An API algorithm is another ants inspired metaheuristic. It is a population-based, cooperative search procedure derived from the Pachycondyla apicalis foraging behavior ants, which are primitive ants that have been studied in the Mexican tropical forest near the Guatemala border [20]. It makes use of ant's colony, which iteratively construct and explore solutions to a combinatorial optimization problem.

Since Monmarché works [20, 21] the API algorithm has been used to solve different optimization problems [22–24].

The Pachycondyla apicalis are primitive ants living in colonies comprise from 20 to 100 individuals and use particular procedure to search their preys. The procedure begins by creating a number of hunting sites distributed uniformly around their nest at approximately 10 m. Starting from the nest, they cover globally a given surface and partition it into many hunting sites and each ant has to perform an exploration of given sites. Because of prey impoverishment or when the nest becomes less comfortable overtime, ants change the nest location periodically.

Using the Pachycondyla apicalis strategy to find their preys an algorithm named the API algorithm has been developed [20]. The API algorithm is based on many parallel local searches collaborating to find a near optimal solution for an objective function.

An effective search technique must not focus the search on only one region of the solution space. It needs to enable a balance between intensification and diversification. In last years, many researches have produced a large number of algorithms that do not fit a single metaheuristic, but they combine various algorithmic ideas, generally originating from artificial intelligence, computer science and operations research. However, to obtain better results, researchers have recently focused on the use of hybrid metaheuristics.

Hybrid metaheuristics are high level procedures that coordinate, generally, two or more metaheuristics to find solutions that are of better quality than those found by only one of them. Hybrid genetic algorithm (HGA), are proposed to enhance the performance of original genetic algorithm.

The present paper deals with the Flow Shop Scheduling Problem with makespan minimization using a hybrid metaheuristic. We propose the adaptation of the hybrid genetic algorithm (HGA) to solve scheduling problem in a flow shop with Cmax minimization with a sensitivity analysis. The remainder of this paper is organized as follows: Sect. 2 describes the flow shop problem formulation, Sect. 3 illustrate the GA principle, Sect. 4 gives the API and the API Fourragement algorithms principles, Sect. 5 describes the Hybrid Genetic Algorithm developed, Sect. 6 gives the sensitivity analysis of the developed algorithms and discusses obtained results and Sect. 7 draws some conclusions and discusses future works.

2 Problem Description

In many manufacturing systems, a number of operations have to be done on all jobs to be processed. When these operations have to be done in the same order and assuming that all machines are set in series the environment is referred to a flow shop.

A Flow Shop Scheduling Problem which can be defined as follows:

There is a number of n independent jobs $(J_1, ..., J_n)$ to be processed on m different machines $(m_1, ..., m_m)$. Each job J_i $(i = 1, ..., n)$ has to be processed on all m_j $(j = 1, ..., m)$ machines for P_{ij} time units (processing time) and the processing of any job can begin only after the completion of the preceding job.

As the problem size increases, exact methods are not computationally practical. For this reason, many researchers have focused on developing heuristics and metaheuristics for NP-Hard problems.

FSSP with makespan minimization can be stated as the problem of finding the minimal processing time such that all the jobs are processed in all the machines.

The makespan can be defined as the time between the beginning of the execution of the first job on the first machine and the completion of the execution of the last job on the last machine.

The assumptions and constraints made for the Flow Shop Scheduling Problem are summarized as:

- There are no precedence constraints among tasks of different jobs;
- Each machine mj can process only one job J_i at a time;
- Each job J_i can be performed only on one machine m_j at a time;
- For all jobs the processing time on each machine is previously known and deterministic;
- The machines sequence is the same for all the jobs, the problem is to find the job sequences on the machines which minimize the makespan;
- Traveling time between consecutive machines is negligible.

The following notations are needed to define the mathematical formulation of the problem:

n: number of jobs
m: number of machines
S_{ij}:   starting time of job i in machine j
P_{ij}:   processing time of job i in machine j
C_{ij}:   completion time of job i in machine j
C_{max}:   completion time of the last job in the last machine

The problem can now be formulated as:

$$f = minC_{max} \tag{1}$$

$$Cm_{ax} = max\{C_{ij}\} \quad \text{for all } i = 1, \ldots, n;\ j = 1, \ldots, m \tag{2}$$

$$C_{ij} = S_{ij} + P_{ij} \qquad \text{for all } i = 1, \ldots, n; \; j = 1, \ldots, m \qquad (3)$$

In our study we consider a 30 machines system with different number of jobs (20, 30 and 50 jobs), that makes we have three different classes of systems.

3 The GA Principle

GA explores a problem space with a population of chromosomes (individuals) and selects chromosomes to be explored iteratively based on their performance. Holland (1975) [1] presented a basic GA in his studies described as follows:

```
          Generate initial population randomly
          Calculate the fitness value of chromosomes
          While termination condition not satisfied
                  Process   crossover   and   mutation   at
chromosomes
                  Calculate    the    fitness    value    of
chromosomes
                  Select the offspring to next generation
```

In this basic form of the genetic algorithm three operators are used to create offspring: selection, crossover and mutation.

- Selection: selects chromosomes (solutions) in the population for reproduction.
- Crossover: The crossover operator is applied on each selected parent and uses usually one cut point and subsequences before and after that point are exchanges between each two selected chromosomes named parents.
- Mutation: The mutation operator is used to diversify the population in the genetic algorithm and prevent the premature convergence of the algorithm. The mutation operator exchanges some genes chosen randomly in the considered chromosome.

4 The Pachycondyla Apicalis Based Metaheuristic (API)

The API algorithm can be viewed as an optimization algorithm performing parallel searches to find best solutions starting from an initial solution (named the nest) and varying its initial solution iteratively.

The API algorithm can be given as follows:

(1) Choose randomly the initial location of the nest N

(2) For each ant a_i, i= 1,2, ..., $ants_{nbr}$

Perform the API-Fourragement

(3) If P_N is reached Then Change the nest location and Reset the memories of all ants

(4) Go to (2) or Stop if a stopping criterion is satisfied.

The API-Fourragement is given as follows:

If a_i has less than p hunting sites in memory Then Create a new hunting site in the neighborhood of N ants Explore this created site

Else

If the previous site exploration was successful Then Explore again the same site

Else Explore a randomly selected site (among the p sites in memory)

Remove from the ants memories all sites which have been explored unsuccessfully more than $P_{local}(a_i)$ consecutive times

Parameters Setting
In this section we present the API algorithm parameters and the signification of each one in our adaptation.

- a_i: Ant
- $ants_{nbr}$: the number of ants used to explore the search space.
- N: The nest. This corresponds, in optimization, to the initial solution of the problem. The random choice of the nest location means a random generation of an initial solution (jobs sequence) of the problem. Hunting sites are also problem solutions initially generated from the nest.
- p: Number of hunting sites in ants memories
- A hunting site: Represents a possible solution of the problem. The number of hunting sites is the number of solutions to be explored by an ant.
- Site exploration: An exploration of a site is to find a new solution in the neighborhood of the considered solution. Then the objective function is evaluated. The exploration will give a new hunting site.
- P_{local}: Represents a parameter which calculates the number of consecutive failures encountered on the same site (this corresponds to calculate the number of unsuccessful explorations of the same solution).

– P_N: Represents the parameter by which the nest changes location. In optimization, this change is equivalent to a reset of the initial condition in the algorithm. P_N is given as follows:

$$P_N = 2 \times (P_{local} + 1) \times p \tag{4}$$

The API algorithm can be viewed as a restarted parallel algorithm which carries out a number of parallel local searches starting from the same point.

Generally, in an optimization problem, the nest and the hunting sites represent admissible solutions.

5   The Hybrid Genetic Algorithm Developed

The HGA proposed hybridizes the GA with the API-Fourragement search procedure. The algorithm begins by generating an initial population randomly. The GA operators are performed exactly like in the basic GA until the mutation operator. In a basic GA selection, crossover and mutation operators are applied to parents in order to generate offspring. Therefore, good performers propagate through the population from generation to the next. Thus, a good exploration of each generation could yield to best solutions. We propose to explore solutions found by the GA (offsprings) using the API-Fourragement algorithm. The main idea is to apply the API-Fourragement on offsprings found by the GA in order to improve them before returning to be evaluated. This can offer to the HAG a good balance between the intensification and diversification search procedures because of its complementary operations.

Before giving the HGA, the solutions encoding, crossover and mutation operators have to be explained.

Solution encoding: In a flow shop scheduling problem, the aim is, generally, to find the best jobs sequence which permits to minimize Cmax. A feasible solution in our study could be viewed as a jobs sequence. Figure 1 shows an example of a possible solution in our study for 10 jobs having to be processed.

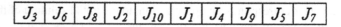

Fig. 1. An example of solution encoding

– Crossover operator: The crossover operator is applied on each two selected parents from the population. It can be done by choosing a crossover point in the chromosome at hand, keeping the left part of the first chromosome and finding in the second chromosome the remaindering genes to complete the new constructed chromosome (offspring) and doing the same procedure on the second parent. Figure 2 gives an illustrating example of the crossover operator applied on two parents.
– Mutation operator: The mutation operator is done by selecting randomly a pair of jobs and swaps them to obtain a new chromosome. An illustrating example of the mutation operator is given in Fig. 3.

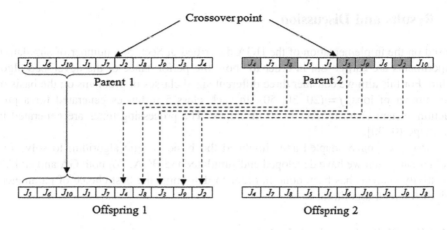

Fig. 2. Example of the crossover operator applied on two selected parents.

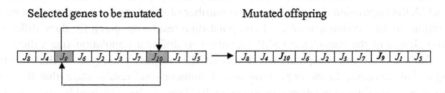

Fig. 3. Example of the mutation operator

The HGA Algorithm is given by the following steps:

```
Generate an initial population of PopSize solutions

Evaluate   each   individual   in   the   population   by
calculating its fitness function (Cmax)

While stopping criterion is not reached

Select two individuals (named parents)

     Perform the crossover operator on the two selected
individuals

     Perform the mutation operator on each obtained
solution (offspring) from the selection operator

Choose randomly one solution among obtained solutions
to be the nest for the API-Fourragement

Consider   all   other   obtained   solutions   (mutated
offsprings) as hunting sites for the API-Fourragement

For each ant ai

Perform the API-Fourragement

Replace parents by obtained offsprings
Output the best solution
```

6   Results and Discussion

Based on the implementation of the HGA described in Sect. 5, a number of simulation experiments are carried out in order to show the performance of the developed algorithm. For this aim we consider three different sized classes of problems on the basis of the number of jobs, $j = (20, 30, 50)$. In each class 5 instances generated for a production system characterized by 30 machines. The processing times are generated in the range [0, 30].

Firstly, we have adapted and simulated the basic genetic algorithm to solve our problem and then we have developed and simulated the HGA. For both GA and HGA a sensitivity analysis has been done in order to determine the best choice of parameters values for each algorithm.

6.1   Sensitivity Analysis of the GA

The number of examined solutions play a main role in population based metaheuristics. In a GA the population size determines the number of solutions to be examined at each iteration. In this section a number of computational results are given for three different sized classes of the considered FSSP according to different population size values.

Table 1 illustrates computational results of the sensitivity analysis of the Genetic Algorithm according to the population size. Computational results show that the GA performs better for a population size equals to 40. This can be explained by the fact that more the explored search space (i.e.: the number of examined solutions) is large better the obtained solutions quality is. But if the population size is higher we may have saturation.

Table 1. Sensitivity analysis on the population size of the GA

	Pop_{Size}	C_{max}		
		10 jobs 20 machines	20 jobs 20 machines	50 jobs 20 machines
Example 1	10	738	1078	1745
	20	727	1054	1730
	30	728	1044	1743
	40	**722**	**1032**	**1720**
	50	724	1037	1729
Example 2	10	787	1071	1726
	20	777	1069	1723
	30	774	1057	1696
	40	**770**	**1052**	**1673**
	50	773	1062	1694
Example 3	10	778	1062	1737
	20	765	1034	1725
	30	739	1024	1714
	40	**732**	**1008**	**1694**
	50	752	1013	1704

(continued)

Table 1. (*continued*)

	Pop_{Size}	C_{max} 10 jobs 20 machines	20 jobs 20 machines	50 jobs 20 machines
Example 4	10	777	1040	1723
	20	786	1040	1715
	30	783	1034	1708
	40	**768**	**1030**	**1687**
	50	772	1036	1688
Example 5	10	820	1085	1734
	20	809	1068	1724
	30	802	1062	1721
	40	**799**	**1041**	**1718**
	50	802	1052	1729

6.2   Sensitivity Analysis of the HGA

To determine a good choice of the HGA parameters, a number of simulations have been done according to the population size.

Table 2. Sensitivity analysis of the HGA

	Number of sites (p)	Pop_{Size}	C_{max} 10 jobs 20 machines	20 jobs 20 machines	50 jobs 20 machines
Example 1	3	10	735	1052	1745
	5	16	731	1043	1734
	7	22	7.30	1039	1726
	9	**28**	**719**	**1021**	**1718**
	11	34	725	1036	1723
	13	40	726	1047	1738
Example 2	3	10	766	1061	1720
	5	16	765	1052	1711
	7	22	760	1052	1696
	9	**28**	**758**	**1043**	**1657**
	11	34	765	1055	1668
	13	40	770	1057	1685
Example 3	3	10	759	1023	1726
	5	16	755	1006	1688
	7	22	741	1001	1689
	9	**28**	**727**	**995**	**1678**
	11	34	744	1002	1696
	13	40	749	1027	1704

(*continued*)

Table 2. (*continued*)

Number of sites (p)		Pop_{Size}	C_{max}		
			10 jobs 20 machines	20 jobs 20 machines	50 jobs 20 machines
Example 4	3	10	785	1021	1705
	5	16	782	1020	1693
	7	22	772	1001	1679
	9	**28**	**761**	**999**	**1671**
	11	34	771	1018	1677
	13	40	773	1022	1679
Example 5	3	10	817	1066	1727
	5	16	806	1065	1717
	7	22	801	1044	1738
	9	**28**	**789**	**1035**	**1694**
	11	34	806	1039	1704
	13	40	809	1057	1710

However, solutions obtained by the GA will be explored by the API-Fourragement the population size of the GA must depend on the number of hunting sites explored by each ant of the API-Fourragement. For this reason we developed the Eq. (5) which gives the population size (Pop_{Size}) according to the number of ants and the number of hunting sites.

$$Pop_{Size} = (p \times ants_{nbr}) + 1. \tag{5}$$

Equation (5) shows that the number of all solutions (Pop_{Size}) to be explored is the number of solutions explored by each ant (p) plus one solution which is the nest. That means the HGA population is built by solutions classified as:

The nest: which is a solution chosen initially in a random way

Hunting sites: which are all other solutions subdivided into $ants_{nbr}$ subpopulations.

The sensitivity analysis results according to the number of hunting sites and consequently to different population sizes is given in Table 2. The number of hunting sites varies from 3 to 13 and therefore the population size varies too.

Computational results show that the solutions quality is better for a number of hunting sites equals to 9.

6.3   Comparison Results of the HGA and the Basic GA

Table 3 shows computational results of both the GA and HGA for best obtained parameters yielding to best C_{max} for each algorithm ($Pop_{Size} = 40$ for the GA and p = 9 for the HGA).

Computational results show that the HGA performs better than the basic GA for the different sized classes of the FSSP considered in term of solutions quality because of its complementary procedures even for relatively high instances.

Table 3. Comparison results of the HGA and the basic GA

	10 jobs and 10 machines		10 jobs and 20 machines		10 jobs and 50 machines	
	GA	HGA	GA	HGA	GA	HGA
Example 1	722	719	1032	1021	1720	1718
Example 2	770	758	1052	1043	1673	1657
Example 3	732	727	1008	995	1694	1678
Example 4	768	761	1030	999	1687	1671
Example 5	799	789	1041	1035	1718	1694

7 Conclusion

In this paper, a hybrid algorithm (HGA) combining GA and API-Fourragement procedure is proposed to solve the flow shop scheduling problem with makespan minimization. The most challenging problem of the GA is to prevent early convergence. The HGA deals with that by the use of the AI-Fourragement search procedure. A sensitivity analysis has been done to determine the best parameters of the algorithm.

The good choice of the algorithm parameters plays a significant role in metaheuristics. In future study, a thorough investigation may be done on this issue.

Experimental results show that induction of the API-Fourragement permits to the HGA algorithm to outperform regular GA in term of solutions quality. Also, this method can be easily extended to solve other FSSP with other criteria which are also hard problems, such as maximum tardiness, total tardiness, etc.

References

1. Holland, J.H.: Adaptation in Natural and Artificial Systems. University of Michigan Press, Ann Arbor (1975)
2. Dorigo, M., DiCaro, G.: The ant colony optimization meta-heuristic. In: Corne, D., Dorigo, M., Glover, F. (eds.) New Ideas in Optimization, McGraw Hill, London, UK, pp. 11–32 (1999)
3. Korošec, P., Šilc, J., Filipič, B.: The differential ant-stigmergy algorithm. Inf. Sci. **192**, 82–97 (1999). http://dx.doi.org/10.1016/j. ins.2010.05.002
4. Kennedy, J., Eberhart, R.: The particle swarm optimization: social adaptation in information processing. In: Corne, D., Dorigo, M., Glover, F. (eds.) New Ideas in Optimization, pp. 379–387. McGraw Hill, London (1999)
5. Karaboga, D., Basturk, B.: A powerful and efficient algorithm for numerical function optimization: artificial bee colony (ABC) algorithm. J. Glob. Optim. **39**(3), 459–471 (2007)
6. Fister, I. Jr., Brest, J., Žumer, V.: Memetic artificial bee colony algorithm for large-scale global optimization. In: IEEE Congress on Evolutionary Computation, pp. 1–8 (2012)
7. Gandomi, A., Yang, X.-S., Talatahari, S., Alavi, A.: Metaheuristic in modeling and optimization. In: Gandomi, A., Yang, X.-S., Talatahari, S., Alavi, A. (eds.) Metaheuristic Application in Structures and Infrastructures, pp. 1–24. Elsevier, Waltham (2013)

8. Yang, X.S.: Optimization and metaheuristic algorithms in engineering. In: Yangetal, X.-S. (ed.) Metaheuristic in Water Geotechnical and Transport Engineering, pp. 1–23. Elsevier, Waltham (2013)

9. Sahab, M., Toropov, V., Gandomi, A.: Traditional and modern optimization techniques – theory and application. In: Gandomi, A.H., et al. (eds.) Metaheuristic Applications in Structures and Infrastructures, pp. 26–47. Elsevier, Waltham (2013)

10. Yang, X.S., Deb, S.: Cuckoo search via levy flights. In: World Congress on Nature & Biologically Inspired Computing (NaBIC2009), pp. 210–214. IEEE Publications (2009)

11. Zhang, Z., Juig, Z.: An improved ant colony for permutation flow shop scheduling to minimize makespan. In: 13th International Conference on Parallel and Distributed Computing, Applications and Technologies (2013)

12. Zhang, J., Tongand, J., Ma, Y.: An effective hybrid ant colony optimization for permutation flow shop scheduling. Open Autom. Control Syst. J. **6**, 62–68 (2014)

13. Ahmadizar, F.: A new ant colony algorithm for makespan minimization in permutation flow shop. Comput. Ind. Eng. **63**, 355–361 (2012)

14. Dorigo, M.: Optimization, learning and natural algorithm. Ph.D. thesis, DEI, Politecnico di Milano, Italy (1992). (in Italian)

15. Dorigo, M., Maniezzo, V., Colorni, A.: The ant system: optimization by a colony of cooperating agents. IEEE Trans. Syst. Man Cybern. B **26**, 29–41 (1996)

16. Holthaus, O., Rajendran, C.: A fast ant-colony algorithm for single-machine scheduling to minimize the sum of weighted tardiness of jobs. J. Oper. Res. Soc. **56**, 947–953 (2005)

17. Marimuthu, S., Ponnambalam, S.G., Jawahar, N.: Threshold accepting and ant-colony optimization algorithms for scheduling m-machine flow shops with lot streaming. J. Mater. Process. Technol. **209**, 1026–1041 (2009)

18. Lin, B.M.T., Lu, C.Y., Shyu, S.J., Tsai, C.Y.: Development of new features of ant colony optimization for flowshop scheduling. Int. J. Prod. Econ. **112**, 742–755 (2008)

19. Rajendran, C., Ziegler, H.: Ant-colony algorithms for permutation flow shop scheduling to minimize makespan/total flow time of jobs. Eur. J. Oper. Res. **155**, 426–438 (2004)

20. Monmarché, N.: Algorithmes de Fourmis Artificielles: Application à la Classification et l'Optimisation. University of François Rabelais, Tours (2000)

21. Monmarché, N., Venturini, G., Slimane, M.: On how Pachycondyla apicalis ants suggest a new search algorithm. Future Gener. Comput. Syst. **16**, 937–946 (2000)

22. Debbat, F., Bendimerad, F.T.: Radiation pattern optimization by apicalis ant algorithm for smart array antennas. Int. J. Sci. Eng. Res. **3**(11) (2012)

23. Houbad, Y., Souier, M., Hassam, A., Sari, Z.: Algorithme API hybride pour la resolution du problem de selection de routages alternatives dans un FMS. In: International Conference on Systems and Processing Information, Guelma, Algeria (2013)

24. Yamina, H., Mehdi, S., Ahmed, H., Zaki, S., Fayçal, B.: An APIm algorithm to solve the scheduling problem in an FMS with presence of breakdowns. Appl. Mech. Mater. **232**, 532–536 (2012)

Optimal Scheduling of Multiproduct Pipeline System Using MILP Continuous Approach

Wassila Abdellaoui[1]([✉]) [iD], Asma Berrichi[1], Djamel Bennacer[2],
Fouad Maliki[1], and Latéfa Ghomri[1]

[1] Department of Electrical Engineering,
University Abou Bekr Blkaid, Tlemcen, Algeria
abdellaouiwgp@gmail.com
[2] Department of Mechanical Engineering,
University Abou Bekr Blkaid, Tlemcen, Algeria

Abstract. To date, the multiproduct pipeline transportation mode has nationally and internationally considerably evolved thanks to his efficiently and effectively of transporting several products. In this paper, we focus our study on the scheduling of a multiproduct pipeline system that receives a number of petroleum products (fuels) from a single refinery source in order to be distributed to several storage and distribution centers (depots). Mixed Integer Linear Programming (MILP) continuous mathematical approach is presented to solve this problem. The sequence of injected products in the same pipeline should be carefully studied, in order to meet market demands and ensure storage autonomy of the marketable pure products in the fuels depots on the one hand and to minimize the number of interfaces; Birth zone of mixture between two products in contact and in sequential flow, which may hinder the continuous operation of the pipeline system, by the necessity of additional storage capacity for this last mixture, that is in no way marketable and requires special processing operations. This work is applied on a real case of a multiproduct pipeline that feeds the western and southwestern region of Algeria with fuels. The obtained results based on the MILP continuous approach give an optimal scheduling of the multiproduct transport system with a minimized number of interfaces.

Keywords: Multiproduct pipeline · Scheduling · Fuels
MILP continuous approach

1 Introduction

Several million barrels of crude oil and oil derivatives such as diesel, gasoline, kerosene, jet fuel and liquid petroleum gases (LPG) are moved daily around the world in imports and exports. These products can be transported with different modes: roads, railroads, vessels, and pipelines among which the latter is the safest and least expensive way to distribute large quantities of energy products from source generally represented by the refineries to distribution terminals located in the different place [1, 2].

Among the best benefits of the pipeline, is its ability to operate 24 h a day, regardless of weather conditions, with much lower operating costs than other inland transport modes [3].

© IFIP International Federation for Information Processing 2018
Published by Springer International Publishing AG 2018. All Rights Reserved
A. Amine et al. (Eds.): CIIA 2018, IFIP AICT 522, pp. 411–420, 2018.
https://doi.org/10.1007/978-3-319-89743-1_36

Pipelines were first utilized by oil transportation companies of crude petroleum and its derivatives, where demand for these petroleum products is high. Oil industries are decided to expand pipeline utilization due to its low operating cost [4].

Products are injected in the pipeline one after the other in batch, in order to be transported to several distribution centers. Generally, we can distinguish two forms of the pipeline; the first one is straight-structured pipeline circulation of batches inside the pipeline, where the flow is established in one direction from source to centers. The second case is where the pipeline transfer products in two directions, called bidirectional pipelines, the last is called tree-structured pipeline because it takes the tree form and at each tree branch called "segment", a depot is located.

In this research, we are interested in the scheduling of multiproduct pipeline take the form of a straight-structured pipeline. Our goal is to find the optimal sequence of the new batches injection inside the pipeline; What allows to satisfy in wanted time daily requests of terminals and to minimize contaminant interface which results between the different batch of product and ensures an autonomy of stock of 20%.

2 Literature Review

Several works on the multiproduct pipeline systems problems have appeared over the last years. Many authors have presented different approaches for scheduling multiproduct pipeline systems in the literature: knowledge-based search techniques and mathematical programming approaches such as Mixed Integer Linear Programming (MILP) used by [5–9] formulations or Nonlinear Mixed Integer Programming (MINLP) formulations. The last can be divided into two approaches: discrete MILP approach and continuous MILP approach [10].

Hane and Ratliff presented a discrete MILP model to transport several products from the refinery to diverse depots; this problem is divided into sub-problems solved by branch and bound method [11].

Rejowski and Pinto were studied a multiproduct system connected a unique refinery to several distribution centers that assure the demand of the local consumer markets. They proposed MILP model based on discrete time for scheduling this system. The model assumed in the beginning that le pipeline is divided into segments and each segment is divided into packs of equal volume, in the second they eliminated this assumption which says that the packs have the same volume [12].

The result of objective function ensures minimization of significant most operational costs, such inventory costs at the refinery and distribution centers, pumping costs and interface costs between adjacent products inside the pipeline, and moreover, the optimal sequence of injected products in the origin of pipeline and also the inventory levels at depot and refinery [1].

A year later, the same author adds special and non-intuitive practical constraints to the original model that can minimize volume between adjacent products (contaminant) inside the pipeline, and after, they included at the first MILP model a set of integer cuts that are based on the demand of depot and pipeline segment initial inventories [13].

Cafaro and Cerdá propose to study the problem treated by [12] with a novel MILP continuous mathematical formulation, which does not require division of pipeline into packs and time discretization. So, it has continuous representation in time and volume. Something special for this presentation, that allows minimizing the number of binary variables and gives the best sequence of injected products into the pipeline with the best costs [14].

Relvas et al. decided to integrate the distribution center operation with the multiproduct pipeline operation in their study. Their objective is to do both, the scheduling of the multi-product pipeline transport and the supply management in the depots. They have applied this model to a real case of a Portuguese oil distribution company [15]. Cafaro and Cerdá studied the problem of scheduling of multiproduct with different assumptions where clients demand was dynamic and they used a multi period moving horizon. They were giving good results appropriate to the real case with very short computational time [16].

Relvas et al. developed a heuristic that can find the optimal sequence of pumped products into the pipeline, they use this heuristic for a short to medium term horizon. MirHassani et al. presented an algorithm for the long-term planning of a simple multiproduct pipeline. The algorithm used a continuous MILP model for a short-term schedule to come to a long-term schedule. The objective function was minimizing the penalty costs of non-use of pipeline capacity, interface costs and cost of no satisfied customer demand [17].

Rejowski and Pinto the authors present a new mixed-integer non-linear programming (MILP) for intermittent multiproduct pipeline scheduling which takes into account several constraints like different flow rates, pipeline works intermittently. etc. This representation gives a good result comparing with Rejowski and Pinto which use a discrete representation of time [18].

Recently, the authors are interested in pipeline networks with multiple origins and destinations have been studied; Like what they did Cafaro and Cerdá Introduce a new tool for optimizing the short-term planning of petroleum product pipelines; Mixed-integer linear programming (MILP) model is expanded to treat pipeline networks with multiple sources, unidirectional flow and a single pipeline between every pair of the adjacent distribution center [19].

3   Problem Statement

In this paper we aim to study the activities scheduling in multiproduct pipeline system of a straight-structured unidirectional pipeline type, connecting a unique origin (refinery) to multiple distribution centers (in the case study we have two centers.

The experiment site is a 168-km multiproduct pipeline linking a refinery in western Algeria to storage and fuel distribution centers. Fuels moved in batches (We note four pure fuel batches, P1, P2, P3 and P4) from the refinery tank farm through pumping station without any physical separation between adjacent products. An area of mixture was established between batches, where this last zone of the mixture progress until reaching the terminal at pipeline's end. The number of mixture depends on the number of initial products injected in the pipeline [20]. Figure 1 shows the physical structure of studied pipeline system.

Fig. 1. Single unidirectional multiproduct pipeline system

The problem purpose is to determine sequence and volumes of new product batches to be pumped in the pipeline, in order to meet market demands and ensure products storage autonomy or the security stock in depots (fixed to 20% of overall storage capacity of each center) with number of interface between adjacent products p and p′ inside the pipeline minimized (Reduced).

3.1 Model Assumptions

(a) All products move in the pipeline without any physical separation between every two products in contact.
(b) The pipeline is always full, so if we think to receive a quantity of products from all the depots, it's necessary to inject the same amount at the origin of the pipeline.
(c) At any new pumping operation, only a unique product i.e. single batch is injected into the pipeline.
(d) Length of the planning horizon is fixe.
(e) Volume between adjacent products in pipeline (contaminant) was fixe.
(f) The scheduling model will meet the demand of products by the depots for daily sales to satisfy customer.

4   Optimization Model

4.1   Objective Function

Problem objective Function is given in Eq. (1) consisted to minimize the interface volume between two adjacent products in the pipeline.

$$min \sum_{i \in I} \sum_{p \in P} \sum_{p' \in P} INV_{i,p,p'} \tag{1}$$

4.2   Constraints

4.2.1   Product Allocation to Batch

$y_{i,p}$ is the binary variable, shows that product p is contained in batch 1 it takes value $y_{i,p} = 1$, if else $y_{i,p} = 0$. And every batch can, at more, take one product so:

$$\sum_{p \in P} y_{i,p} \leq 1 \qquad \forall i \in I^{new} \tag{2}$$

The new batch i will be injected after batch i − 1 so if batch i − 1 take any product $\sum_{p \in P} y_{i-1,p} = 0$ the batch i was not injected.

$$\sum_{p \in P} y_{i,p} \leq \sum_{p \in P} y_{i-1,p} \ \forall i \in I^{new} \tag{3}$$

4.2.2 Batch Sequencing
The injection of a new batch $i \in I^{new}$ in the pipeline should start after the end of injected batch $i − 1$.

$$s_i \geq s_{i-1} + L_{i-1} \ \forall i \in I^{nouveaunew} (i \geq 2) \tag{4}$$

4.2.3 Interface Volume Between Two Successive
Inside the multiproduct pipeline, there is no physical separation between different products, so we record certain volume of intermixing between the two adjacent batches which is assumed a constant value and it is presented with $mix(p,p')$ The continuous variable $INV_{i,p,p'}$ that presents interface volume between batches i and i + 1 take the value of $mix(p,p')$, if product p was located in batch i and product p' was located in batch i + 1.

$$\sum_{i \in I} \sum_{p \in P} \sum_{p' \in P} INV_{i,p,p'} \geq mix(p,p') * (y_{i,p} + y_{i+1,p'} - 1) \ \forall i \in I, i < |I|, p, p' \in P \tag{5}$$

4.2.4 Inventory in Distribution Center
The inventory of products in distribution center j at the end of injection batch i was equal to the inventory product at the end of pumped batch i − 1 $nvs_{p,j,i-1}$ by adding sum of product volume transferred to the distribution center during pumped batch i ($vmp_{h,p,j,i}$ for depot 1 and $vsq_{h,p,j,i}$ for depot 2) minus quantity delivery to clients.

$$nvs_{p,j,i} = vint_{p,j} + \sum_{h \in I, h \leq i} vmp_{h,p,j,i} - vom_{p,j,i}$$
$$\forall p \in P, j \in J(j < |J|), i \in I^{nouveau} (i = first \ new \ batch) \tag{6}$$

$$nvs_{p,j,i} = nvs_{p,j,i-1} + \sum_{h \in I, h \leq i} vmp_{h,p,j,i} - vom_{p,j,i}$$
$$\forall p \in P, j \in J(j < |J|), i \in I^{new} \tag{7}$$

To ensure that the level inventory in distribution center was grater than or equal to the stock of security, we use the following constraint, where $ss_{p,j}$ presents stock of security that is fixed at 20% of the overall stock capacities of each dept.

$$nvs_{p,j,i} \geq ss_{p,j} \quad \forall \, p \in P, j \in J, i \in I^{new} \tag{8}$$

5   Result and Discussion

5.1   Given

These data is harvest according to our real case study (Tables 1, 2, 3 and 4).

Table 1. Daily demand of depots

Product	Daily demand[m^3]	
	Depot 1	*Depot 2*
P1	1200	3000
P2	400	800
P3	80	150
P4	-	150

Table 2. Tanks storage capacities, products inventories and products security inventories

Product	Level inventory [m^3]					
	Depot 1			*Depot 2*		
	Capacity	Initial inventory	Security inventory	Capacity	Initial inventory	Security inventory
P1	6000	814	1200	22000	5572.4	4400
P2	1700	809	340	9500	3394.6	1900
P3	450	196	90	1000	996	200
P4	-	-	-	5000	3284.7	1000

Table 3. Interface volume and possible sequences between subsequent products inside the pipeline, p and p′

Product	Volume interface [m^3]			
	P1	*P2*	*P3*	*P4*
P1		28	30	
P2	28			0
P3	30			
P4	30			

Table 4. Initial batches volume inside pipeline at t = 0

	Initial batchs inside pipeline
	Volume [m³]
Batch01 (p2)	520
Batch01 (p1)	9580

5.2 Analyses Result and Discussion

The value of function objective was 320 m³ presented the total of interface mixture result between adjacent product inside the pipeline. They presented 2.30% of total volume, so we can say that the result was good more than can be satisfied the daily demand of client. The optimal sequence of the new batch injected inside the pipeline that can minimize the objective function was:

$$p3^{(1)} - p1^{(2)} - p2^{(3)} - p1^{(4)} - p3^{(5)} - p1^{(6)} - p2^{(7)} - p1^{(8)}$$

This sequence can minimize the number of an interface between adjacent product p and p′ and therefore minimize the objective function by satisfying the sales forecasts of centers and finally customer demands.

Table 5 shows the volumes of a new batch injected inside the pipeline satisfying the centers demands (depot 1 and 2) and ensure level tanks higher or equal than the stock of security.

Table 5. In volume of a new batch injected inside the pipeline

No of batch	Volume of batch [m3]			
	p1	p2	p3	p4
Batch 1			386	
Batch 2	4776..5			
Batch 3		8298.4		
Batch 4	15000			
Batch 5			1029.1	
Batch 6	13710			
Batch 7		10000		
Batch 8				2280

In the Tables 6 and 7 we can see that tanks of all product at each depot (depot 1 and depot 2) at the end injection of the new batch; The level inventory was more than the stock of security so we assure that the probability of no satisfied the demands of clients will be decrease so always the demands of distribution centers was satisfied.

Table 6. Level tanks in the depot 1 at the end of injection of the new batch

Batch	Tank $p1$		Tank $p2$		Tank $p3$	
	Volume [m³]	%	Volume [m³]	%	Volume [m³]	%
Initial	814	13.6	809	47.6	196	43.6
Batch 1	1200	20	809	24.1	116	25.78
Batch 2	2414.9	40.24	409	24.06	116	25.78
Batch 3	4800	80	400	23.53	347.5	83.23
Batch 4	2400	40	1480	87.06	187.55	41.67
Batch 5	1200	20	1080	63.53	107.55	23.9
Batch 6	4256	70.93	680	40	426.67	94.81
Batch 7	3200	53.33	340	20	213.33	47.41
Batch 8	1600	26.67	1260		74.12	106.67	23.71

Table 7. Level tanks in the depot 2 at the end of injection of the new batch

Batch	Tank $p1$		Tank $p2$		tank $p3$		Tank $p3$	
	Volume [m³]	%	Volume [m³]	%	Volume [m³]	%	Volume [m³]	%
Initial	5572.4	25.3	3394.6	35.7	996	99.6	3284.7	65.7
Batch 1	4400	20	2594.6	27.31	1000	90.6	3284.7	62.6
Batch 2	4400	20	3100.6	32.64	905.55	90.6	3134.7	53.7
Batch 3	5410.7	24.59	2300.6	24.22	755.55	90.6	2984.7	53.7
Batch 4	6370.9	28.96	6700	70.53	500	45.6	2684.7	53.7
Batch 5	4400	20	5900	62.11	350	100	2534.7	53.7
Batch 6	10455	47.52	5100	53.68	200	55	2384.7	47.7
Batch 7	11402	51.83	3800	40	400	55	2384.7	38.7
Batch 8	5750.9	26.14	1900	20	200	40	1784.7	35.7

6 Conclusions

Scheduling of a unidirectional multiproduct pipeline connecting a single refined to multiple distribution terminals is carried out with The MILP continuous approach. The MILP continuous can minimize the number of variable discussion compared to the MILP discrete and gives best results in the multiproduct pipeline problem scheduling case.

The MILP continuous was applied on the Algerian multiproduct pipeline scheduling. The pipeline in question links a refinery in western Algeria to two storage and fuel distribution centers. It has be seen that the use of MILP continuous approach gives the best result by taking into account the different operating conditions of the multiproduct pipeline which is served as experiment site.

The obtained results present an optimal solution that gives the sequence of new batches to be introduced in the pipeline. The flow configuration has reduced the number of interfaces between products p and p' in contact and in a sequential flow, unlike to the current planning system of the company which is not based on recognized models.

Furthermore, we were able to quietly ensure autonomy storage that exceeds the required safety stock set at 20%.

Finally, our work based on MILP continuous approach can be considered as a decision support tool, to be called for use in the case of planning and scheduling of multiproduct pipeline use the MILP continuous approach for solve this type of problems.

References

1. Mostafaei, H., Ghaffari-Hadigheh, A.: A general modeling framework for the long-term scheduling of multiproduct pipelines with delivery constraints. Ind. Eng. Chem. Res. **53**(17), 7029–7042 (2014).
2. LNCS Homepage. http://www.aopl.org. Accessed 12 Dec 2001
3. Cafaro, V., Cafaro, D., Mendez, C.A., Cerdá, J.: Detailed scheduling of oil products pipelines with parallel batch input at intermediate source. Chem. Eng. Trans. **32**, 1345–1350 (2013)
4. Hobson, G.D., Pohl, W.: Modern Petroleum Technology, 5th edn. Wiley, England (1982)
5. Cafaro, V.G., Cafaro, D.C., Méndez, C.A., Cerdá, J.: Detailed scheduling of operations in single-source refined products pipelines. Ind. Eng. Chem. Res. **50**(10), 6240–6259 (2011)
6. Cafaro, V.G., Cafaro, D.C., Méndez, C.A., Cerdá, J.: Detailed scheduling of single-source pipelines with simultaneous deliveries to multiple offtake stations. Ind. Eng. Chem. Res. **51** (17), 6145–6615 (2012)
7. Mostafaei, H., Castro, P.M., Ghaffari-Hadigheh, A.: A novel monolithic MILP framework for lot-sizing and scheduling of multiproduct treelike pipeline networks. Ind. Eng. Chem. Res. **54**(37), 9202–9221 (2015)
8. Zaghian, A., Mostafaei, H.: An MILP model for scheduling the operation of a refined petroleum products distribution system. Oper. Res. - Int. J. (ORIJ) **16**, 513–542 (2016)
9. Castro, P.M., Mostafaei, H.: New continuous-time scheduling formulation for multiproduct Pipelines. Comp. Aided Chem. Eng. **40**, 1381–1386 (2017)
10. Cafaro, D.C., Cerdá, J.: Efficient tool for the scheduling of multiproduct pipelines and terminal operations. Ind. Eng. Chem. Res. **47**, 9941–9956 (2008)
11. Hane, C.A., Ratliff, H.D.: Sequencing inputs to multi-commodity pipelines. Ann. Oper. Res. **57**, 73–101 (1995)
12. Rejowski, R., Pinto, J.M.: Scheduling of a multiproduct pipeline system. Comput. Chem. Eng. **27**, 1229–1246 (2003)
13. Rejowski, R., Pinto, J.M.: Scheduling of a multiproduct pipeline system. Comput. Chem. Eng. **28**, 1511–1528 (2004)
14. Cafaro, D.C., Cerdá, J.: Optimal scheduling of multiproduct pipeline systems using a non-discrete MILP formulation. Comput. Chem. Eng. **28**, 2053–2068 (2004)
15. Relvas, S., Matos, H.A., Barbosa-Povoa, A.P.F.D., Fialho, J., Pinheiro, A.S.: Pipeline scheduling and inventory management of a multiproduct distribution oil system. Ind. Eng. Chem. Res. **45**, 7841–7855 (2006)
16. Cafaro, D.C., Cerdá, J.: Dynamic scheduling of multiproduct pipelines with multiple delivery due dates. Comput. Chem. Eng. **32**, 728–753 (2008)
17. MirHassani, S.A., Moradi, S., Taghinezhad, N.: Algorithm for long-term scheduling of MUI product pipelines. Eng. Chem. Res. **50**(24), 13899–13910 (2011)

18. Rejowski, R., Pinto, J.M.: A novel continuous time representation for the scheduling of pipeline systems with pumping yield rate constraints. Comput. Chem. Eng. **32**, 1042–1066 (2008)
19. Cafaro, D.C., Cerdá, J.: Short-term operational planning of refined products pipelines. Optim. Eng. **18**, 241–268 (2016)
20. Bennacer, D., Saim, R., Abboudi, S., Benameur, B.: Interface calculation method improves multiproduct transport. Oil Gas **114**(12), 74–80 (2016)

Wireless Communications and Mobile Computing

Terrain Partitioning Based Approach for Realistic Deployment of Wireless Sensor Networks

Mostefa Zafer[1]($\boxtimes$), Mustapha Reda Senouci[2]($\boxtimes$), and Mohamed Aissani[2]

[1] Ecole nationale Supérieure d'Informatique,
BP 68M, 16309 Oued-Smar, Alger, Algérie
m_zafer@esi.dz
[2] Ecole Militaire Polytechnique, BP 17, 16046 Bordj El-Bahri, Alger, Algérie
mrsenouci@gmail.com, maissani@gmail.com

Abstract. This paper addresses the NP-hard problem of deploying wireless sensor networks on 3D terrains. On the contrary to previous works that place the sensors without any analysis of the terrain, we propose a two-phase solution based on terrain partitioning. The main idea is to estimate the number of sensors to be used and simplify the sensors deployment by partitioning the terrain according to topographic criteria. Simulation results based on real-world terrains confirm the efficiency of our solution in terms of coverage quality.

Keywords: Wireless Sensor Networks · 3D terrain · Deployment

1 Introduction

A Wireless Sensor Network (WSN) is a special type of ad hoc networks, consisting of a set of sensors deployed in a region of interest (RoI) to monitor a target event. Each sensor is battery-powered and has a sensing unit, a processing unit, and a wireless communication interface. The primary task of a sensor is to take measurements related to the target event and to communicate them wirelessly to a collection point (sink), using a one-hop or a multi-hop routing protocol [1].

For a WSN to be fully functional, at least two conditions must be met: coverage and connectivity [1]. Coverage suggests, in its simplest form, that each point of the RoI is covered by at least one sensor. This gives the network the ability to detect, at any point, the target event occurrence. Connectivity implies the existence, for each sensor, of at least one path connecting it to the sink.

To satisfy the two aforementioned conditions, it is necessary that the sensors must be well distributed in the RoI. The usually used solution consists in deploying the sensors in a random but very dense manner [1]. This practice is recommended in some cases, especially when the RoI is inaccessible. Nevertheless, for a three-dimensional RoI, characterized by a complex topography (e.g. mountainous areas), this random deployment produces many coverage holes, which

A. Amine et al. (Eds.): CIIA 2018, IFIP AICT 522, pp. 423–435, 2018.
https://doi.org/10.1007/978-3-319-89743-1_37

are generally hidden parts of the RoI with very low sensor density; and therefore a very low coverage quality, which gravely impairs the WSN performances.

The limitations of the random deployment of 3D WSNs, notably in terms of coverage, have been highlighted in several works [2–4] that recommended the use of a deterministic sensor placement. In this latter, the sensors' positions are precomputed, while taking into consideration not only the factors related to the sensors characteristics but also the topography of the RoI. This requires first a reliable representation of the RoI and a rigorous modeling of its influence on the network performances.

The problem of WSNs deployment on a 3D terrains, proved NP-hard [5,6], has been the subject of several researches [2–15] that have proposed, under different hypotheses and formulations, approximates solutions based on heuristics or meta-heuristics, aiming to find the best nodes positions, that meet the network objectives. However, these works treat the entire RoI in the same way, without taking into account the differences in topographic complexity between its different sub-regions. To contribute in removing this weakness and then improve coverage and connectivity, we propose a new deployment approach, in which the novelty is to partition the RoI into sub-regions characterized by simple topographies, to guide the deployment of the network composed of sensors and relays. In this work, we tackle the part related to the sensors deployment in order to maximize coverage. For the relays deployment, which are used to overcome the connectivity problem, we just give an indication that will be detailed and evaluated in a future work.

The rest of this paper is organized as follows. The related works and their limits are summarized in Sect. 2. The proposed approach is described in Sect. 3. The coverage quality produced by the proposed approach is evaluated by simulation in Sect. 4. The conclusion of this work is given in Sect. 5.

2   Related Works

The main difference between the existing approaches for the problem at hand lies in the terrain representation. As such, most approaches [2–4,7–15] adopt the matrix model, in which the terrain is represented by a matrix, where the value of each cell corresponds to the elevation of the place that it represents on the terrain. Another digital model, used by other approaches [5,6,16–18], consists in representing the terrain by triangles network form. This model, called TIN (Triangulated Irregular Network), can be deduced from the matrix model by using triangulation methods. Few simplistic approaches [19–21] adopt a mathematical model, where the coordinates of each point are calculated directly according to a mathematical formula.

The aforementioned approaches consider different assumptions and models to represent the sensor capabilities. The sensors can be considered homogeneous or heterogeneous, directional or omnidirectional, mobile or static. Their coverage capacities can be represented according to several parameters, with binary or probabilistic models. The parameters considered include, besides sensors-related ones, those related to the RoI, in particular its topography that is introduced, by the most of these approaches, via the visibility condition, which is verified in different ways, depending on the adopted terrain model [22]. Also, these approaches address various objectives such as connectivity [11,19,20], sensors stealth [16,17], and coverage that is the primary objective for all of these approaches.

In the literature, different heuristics and meta-heuristics are employed to find the best positions (and orientations) of the sensors to meet the planned objectives. In the sequel, these approaches are classified according to the adopted terrain model, and described in terms of assumptions and resolution schemes.

2.1 Deployment Approaches Based on a Mathematical Terrain Model

All the algorithms of this class are based on the binary coverage model. To improve the quality of the random deployment in terms of coverage and connectivity, two sensors redistribution algorithms have been proposed in [19,20], where the terrain is modeled as a cone, and sensors are assumed to be homogeneous and omnidirectional. The algorithm proposed in [19] is based on the virtual forces technique, under the constraint that the sensors move only on the surface of the terrain. In [20], the proposed solution consists in moving the sensors to positions situated on a spiral curve surrounding the terrain.

Assuming that the sensors are directional, the SA (Simulated Annealing) algorithm is used in [21] to relocate and reorient the sensors, initially deployed in a uniform manner, with the aim of improving the coverage.

2.2 Deployment Approaches Based on a Matrix Terrain Model

Some deployment solutions of this class adopt the binary coverage model, assuming that the sensors are omnidirectional [4,7,9] or directional [8]. To maximize coverage, the SA is implemented in [7], in both centralized and distributed manners. For the same goal, the S-GA (Simple Genetic Algorithm) and the CMA-ES (Covariance Matrix Adaptation-Evolution Strategy) algorithms are used in [4] and [8], respectively. In [9], the Voronoï diagram reinforced by the use of mobile nodes is made more realistic by taking into account the presence of obstacles. Other solutions adopt the probabilistic coverage model and the sensors are supposed to be omnidirectional [2,3,10–12] or directional [13–15].

To maximize coverage, the S-GA and SS-GA (Steady State Genetic Algorithm) algorithms are used in [2] where they were reinforced by the 2D-DWT (Discrete Wavelet Transform) transformation used to locate the coverage holes. For the same goal, this same transformation is adopted to reinforce the CSO (Cat Swarm Optimization) and ABC (Artificial Bee Colony) algorithms, used in [3] and [10], respectively.

In [11], in addition to coverage, connectivity is taken into account by using the link evaluation model adopted in [23]. To maximize these two criteria, the S-GA algorithm, reinforced by the 2D-DWT transformation, is used. The authors in [12] proposed a greedy algorithm to achieve k-coverage. Also, the work done in [8] is revised in [13–15] by adopting a probabilistic coverage model. In [13,14], SA and CMA-ES are the main methods used to maximize coverage. In [15], the PSO (Particle Swarm Optimization) method, consolidated by the virtual force technique, is used to maximize coverage of critical points on the terrain.

2.3 Deployment Approaches Based on a TIN Terrain Model

Assuming that the sensors are homogeneous and omnidirectional and by adopting a binary coverage model, a greedy algorithm is proposed in [5,6] based on the partitioning of the terrain in a regular grid, whose cells are smaller than the sensors coverage range. This algorithm consists in choosing, in a progressive (iterative) way, the cells that should host sensors in order to maximize coverage. With the same assumptions and by adopting a probabilistic coverage model, another solution is proposed in [18]. This latter is based on the Centroidal Voronoï Tessellation adapted to 3D terrain and guided by Ricci one-to-one mapping, with the aim to ensure coverage.

Table 1. A comparison between various deployment approaches.

Ref	Objectives	Resolution approach	RoI model	Deployment parameters		
				Sensor type	Visibility	Coverage model
[19]	Coverage + Connectivity	Virtual forces	Mathematical	Omnidirectional	✗	Binary
[20]	Coverage + Connectivity	Computational geometry	Mathematical	Omnidirectional	✗	Binary
[21]	Coverage	SA	Mathematical	Directional	✗	Binary
[7]	Coverage	SA	Matrix	Omnidirectional	✓	Binary
[9]	Coverage	Voronoï diagram	Matrix	Omnidirectional	✗	Binary
[4]	Coverage	S-GA	Matrix	Omnidirectional	✓	Binary
[8]	Coverage	CMA-ES	Matrix	Directional	✓	Binary
[2]	Coverage	S-GA + SS-GA	Matrix	Omnidirectional	✓	Probabilistic
[3]	Coverage	CSO	Matrix	Omnidirectional	✓	Probabilistic
[10]	Coverage	ABC	Matrix	Omnidirectional	✓	Probabilistic
[11]	Coverage + Connectivity	S-GA	Matrix	Omnidirectional	✓	Probabilistic
[12]	k-Coverage	Greedy Algorithm	Matrix	Omnidirectional	✗	Probabilistic
[13]	Coverage	SA + CMA-ES	Matrix	Directional	✓	Probabilistic
[15]	Coverage	PSO	Matrix	Directional	✓	Probabilistic
[14]	Coverage	SA + CMA-ES	Matrix	Directional	✓	Probabilistic
[18]	Coverage	Centroidal Voronoï Tessellation	TIN	Omnidirectional	✗	Probabilistic
[5,6]	Coverage	Greedy Algorithm	TIN	Omnidirectional	✗	Binary
[16,17]	Coverage + Sensors stealth	SS-GA	TIN	Directional	✓	Probabilistic

By adopting a probabilistic coverage model and assuming that the sensors are directional, the authors in [16,17] used the SS-GA algorithm to find the minimum number of sensors, their positions and their orientations, to maximize coverage while ensuring sensors stealth. Table 1 provides a comparative summary of the aforementioned deployment approaches.

It is worth to mention that most of the above-discussed approaches focused on the coverage of the RoI without considering other deployment-related aspects, such as connectivity, which is as important as coverage. In addition, some of these approaches are based on a very simplified and unrealistic terrain model, where its effective influence on coverage and connectivity has not been well taken into account. This observation concerns in particular works based on a mathematical terrain model. Also, the complexity of the RoI is an important factors to be taken into consideration to estimate the number of sensors to be deployed. This issue has not been addressed in these approaches. In addition, the performance results presented in these works show that the most effective approaches are those that use digital models to represent the terrain and exploit meta-heuristics to find better positions of the sensors. However, their weakness lies in the fact that they treat the entire RoI in the same way, without taking into account the differences in topographic complexity between the different sub-regions of the RoI.

3   The Proposed Approach

In order to face the topography challenges and simplify the deployment of the network, consisting of sensors and relays, our key idea is to partition the RoI according to the visibility criterion. Our approach is composed of three phases described as follows. The first phase consists of partitioning the RoI into relatively simple topography sub-regions by using a simple heuristic designed for this purpose and based on the visibility analysis. Thus, in each built sub-region, the visibility factor is less pronounced and its influence on detection and communication operations is greatly degraded. The second phase is intended for the deployment of the sensors on each sub-region, with the aim of maximizing its coverage, by using a method designed for this purpose. The first and second phases of our deployment approach are depicted on Fig. 1.

The third and final phase is intended to consolidate the network connectivity by using relays that will be placed at positions with wide visibility, located on the crest lines separating the constructed sub-regions. Our goal is to build a 2-tier architecture for the network, whose advantage lies in the fact that the sensors do not participate in data routing, which increases their lifetimes. Visibility, distance and load balancing are the most important factors that will be taken into consideration when building this architecture. We recall that this paper focuses only on the first and the second phases. The third phase is the subject of an upcoming work.

(a) Phase 1   (b) Phase 2

Fig. 1. First and second phases of our deployment approach.

3.1 RoI Partitioning

In this section, we describe our approach for partitioning the RoI into sub-regions with relatively simple topographies, which should facilitates the deployment of sensors on each of these sub-regions. We assume that the RoI, denoted $\mathcal{T}$, is represented by a TIN model that is composed of n triangles $t_i (1 \leq i \leq n)$, thus $\mathcal{T} = \{t_i, 1 \leq i \leq n\}$. To construct a partition $\mathcal{P}$ of $\mathcal{T}$, we apply an iterative fusion of triangles based on visibility analysis. The number of sub-regions constructed, denoted $\|\mathcal{P}\|$, is not fixed beforehand, and it is strongly depended to the RoI complexity. Each constructed sub-region, denoted $\mathcal{R}_k (1 \leq k \leq \|\mathcal{P}\|)$, is composed of one or more triangles and each triangle belongs only to one sub-region. To explain this partitioning heuristic, we define the following notions.

Definition 1 (Intervisibility triangle-triangle). *Two triangles t_i and t_j are considered intervisible if and only if they are adjacent; that is to say that they share a common edge, and their non-common vertices are intervisible. Therefore, any point $p \in t_i$ is visible to any other point $q \in t_j$.*

If the visibility between two points p and q is verified by the binary function $v(p, q)$ and the adjacency between two triangles t_i and t_j is verified by the binary function $adj(t_i, t_j)$, then the visibility between t_i and t_j is verified by the binary function $\mathcal{V}_\triangle(t_i, t_j)$ where $\mathcal{V}_\triangle(t_i, t_j) \iff adj(t_i, t_j) \wedge (\forall p \in t_i, \forall q \in t_j : v(p, q))$.

The visibility between two triangles (see Fig. 2) means that the coverage of one of the triangles could be ensured by the sensors deployed on the other, hence the opportunity to consider both as a single entity when deploying the sensors.

Fig. 2. Intervisibility triangle-triangle $(\mathcal{V}_\triangle(t_1, t_2) = 0, \mathcal{V}_\triangle(t_3, t_4) = 1)$.

Definition 2 (Intervisibility triangle-sub-region). *A triangle t_i is considered visible to a sub-region $\mathcal{R}_k$ if and only if it does not belong to $\mathcal{R}_k$ and it is visible to a triangle $t_j \in \mathcal{R}_k$. This relation is modeled by the binary function $\mathcal{V}_\mathcal{R}(t_i, \mathcal{R}_k)$ where $\mathcal{V}_\mathcal{R}(t_i, \mathcal{R}_k) \iff (t_i \notin \mathcal{R}_k) \wedge (\exists t_j \in \mathcal{R}_k : \mathcal{V}_\triangle(t_i, t_j))$.*

Similarly, the visibility between a triangle t_i and a sub-region $\mathcal{R}_k$ means that the coverage of t_i can be ensured by the sensors deployed in $\mathcal{R}_k$, hence the opportunity to consider them as a single entity when deploying the sensors.

Definition 3 (Free triangle). *A triangle t_i is considered free if and only if it is not assigned to any sub-region.*

Definition 4 (Front of a sub-region). *The front of a sub-region $\mathcal{R}_k$ is the set of free triangles visible to $\mathcal{R}_k$. If the set of free triangles is represented by Ω, then the front of $\mathcal{R}_k$ is $\mathcal{F}(\mathcal{R}_k) = \{t_i \in \Omega / \mathcal{V}_\mathcal{R}(t_i, \mathcal{R}_k) = 1\}$.*

Description. Initially, the set of free triangles Ω is equal to the set of triangles composing the RoI $\mathcal{T}$ (all the triangles of $\mathcal{T}$ are free). We assign to the first sub-region to be constructed, denoted $\mathcal{R}_1$, a triangle t arbitrarily selected from Ω ($t \longleftarrow rand_\triangle(\Omega)$). The sub-region $\mathcal{R}_1$ will be extended in an iterative manner, adding to it, at each iteration, the set $\mathcal{F}(\mathcal{R}_1)$. When there is no free triangle to add to $\mathcal{R}_1$; that is to say that $\mathcal{F}(\mathcal{R}_1) = \emptyset$, the construction of $\mathcal{R}_1$ is completed, and the construction of a new sub-region in the same way is launched. The sub-region construction algorithm stops when the set of free triangles becomes empty ($\Omega = \emptyset$). Obviously, the number of sub-regions produced by this heuristic depends on the complexity of the topography. In the ideal case, we will have a single sub-region including all the n triangles of $\mathcal{T}$. In extremely complex cases, we will have a number of sub-regions equal to the number of triangles of $\mathcal{T}$. It is easy to confirm that the temporal complexity of this heuristic is $\mathcal{O}(n^2)$, but it can be reduced to $\mathcal{O}(n \times log(n))$. Indeed, adding a triangle to a sub-region requires confirming that it is still free, by checking its presence in Ω, which is done in a sequential way in our basic heuristic, and can be improved by adopting a dichotomy search, under the condition that the triangles in Ω are sorted (by their numbers for example). The partitioning produced by our heuristic has two interesting characteristics: (i) two triangles lying in two different sub-regions are necessarily not inter-visible; and (ii) for every point in a sub-region, there are necessarily other points in that sub-region, which are visible to it.

3.2 Sensors Deployment

This section describes the sensors deployment phase, which includes the formulation of the desired objective according to the considered parameters and the adopted assumptions, an estimate of the number of sensors to be used, as well as the design of a resolution method, allowing to select the appropriate positions of these sensors.

Algorithm 1. RoI partitioning

Input: $\mathcal{T} = \{t_i/1 \leq i \leq n\}$
Output: $\mathcal{P}$
$\mathcal{P} \longleftarrow \emptyset; \; \Omega \longleftarrow \mathcal{T}; \; i \longleftarrow 0;$
while $(\Omega \neq \emptyset)$ **do**
    | $i \longleftarrow i+1;$
    | $t \longleftarrow rand_\triangle(\Omega);$ /* random selection of a free triangle   */
    | $\mathcal{R}_i \longleftarrow \{t\}; \; \Omega \longleftarrow \Omega - \{t\};$
    | **while** $(\mathcal{F}(\mathcal{R}_i) \neq \emptyset)$ **do**
    | | $\mathcal{R}_i \longleftarrow \mathcal{R}_i \cup \mathcal{F}(\mathcal{R}_i); \; \Omega \longleftarrow \Omega - \mathcal{F}(\mathcal{R}_i);$
    | **end**
    | $\mathcal{P} \longleftarrow \mathcal{P} \cup \mathcal{R}_i;$
end

Hypotheses and Objective. We aim to achieve the minimum coverage (1-coverage) level of the RoI $\mathcal{T}$, wherever every point must be covered by at least one sensor. For simplicity, we assume that the sensors are homogeneous and omnidirectional. Also, we adopt the binary coverage model, where the coverage capacity of the sensors is influenced by the distance and the terrain topography, which is introduced via the visibility factor. Thus, a point p_i is considered to be covered by a sensor s_j if and only if: (1) p_i and s_j are intervisible and (2) p_i is within the coverage range of s_j. This last condition is examined by the binary function $\mu_d(p_i, s_j)$, defined by Eq. 1, where r_s represent the coverage range of the sensors and $d(p_i, s_j)$ the Euclidean distance between p_i and s_j.

$$\mu_d(p_i, s_j) = \begin{cases} 1 & \text{if } d(p_i, s_j) \leq r_s \\ 0 & \text{otherwise} \end{cases} \tag{1}$$

Similarly, the visibility between p_i and s_j is modeled by a binary function $\mu_v(p_i, s_j)$, calculated by the Bresenham's algorithm [22]. Thus, the coverage of p_i by s_j is modeled by the binary function $\mathcal{C}(p_i, s_j) = \mu_d(p_i, s_j) \times \mu_v(p_i, s_j)$.

In the sequel, we describe our approach for the deployment of sensors on an arbitrary sub-region $\mathcal{R}_k \in \mathcal{P}$, constructed by the above-described partitioning phase. The same deployment approach is applied for the other sub-regions.

Deployment of Sensors on a Sub-region $\mathcal{R}_k$. We assume that the sensors deployed on $\mathcal{R}_k$ only contribute to the coverage of $\mathcal{R}_k$. In the presence of $\mathcal{N}_k$ sensors in $\mathcal{R}_k$, the coverage of a point $p_i \in \mathcal{R}_k$, is modeled by the function $Cov(p_i, \mathcal{N}_k) = \max_{1 \leq j \leq \mathcal{N}_k} \mathcal{C}(p_i, s_j)$.

We consider by approximation that the coverage $Cov_\triangle(t_i, \mathcal{N}_k)$ of a triangle $t_i \in \mathcal{R}_k$ is the mean of the coverage value of its vertices p_1^i, p_2^i and p_3^i (Eq. 2). The objective is therefore to find the number $\mathcal{N}_k$ and the positions of the sensors to be deployed in the sub-region $\mathcal{R}_k$, to maximize its coverage quality $Cov_\mathcal{R}(\mathcal{R}_k, \mathcal{N}_k)$ calculated according to Eq. 3, where $\|\mathcal{R}_k\|$ represents the number of triangles composing the sub-region $\mathcal{R}_k$.

$$Cov_\triangle(t_i, \mathcal{N}_k) \approx \frac{1}{3} \times \sum_{j=1}^{3} Cov(p_j^i, \mathcal{N}_k) \tag{2}$$

$$Cov_\mathcal{R}(\mathcal{R}_k, \mathcal{N}_k) = \frac{1}{\|\mathcal{R}_k\|} \times \sum_{t_i \in \mathcal{R}_k} Cov_\triangle(t_i, \mathcal{N}_k) \tag{3}$$

Estimation of the number of sensors. The number of sensors $(\mathcal{N}_k)$ to be deployed on the sub-region $\mathcal{R}_k$ to ensure its coverage depends on the coverage range r_s of the sensors, the surface area of $\mathcal{R}_k$, and the arrangement of the triangles of $\mathcal{R}_k$. For simplicity, we take into account only the first and the second factors. Therefore, we compute $\mathcal{N}_k$ by dividing the surface of $\mathcal{R}_k$, which is equal to the sum of the surfaces of all its triangles, on the surface covered by a sensor in an ideal (flat) terrain. The number $\mathcal{N}_k$ is given by Eq. 4, where $\|t_i\|$ represents the surface of the triangle t_i and $\mathcal{I}$ denotes the integer part function.

$$\mathcal{N}_k = \mathcal{I}\left(\frac{\sum_{t_i \in \mathcal{R}_k} \|t_i\|}{\pi \times r_s^2}\right) + 1 \tag{4}$$

Algorithm 2. Sensors deployment on sub-region $\mathcal{R}_k$

Input: X_r
Output: X
$X \longleftarrow X_r$; $i \longleftarrow 0$;
while $i < i_s$ do
$\quad$ $j \longleftarrow rand_\mathbb{N}\{1, ..., \mathcal{N}_k\}$; /* random selection of a sensor s_j $\qquad$ */
$\quad$ $X_i \longleftarrow X_r + rand_{s_j}(\mathcal{R}_k)$ /* generation of X_i from X_r by randomly
$\qquad$ changing the position of s_j in $\mathcal{R}_k$ $\qquad$ */
$\quad$ if $Cov_\mathcal{R}(X_i) > Cov_\mathcal{R}(X_r)$ then
$\quad\quad$ $X_r \longleftarrow X_i$;
$\quad\quad$ if $Cov_\mathcal{R}(X_i) > Cov_\mathcal{R}(X)$ then
$\quad\quad\quad$ $X \longleftarrow X_i$;
$\quad\quad$ end
$\quad$ else
$\quad\quad$ $\zeta \longleftarrow rand_\mathbb{R}[0, 1]$; /* generating a random number $\qquad$ */
$\quad\quad$ if $\zeta < \mathcal{T}(i)$ then
$\quad\quad\quad$ $X_r \longleftarrow X_i$;
$\quad\quad$ end
$\quad$ end
$\quad$ $i \longleftarrow i+1$;
end

Deployment of sensors. In order to maximize the coverage of $\mathcal{R}_k$ by the $\mathcal{N}_k$ sensors, we use the simulated annealing to determine the positions to be occupied by the sensors. Thus, a possible solution to our problem is coded in the form of a vector containing the positions of the $\mathcal{N}_k$ sensors. The method is based on the follow-up of the evolution of a current solution, that we note it X_r. Let $\mathcal{C}ov_{\mathcal{R}}(X_r)$ be the coverage quality $\mathcal{C}ov_{\mathcal{R}}(\mathcal{R}_k, \mathcal{N}_k)$ generated by X_r. Initially, X_r is generated in an arbitrary manner, where the sensors occupy random positions in $\mathcal{R}_k$. Thereafter, X_r evolves in several iterations. At an iteration i, a new solution X_i is generated from X_r, by randomly changing the position of an arbitrary sensor in X_r ($X_i \longleftarrow X_r + rand_{s_j}(\mathcal{R}_k)$), under the constraint that its new position is always in $\mathcal{R}_k$. Let $\mathcal{C}ov_{\mathcal{R}}(X_i)$ be the coverage quality $\mathcal{C}ov_{\mathcal{R}}(\mathcal{R}_k, \mathcal{N}_k)$ produced by X_i. The solution X_r is updated to X_i ($X_r \leftarrow X_i$) if X_i improves the coverage quality; that is to say that $\mathcal{C}ov_{\mathcal{R}}(X_i) > \mathcal{C}ov_{\mathcal{R}}(X_r)$. In the contrary case, this update ($X_r \leftarrow X_i$) is carried out with a probability $\mathfrak{T}(i)$ calculated according to Eq. 5. This probability is impaired as a function of the number of iterations performed, which is limited to a threshold value i_s. The best value taken by X_r during all its evolution on the i_s iterations is retained as a solution to the problem.

$$\mathfrak{T}(i) = 2^{-\left(\frac{2i}{i_s}+1\right)} \tag{5}$$

The total number $\mathcal{N}$ of sensors to be used for the terrain $\mathcal{T}$, and its coverage quality, are computed using Eqs. 6 and 7, respectively.

$$\mathcal{N} = \sum_{k=1}^{\|\mathcal{P}\|} \mathcal{N}_k \tag{6}$$

$$\mathcal{C}ov_{\mathcal{T}}(\mathcal{T}, \mathcal{N}) = \frac{1}{\|\mathcal{P}\|} \times \sum_{k=1}^{\|\mathcal{P}\|} \mathcal{C}ov_{\mathcal{R}}(\mathcal{R}_k, \mathcal{N}_k) \tag{7}$$

It should be noted that the sensor deployment phase can be easily upgraded, by eliminating the assumption that the sensors are omnidirectional and replacing the binary coverage model with a probabilistic model, which is more realistic. Whatever the choices adopted concerning the coverage model, the main contribution proposed in this paper, which consists in preceding the phase of deployment of the sensors by a phase of partitioning the terrain according to topographic criteria, remains intact.

4   Performance Evaluation

To evaluate the coverage quality produced by our approach, we have chosen four real RoIs of size $1200\,\mathrm{m} \times 1200\,\mathrm{m}$ with highly distinguishable and progressive complexities (RoI1 < RoI2 < RoI3 < RoI4). Each of these RoI is represented by a TIN model, built by the application of a Delaunay triangulation on altimetry data of these terrains, retrieved via "http://www.zonums.com/gmaps/terrain. php?action=sample". We set the coverage range of each sensor (r_s) to $30\,\mathrm{m}$, and

Table 2. Simulation results.

RoI	# Sensors	Coverage quality		
		Without partitioning	With partitioning	Gain
1	723	0.7460	0.8537	+0.1077
2	779	0.7187	0.8263	+0.1076
3	897	0.6832	0.8038	+0.1206
4	1081	0.6325	0.7923	+0.1598

we apply Eq. 6 to determine the number of sensors to use for each terrain, where the results obtained are shown on Table 2.

For the simulated annealing used for sensor deployment, we set the maximum iteration count (i_s) to 100. We note that the terrain partitioning heuristic and the simulated annealing are implemented on Matlab. The main objective is to analyze the impact of the partitioning phase on the efficiency of the simulated annealing in terms of coverage quality. Thus, for each RoI, the deployment of the sensors by the simulated annealing is carried out according to two scenarios. In the first scenario, all the $\mathcal{N}$ sensors are deployed on the entire RoI, without partitioning it. In the second scenario, the RoI is partitioned using the partitioning heuristic, after which the $\mathcal{N}_k$ sensors allocated to each sub-region $\mathcal{R}_k$ are deployed on it. The coverage quality produced by the simulated annealing implemented for the four RoIs according to these two scenarios (with and without partitioning) is summarized in Table 2.

The obtained results indicate the following. (1) *the number of sensors increases according to the complexity of the RoI.* This is justified by the fact that this complexity reflects the number of peaks. Each peak contributes to the increase of the triangles surfaces of the TIN model, and consequently to the increase of the surface of RoI, used as a factor to calculate the number of sensors; (2) *the coverage quality produced by the SA, in both scenarios (with and without partitioning), degrades according to the RoI complexity, despite the increase in the number of sensors.* This is justified by the fact that this complexity decreases in some way the factor of inter-visibility (notably for the third and the fourth RoI), which obviously has an influence on the coverage quality produced by SA, especially since it was launched with the same number of iterations (set at 100) for the four RoI; (3) *the quality of coverage is improved by the partitioning phase, and this improvement becomes more significant for the third and fourth RoI.* This is justified by the fact that this partitioning has already been based on the visibility analysis, in such a way that the increase in RoI complexity implies first and foremost the increase in the number of sub-regions constructed, which has allowed to minimize the influence of the terrain complexity on the visibility criterion in each constructed sub-region. Also, this partitioning allowed to subdivide the global problem, which consists of deploying the total number of sensors over the entire RoI, to very small sub-problems, where each one consists of deploying on each constructed sub-region, only the number of sensors allocated to it.

5   Conclusion

Examining previously proposed solutions for the deployment of WSNs on realistic terrains allowed us to remark that these solutions are "blind", because they proceed to the deployment of the sensors without analyzing the terrain and without taking into account the topographical differences that may exist between its different sub-regions. From this observation, in this paper, we have proposed a new deployment approach, which consists of partitioning the terrain into simple topography sub-regions, to simplify the estimation of the number of sensors to be used and guide their deployment. The obtained simulation results confirm the efficiency of our approach in terms of coverage quality. As a future work, we aim to design a relay placement algorithm to ensure the connectivity between the constructed sub-regions, and a logical topology management algorithm in order to maximize the network lifetime.

References

1. Senouci, M.R., Mellouk, A.: Deploying Wireless Sensor Networks: Theory and Practice. Elsevier, Amsterdam (2016)
2. Unaldi, N., Temel, S., Asari, V.K.: Method for optimal sensor deployment on 3D terrains utilizing a steady state genetic algorithm with a guided walk mutation operator based on the wavelet transform. Sens. J. **12**, 5116–5133 (2012)
3. Temel, S., Unaldi, N., Kaynak, O.: On deployment of wireless sensors on 3D terrains to maximize sensing coverage by utilizing cat swarm optimization with wavelet transform. IEEE Trans. Syst. Man Cybern. Syst. **44**, 111–120 (2014)
4. Ko, A.H.-R., Gagnon, F.: Process of 3D wireless decentralized sensor deployment using parsing crossover scheme. EACI **11**, 89–101 (2015)
5. Zhao, M.-C., Lei, J., Wu, M.-Y., Liu, Y., Shu, W.: Surface coverage in wireless sensor networks. In: IEEE INFOCOM, pp. 109–117 (2009)
6. Kong, L., Zhao, M.-C., Liu, X.-Y., Lu, J., Liu, Y., Wu, M.-Y., Shu, W.: Surface coverage in sensor networks. IEEE TPDS **25**, 234–243 (2014)
7. Veenstra, K., Obraczka, K.: Guiding sensor-node deployment over 2.5D terrain. In: IEEE ICC, London, pp. 6719–6725, June 2015
8. Akbarzadeh, V., Ko, A.H.-R., Gagné, C., Parizeau, M.: Topography-Aware Sensor Deployment Optimization with CMA-ES. In: Schaefer, R., Cotta, C., Kołodziej, J., Rudolph, G. (eds.) PPSN 2010. LNCS, vol. 6239, pp. 141–150. Springer, Heidelberg (2010). https://doi.org/10.1007/978-3-642-15871-1_15
9. Doodmana, S., Afghantoloee, A., Mostafavi, M.A., Karimipour, F.: 3D extention of the VOR algorithm to determine and optimize the coverage of geosensor networks. In: ISPRS, Tehran, Iran, pp. 103–108, November 2014
10. Hang, Y., Xunbo, L., Zhenlin, W., Wenjie, Y., Bo, H.: A novel sensor deployment method based on image processing and wavelet transform to optimize the surface coverage in WSNs. Chin. J. Electron. **25**, 495–502 (2016)
11. Unaldi, N., Temel, S.: Wireless sensor deployment method on 3D environments to maximize quality of coverage and quality of network connectivity. In: Proceedings of WCECS, San Francisco, USA, October 2014
12. Song, T., Gong, C., Liu, C.: A practical coverage algorithm for wireless sensor networks in real terrain surface. Int. J. Wirel. Mob. Comput. **5**, 671–680 (2012)

13. Akbarzadeh, V., Gagné, C., Parizeau, M., Argany, M., Mostafavi, M.A.: Probabilistic sensing model for sensor placement optimization based on line-of-sight coverage. IEEE ToIM **62**, 293–303 (2013)
14. Akbarzadeh, V., Gagné, C., Parizeau, M., Mostafavi, M.A.: Black-box optimization of sensor placement with elevation maps and probabilistic sensing models. In: IEEE International Symposium, pp. 89–94 (2011)
15. Tam, N.T., Thanh, H.D., Son, L.H., Le, V.T.: Optimization for the sensor placement problem in 3D environments. In: 12th International Conference on Networking, Sensing and Control, Taipei, Taiwan, pp. 327–333 (2015)
16. Topcuoglu, H.R., Ermis, M., Sifyan, M.: Positioning and utilizing sensors on a 3D terrain Part I: theory and modeling. IEEE Trans. Syst. Man Cybern. Part C Appl. Rev. **41**, 376–382 (2011)
17. Topcuoglu, H.R., Ermis, M., Sifyan, M.: Positioning and utilizing sensors on a 3D terrain Part II: solving with a hybrid evolutionary algorithm. IEEE Trans. Syst. Man Cybern. Part C Appl. Rev. **41**, 470–480 (2011)
18. Jin, M., Rong, G., Wu, H., Shuai, L., Guo, X.: Optimal surface deployment problem in wireless sensor networks. In: IEEE INFOCOM, pp. 2345–2353 (2012)
19. Boufares, N., Khoufi, I., Minet, P., Saidane, L.: 3D surface covering with virtual forces. In: PEMWN, Hammamet, Tunisia, pp. 103–108, November 2015
20. Kim, K.: Mountainous terrain coverage in mobile sensor networks. IET Commun. **9**(5), 613–620 (2015)
21. Xiao, F., Yang, X., Yang, M., Sun, L., Wang, R., Yang, P.: Surface coverage algorithm in directional sensor networks for three-dimensional complex terrains. Tsinghua Sci. Technol. **21**, 397–406 (2016)
22. Proctor, M.D., Gerber, W.: Line-of-sight attributes for a generalized application program interface. Def. Model. Simul. Appl. Method. Technol. **1**(1), 43–57 (2004)
23. De Marco, G.: MOGAMESH: a multi-objective algorithm for node placement in wireless mesh networks. In: 6th International Symposium on Wireless Communication Systems, pp. 388–392, September 2009

Hybrid Acknowledgment Punishment Scheme Based on Dempster-Shafer Theory for MANET

Mahdi Bounouni[1,2]($\boxtimes$) and Louiza Bouallouche-Medjkoune[1]

[1] LaMOS Research Unit, Faculty of Exact Sciences,
University of Bejaia, 06000 Bejaia, Algeria
Bounouni@gmail.com, louiza_medjkoune@yahoo.fr
[2] Faculty of Law and Political Sciences, University of Setif 2, Setif, Algeria

Abstract. In this paper, we cope with malicious nodes dropping packets to disrupt the well-functioning of mobiles ad hoc networks tasks. We propose a new hybrid acknowledgment punishment scheme based on Dempster Shafer theory, called HAPS. The proposed scheme incorporates three interactive modules. The monitor module monitors the behaviour of one-hop nodes in the data forwarding process. The reputation module assesses the direct and the indirect reputation of nodes using Dempster Shafer theory, which is a mathematical method, that can aggregate multiple recommendations shared by independent sources, while some of these recommendations might be unreliable. Since recommendations exchange between nodes consumes resources, a novel recommendation algorithm has been incorporated to deal with false dissemination attack and to minimize the recommendation traffic. The exclusion module punishes nodes regarded as malicious. The simulation results show that HAPS improves the throughput and reduces the malicious dropping ratio in comparison to existing acknowledgment scheme.

Keywords: Mobile ad hoc network · Security · Cooperation
Dempster Shafer theory · Uncertainty

1 Introduction

Mobile ad hoc network (MANET) is a collection of wireless mobile nodes that are able to perform the network tasks without requiring a fixed infrastructure or centralized administration. The communication between nodes follows a multi-hop approach. This approach depends on the assumption that all mobile nodes cooperate. Nevertheless, this assumption cannot be ensured due to the MANET features including the distributed nature, resource constraint of nodes [1]. These features make MANET vulnerable to selfish and malicious nodes. Selfish nodes

© IFIP International Federation for Information Processing 2018
Published by Springer International Publishing AG 2018. All Rights Reserved
A. Amine et al. (Eds.): CIIA 2018, IFIP AICT 522, pp. 436–447, 2018.
https://doi.org/10.1007/978-3-319-89743-1_38

may refuse to relay packets for other nodes to preserve their resources. On the other hand, malicious nodes may drop all packets passing through them in order to disrupt the functioning of the network's activities. Therefore, to improve the network performance, it is critical to cope against the selfish and malicious behavior.

In the literature, one can categorize two types of related works dealing with selfish and malicious nodes dropping packets: credit-based schemes [3–7] and reputation-based schemes [2,5,8–10,12,13,18–21,24]. The goal of incentive-based schemes consists of encouraging nodes to relay packets for the benefits of other nodes by using credit. Node earns credits by relaying packets for other nodes and loses credits to send their packets. In the reputation-based schemes, each node monitors its one-hop nodes and computes their reputation values according to their behaviour. Almost of the reputation-based schemes use the watchdog technique [2] for the monitoring. However, this technique presents several feebleness as reported in [2,11]. To deal with this feebleness, the acknowledgment technique is proposed in [12]. This technique permits to expand the range of neighbours monitoring to the two-hop by introducing a new kind of packet called TWOACK packet.

One of the recent scheme employing the acknowledgment technique is EAACK scheme [13]. EAACK can detect and punish malicious links. EAACK can effectively resolve some feebleness of the watchdog technique. However, EAACK is still vulnerable to other threats. (1) When nodes move faster, their neighbourhoods change often and therefore, malicious nodes have a several chances to drop more packets. Because, each new neighbour for malicious node forms a potential malicious link. This threat is inherited from TWOACK scheme [12], since TWOACK scheme can detect only malicious links and EAACK is based on TWOACK. (2) All requests initiated by malicious nodes are still relayed because the purpose is to relive malicious nodes from relaying data packets instead of punishing them.

To address the above threats, we propose a hybrid acknowledgment punishment scheme based on Dempster Shafer theory [14,15,23], called HAPS. HAPS scheme aims to enhance the performance of EAACK [13] by punishing malicious nodes more severely. HAPS is structured around three interactive modules: monitor, reputation, exclusion. The monitor module monitors the behavior of one-hop nodes in the data forwarding process. The reputation module computes the direct and indirect reputation values of neighbour nodes based on the information provided by the monitor module and the recommendations shared between nodes. We propose a new combination algorithm based on Demspter Shafer theory [14,15] to compute the direct reputation value of the node. Thus, HAPS enables nodes to share their recommendations about other nodes, but only when it is necessary, and the combination of different recommendations is done based on Dempster Shafer Theory. The exclusion module punishes all nodes having reputation values smaller than the reputation threshold.

The remainder of this paper is organized as follows. In the Sect. 2, We present some preliminaries on Dempster Shafer Theory. Section 3 is devoted to the adversarial model. In Sect. 4, we present our proposed scheme (HAPS). In Sect. 5, we examine the performance of HAPS via simulation and finally conclude the paper.

2   Preliminaries on Dempster-Shafer Theory

Dempster Shafer theory of evidence [14,15] is a mathematical method, handling the uncertainty and the subjective judgment. This method is especially efficient in situation when there is a need to aggregate multiple evidences shared by independent sources while some shared evidences might be unreliable, imprecise or incomplete/ambiguous. Let $\varphi = \{A_1, .., A_n\}$ be a finite set of mutually exclusive and exhaustive hypotheses denoted as the frame of discernment, where A_i are the individual hypotheses [22]. 2^φ denotes the possible subsets (or power set) of φ. In this section, we outline some basic concepts of Dempster Shafer theory.

Definition 1 [15]: A basic probability assignment function (BPA) or a mass function m is a function that assigns to each subset of ϕ a quantity of belief which is a number between 0 and 1. m is defined from $2^\varphi \rightarrow [0,1]$ and satisfying the following two constraints:

$$m\,(\emptyset) = 0 \ et \ \sum_{A \in \varphi} m(A) = 1 \tag{1}$$

Definition 2 [15]: Let $m : 2^\varphi \rightarrow [0,1]$ be a mass function. The belief function $bel : 2^\varphi \rightarrow [0,1]$ related to the mass function m over φ is defined as follows

$$bel\,(A) = \sum_{B \in A} m(B) \tag{2}$$

$bel(A)$ corresponds to the total of belief given to the hypotheses A.

Definition 3 [15]: Dempster's rule of combination permits to combine independent evidences issued from independent sources by applying the orthogonal sum $\oplus$. Given two mass functions m_1 and m_2 over the same frame of discernment φ. According to the Dempster's rule of combination, m_1 and m_2 can be combined into a new mass function $m : 2^\varphi \rightarrow [0,1]$ as follows:

$$m\,(C) = m_1\,(A) \bigoplus m_2\,(B) = \frac{\sum_{A_i \cap B_j = C} m_1\,(A_i) m_2\,(B_j)}{1 - K_{12}} \tag{3}$$

Where $K_{12} = \sum_{A_i \cap B_j = \emptyset} m_1\,(A_i)\,m_2(B_j)$. $m\,(c)$ represents the mass function of the combined evidence and K_{12} reflects the amount of conflicts between m_1 and m_2.

According to the Dempster's rule of combination, we can combine n evidences as follows:

$$m_1 \bigoplus m_2 \cdots \bigoplus m_n\,(C) = \frac{\sum_{C_1 \cap \cdots \cap C_n = C} m_1\,(C_1) m_2\,(C_2) \ldots m_n(C_n)}{1 - K_{1\ldots n}} \tag{4}$$

Where $K = \sum_{C_1 \cap \cdots \cap C_n = \emptyset} m_1\,(C_1)\,m_2\,(C_2) \ldots m_n(C_n) < 1$.

3   Adversarial Model

According to their purposes, nodes may behave maliciously in order to degrade the network performance. In our paper, we suppose that malicious nodes may launch: (1) Black hole attack by dropping all data packets passing through them. (2) False dissemination attack by sharing fake recommendation to falsely improve or degrade the reputation value of the malicious or honest node, respectively.

4   The Proposed HAPS Scheme

HAPS scheme is structured around three interactive modules: monitor, reputation, exclusion.

4.1   Monitor Module

This module monitors the behaviour of one-hop nodes in the data forwarding process. HAPS employ the monitoring technique proposed in the EAACK scheme [13]. EAACK scheme is the result of the combination of three modes: ACK, S-ACK and MRA. In this paper, we implement only ACK and S-ACK modes. In the ACK mode, the destination node should send back an ACK packet to the source node for every data packet received. The S-ACK mode is similar to the TWOACK scheme. In the S-ACK mode, a new kind of packet called S-ACK is used. Each node forwarding data packets should send an S-ACK packet to the two-hop node in the opposite direction of the forwarding path.

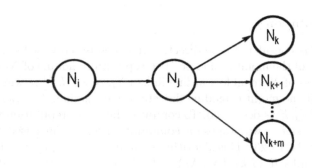

Fig. 1. Monitoring scenario

To illustrate the functioning of this technique, let $p = \{N_s, ..., N_i, N_j, N_k, ..., N_d\}$ the selected forwarding path, $<N_i, N_j, N_k> \in p$ a triplet of nodes taken as an example (see Fig. 1). $List_{ID}$ denotes the list of ID of data packets sent or forwarded waiting to be acknowledged. The source N_s sends data packets to the destination N_d through the path p. In the startup, the ACK mode is employed. In this mode, N_s adds the ID of each data packet sent D to $List_{ID}$.

Each ID is maintained for θ second. Upon reception of D at N_d, it should send back an ACK packet to N_s. For each ACK packet relayed by all nodes $N_i \in p$, the monitor module of N_i registers a good action through the link (N_j, N_k). If N_s receives an ACK packet before θ expires, which means that there are no malicious actions along the path p, it removes the ID of D from $List_{ID}$. Otherwise, N_s switches to the S-ACK mode. N_i adds the ID of each data packet forwarded D to $List_{ID}$. Each ID is maintained for ϑ second. N_j Will forward D to N_k if it behaves cooperatively. Once the packet D reaches N_k , it should send back an S-ACK packet N_i if it does not behave maliciously. If N_i receives S-ACK packet before ϑ expire, it deletes the ID of D from $List_{ID}$ and registers a good action against the link (N_j, N_k). Otherwise, if N_i does not receive S-ACK packet after ϑ expires, N_i removes the ID of D from $List_{ID}$ and registers a bad action against the link (N_j, N_k). The same process is repeated for each triplet of nodes along p. This process is repeated until N_s receives a switch packet from N_d, which means that p is a safer path. Therefore, N_s switches to the ACK mode.

4.2   Reputation Module

This module assesses and manages the reputation values of one-hop nodes. The reputation is classified into three types: direct, indirect and final. This module maintains four parameters ranging from 0 and 1: the reputation value of each monitored link $(N_j, N_k) \in FL_i^j$ denoted by $R_i(j, k)$ where FL_i^j denotes the set of forwarding links in which N_j is involved, The direct, indirect and final reputation values denoted by $DR_i^j(t)$, $IR_i^j(t)$ and $FR_i^j(t)$, respectively, where denotes the time of the computation of the reputation value.

Direct Reputation

A reputation is considered type direct, if it is computed based only on the recommendation of the monitor module. The reputation module of N_i evaluates the trustworthiness of N_j in all forwarding links $(N_j, N_k) \in FL_i^j$ in which is involved (see Fig. 1). If the monitor module detects a good action, $R_i(j, k)$ is increased. Otherwise, $R_i(j, k)$ is decreased. To compute the direct reputation value $DR_i^j(t)$ of N_j at time slot t, we propose a combination algorithm based on Demspter Shafer theory [14,15,23]. This algorithm combines and aggregates the reputation values of all links $R_i(j, k), (N_j, N_k) \in FL_i^j$ to come up to a single reputation value of N_j. The proposed algorithm functions as follows. In the startup, $R_i(j, k)$ of each link $(N_j, N_k) \in FL_i^j$ is initialized to $neutral_v$ and it is updated according to the action detected by the monitor module. We consider two exclusive and exhaustive hypothesis that construct the frame of discernment $\varphi = \{C, \overline{C}\}$ where C means that the node N_j is cooperative and $\overline{C}$ means that the node N_j is uncooperative. The power set 2^φ consists of four elements: $\emptyset$, $C = cooperative$, $\overline{C} = uncooperative$ and hypothesis $U = \varphi$ (N_j is either cooperative or uncooperative which represents the uncertainty). In this scenario, the reputation module of N_i perceives the reputation of each link $(N_j, N_k) \in FL_i^j$ as recommendation provided by N_k. The reputation module determines the state of the node N_j in the link (j,k) according to $R_i(j,k)$. If reputation module states that N_j is cooperative through the link (j,k), which means that $R_i(j,k) \geq neutral_v$, The BPA of N_k is:

$$
\begin{aligned}
m_k(C) &= R_i(j,k) \\
m_k(\overline{C}) &= 0 \\
m_k(U) &= 1 - R_i(j,k)
\end{aligned}
\tag{5}
$$

If reputation module states that N_j is uncooperative through the link (j,l), which means that $R_i(j,k) < neutral_v$, the BPA of N_l is

$$
\begin{aligned}
m_l(C) &= R_i(j,l) \\
m_l(\overline{C}) &= neutral_v - R_i(j,l) \\
m_l(U) &= 1 - neutral_v
\end{aligned}
\tag{6}
$$

The direct reputation value $DR_i^j(t)$ is computed by combining all recommendations collected from all links $(N_j, N_k) \in FL_i^j$ by applying the Dempster rule of combination. $neutral_v - R_i(j,l)$ value reflects the degree of maliciousness of the link (j,l). The rational of this algorithm is that: a malicious node should compromise multiple forwarding links (multiple forwarding paths) to achieve its purpose that consists on disrupting the data forwarding process. Therefore, it is involved in multiple bad actions that cause the degradation of its direct reputation value. On the other hand, an honest node is involved in more good actions than bad actions; therefore, it can improve its direct reputation value.

Indirect Reputation

A reputation is considered indirect, if it is computed based only on the recommendations shared between neighbours. In HAPS, this reputation is calculated and used only when there is need. The exchange method is done only when it is necessary, especially when a particular neighbour needs to send its packets. The goal is to improve the accuracy of the computation of nodes reputation and to minimize the recommendations traffics. When a node termed requestor needs to relay its packets through neighbours, all neighbours exchange their computed direct reputation values about this requestor. After that, they compute its indirect reputation value by aggregating all received recommendations using Dempster Shafer theory. Note that recommendations from nodes regarded as malicious are ignored.

When N_i receives a RREQ packet, it checks whether the requestor N_j is a neighbour $(N_j \in NG_i)$. If the requestor N_j is not neighbour, N_i simply forwards the RREQ packet. Else, N_i shares its recommendation about N_j in the neighbourhood and set the timer T_r. To prevent malicious nodes from colluding with other nodes or from manipulating the reputation values of some nodes, node N_i accepts only recommendation received before T_r expire. In order to compute the indirect reputation value of N_j, N_i aggregates all received recommendations using Dempster Shafer Theory. The recommendation of N_i about N_j is one among the set {cooperative-uncooperative}. Therefore, the frame of discernment is $\varphi = \{cooperative, uncooperative\}$. For instance, the reputation value of N_k at N_i is $DR_i^k(t)$. If N_k states that N_j is cooperative, the BPA of N_k is [16]:

$$m_k(C) = DR_i^k(t)$$
$$m_k(\overline{C}) = 0$$
$$m_k(U) = 1 - DR_i^k(t) \tag{7}$$

If N_k claims that N_j is uncooperative, the BPA of N_k is:

$$m_k(C) = 0$$
$$m_k(\overline{C}) = DR_i^k(t)$$
$$m_k(U) = 1 - DR_i^k(t) \tag{8}$$

The indirect reputation value of N_j is obtained after combining all recommendations using the Dempster's rule of combination. According to Dempster Shafer Theory features, the relevance of a recommendation depends on the reputation value of the recommender, which permit to get a reliable reputation value. This feature make our approach resilient to false recommendation dissemination. Thus, our approach can cope with collusion attack that occurs when a group of malicious nodes provides fake recommendations about an honest node. Because, the time T_rec between the diffusion and the combination of recommendations is very low. Then, these nodes have not sufficient times to collude.

Illustration Example:

Let assume three nodes N_k, N_l and N_f with reputation values 0.4, 0.45 and 0.9 at N_i, respectively. They share their recommendations about N_j. N_k claims that N_j is cooperative, N_l and N_f claims that N_j is uncooperative. Hence, the mass function are:

$$m_k(C) = 0.9,\ m_k(\overline{C}) = 0,\ m_k(U) = 0.1$$
$$m_l(C) = 0,\ m_l(\overline{C}) = 0.3,\ m_l(U) = 0.7$$
$$m_f(C) = 0,\ m_f(\overline{C}) = 0.3,\ m_f(U) = 0.7$$

The reputation module of N_i combines m_k and m_l as follows:

$$K_{kl} = m_k(C)\, m_l(\overline{C}) + m_k(\overline{C})\, m_l(C) = 0.27$$
$$m_{kl}(C) = m_k \bigoplus m_l(C) = \frac{m_k(C)\, m_l(C) + m_k(C)\, m_l(U) + m_k(U)m_l(C)}{1 - K_{kl}} = 0.863$$
$$m_{kl}(\overline{C}) = m_k \bigoplus m_l(\overline{C}) = \frac{m_k(\overline{C})\, m_l(\overline{C}) + m_k(\overline{C})\, m_l(U) + m_k(U)m_l(\overline{C})}{1 - K_{kl}} = 0.041$$
$$m_{kl}(U) == m_k \bigoplus m_l(U) = \frac{m_k(U)\, m_l(U)}{1 - K_{klf}} = 0.0958$$

The obtained m_{kl} is combined with m_f as:

$$K_{klf} = m_{kl}(C)\,m_f\,(\overline{C}) + m_{kl}\,(\overline{C})\,m_f\,(C) = 0.258$$

$$m_{klf}(C) = m_{kl}\bigoplus m_f(C) = \frac{m_{kl}(C)\,m_f(C) + m_{kl}(C)\,m_f(U) + m_{kl}(U)m_f(C)}{1 - K_{klf}} = 0.815$$

$$m_{klf}(\overline{C}) = m_{kl}\bigoplus m_f(\overline{C}) = \frac{m_{kl}(\overline{C})\,m_f(\overline{C}) + m_{kl}(\overline{C})\,m_f(U) + m_{kl}(U)m_f(\overline{C})}{1 - K_{kl}} = 0.094$$

$$m_{klf}(U) == m_{kl}\bigoplus m_f(U) = \frac{m_{kl}(U)\,m_f(U)}{1 - K_{klf}} = 0.0905$$

From this result, the indirect reputation value $IR_i^j(t)$ of node N_j is 0.81.

Final reputation

After obtaining $DR_i^j(t)$ and $IR_i^j(t)$, the final reputation value $FR_i^j(t)$ is computed by combining $DR_i^j(t)$ and $IR_i^j(t)$ with the following equation:

$$FR_i^j(t) = \delta * DR_i^j(t) + (1 - \delta) * IR_i^j(t). \tag{9}$$

Where $\delta(0 < \delta < 1)$ determines the relevance of direct reputation compared to the indirect reputation.

4.3   Exclusion Module

The exclusion module is responsible for punishing malicious nodes. It considers a node having final reputation value smaller than the threshold as malicious. This module puts the detected node in its black list $Black_{list}$, and it sends a misbehaving report to the source node of data packets to proceed to its punishment. All nodes forwarding, receiving or overhearing the misbehaving report put the detected node in their $Black_{list}$ and proceed to its punishment. The punishment consists on: (1) invalidating all forwarding paths involving this node and evicting to route their data packets through this node. (2) Refusing to forward packets initiated from this node by discarding all its RREQ packets generated.

5   Performance Evaluation

In this section, we conduct a series of simulation experiments to examine the performance efficiency of the HAPS scheme using the network simulator NS-2.34. We evaluate the effectiveness of the HAPS scheme on the exclusion of malicious nodes dropping data packets in comparison to EAACK [13]. We simulate 40 mobile nodes deployed within an area of $700 * 700$ m. The number of malicious nodes varies from 2 to 12. The rest of simulation parameters are shown in Table 1. The following two metrics are used to examine the efficiency of HAPS:

Table 1. Malicious scenario parameters

Parameter	Value
Number of node	40
Routing protocol	DSR [17]
Simulation area	700 m × 700 m
Transmission range	250 m
Node speed	2 m/s, 4 m/s, 6 m/s, 8 m/s and 10 m/s
Pause time	0 s
Number of malicious nodes	2, 4, 6, 8, 10, 12
Number of CBR	20 connections
Simulation time	600 s

- **Average throughput (Kbps)** represents the total size of data packets that successfully reached their destination over the simulation times.
- **Malicious dropping ratio** refers to the ratio between the total numbers of data packets dropped by malicious nodes to the total numbers of data packets sent.

All plotted results are obtained after averaging the result of 20 simulation runs.

5.1   Average Throughput (Kbps)

Figure 2 shows the average throughput of HAPS and EAACK under varying the number of malicious nodes. We can observe that as the number of malicious nodes increases, the average throughput of two schemes decreases. However,

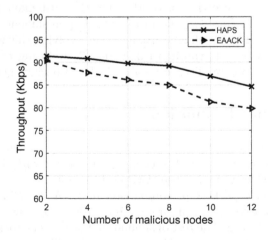

Fig. 2. Average throughput vs Number of malicious nodes

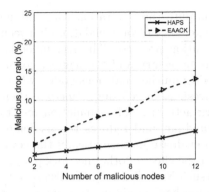
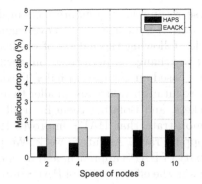

Fig. 3. Malicious dropping ratio vs Number of malicious nodes

Fig. 4. Malicious dropping ratio vs Nodes speeds

the obtained results indicate that HAPS improves the average throughput much more than EAACK. This improvement is due to the fact that HAPS can identify and isolate malicious nodes instead of malicious links. Therefore, using HAPS scheme, the established paths between each pair of nodes is more reliable.

5.2 Malicious Dropping Ratio

Figure 3 depicts the malicious dropping ratio of HAPS and EAACK as a function of the number malicious nodes. The results show that the malicious dropping ratio increases as the number of malicious nodes increases. But, the malicious dropping ratio with HAPS increases more gently than with EAACK. This is because AASC penalizes malicious nodes more effectively and severely compared EAACK that is able to isolate only malicious links.

Figure 4 plots the malicious dropping ratio across varying the node speed. In this scenario, the network contains 4 malicious nodes. From this figure, we can observe that HAPS has a lower malicious dropping ratio in all cases compared to EAACK. This gap is more apparent when the nodes move faster. Because, when the nodes move faster, their neighbourhoods change frequently. As EAACK isolates only malicious links, each new neighbour for malicious nodes forms a new opportunity (malicious link) to drop more packets. Therefore, HAPS is more resilient to topology changes.

6   Conclusion and Future Work

In this paper, we have proposed HAPS, which is a novel acknowledgment Punishment Scheme aiming to detect and punish malicious nodes dropping data packets. In HAPS scheme, the reputation values of all links in which node is involved, are perceived as recommendations. Using Dempster Shafer Theory, these recommendations are combined to compute the reputation value of the node. HAPS incorporates a novel manner to exchange recommendations between nodes following the nature of on-demand routing protocol. The recommendations exchange is performed only when it is necessary and the aggregation is done based on Dempster Shafer Theory. HAPS punishes malicious nodes, whose the reputation values are smaller than the threshold by refusing to forward their packets, and isolating them from all network activities. The simulation results demonstrate that HAPS improves the throughput and reduces the malicious dropping ratio. As future work, We plan to evaluate mathematically the complexity of HAPS approach, and by simulation the effect of other network parameters in the effectiveness of HAPS approach (such Node density). We also plan to extend the HAPS scheme with an incentive mechanism aiming to cope against the selective packet dropping attack, while to motivate the cooperation of selfish nodes.

References

1. Peng, S.C., Wang, G.J., Hu, Z.W., Chen, J.P.: Survivability modeling and analysis on 3D mobile ad-hoc networks. J. Cent. South Univ. Technol. **18**(4), 1144–1152 (2011)
2. Marti, S., Giuli, T.J., Lai, K., Baker, M.: Mitigating routing misbehavior in mobile ad hoc networks. In: Proceedings of the 6th Annual International Conference on Mobile Computing and Networking, pp. 255–265 (2000)
3. Buttyan, L., Hubaux, J.P.: Stimulating cooperation in self-organizing mobile ad hoc networks. Mob. Netw. Appl. **8**(5), 579–592 (2003)
4. Zong, C., Yang, S.: Sprite: a simple, cheat-proof, credit-based system for mobile ad-hoc networks. In: Twenty-Second Annual Joint Conference of the IEEE Computer and Communications (INFOCOM), pp. 1987–1997 (2003)
5. Bansal, S., Baker, M.: Observation-based cooperation enforcement in ad hoc networks. arXiv preprint cs/0307012 (2003)
6. Zhu, H., Lin, X., Lu, R., Fan, Y., Shen, X.: Smart: a secure multilayer credit-based incentive scheme for delay-tolerant networks. J. IEEE Trans. Veh. Technol. **58**(8), 4628–4639 (2009)
7. Mahmoud, M.E., Shen, X.: FESCIM: fair, efficient, and secure cooperation incentive mechanism for multihop cellular networks. J. IEEE Trans. Mob. Comput. **11**(5), 753–766 (2012)
8. Buchegger, S., Le Boudec, J.Y.: Performance analysis of the CONFIDANT protocol. In: Proceedings of the 3rd ACM International Symposium on Mobile Ad Hoc Networking and Computing, pp. 226–236 (2002)
9. Michiardi, P., Molva, R.: Core: a collaborative reputation mechanism to enforce node cooperation in mobile ad hoc networks. In: Jerman-Blažič, B., Klobučar, T. (eds.) Advanced Communications and Multimedia Security. IFIP AICT, vol. 100, pp. 107–121. Springer, Boston (2002). https://doi.org/10.1007/978-0-387-35612-9_9
10. He, Q., Wu, D., Khosla, P.: SORI: a secure and objective reputation-based incentive scheme for ad-hoc networks. In: Wireless Communications and Networking Conference (WCNC), pp. 825–830 (2004)

11. Yau, P.W., Mitchell, C.J.: Reputation methods for routing security for mobile ad hoc networks. In: Proceedings of the Mobile Future and Symposium on Trends in Communications, pp. 130–137 (2003)

12. Balakrishnan, K., Deng, J., Varshney, P.K.: TWOACK: preventing selfishness in mobile ad hoc networks. In: Wireless Communications and Networking, pp. 2137–2142 (2005)

13. Shakshuki, E.M., Kang, N., Sheltami, T.R.: EAACK—a secure intrusion-detection system for MANETs. IEEE Trans. Industr. Electron. **60**(3), 1089–1098 (2013)

14. Shafer, G.A.: Mathematical Theory of Evidence. Princeton University Press, Princeton (1976)

15. Sudkamp, T.: The consistency of Dempster-Shafer updating. Int. J. Approx. Reason. **7**(1), 19–44 (1992)

16. Chen, T.M., Venkataramanan, V.: Dempster-Shafer theory for intrusion detection in ad hoc networks. IEEE Internet Comput. **9**(6), 35–41 (2005)

17. Johnson, D.B., Maltz, D.A.: Dynamic source routing in ad hoc wireless networks. In: Imielinski, T., Korth, H.F. (eds.) Mobile Computing. The Kluwer International Series in Engineering and Computer Science, vol. 353, pp. 153–181. Springer, Boston (1996). https://doi.org/10.1007/978-0-585-29603-6_5

18. Liu, K., Deng, J., Varshney, P.K., Balakrishnan, K.: An acknowledgment-based approach for the detection of routing misbehavior in MANETs. IEEE Trans. Mob. Comput. **6**(5), 536–550 (2007)

19. Sheltami, T., Al-Roubaiey, A., Shakshuki, E., Mahmoud, A.: Video transmission enhancement in presence of misbehaving nodes in MANETs. Multimedia Syst. **15**(5), 273–282 (2009)

20. Sun, H.M., Chen, C.H., Ku, Y.F.: A novel acknowledgment-based approach against collude attacks in MANET. Expert Syst. Appl. **39**(9), 7968–7975 (2012)

21. Djahel, S., NaïtAbdesselam, F., Zhang, Z., Khokhar, A.: Defending against packet dropping attack in vehicular ad hoc networks. Secur. Commun. Netw. **1**(3), 245–258 (2008)

22. Campos, F., Cavalcante, S.: An extended approach for Dempster-Shafer theory. In: Information Reuse and Integration, pp. 338–344 (2003)

23. Bloch, I.: Some aspects of Dempster-Shafer evidence theory for classification of multi-modality medical images considering partial volume effect. Pattern Recogn. Lett. **17**(8), 905–919 (1996)

24. Bounouni, M., Bouallouche-Medjkoune, L.: A hybrid stimulation approach for coping against the malevolence and selfishness in mobile ad hoc network. Wirel. Pers. Commun. **88**(2), 255–281 (2016)

Efficient Broadcast of Alert Messages in VANETs

Hadjer Goumidi[1,2]([✉]), Zibouda Aliouat[1,2], and Makhlouf Aliouat[1,2]

[1] University Ferhat Abbas Setif 1, 19000 Setif, Algeria
{hadjer.goumidi,zaliouat,maliouat}@univ-setif.dz
[2] Laboratory of Networks and Distributed Systems,
University Ferhat Abbas Setif 1, 19000 Setif, Algeria

Abstract. VANET networks consist of several vehicles communicate with each other or with road fixed stations in order to offer a secure collaborative driving environment with the help of broadcasting emergency messages. These messages play a significant role in road safety by warning vehicles of any potential danger. The work reported in this paper is to devise an efficient broadcast protocol fulfilling alert message requirements in the VANETs context. Reacting in time to these urgent events may be important for crucial situations like road accidents.

Performance evaluation of our proposal EBAM is carried out through simulations with combination of VanetMobiSim and Network Simulator 2 and obtained results, in terms of loss rate, delay and reception rate of urgent messages, outperform those given by the two known referred protocols.

Keywords: Efficient broadcast · Emergency message
Dependable delivery · VANETs

1 Introduction

The technological development that has seen the world of today affects all areas, particularly the communication area which is experiencing a considerable evolution by the advances of wireless technology which has partially solved the problem of road accidents. Our work is coming in this context. In order to solve the problem of vehicle crashes, researchers think to exploit this technology enabling vehicles to establish direct (or indirect) links between them for being aware of potential critical situations like imminent accident or dangerous obstacles. This enabling technology is represented by VANET (Vehicular Ad hoc NETwork). Indeed, one of the promising applications of these networks is to allow vehicles equipped with specific sensors to detect the immediate environment, and alert drivers of vehicles around early in case of accident risks.

VANET as comfort communication can be made by two means: Periodic Safety Message (Referred in the sequel as Beacon) and Event Driven Message

© IFIP International Federation for Information Processing 2018
Published by Springer International Publishing AG 2018. All Rights Reserved
A. Amine et al. (Eds.): CIIA 2018, IFIP AICT 522, pp. 448–459, 2018.
https://doi.org/10.1007/978-3-319-89743-1_39

(referred as Emergency Message). Both message types share only one control channel. The Beacon messages are messages about status of sender vehicle. Status information includes position, speed, heading towards, etc ... about the sender. Beacons provide resent or latest information of the sender vehicle to all present vehicles in the network, which help to know the position in the current network and anticipate the movement of vehicles.

Beacon messages are sent antagonistically to neighboring vehicles every 10 s. Emergency Messages are messages sent by a vehicle that detects a potential dangerous situation on the road; this information should be dispersed as alarm or alert to other vehicles pointing out a prospective danger that could affect the incoming vehicles. VANET is a high mobile or volatile network where nodes are kept changing their positions, moving in speed, which means that these vehicles may be get influence, even if they are very far from the danger. They can reach a danger in short time, so a fraction of second should be very important to avoid a danger [5]. Because of the importance of these messages exchanged between vehicles, it is necessary to ensure that these messages attain the largest number of vehicles in a minimum delay. These emergency messages are delivered in broadcasting way, where all the vehicles within the coverage area of the sender should receive the messages. The communication range of sender is not enough, due to the fading effects and attenuation; it hardly reaches a 1000 m (which is the DSRC communication range). Critical information should be received by vehicles, in time, far from the danger for avoiding risks. The prospect of message reception can reach 99% in short distances and can be decreased up to 20% at half of the communication range [7]. Therefore, it is necessary to provide a technique for raising the emergency message reception with high reliability and availability. The dissemination of the emergency messages poses several problems such as collisions, the problem of broadcast storm, problem of the hidden node etc. Several works such as PCBB, CBB, and EMDV... have been proposed to solve the problem of broadcasting emergency messages in VANETs.

2   Related Works

In [2], authors proposed a Dynamic Search-Assisted Broadcast (DSAB) protocol to efficient broadcast of emergency messages in VANETs. It based on controlling dynamically the transmission power to estimate the vehicle density and the use of n-way search mechanism to find the farthest vehicle in VANETs. The problem of this protocol is the choose of one vehicle which is the farthest one to rebroadcast the message because in VANET, vehicles always change their position and the farthest vehicle may do not receive the message, it may become out of range transmission when sending the message or it cannot receive the message because of the channel problems. A Distribution-Adaptive Distance with Channel Quality (DADCQ) protocol for efficient multi-hop broadcast of VANET is presented in [6]. It combines local spatial distribution information and other factors with the distance method heuristic to select rebroadcasting nodes. Authors in [4] presented new distance-aware safety-related message broadcasting algorithm to enhance the propagation distance in the latency time of safety related message broadcasting. It generates the lenghts of backoff times by calculating the distances between the source node and its forwarding node. The farthest forwarding vehicle has the highest probability to forward messages. In [9], authors proposed an algorithm for data dissemination mechanism for motorway environment in VANETs. This protocol is based on an improved K-nearest neighbors prediction to predict the status of the nodes, and on construction of a forwarder set to avoid the broadcast storm problem. Limited Area-Based (LAB) scheme [3] adjusts the size of an area that contains the broadcasting nodes for obtaining a faster propagation time. Authors in [8] presented a counting-based broadcast model of emergency message dissemination in VANETs which is based on counting all possible cases of contention window assignments to all the vehicles simultaneously receiving a broadcast message. In [5], authors proposed Pso Contention Based Broadcast (PCBB) protocol that allows vehicles to increase the reception percentage of emergency information. In this protocol the network must be divided into several segments to help the sender to determine the next forwarder of the emergency message which is the furthest vehicle in the last non-empty segment.

The MIN is the lower bound of the last non-empty segment it calculated using the fitness function [5].

This protocol has a serious problem because the MIN calculation by fitness function is not always correct. In some cases, the MIN is greater than the max, this algorithm can be applied only in dense areas where the number of vehicles is large and the distance between them is small.

In [1], the authors proposed that the reception probability of the broadcast message instantly decreases at a distance greater than 66% of the transmission range, then they choose the forwarder as the node that has a distance equal to 66% of the transmission range.

To resolve the problem of the Protocol PCBB, we tried to apply this algorithm which chooses the node that has a distance equal to 66% of the transmission range of the transmitter as forwarder

When we choose the node 66%, not the max (the farthest), the message takes a great reception time than that of the max. This algorithm does not use segmentation, so the forwarder has not potential nodes, and if the message fails to arrive to the forwarder it will be lost.

3   The Proposed Protocol

As a first step in our approach, we will present the EBAM (Efficient Broadcast of Alert Messages) protocol that ensures safe transmission of alert messages. From the disadvantages noted during the study of the protocols proposed in the specialized literature, we were able to identify an approach to avoid some problems detected in these protocols. The approach also assured high reliability avoiding collisions, decreasing the channel load with minimal cost and low response time. EBAM protocol can be used in all areas (dense and sparse networks) contrary to PCBB algorithm which can be applied only in dense areas. (Table 1)

Table 1. Value notations table

Notation	Value
Fij	The forwarder of node (i)
Vi	The neighbors of node (i)
$Dist_{a,b}$	The distance between node a and b
Bmin(i)	The lower bound (min) of node (i)
Del	The delay of message defined by the sender
D	The delay of message
m	Message
T(i)	Emergency message table of node (i)
Pi	Potential node for node (i)
CT	Contention time
tt	Transmission time
t1	The duration of message transmission to the forwarder
t2	The duration of ACK transmission from potential node to forwarder
LR	Loss rate
Msgs	Messages sent
Msgr	Messages received
MsgL	Messages lost

The main idea of our protocol is to transmit the message to the farthest node within the transmission range. This node is defined as a forwarder. If the forwarder does not receive the message we resort to using potential nodes, such as the min of these nodes represents 60% of the farthest node distance.

We choose 60% as a lower bound because in [1] authors suppose that the distance 66% is the most appropriate to increase the reception rate.

Our approach has four steps:

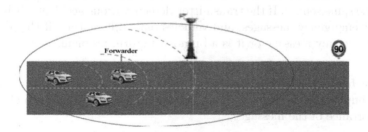

Fig. 1. Case of RSU further than forwarder.

3.1 Preparing to Send

Each node n_i knows the position of its neighbor nodes because of periodic security messages that are transmitted every second (beacon), allowing to calculate the distance between nodes to select the forwarder.

The selection of forwarder inspired from the PCBB protocol, which is the furthest vehicle in the last non-empty segment by using the formula (1):

$$F_{i,j} = j/\max(Dist_{i,k}); k \in V_i \tag{1}$$

The RSU sends its location to all nodes in its transmission range periodically, and each vehicle can calculate the distance between it and the RSU, for deciding which will receive the emergency message (forwarder or the RSU) by using the formula (2):

$$Dist_{i,j} = \min(Dist_{i,RSU}, Dist_{i,F_{i,j}}) \tag{2}$$

In sparse area, we can found the case when the RSU is within transmission range of node (i) and it is farther than the forwarder as explain in example of Fig. 1. So, each vehicle must check if the RSU is within its transmission range, it will receive the message, else it sends the message to forwarder.

We calculate the min by using formula (3):

$$Bmin_i = Dist_{i,F_{i,j}} \times 60\% \tag{3}$$

When a vehicle detects an emergency (e.g. accident), it creates an alert message *ADV-ALERT*, the format of this message is represented in Table 2. The emergency message contains:

Source Identifier. To avoid the problem of broadcast storm, we use this identifier which prevents the sender to retransmit the same message.

Message Identifier. We use this identifier to differentiate messages transmitted by the same sender. If the transmitter detects various accidents, it must send a various emergency messages, and vehicles must differentiate if this message is a new emergency message or it is a broadcast storm problem.

Priority Identifier. This identifier is used to classify the messages according to their importance in order to make it easier for the receiving vehicle to know the importance of the messages.

Forwarder ID. We select the forwarder as the last vehicle in the last non-empty segment to avoid collisions and to transmit the message very quickly.

Transmission Time and Delay. The emergency message requires a short time, and the transmission of the emergency message after this time occupies the channel and loads the network without interest. The sender defines its message transmission time and the delay (Del) of this message.

Position. The absence of characterization of the emergency event causes network load problem and can cause collisions leading to message loss and delay in reception. To solve this problem, the vehicle that detects the event sends the position of that event using GPS.

Min. It is necessary to send the min in the message, for each vehicle that receives this message to know if it is a potential node or not.

Data. The emergency message must contain important data such as:

- If there are victims or wounded persons.
- The nature of the urgent case (accident, avalanche...).

Table 2. Format of the emergency message.

SourceID	MessageID	PriorityID	ForwarderID	Transmission time	Delay	Position	Min	Data

If a vehicle detects an event, it checks whether the same location exists in the emergency message table, if so, it does not creates the message, otherwise it creates the message and it adds the location of this event.

The use of the event position can help the emergency vehicles (ambulances) to intervene more quickly.

3.2   The Sending Step

After the creation of the emergency message *ADV-ALERT* and the selection of the forwarder, the sender broadcasts the message in the network to alert other vehicles about any potential danger.

3.3   Receiving Step

Each vehicle existing in the transmission range of the sender that receives the message *ADV-ALERT*, checks if it is a potential node by using formula (4):

$$S = Dist_{i,j} - Bmin_j; \begin{cases} j \in P_i, & S \geqslant 0 \\ j \in V_i, & S < 0 \end{cases} \tag{4}$$

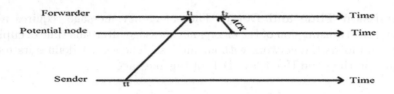

Fig. 2. Contention time illustration.

The vehicles running in the last segment (potential nodes) initiate a contention time for waiting the forwarder to transmit a reply by using the formula (5) as explained in the example of Fig. 2.

$$CT = tt + t_1 + t_2 + \varepsilon \tag{5}$$

Then they detect the channel if no reply has been transmitted from the forwarder or other vehicles. It means that the forwarder has not received the message, so the first vehicle from the potential vehicles set that transmits the reply is defined as a forwarder. Once the reply is transmitted, the other vehicles stop the transmission to avoid collisions. Figure 3 depicts the receiving step.

Before sending the reply, the forwarder must check if it saved the *ADV-ALERT* in its emergency message table for sending the reply and for transmitting the *ADV-ALERT* to its forwarder. As shown in Fig. 4, it must check:

Priority ID: The forwarder checks the importance of the message if it is a big number, it does not save it, if none it checks the other indices (if they are acceptable) it saves the message in the table according to its priority level.

Transmission Time and Delay: In general, the alert message is urgent and it does not take much time. Each vehicle receiving this message must calculate Dm using formula (6). To decide saving the message or not, in its emergency message table for the retransmission, the node uses the formula (7):

$$D_m = Tr - Te \tag{6}$$

$$h = Del - D_m, \begin{cases} m \in T[i], & h \geqslant 0 \\ m \notin T[i], & h < 0 \end{cases} \tag{7}$$

After each 30 s, each vehicle must update its table, the messages that exceed the delay will be deleted.

Source ID and Message ID are used to check if the message is transmitted before or not. If the forwarder found that the *ADV-ALERT* message has the same Source ID of a message stored in the emergency message table, it should check its message ID: if it found the same message ID then it does not save the message, otherwise the message will be saved.

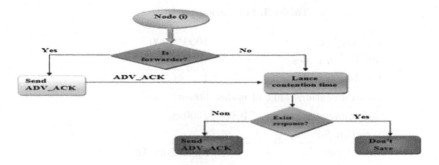

Fig. 3. Alert message reception flowchart.

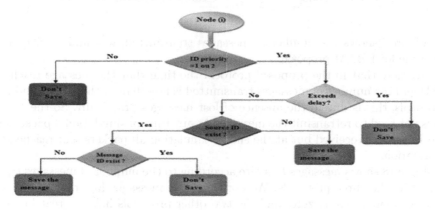

Fig. 4. Flowchart for checking message saving.

3.4 Rebroadcasting Steps

After saving the *ADV-ALERT* in emergency message table, The forwarder must broadcast it in its transmission range to its forwarder and follow the same steps defined previously.

4 Simulation

4.1 Simulation Parameters

The simulation parameters are shown in Table 3.

Figure 5 shows the number of messages transmitted, lost and received over the time for the PCBB protocol. We note that only one message is received the rest of messages are lost (the Min is greater than the max) until the end of the simulation.

Figure 6 shows the number of messages transmitted, lost and received over the time for the 66% protocol. It is clear that the number of lost messages is great, only 8 messages among 307 sent messages are received and the rest is lost. This loss is due to the absence of potential nodes.

Table 3. Simulation parameters.

Topology (m^2)	1000 * 1000
Number of nodes	100
Number of RSUs	1
Communication range of nodes	300 m
Communication range of RSU	1000 m
Speed (m/s)	6
MAC layer	MAC 802_ 11
Simulation time (s)	300 s
Mobility model	Random following IDM_ LC

Figure 7 shows the number of messages transmitted, lost and received over the time for EBAM proposal.

We note that in the proposed protocol the time that the message reach the RSU and the number of messages transmitted is less than in the PCBB and 66% protocols, this is due to the absence of lost messages practically. So the vehicle does not need to retransmit the message. At any time of simulation a packet sent can be not yet received but at the end of simulation all packets sent reached the destination.

Figure 8 shows messages loss rate according to the number of messages transmitted in the three protocols. We note that the messages loss rate in our protocol is approximately zero, and the two other protocols have a great loss rate

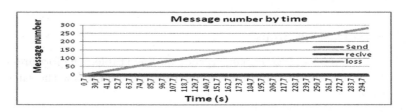

Fig. 5. Number of messages transmitted, received and lost in the PCBB protocol.

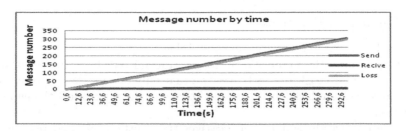

Fig. 6. Number of messages transmitted, received and lost in the 66% protocol

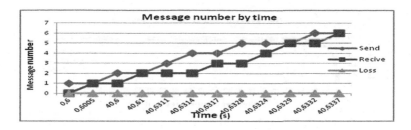

Fig. 7. Number of messages transmitted, received and lost in EBAM protocol.

compared to the EBAM, but the loss rate of the PCBB protocol is greater than in the 66% protocol.

Figure 9 shows the reception delay of alert messages according to the three protocols. We calculate the reception delay of alert message by using the expression (7). From Fig. 9, we note that in EBAM, a message is received after 40.63 s from its transmission, but in the other two protocols the message does not arrive at the RSU (the simulation time is exhausted and the message did not achieve its destination). The loss rate of the three protocols is calculated by (8):

$$Msg_L = \sum Msg_s - \sum Msg_r; \qquad LR = \frac{Msg_L}{\sum Msg_s} \qquad (8)$$

4.2   Evaluation of Protocol Performance

For evaluating our protocol performance, we tested various scenarios by changing the parameters.

Figure 10 represents the reception rate of the alert messages compared with the number of nodes according to the number of RSUs. In this figure we change the number of nodes and RSUs and the topology; we use, 50, 100, 200, 350 and 500 nodes, with 0, 1 or 2 RSUs in topology of 2000 * 2000 (m²).

We remark that the reception rate is almost 20% when nodes broadcast the message between them. When one RSU broadcasts it, the reception rate is increased up to 80%, and in the case where two RSUs broadcast the message, the reception rate reach approximately 100%(the RSUs almost cover all the Topology).

Figure 11 represents the simulation result of reception delay in second compared with the number of nodes. In this scenario we change only the number of nodes we use, 50, 100, 200, 350 and 500 nodes.

The reception delay is decreased with the increase of the number of nodes. This is due to the number of vehicles that transmit the message. When the number of vehicles becomes important, the message takes less reception time.

Figure 12 represent the simulation result of reception delay in second related to the changing in speed. In this figure we change the speed parametre; we test speed of 6, 12, 20, 40 and 100 m/s with fixed number of nodes (100 nodes).

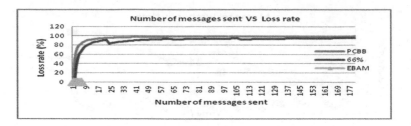

Fig. 8. Messages loss in the three protocols

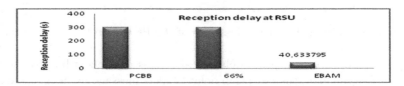

Fig. 9. Reception delay according to three protocols.

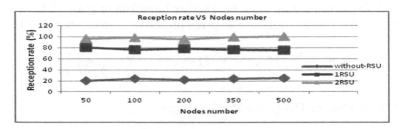

Fig. 10. Reception rate with 1 RSU, 2 RSU and without RSU.

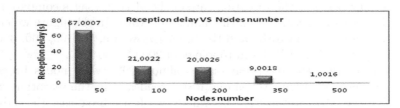

Fig. 11. Reception delay according to number of nodes.

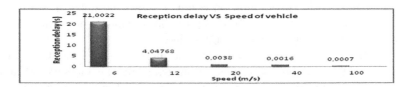

Fig. 12. Reception delay vs speed.

The reception delay decreased up with the increase of speed due to the speeds of the vehicles transmitting the message. When the speed of the vehicles is important the message takes less time to be received.

5 Conclusion

In this paper, we studied the problem of efficient emergency message transmission in VANETs in order to propose a solution respecting dependable communications and specific time constraints of VANET environments. The EBAM proposed protocol showed an appreciable improvement of successfully delivered urgent messages in accordance with the real time constraint. This enhanced result obtained even at high network density and high node velocities is of great importance in order to well-timed reacting to crucial events (Like road accidents) by rescue services.

As a future work, it is desirable to increase the improvement of the proposal, by considering the vehicle directions using GPS.

References

1. Batish, S., Kakria, A.: Efficient broadcasting protocol for vehicular ad-hoc network. Int. J. Comput. Appl. **49**(23), 38–41 (2012)
2. Chen, Y.D., Shih, Y.P., Shih, K.P.: An emergency message dissemination protocol using n-way search with power control for vanets. In: 2015 IEEE International Conference on Communications (ICC), pp. 3653–3658. IEEE (2015)
3. Kang, M., Virdaus, I.K., Shin, S., Lee, C.G.: Limited area broadcast for warning message delivery over vehicular ad-hoc networks. EURASIP J. Wirel. Commun. Network. **1**, 1–16 (2016)
4. Li, X., Hu, B., Chen, H., Ye, J.: A distance-aware safety-related message broadcasting algorithm for vehicular networks. Int. J. Distrib. Sens. Netw. **10**(2), 1–11 (2014). Article No. 139857
5. Samara, G., Alhmiedat, T.: Intelligent emergency message broadcasting in VANET using PSO. World Comput. Sci. Inf. Technol. J. (WCSIT) **4**(7), 90–100 (2014). ISSN: 2221-0741
6. Slavik, M., Mahgoub, I.: Spatial distribution and channel quality adaptive protocol for multihop wireless broadcast routing in VANET. IEEE Trans. Mob. Comput. **12**(4), 722–734 (2013)
7. Torrent-Moreno, M., Jiang, D., Hartenstein, H.: Broadcast reception rates and effects of priority access in 802.11-based vehicular ad-hoc networks. In: Proceedings of the 1st ACM International Workshop on Vehicular Ad Hoc Networks, Philadelphia, pp. 10–18. ACM, October 2004
8. Virdaus, I.K., Kang, M., Shin, S., Lee, C.G., Pyim, J.Y.: A counting-based broadcast model of emergency message dissemination in VANETs. In: Ninth International Conference on Ubiquitous and Future Networks (ICUFN), pp. 927–930. IEEE (2017)
9. Yang, Y., Liu, Q., Gao, Z., Qiu, X., Rui, L., Li, X.: A data dissemination mechanism for motorway environment in VANETs. EURASIP J. Wirel. Commun. Netw. **93**(1), 1–11 (2015)

Improved Sensed Data Dependability and Integrity in Wireless Sensor Networks

Zibouda Aliouat and Makhlouf Aliouat[(✉)]

Networks and Distributed Systems Laboratory, Computer Science Department,
Faculty of Sciences, University Ferhat Abbas Sétif 1, Sétif, Algeria
{zaliouat,maliouat}@univ-setif.dz

Abstract. The mainspring of a Wireless Sensor Network (WSN) is to capture data from an interesting deployment area and sending them out to the end user. However, one can ask to what extent the end user can have confidence in the use of these data? Especially when these collected data are employed in a crucial application that definitely excludes wrong data use. Thereby, WSN mission success is basically dependent on trustworthy of the data delivery process to the end user. To reach this goal, some obstacles, related to malicious node behavior or failure nodes, must be avoided or tolerated. Therefore, we propose, in this paper, a scheme to improve the dependability of the data sensed by sensor nodes in one hand and the reliable communication of these data to the sink in other hand. The proposal is based on a fault tolerant sensing process and the resilience to malware threat on the transmitted data from nodes to sink. The proposal was integrated to the well know Leach protocol and the performance evaluation, carried out on NS2 simulator, showed convincing results in terms of energy conserving, received data rate and node failure occurrence and attack detections.

Keywords: WSN · Dependable data sensing · Trusted data aggregation
Secure data transmission

1 Introduction

Dependable data sensing is of paramount importance in WSNs. Indeed, what should be the consequences when wrong data are used, particularly in deployed critical applications? To make suitable decision from application outcomes, the end users have to rely on consistent information gathered by sensor nodes from the area of interest. Otherwise, not only the application outcomes should be erroneous, but they could lead to a disaster if they should be used to generate a final critical decision.

Data should be correct since the time they are captured until the time they should be used by the WSN controller. To this end, data must be free from alterations purposely induced by an adversary or due to a sensor failed component or resulting from the unpredictable wireless environment.

Prospective wrong data received by an end user, after data aggregation process [1], may have been corrupted from diverse ways such as: a malfunctioning of node sensing unit, any impairment during transmission or alteration of data integrity caused by malware or intruder at any node acting as router located along the routing path. This is

A. Amine et al. (Eds.): CIIA 2018, IFIP AICT 522, pp. 460–471, 2018.
https://doi.org/10.1007/978-3-319-89743-1_40

for what is devoted the work reported in this paper; more precisely, coping with the tricky issue of node fault tolerance in case of unit sensing failure and security issue related to data integrity in case of network attack by intruders. The two problems have been treated separately in the specialized literature; they should be conjointly considered in order to reach end user confidence requirements in consuming collected data. We point out that node failures concern any flaws in correct data forwarding but more precisely incorrect data sensing generated by lack of energy or fault occurrence in sensing unit.

In the sequel, the paper is organized as follows: in Sect. 2, we give a related work dealt with the data aggregation issue; Sect. 3 presents our proposals for dependable data sensing and aggregation and performance evaluation of our proposals. We conclude the paper by a conclusion and future work.

2   Related Works

Several approaches have been proposed for reliably aggregating data in WSNs or securing them. These approaches coped only with routing problems like adversary attacks or node malfunctioning to forward data they have in charge.

De Cristofaro et al. [2] have proposed FAIR (Fuzzy-based Aggregation providing In-network Resilience for real-time Wireless Sensor Networks) an algorithm for robust data aggregation in real time. Witness nodes are used to confirm the outcomes of the aggregation process. The protocol seems to be robust but is only suitable for small WSNs and suffers from security lack and time overhead.

Wang et al. [3] have proposed EESSDA (Energy-Efficient and Scalable Aggregation Secure Data) a protocol using secure channel data encryption scheme. EESSDA uses several steps: creation of an aggregation tree rooted by the sink, creation of a secure channel between children and parents sharing a common key. Any node waits a certain time to receive data from its children, aggregates them with its own data and sends the new result to its parent. The protocol is costly and suffers from lack of security, because a compromised node near BS may jeopardize the aggregates confidentiality.

Shivakumar et al. [4] proposed ERDRA (Efficient and Reliable Data Routing for In-Network Aggregation in Wireless Sensor Networks) a protocol building a routing tree with the shortest path connecting all nodes for reliable data aggregation. The tree is used by coordinators, elected among nodes having detected a new event, to collect data and sending them along a reliable new path. Weaknesses: Compromising a channel involves to intercept all messages, time overhead.

In [5] Jose et al. have proposed a data aggregation protocol, ensuring confidentiality, authentication and freshness, suitable for the critical time applications. The message authentication is obtained via a key pair and a secret identification of each node. The protocol uses a tree aggregation including terminal nodes, parent nodes and sink. The protocol suffers from possible internal attacks, CPU overload and an unsafe key management mechanism.

Zhu et al. [6] proposed ECIPAP (Efficient Confidentiality and Integrity Preserving Aggregation Protocol), an effective protocol ensuring confidentiality and aggregated data integrity. A node getting sensed data sends them to its parent with a Mac. Each parent node aggregates the received value with its own and sends the result to the upper level until reaching the sink. The latter decrypts the received message and broadcasts the aggregated data allowing each node to verify if its own data have been added. Protocol weaknesses: Possible internal attacks and time overhead.

Lathamanju et al. [7] proposed a secure data aggregation algorithm improving network life time and ensuring safety and security via Diffie-Hellman algorithm. Nodes are classified as friends or malicious. When a node wants to send data, it initiates a request message to route. Each intermediate node forwards the request if the sender is not malicious. When the destination node receives the route request, it sends an ACK response and its public key. The source node evaluates the best route with the largest number of friend nodes to send data. Protocol Weaknesses: Time overhead and resources waste.

BabuKaruppiah et al. [8] have proposed NADSPSD (A Novel Approach to Detect the Shortest Path for Secure Data Aggregation using Fuzzy Logic in WSNs) an effective technique to detect the shortest path using fuzzy logic to secure the data aggregation. It is based on trust and residual energy of a node. The selection of the best route to send aggregated data is based on the combination of the path length, the available energy level and the node reputation. Protocol weakness: Cost and lack of security.

In [9], Jia et al. have proposed MDRN (Minimum Distance Redundant Nodes) a fault recovery protocol which is deployed on the receiver node that has knowledge of node locations and failed ones. By choosing an appropriate number of redundant nodes, the algorithm will provide an accurate recovery. The proposal is feasible and effective to deal with the coverage hole issue caused by failed node. The protocol is unable to deal with multiple faults and requires a significant nodes redundancy.

3   Proposed Protocols

WSNs are typically deployed in harsh unattended environment. The inherent node vulnerabilities may lead to an unauthorized modification of sensed data. To overcome this problem, we present a solution offering safe data aggregation and sensing fault tolerance capability.

3.1   Description of the First Proposal

The different proposals work with the following hypothesis: - A WSN with N sensor nodes organized in clusters with a Cluster Head (CH) as manager and aggregator for each cluster. Each node i has a unique identifier Idi. The Base Station (BS) is assumed to be robust and reliable, with inexhaustible resources. - A compromised node may send corrupted data to the BS.

Notations: Emin: minimal energy amount, Round: period for changing aggregators. Frame: temporal interval for data sensing, Vmax: maximal value, R: node communication range.

The first proposal named LEACH-FD (Low Energy Adaptive Clustering Hierarchy Fault Discovering) is a version improving the well-known LEACH Protocol [10] by including a filtering to eliminate incorrect data before their aggregation.

Proposed Protocol Steps: The algorithm takes place in rounds having approximately the same interval of time determined in advance. Each round consists of an initialization phase and a transmission phase. The general proposed algorithm is the following:

- Selecting of ClusterHeads (CHs) as done in LEACH protocol.
- Request dissemination by CHs to all nodes.
- A node may join a CH by sending it a join request.
- Formation of TDMA-schedule by CHs and Phase of data transmission to the CHs.
- Filtering received data and aggregation and sending them to the base station.

The different phases are described hereafter:

Initialization Phase: This leads to the clusters formation by electing a Cluster-Head (CH) for each cluster and establishing nodes channel access strategy within each cluster. This phase begins by local decision making to become cluster-head. Each node Ni chooses a random number rn, if $rn \leq t$, the node Ni becomes cluster-Head as in [9]. When elected, a CH must inform its neighboring nodes of its new rank. The cluster member nodes managed by a CH are those nodes having joined CH according to the signal strength of the rank notification message sent out by CH. Each member node has to inform a CH of its decision to be membership. After that, the communications within a cluster can be made according to the TDMA communication protocol. For this, each CH establishes a TDMA schedule for its members.

Transmission Phase: Data transmission operation is repeated at each frame, where nodes send data to their respective CH once a frame during their own allocated slots. Outside their slots, nodes get into sleeping mode for conserving their energy [11].

Aggregation Step: We added an important preliminary step to the aggregation operation, in which, after receiving data sensed by all nodes, each aggregator has to locally perform a filtering of received data to eliminate potential erroneous sensed data before the final aggregation process. A suitable detection algorithm of outliers should detect most of the errors and the number of false positive must be as small as possible. It uses the median, which is classified statistically among the robust features for detecting outliers [12].

Simulation Results: For the sake of limited space, the details of simulation are given only for the third proposal. Therefore, the conducted simulations showed that the proposal works only if each cluster includes more than two sensors nodes in order to be able to compare the values sent out from a cluster. Ignoring the incorrect values is not enough, but we must determine the causes in order to avoid using them in the future. When a node Ni sends an abnormal value compared to other nodes Nj, $j \neq i$, this does not necessarily imply that Ni is failed. We can find a case where an event is triggered at a node level but other nodes have not yet discovered it. Therefore, we must add a method to ensure that the value sent is correct without waiting for the next frame since in the critical systems, delays in event discovery are crucial.

3.2 Description of the Second Proposal

We present a new protocol FDP (Fault Discovering Protocol), that uses neither the centralized approach nor the only distributed one but it combines the two. That is, the aggregators selection is done by the BS to ensure a global view of the network and a good energy management, but each node chooses its aggregator alone to reduce the load on the BS. This proposal is trust based where nodes history is used to decide if a node is faulty or not. The BS maintains a table of confidence in which each node has a certain degree of confidence. This value may change if sensed data by a node are not correct. It always uses sensed data filtering at the aggregators' level, but it adds a procedure to determine if a sensor is really faulty.

Proposed Algorithm: This algorithm is also carried out in rounds. Each round consists of an initialization phase, transmission phase and verification phase see Fig. 1.

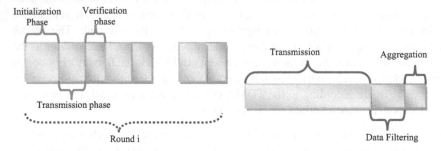

Fig. 1. 2nd proposal phases.   **Fig. 2.** Transmission phase.

Initialization Phase: This phase is composed of 3 sub-phases: aggregators selection, clusters formation and scheduling. At the beginning of each round, each node checks its residual energy if it is greater than a threshold value *Emin,* then it sends to BS a message containing its location X, Y and its energy level E.

- Aggregators Selection: In this phase, BS will retrieve all data sent by nodes and then will choose nodes with maximum value of confidence as aggregators. If there are two nodes with same value, the node with maximum energy will be chosen. Also, in case of equality, the node with the smallest index is selected. After that, BS will broadcast a message including the CH identities and their coordinates.
- Clusters formation: Upon reception of the information message sent by the BS, if the receiver node is a CH, it will do nothing, otherwise (standard node) it chooses from the list of the aggregator nodes the node closest to it as its leader. The node must inform its leader by sending it a JOIN message (belonging) containing the identity of the member node.

An Aggregator receiving JOIN message adds the sender to a list containing all member nodes. After that, any aggregator builds a TDMA-schedule and sends it to its members. Each one will know as well, its position in this TDMA-schedule and the data transmission time slot to its aggregator. The TDMA-schedule allows avoiding collisions between nodes which minimizes nodes energy consumption. This energy savings is also enhanced by node going to sleep out of its own time slot.

Transmission Phase: This phase also is decomposed in frames whose size is larger than that of the frame of the first proposal, because each frame contains 3 sub-phases: transmission, data filtering and accepted data aggregation (Fig. 2).

- Data Transmission: It is decomposed into several time slots, a slot in a frame is dedicated to a single node during which it transmits its data to the aggregator. Each node sends a message containing the sensed value and its residual energy.
- Aggregation Phase: When the aggregator receives data from a node Ni, it checks if the energy of Ni is less than $Emin$, if so, it will ignore the received value otherwise it will store it in a list. At the end of each frame, the aggregator will count the number of values in its list, if this number is greater than 3 then it will filter the data using the procedure used in the first proposal but this time when it finds an abnormal value it will send an alert message to the BS containing the node identity that has captured the incorrect value.

Verification Phase: It contains two sub-phases: data verification, data confirmation.

- Data Verification: This sub phase is carried out when the BS receives an alert message containing the ID of a node; it will send a request for that node to re-capture data in its proximity according to radius R.
- Confirmation phase: When a node receives a confirmation message, it will sense again and return the value captured for the second time. When the BS receives a confirmation message, it will compare the data received and if there is a difference between the data of the nodes, so the node is considered faulty. BS will decrement trust value of a node having sent incorrect values, if this value reaches 3 then that node is declared faulty. These operations are repeated until the round end.

The initialization phase is repeated at each round up to the simulation end caused either by the exhaustion of the simulation time set to 3600 s or by the condition "the number of remaining nodes in WSN is less than the number of aggregators" but the confirmation phase is made only in case of a problem.

Simulation Results: The simulation results show that we cannot surely conclude if the sensor node that sent incorrect values is failed or compromised. Also the FDP protocol does not cover the case of failed aggregator.

3.3 Description of the Third Proposal

This proposal FDAP (Fault Discovering and Attack Protocol) is to enhance the second proposal to which is added an encryption mechanism for node authentication.

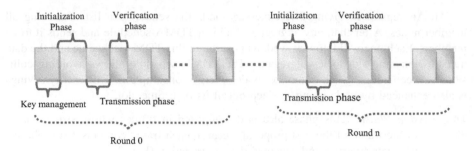

Fig. 3. Phases of the 3rd proposition.

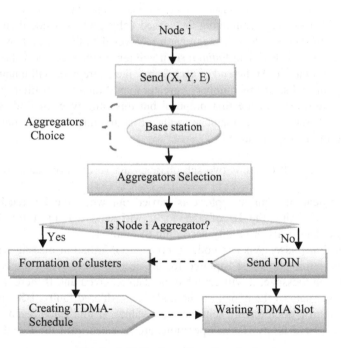

Fig. 4. Initialization phase flowchart.

Proposed Algorithm: Each round consists of an initialization phase (Fig. 4), a transmission phase (Fig. 5) and a confirmation phase (Fig. 6); this is before the phase of key management (Fig. 3). The main phases of the third proposal are as follows:

Key Management and Authentication Nodes: Key management is carried out a time the deployment is achieved and before nodes are operating as sensors. So, BS assigns to each node Ni a unique identifier IDi and a symmetrical key that BS shares with the node Ni and a large integer value M is also preset, where key > M. The other phases remain the same until the confirmation phase of detecting erroneous data. The verification phase remains the same except that it is added a technique of cryptography to ensure the nodes authentication:

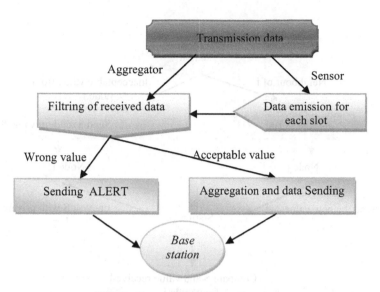

Fig. 5. Transmission phase flowchart.

- *Detection of compromised node:* When BS receives an alert message containing the ID of a node; it will send it a message containing a random value S asking it to encrypt the value using its private key. The BS will also send to the neighbors of this node a confirmation message in the same way as in the previous proposal. A node *Nj* which receives a confirmation message of its identity, uses the following expression to encrypt the received value: C = ENC (value received) = value received + key(Nj) MOD M. Then it sends a confirmation message containing the value C. When the base station receives this message, it will decrypt it using this expression: d = Dec(C) = C key(Nj) MOD M. Finally BS compares D and S if they are equal then the authentication is ensured otherwise the node Nj has been compromised.
- *Detection of faulty aggregator:* The base station can receive final erroneous aggregate values due to a faulty aggregator, so we add a procedure to cover this case. The BS will check the values received from the aggregators if it finds a value superior to *Vmax*, it will trigger the verification procedure. But now, it does not send to the nodes in the vicinity of the aggregator but to all members of its cluster. BS will then compare the new received values with those of the aggregator, if there is a large difference, BS will decrement confidence value of the aggregator. BS will then play the role of the aggregator during the remaining time of the current round.

3.4   Proposal Performance Evaluation

The performance evaluation of our final proposal is carried out through the well known simulator NS2 according to the network parameters summarized in the Table 1 hereafter. The metrics of performance evaluation used are detected anomalies, life time duration, energy consumed and the data amount received by the base station. The three

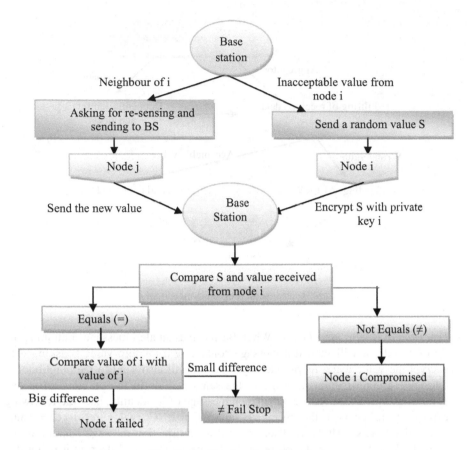

Fig. 6. Verification phase flowchart

Table 1. Parameters used in the simulation process

Parameters	Values	Parameters	Values
Size of the network	1000 m * 1000 m	Round time duration	30 s
Base station position	50, 175	Initial energy amount	2 J
Clusters number	5	Initial trust value	6
Sensor nodes number	100	Frame	3 s
Simulation time	3600 s	Round	30 s
		Radius R	10 m

algorithms are used to simulate the same scenario designed to sense the temperature in a forest and monitor fires (The scenario contains a failed node, an intruder and a fire). Table 2 shows the anomalies recorded by sensor nodes according to the three proposals. As a result, the first proposal detects no anomalies while the 2nd proposal detects only failures and fires but the 3rd proposal detects all the assumed anomalies (failures, fires and attacks).

Table 2. Anomalies detected during simulation operation

Proposals	Failure	Attack	Fire
1st Proposal	No	No	No
2nd Proposal	Yes	No	Yes
3rd Proposal	Yes	Yes	Yes

Comparison Between Different Proposal Versions: Figure 7 showed the energy consumed as a function of simulation time in different proposals. We note a great improvement obtained by FDP and FDAP regarding to Leach-FD version. Therefore, the strategy of aggregator's selection we proposed gives lower energy consumption which is vital for WSNs to last longer until achieving their missions. Also, we note that the use of cryptography in the third protocol does not consume a lot of energy. Figure 8 shows the network life time duration ensured par different proposals. It represents the number of survival nodes during the entire simulation. From the curves, we note that FDP and FDAP allow ensuring a survival of all nodes until point of time 320 and 350, and maintain a functionality of all nodes during this period and prolong the life time of the WSN, while the life time duration of nodes in improved leach-FD is limited to time 200 only. We justify this difference by the new technique of selecting the aggregators used by FDP and FDAP.

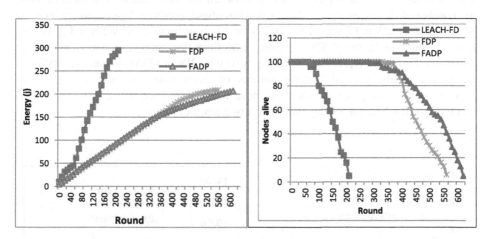

Fig. 7. Energy consumed by nodes.   **Fig. 8.** Alive nodes number function of time.

Figure 9 represents the amount of data received by BS, which is expressed as a function of the number of messages sent by nodes to BS. The curves show that FDP and FDAP, ensures data transmission up to time 600 with a number of messages equals 11800. On the other hand, the transmission of data in leach-FD stops at time 210 with almost 17500. Therefore, no remarkable improvement regarding the amount of data received by BS is noted. The cause of data reduction is the elimination of data of the faulty sensors and the nodes that have a minimum energy because we are interested in the quality of the data more than the quantity.

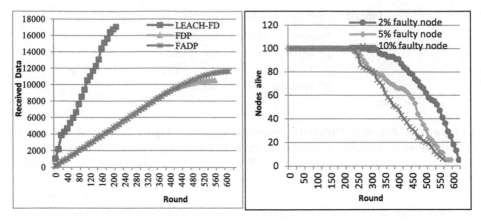

Fig. 9. Data amount received by BS.   **Fig. 10.** Energy consumed in FDAP.

Proposals Behavior in Presence of Faults: The results obtained after variation of the number of failed nodes are illustrated by the Figs. 10, 11 and 12 with 2%, 5% and 10%. of failed nodes. Figure 10 represents the energy consumption as a function of simulation time in FDAP in the three cases: 2%, 5% and 10% of faulty nodes. The curves show a difference in energy consumption between the three cases. When a node fails, the number of messages, during the verification phase, increases leading to the increase of energy dissipation. Figure 11 represents the life time duration of WSN in FDAP. We note a reduction in the WSN life time each time the number of failed nodes has increased. Figure 12 represents the data received by BS in FDAP in the three different cases. The results showed that the number of failed nodes impacts negatively the performance of the protocol where we note a big difference in the amount of data received by BS because of the elimination of incorrect values.

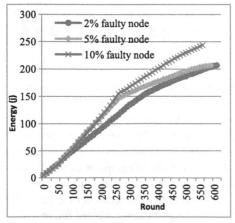

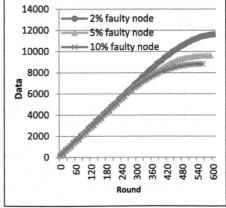

Fig. 11. Nodes life time duration in FDAP.   **Fig. 12.** Data amount received by BS in FDAP.

4   Conclusion

Since the decision taken by the end user of a WSN greatly depends on the data sensed by nodes from an area of interest, these data which have undergone an aggregation process, must be as dependable as possible to ensure the success of the network mission. That is why our protocol is designed to transmit only reliable data from healthy nodes avoiding so those that are faulty or compromised by intruders. The results of simulation carried out via the well known Network Simulator NS2 shown that our third proposal is able to detect all the considered anomalies like failure node or node attack. The convincing results have been obtained in an efficient way in terms of energy consumption and network life time duration. To be complete, we intend to include in our protocol the robustness of the sink to cope with both fault occurrences and attacks.

References

1. Ramar, C., Rubasoundar, K.: A survey on data aggregation techniques in WSNs. Int. J. Mob. Netw. Des. Innov. **6**(2), 81–91 (2015)
2. De Cristofaro, E., Bohli, J.M., Westhoff, D.: FAIR: fuzzy-based aggregation providing in-network resilience for real-time wireless sensor networks. In: Proceedings of ACM WiSec 2009, Zurich, Switzerland, 16–18 March 2009 (2009)
3. Wang, T., Qin, X., Liu, L.: Energy-efficient and scalable aggregation secure data a secure protocol. In: Distributed Sensor Networks, January 2013
4. Shivakumar, R., Jagadeesha, J., Babu, A., Rashmi, K.R.: An efficient and reliable data routing for in-network aggregation in wireless sensor network. Int. J. Emerg. Technol. Eng. **2**(4) (2015)
5. Jose, J., Princy, M., Jose, J.: PEPPDA: power efficient privacy preserving data aggregation for wireless sensor networks. In: ICE-CCN, March 2013
6. Zhu, L., Yang, Z., Xue, J., Guo, C.: An efficient confidentiality and integrity preserving aggregation protocol in wireless sensor networks. Int. J. Distrib. Sens. Netw. **10**(?), 1–8 (2014)
7. Lathamanju, R., Senthilkumar, P.: CRSR algorithm: a secure data aggregation algorithm in WSN. Int. J. Adv. Res. Electron. Commun. Eng. (IJARECE), **2**(9), 781–789 (2013)
8. BabuKaruppiah, A., Kannadhasan, S.: A novel approach to detect the shortest path for secure data aggregation using fuzzy logic in WSNs. Int. J. Eng. Comput. Sci. **2**(2), 506–510 (2013)
9. Jia, S., Bailing, W., Xiyuan, P., Jianfeng, L., Cheng, Z.: A recovery algorithm based on minimum distance redundant nodes in fault management in WSNs. Int. J. Control Autom. **6**(2), 175–184 (2013)
10. Heinzelman, W., Chandrakasan, A., Balakrishnan, H.: Energy-efficient communication protocol for wireless micro sensor networks. In: Proceedings of the 33rd Hawaii International Conference on System Sciences (2000)
11. Aliouat, Z., Aliouat, M.: Improved WSN capabilities through efficient duty-cycle mechanism. In: Yao, L., Xie, X., Zhang, Q., Yang, L.T., Zomaya, A.Y., Jin, H. (eds.) APSCC 2015. LNCS, vol. 9464, pp. 268–277. Springer, Cham (2015). https://doi.org/10.1007/978-3-319-26979-5_20
12. Titouna, C., Aliouat, M., Gueroui, M.: FDS: fault detection scheme for wireless sensor networks. Wirel. Person. Commun. **86**(2), 549–562 (2016)

Impact of Clustering Stability on the Improvement of Time Synchronization in VANETs

Khedidja Medani[1,2]($\boxtimes$), Makhlouf Aliouat[1,2], and Zibouda Aliouat[1,2]

[1] Department of Computer Science, Faculty of Sciences,
Ferhat Abbas University, Setif, Algeria
{khadidja_medani,maliouat,zaliouat}@univ-setif.dz
[2] LRSD laboratory, Faculty of Science, Computer Science Department,
UFAS1, Setif, Algeria

Abstract. Vehicular Ad Hoc NETworks (VANETs) has been developed to be, as a part of the Internet of Vehicles (IoVs), one of the most promoting technologies in the near future. Extensive researches are focusing on developing appropriate solutions in order to provide better Quality of Services (QoS) to these latter. This paper deals with network performance improvement based on clustering mechanism to reduce the number of communications caused by the high frequency of arrival/departure of vehicles while maintaining relevant clock synchronization solutions adequately operating. Thus, we attempt to adopt certain known clustering algorithms to our previous solution of time synchronization in VANETs namely OTRB (Offsets Table Robust Broadcasting) in order to evaluate the impact of a hierarchical communication on time synchronization process. An analytical study with a comparison on the influence of the clustering mechanism is given. The comparison takes place using simulation software NS2 (Network Simulator 2) and VanetMobiSim (VANET Mobility Simulator). The performance parameters include the average of arrival/departure of nodes and the number of isolated nodes. Simulation results reveal that an appropriate method should reduce the overhead of re-clustering and lead to an efficient network coverage with the best time synchronization rate.

Keywords: Intelligent transportation systems · VANETs
Clustering · Time synchronization performance

1 Introduction

The emergent Vehicular ad hoc networks (VANETs) tend to improve road safety and users infotainment during displacement. To this end, nodes in VANETs, as moving vehicles, exchange their information using wireless ad hoc communications. Due to the nature of the medium access, vehicles communicate directly with one another or with the available infrastructure using single hop communication. To reach a further destination, multi-hop communications can be established using relay nodes. However, the constraints created by the intrinsic characteristic of VANETs, such as the high density of nodes, make data routing a challenging task. To handle such a limitation, data should be spread among an optimized route to ensure reliable delivery with an optimized communication overhead. This can take place by splitting the network into small groups, clusters. For each cluster, a node is elected as a coordinator to communicate data with its members on the one hand and to spread the information to other clusters on the other hand. In addition, several issues depend on a strong assumption of reliable clustering mechanism.

The standard IEEE 802.11 p has been proposed to deal with VANETs safety application. However, due to the high density of nodes, the standard suffers predictability, fairness, low throughput and high collision rate [1]. In addition, the high rate of transmitted data, such as safety packet dissemination, make network prone to face high congestion. So, collisions, especially when large transmission scope is used, lead to packet loss rate increases. To deal with these drawbacks and to improve reliability, many researchers suggest network partitioning into small clusters in order to allow better shared medium access control. Where clustering is used to limit the channel contention and provide fair channel access. Furthermore, clustering algorithms allow power control and packet delivery enhancement.

Time synchronization in VANETs is crucial. Several proposed algorithms, whatever derived from MANETs or not, rely on some sort of logical clustering view [2]. The aim is to provide local time synchronization among the neighboring nodes in order to reach global time synchronization. For instance, in HCS [3] and OTRB [2] protocols, nodes synchronize their clocks by the aid of a randomly selected node, which is considered as a reliable source of time. The time synchronization solution performances are related to the selected node. This elected node can be seen as the coordinator of its neighbors. Such protocols rely on the assumption that the algorithm provides reliable clustering with a minimized number of in/out moving vehicles.

In order to enhance its performances, the present paper carries a study case on the influence of clustering algorithms in time synchronization using OTRB protocol. OTRB algorithm requires the organized clusters to be stable for a period of time enough to achieve the time synchronization. The stability parameters, which refer to the average lifetime of the cluster, is related to the synchronization round period in OTRB. In turn, this latter is related to the synchronized group scale and the desired time precision [2]. Different clustering algorithms in the literature take only into account the distance between nodes and their directions as parameters to form the clusters. However, vehicles displacement is related to the road infrastructure. Even vehicles move roughly with same velocities starting from a nearby zone, they can be found in further ways after a while. Therefore, other parameters, like the velocity of nodes, should be considered in order to deal with the high rate of arrival/departure of nodes.

The main contribution of this paper is to explore the impact of clustering on time synchronization process. To this end, we adopt an appropriate clustering algorithm to our previous solution OTRB. The best clustering algorithm should improve the synchronization rate and enhance the communication overhead.

The rest of the paper is organized as follow: Sect. 2 introduces the main clustering algorithm concepts and their requirement in VNAETs. While Sect. 3 gives an overview of the proposed solution of clustering issue. Section 4 gives an overview of OTRB protocol. Finally, in Sect. 5, a comparison of the implemented clustering protocols under OTRB solution is discussed in order to evaluate its performance. Finally, Sect. 6 concludes the paper.

2   Basic Clustering Notions and Requirements

The clustering is defined by the grouping of similar object items into clusters. In [4], a mathematical definition is given as follows: let $X \in R^{m \times n}$ a set of items representing a set of m objects x_i in R^n. The goal of the clustering algorithm is to divide X into K groups C_k. Each one is called a cluster. Where all objects that belong to the same cluster are more alike than objects in different clusters. The result of the algorithm is an injective mapping $X \rightarrow C$ of object items X_i to clusters C_k.

The clustering can be exclusive, overlapping or fuzzy. In case of exclusive clustering, one object (node) can be found in one and just one cluster. When, in overlapping clustering, the node can belong to more than one cluster (>=1). Whereas, in fuzzy clustering, for each node is assigned a membership weight. Therefore, the node belongs to the cluster according to its associated weight. Fuzzy algorithms can be used to improve exclusive clustering by avoiding the arbitrary node assignment. In the case, the node will be assigned to the cluster within its membership weight is highest.

In VANETs, clustering is the act of grouping the vehicles into multiple clusters (groups of vehicles). In which, vehicles belonging to the same cluster have the same properties, such as velocity, position and direction.

Under each cluster structure, there is one node acting as a coordinator of the group. It is the cluster-head node (CH). The remaining nodes are classified to cluster-member (CM) and/or cluster-gateway (GW) nodes. In term of communication, a cluster-gateway node is able to communicate with all the clusters in which it belongs. Consequently, this latter represents the relay node of its clusters. In other words, it is the relationship between the neighboring clusters.

It is announced in [5] that "The clusters should reflect some mechanism at work in the domain from which instances or data points are drawn, a mechanism that causes some instances to bear a stronger resemblance to one another than they do to the remaining instances." Regarding to vehicular environments, several parameters have been considered, such as node density, position, speed and direction. These parameters define the requirement that should be employed by any clustering algorithm. From which, we define the following that affects clustering efficiency and network performance [6,7] :

1. Cluster transmission overhead: The cluster transmission overhead refers as to the average number of packets exchanged to maintain the cluster structure. The more the transmission overhead is less, the more the clustering algorithm is desired.
2. Cluster convergence: Cluster Convergence refers as to the time needed for all the nodes in order to join a cluster. The clustering algorithm that takes large convergence time is less suitable.
3. Cluster connect time: this parameter refers as to the rate of connection of the vehicle to one cluster. The highest connect time rate is given, the more suitable clustering algorithm.
4. Cluster stability: the cluster stability refers as to the average lifetime of the cluster. The best clustering algorithm goes after high stability.

3   Related Work

For the purpose of supporting Quality of Service (QoS) in VANETs networking, stable clustering of the moving vehicles is required. The design of clustering algorithms enables organized medium access control and simplified data communication task. Several works in MANETs literature focus on improving the generated communication overhead by minimizing the number of clusters.

In [8], Gerla et al. propose to select the node having the highest number of neighbors as a CH in its vicinity. The proposed scheme minimizes the number of clusters. On the other hand, it deteriorates the network throughput due to the high rate of CMs within a cluster. This technique is not suitable for VANETs environment characterized by the high density of its nodes.

Other works, such as [9] selects the node with the lowest ID in its vicinity to be a CH of non-overlapping cluster. This protocol does not take nodes' mobility character into consideration. Then, CH selection process has to be frequently invoked especially in case of highly networks topology change, such as in VANETs. This will increase the cluster transmission overhead, which in turn reduces the network performances.

Su and Zhang [10] propose their algorithm which is a time and size based CH selection. The node that succeeds on sending its invite-to-join packet first and has more neighboring nodes is selected to be a CH. As the vehicles move in and out of the cluster frequently, there is high frequent of arrival and departure of nodes. Thus, maintaining the cluster stability becomes a hard task. Whereas, the mobility of vehicle in VANET systems is a major challenge to ensure clustering stability. Many researches in the current literature attempt to deal such a limitation.

In [11] the authors adopted the idea of [12]. In which, a node will announce itself as a CH in case it does not receive any Hello packet from its vicinity. Nevertheless, the node is allowed to be a CM in at least one cluster. The authors in the presented algorithm permit the merging of two adjacent clusters if their CHs come in direct communication range. The new CH is the one with the suitable weighed factor. This factor is calculated for each CH considering the mobility, connectivity and distance factors. However, CMs can be found out of the new elected CH, thus the possibility of frequent disconnection between the communicating nodes.

Kayis and Lichuan in [13] a speed-based classification in order to partition the vehicles into clusters. In which, seven groups of velocities are defined based on the min and the max speed boundaries. The "First Declaration Wins Rule" is adopted to select the set of the CHs. However, the nodes speed changes from node to node and can also change over the time. Therefore, nodes can frequently leave their group to join the more suitable. The algorithm may exhibit a frequent change in the network topology due to high cluster change rate. In addition, the CH can easily lose the connection with its member nodes.

A heuristic position-based clustering algorithm is proposed in [14]. The cluster formation is obtained by the partition of the network based on the geographical position and the priority associated with each vehicle. Each node calculates its priority value using its node ID, the current time and the eligibility value. The eligibility value increases with the travel time of the node and decreases when the node speed deviates largely from the average speed. The node with the highest priority in all its one-hop neighbors and one of its two-hop neighbors is elected to be a CH. Also, in the proposed technique, in order to optimize the cluster size, a maximum distance between the CH and its members is predefined. However, simulation results show that the formation of small cluster size increases the cluster reconfiguration rate. On the other hand, large communication range decreases the cluster transmission efficiency.

Wu et al. [15] have proposed the Type-based Cluster-forming Algorithm (TCA) in order to reduce the update frequency of CHs in emergency ad hoc networks. Each node is assigned a stability factor, S, which is updated frequently. The node having a lower value of S has low mobility and reliable connectivity. The node with the lower stability factor is more likely to be elected as a CH. However, because of the assumption that the nodes have slow speeds, this algorithm generates more communication overhead for the cluster updating. This makes the algorithm not suitable to be implemented in VANETs environment.

A distributed CH selection is proposed in [16] to dynamically organize the network into clusters. The novel proposed algorithm is speed and distance based. The CH is selected according to its relative speed and the distance within its neighbors. Also, fuzzy logic inference system is used for predicting the future speed and position of all the CMs. As the maintenance phase is adaptable to the drivers' behavior, the proposed algorithm is highly stable. However, the distributed cluster overhead causes the number of messages received/transmitted to be decreased [6]. Also, the algorithm is implemented under medium vehicle density and low-speed conditions.

Rawashdeh and Mahmud [17] attempts to deal with the clustering issue with stability on highway VANET environments. The authors apart from the assumption that "the existence of VANETs nodes in the same geographic proximity does not mean that they exhibit the same mobility pattern", and try to improve the stability by making the network topology less dynamic. In addition to the velocity and the position, the proposed approach takes the location and the velocity difference between the nodes into consideration. Each node runs the algorithm in fully distributed manner. First, a separation of the vehicles into highly mobile and low mobile groups is made. The network partition should provide a minimized number of clusters and ensure that all the CMs are stable with respect to one another. Then, the set of nodes with the highest suitable value are elected as CH nodes. Each node calculates its suitable value, which is calculated based to the velocity and the distance with regard to its stable neighbors. The closer velocity to the average velocities of its stable neighbors the node has, the highest suitable value is given. Simulation results show that the proposed algorithm improves the average cluster lifetime, which increases the stability of network topology.

In [18] a new clustering algorithm in vehicular ad hoc networks is presented. The proposed scheme, called Hierarchical Clustering Algorithm (HCA), provides network two-hop clustering in fully distributed randomized fashion. Three roles can be defined to form the network hierarchy, in which a node can be a cluster head (CH), cluster relay (CR) or slave node. The CR nodes relay the transmitted data between the CHs and the slaves. The algorithm forms the clusters running four steps; cluster Relays Selection, ClusterHead Selection, Cluster Formation and Scheduling and Cluster Maintenance. Where it debates fast clusters construction in the first third steps and leave the mobility handling to the maintenance phase. However, the algorithm improves the clustering stability at the price of transmission efficiency and the average of clustering overhead. Also, the high rate of inter-cluster interferences causes frequent cluster changes and message loss due to message collisions. Another clustering algorithm for VANETs is proposed by Hassan abadi et al. in [19]. The authors have presented the Affinity PROpagation for VEhiclar networks (APROVE) protocol which attempts to produce high stability clustering taking into account the mobility parameter of nodes. In the proposed protocol, the CH selection is based on nodes' interdistance. The closest node to its neighbors is selected. This scheme suffers long cluster convergence time in order to make all the nodes belong to a cluster.

In [20], the authors rely on neighbor vehicles' mobility to maintain VANETs stable clustering in rounds. In which, the node with the lowest neighbor vehicles mobility is elected to be the CH in its vicinity. Each node is assumed to have a unique identity, ID and maintain a neighbor vehicles table, which includes, for each neighbor, its ID and the neighbor vehicles mobility value. The neighbor vehicle table will be updated at each round and the neighbor vehicle mobility is calculated considering the number of vehicles entering/leaving the service convergence of the node.

In [21] a new oriented VANETs clustering scheme for urban city scenario is presented. In order to improve clustering stability, the proposed scheme considers the vehicles direction, position and the link lifetime estimation. Each node is able to estimate its speed and distance relative to its one-hop neighbors. The nearest vehicle into the central geographical position is likely chosen as a CH. The algorithm defines a safe distance threshold, D which D is less than the transmission range. The vehicles within this value are considered having more stable links with the CH. Therefore, the CMs are selected by the CH from its one-hop neighbors, in which the cluster size is defined by $L \leq 2D$.

An analytical study that evaluates the impact of the clustering mechanism on the QoS of VANET systems implementation must be carried out. By way of illustration, we analyze the influence of the clustering protocol performed while maintaining proper clock synchronization using OTRB protocol [2]. The analytical study is carried out using network simulator (NS2) to compare the impact of the rate of arrival/departure nodes on the performance of OTRB protocol. To this end, the following sections demonstrate the behavior of OTRB protocol to maintain synchronized clocks in VANET systems. Then the simulation results are presented.

4   OTRB Protocol Overview

For the purpose of time synchronization in distributed VANET environments, the OTRB protocol is proposed in [2] by Medani et al. The main idea is to set a number of nodes, named transporter nodes, to spread the time information over the entire network. For each transporter node, the time information broadcasted consists of the offsets related to all its neighbors. However, the transporter node is supposed a reliable source of time.

The OTRB protocol performs time synchronization relying on pair-wise time synchronization mechanism [22]. Initially, each transporter node broadcasts a synchronization packet to its neighbors. Upon receiving this synchronization packet, neighboring nodes respond by their time values as shown in Fig. 1. Then, the transporter node calculates the clock offset relative to each neighbor i and broadcasts the estimated values, as a time table, allowing all of them to synchronize to one another.

The transporter node selection phase has an important impact on OTRB performance in terms of time synchronization rate and communication overhead. Thus, the transporter node should have a higher stability relative to its neighbors. That is, a node leaving its synchronized group before the synchronization will be achieved, generates more synchronization requests. In this case, time synchronization maintenance requires more communication overhead. The more the overhead is generated, the more the communication collisions. In turn, a high rate of collisions decreases the time synchronization rate. On the other hand, the transporter node selection phase should be as fast as possible in order to deal with the time synchronization requirements, delay latency for example.

It is stated in [2] that each transporter node with its synchronized group can be seen as a cluster. Where the transporter node represents the CH and its neighbors represent the CMs (see Fig. 2). Each node can join more than one transporter node. In the rest of this paper, we refer to the transporter node as the CH, and to the synchronized group as the CMs.

To show the influence of the clustering algorithm on OTRB performance, in the next sections, we present the simulation and the comparison of three clustering algorithms (CBLR [12], LID [9] and NMCS [20]) working under OTRB protocol.

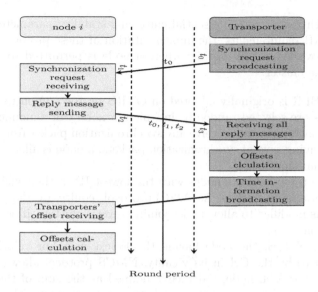

Fig. 1. OTRB protocol process [2].

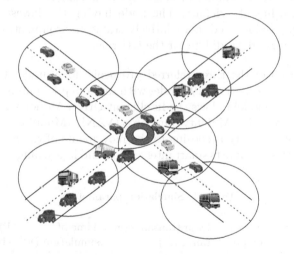

Fig. 2. Network architecture in OTRB [2].

5   Simulation Results

The proposed time synchronization OTRB protocol relies on a strong assumption of stable clustering mechanism. To evaluate its performances and demonstrate its better quality of functioning, we adopt three clustering protocols CBLR [12], LID [9] and NMCS [20] to work under the referred OTRB and the behavior of the protocol is observed for each one. The time synchronization rate and the average communication overhead needed to accomplish the synchronization process are measured with regard to the essential clustering stability parameters. In order to meet OTRB requirements, the implementation of these protocols is adapted to maintain overlapping clustering; where a node is permitted to be a CM in more than one cluster.

1. CBLR: CBLR is originally adopted on OTRB protocol upon its creation. A set of CHs are selected randomly, in which, a node will announce itself as a CH if this latter does not receive any synchronization packet from its vicinity. When receiving several synchronization packets, a node is allowed to be CMs in more than one cluster.
2. LID: LID protocol selects nodes with the lowest ID in their vicinity as CHs of non-overlapping clusters. As OTRB protocol requires overlapping clusters, LID is thus modified to allow nodes joining several clusters, if possible, in our contribution.
3. NMCS: In NMCS, the node having the lowest neighbor vehicles mobility is selected to be the CH in its vicinity. NMCS protocol allows overlapping clustering. For each node, a degree is defined as the sum of the number of nodes that have left/arrived the node's transmission range. The sum is then divided by the total number of neighbors to normalize the degree of node's mobility regarding its vicinity. The node having the lowest value for this metric is supposed located in a relatively stable environment, indicating that such node is a good candidate for the CH role.

The simulation of the three referred protocols holds under the parameters shown in Table 1. A scenario of 199 nodes with random positions is initially generated. The mobility model follows the Intelligent Driver Model with Lane Changing (IDM_LC) of VANET Mobility Simulator (VANETMobiSim), in which, vehicles regulate their velocity according to the movements of the neighboring vehicles. This model also supports smart intersection and lane changing management.

Table 1. Simulation parameters

Topology $(m * m)$	Nodes number	Data transmission range (m)	Time of simulation (s)	Physical channel
3400 * 4900	199	300	200	IEEE 802.11b

In order to evaluate OTRB performances, we observe the changes in the synchronization rate and the communication overhead for each clustering protocol implemented under OTRB. On the other hand, to shed the light on the influence of the clustering protocol stability on OTRB performance, the average of arrival/departure and isolated nodes are captured in parallel.

As shown in Fig. 3, NMCS protocol reduces the average of the nodes moving into new clusters during the synchronization process. That is NMCS protocol

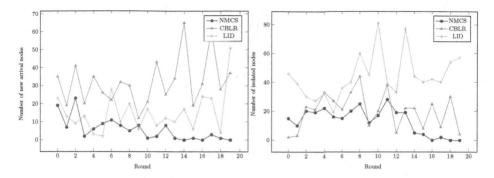

Fig. 3. Average of new arrival nodes per round.

Fig. 4. Number of isolated nodes per round.

favors nodes located in a stable environment relative to their vicinities to be CH nodes. In contrast, CBLR and LID protocols do not take into account nodes' mobility while performing the clustering operation. In addition, relying on the lowest ID and the random metrics to select the CH nodes in LID and CBLR respectively multiplies the risk of having so much number of CHs and on the other hand a high number of isolated nodes (see Fig. 4).

Actually, the clustering stability metrics, arrival/departure, and isolated nodes affect the performance of OTRB protocol. Table 2 demonstrates that the more stable clustering protocol, the more OTRB performances are improved.

Figure 5 shows that the synchronization rate of OTRB using NMCS outperforms OTRB using the two other protocols. This is due to the improvement in the average of arrival/departure nodes which reveals successful clock synchronization process in the overall of nodes. Therefore, the number of messages generated in order to accomplish the synchronization process is minimized. On the other hand, the more the number of isolated nodes, the more supplementary synchronization requests are generated. For that, using NMCS under OTRB protocol improves the communication overhead compared to CBLR and LID protocols under OTRB (see Fig. 6).

Table 2. Comparison.

Implemented protocol	Arrival/departure nodes	Isolated nodes	Synchronization rate (s)	Communication overhead
CBLR under OTRB	High	Low	Medium	High
LID under OTRB	Medium	Medium	Medium	High
NMCS under OTRB	Low	Medium	Improved	Improved

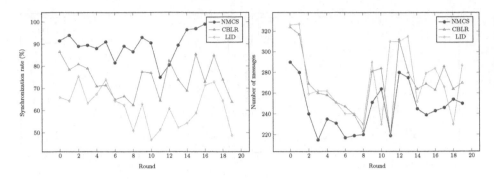

Fig. 5. Synchronization rate comparison.   **Fig. 6.** Communication overhead comparison.

6   Conclusion

Because time synchronization is an essential attribute of VANET networks, because VANETs are included in a larger area: the IoV which itself is part of the largest networks: the IoT, then the quality of service of the synchronization process must fulfill the expected objective in accordance with network importance. The paper is dealing with the impact of the well know clustering nodes organization on time synchronization ORTB protocol in order to improve the latter capabilities. Several known clustering algorithms have been used by ORTB and the comparative analysis of simulation results reveals that better performance, in term of synchronization rate and average communication overhead, is offered by the high stability clustering algorithm. However, multi-metric algorithms can be used to further enhance the clustering stability and so the network performance. This can be the subject of future work.

References

1. Yin, J., ElBatt, T., Yeung, G., Ryu, B., Habermas, S., Krishnan, H., Talty, T.: Performance evaluation of safety applications over DSRC vehicular ad hoc networks. In: Proceedings of the 1st ACM International Workshop on Vehicular ad hoc Networks, pp. 1–9. ACM (2004)
2. Medani, K., Aliouat, M., Aliouat, Z.: Fault tolerant time synchronization using offsets table robust broadcasting protocol for vehicular ad hoc networks. AEU-Int. J. Electron. Commun. **81**, 192–204 (2017)
3. Sam, D., Velanganni, C., Evangelin, T.E.: A vehicle control system using a time synchronized hybrid VANET to reduce road accidents caused by human error. Veh. Commun. **6**, 17–28 (2016)
4. Graepel, T.: Statistical physics of clustering algorithms. Technical report, 171822 (1998)
5. Chakrabarti, S., Cox, E., Frank, E., Güting, R.H., Han, J., Jiang, X., Kamber, M., Lightstone, S.S., Nadeau, T.P., Neapolitan, R.E., et al.: Data Mining: Know It All. Morgan Kaufmann, Burlington (2008)

6. Bali, R.S., Kumar, N., Rodrigues, J.J.P.C.: Clustering in vehicular ad hoc networks: taxonomy, challenges and solutions. Veh. Commun. **1**(3), 134–152 (2014)
7. Cooper, C., Franklin, D., Ros, M., Safaei, F., Abolhasan, M.: A comparative survey of vanet clustering techniques. IEEE Commun. Surv. Tutorials **19**(1), 657–681 (2017)
8. Gerla, M., Tsai, J.T.-C.: Multicluster, mobile, multimedia radio network. Wireless Netw. **1**(3), 255–265 (1995)
9. Lin, C.R., Gerla, M.: Adaptive clustering for mobile wireless networks. IEEE J. Sel. Areas Commun. **15**(7), 1265–1275 (1997)
10. Hang, S., Zhang, X.: Clustering-based multichannel MAC protocols for QoS provisionings over vehicular ad hoc networks. IEEE Trans. Veh. Technol. **56**(6), 3309–3323 (2007)
11. Guenter, Y., Wiegel, B., Großmann, H.P.: Medium access concept for VANETs based on clustering. In: 2007 IEEE 66th Vehicular Technology Conference, VTC-2007 Fall, pp. 2189–2193. IEEE (2007)
12. Santos, R.A., Edwards, R.M., Seed, N.L.: Inter vehicular data exchange between fast moving road traffic using an ad-hoc cluster-based location routing algorithm and 802.11 b direct sequence spread spectrum radio. In: PostGraduate Networking Conference (2003)
13. Kayis, O., Acarman, T.: Clustering formation for inter-vehicle communication. In: 2007 IEEE Intelligent Transportation Systems Conference, ITSC 2007, pp. 636–641. IEEE (2007)
14. Wang, Z., Liu, L., Zhou, M., Ansari, N.: A position-based clustering technique for ad hoc intervehicle communication. IEEE Trans. Syst. Man Cybern. Part C (Appl. Rev.) **38**(2), 201–208 (2008)
15. Wu, H., Zhong, Z., Hanzo, L.: A cluster-head selection and update algorithm for ad hoc networks. In: 2010 IEEE Global Telecommunications Conference (GLOBECOM 2010), pp. 1–5. IEEE (2010)
16. Hafeez, K.A., Zhao, L., Liao, Z., Ma, B.N.-W.: A fuzzy-logic-based cluster head selection algorithm in VANETs. In: 2012 IEEE International Conference on Communications (ICC), pp. 203–207. IEEE (2012)
17. Rawashdeh, Z.Y., Mahmud, S.M.: A novel algorithm to form stable clusters in vehicular ad hoc networks on highways. EURASIP J. Wirel. Commun. Netw. **2012**(1), 15 (2012)
18. Dror, E., Avin, C., Lotker, Z.: Fast randomized algorithm for 2-hops clustering in vehicular ad-hoc networks. Ad Hoc Netw. **11**(7), 2002–2015 (2013)
19. Hassanabadi, B., Shea, C., Zhang, L., Valaee, S.: Clustering in vehicular ad hoc networks using affinity propagation. Ad Hoc Netw. **13**, 535–548 (2014)
20. Kwon, J.-H., Chang, H.S., Shon, T., Jung, J.-J., Kim, E.-J.: Neighbor stability-based VANET clustering for urban vehicular environments. J. Supercomputing **72**(1), 161–176 (2016)
21. Ren, M., Khoukhi, L., Labiod, H., Zhang, J., Vèque, V.: A mobility-based scheme for dynamic clustering in vehicular ad-hoc networks (VANETs). Veh. Commun. **9**, 233–241 (2016)
22. Sundararaman, B., Buy, U., Kshemkalyani, A.D.: Clock synchronization for wireless sensor networks: a survey. Ad hoc Netw. **3**(3), 281–323 (2005)

Internet of Things and Decision Support Systems

Enhancement of IoT Applications Dependability Using Bayesian Networks

Yasmine Harbi$^{(\boxtimes)}$, Zibouda Aliouat, and Sarra Hammoudi

LRSD Laboratory, Computer Science Department,
Ferhat Abbas University, Setif, Algeria
harbi_yas@yahoo.com, {zaliouat,hammoudi_sarah}@univ-setif.dz

Abstract. Sensors play a vital role in Internet of Things (IoT) monitoring applications. However, the harsh nature of the environment influences negatively on the sensors reliability. They may occasionally generate inaccurate measurements when they are exposed to high level of
humidity, temperature, etc. Forwarding incorrect data to the base station may cause disastrous decision-making in critical applications. To
address this problem, we proposed a new fault-tolerant mechanism based
on Bayesian Networks that carefully solve such a problem. The mechanism is called Reliability of Captured Data (RCD); it enables sensors
to both detect and recover faulty data. To show the performance of our
algorithm, we compared this latter with another recent algorithm called
Fault Detection Scheme (FDS) using Network Simulator 2 (NS2). The
results showed that our strategy gives remarkable enhancements in terms
of fault detection accuracy and fault alarm rate.

Keywords: Internet of Things · Wireless Sensor Networks
Smart sensors · Fault tolerance · Bayesian Networks · Reliability

1 Introduction

The IoT refers to the next generation of the Internet [1]. It cuts across many critical applications. Among of these applications are: process automation applications, infrastructure monitoring networks, process control networks, smart building, patient monitoring, oil and gas industry [1–4].

The IoT is empowered by the proliferation of a myriad number of miniature
sensors and actuators. Sensors can efficiently monitor critical environments in
diverse domains.

Fault tolerance is one of the critical problems in IoT applications. The problem of missing sensor node and data are inevitable in Wireless Sensor Networks.
Failure problems are due to various factors such as power depletion, environmental impact and dislocation of sensor nodes [5]. In such a case, sensors may
generate incorrect measurements, cause false alarms and degrade the reliability
of collected data [6]. Therefore, it is extremely important to ensure the accuracy

© IFIP International Federation for Information Processing 2018
Published by Springer International Publishing AG 2018. All Rights Reserved
A. Amine et al. (Eds.): CIIA 2018, IFIP AICT 522, pp. 487–497, 2018.
https://doi.org/10.1007/978-3-319-89743-1_42

of sensed data before decision-making process. A wide variety of classification-based approaches including Bayesian Networks (BNs), Neural Networks (NNs), and Hidden Markov Models (HMMs) can be used to detect inaccurate data.

To address such a problem, this paper presents a new fault-tolerant mechanism, called Reliability of Captured Data (RCD) that is able to achieve higher data reliability and accuracy. The RCD makes sensors more intelligent, it enables them to efficiently detect the faulty nodes by analyzing the measured data. The algorithm is also able to recover the unreliable data before forwarding it to the base station. The accuracy of the faulty node detection reflects in the diagnosis of the sensed data using Bayesian Networks.

The Bayesian Network is a probabilistic model which enables the sensors to locally predict their current readings based on their own past readings and the current readings of their neighbors [7,8]. The choice of the Bayesian Networks helps to accurately detect the faulty nodes since its rate of misclassification is very low compared to NN and HMM [9]. In addition, the BN requires fewer parameters than the HMM to represent the same information, and provides the spatial-temporal correlation between the sensors [10,11].

To show the performance of our proposal, we have conducted a several series using Network Simulator 2 (NS2). We compared our strategy to a recent one which is called Fault Detection Scheme (FDS) [13]. The results show that our mechanism achieves high amelioration in terms of fault detection accuracy and false alarm rate.

The rest of this paper is organized as follows. Section 2 presents the related works in the context of fault tolerance. Section 3 describes the concept of Bayesian classifier. Section 4 discusses the proposal and the simulation results are presented in Sect. 5. Section 6 concludes our work.

2   Related Works

Failure detection and recovery is a crucial task in IoT since sensors are deployed in harsh and unattended environments.

Yuvaraja and Sabrigiriraj in [12] designed a fault detection and recovery scheme for routing protocols in WSN. The sink periodically broadcasts agent packets to all its neighbors. The non faulty nodes reply with an acknowledgment. The agents form a query path towards the dead or faulty node. To decrease the broadcasting cost overhead, the receiving nodes apply the Random Decision Function (RDF) to be able to take a decision to whether they forward the agent packet or not in order to detect the dead or faulty nodes. Once the faulty nodes is detected, the connectivity is restored by replacing faulty node with block movement. The technique is called Least-Disruptive Topology Repair (LeDiR). The shortcoming of this paper is when applying the RDF function risks the agent packets to not arrive at the final nodes of the network. Therefore, sink may consider it as faulty nodes.

Titouna et al. in [13] proposed Fault Detection Scheme (FDS) which consider both battery power and sensed data to identify faulty nodes. The FDS is performed in two-levels where sensor nodes detect outlier (faulty data) using a Bayesian model, then transmit the decision to Cluster Head (CH) which verifies only sensor nodes that have a primary decision as faulty. This second verification is based on similarity between primary decisions of cluster's members. The drawback of this approach is with a high number of faulty nodes, the process of detection is very slow and false alarm rate reaches a higher value.

The authors in [14] presented an extension of previous work [13] named Distributed Fault-Tolerant Algorithm (DFTA) in order to recover faulty nodes in WSN. The main idea is to eliminate the faulty node from the network and replace it with a switch-off node which has a higher degree of connectivity and belongs to the same cluster. Sleeping nodes are selected by CH after deployment of the network. The sleeping nodes neighbors' number must be less than the average number of cluster's members. After detecting faulty node, the CH sends a wakeup message to all sleeping nodes. These latter update their routing table and send the number of hops to reach the faulty node to CH. Finally, the CH chooses the appropriate recovery node that has fewer hops to reach the faulty one. The shortcoming of this strategy is that during the recovery phase, the CH drains too much energy since it sends three types of messages. Which overtime may decrease the network's lifetime.

Yuan et al. in [15] proposed a Distributed Bayesian Algorithm (DBA) which can efficiently detect faulty readings in WSN using Bayesian networks. The sensor nodes exchange their reading with neighbors and calculate the probability of fault. This latter can be incorrect if the most of the neighbors are faulty. Then, the faulty node will consider itself as good node, or the good node may consider itself as faulty one. To avoid such a case, the probability of fault must be adjusted by border nodes. If a node has two neighbors that have a low probability of fault, but have different status (i.e. one is faulty and other is good), then it will be considered as border node. The drawback of this algorithm is that, in dense networks exchanging with all neighbors will increase the energy consumption and decrease the network's lifetime. Also, the algorithm does not take into account the overhead cost.

Sutagundar et al. in [16] proposed a fault-tolerant approach based on static and mobile agents which reside in all nodes of the network. This approach is developed at three levels: node level, CH level and Sink level. At node level, the Fault Tolerant Agent (FTA) checks whether the reading is in the reference range. Then, it discards the faulty readings. The Data Transfer Agent (DTA) transmits correct readings to CH. This latter compares received data and computes the average of all corrected data, then sends it to sink through DTA. The Sink Manager Agent (SMA) broadcasts periodically "alive messages" to all nodes and duplicate sink. If the principal sink fails, the duplicate one replaces it. The shortcoming of this mechanism is that in agent based approaches the network experiences a significant amount of delay to retrieve the state of the node, because it needs to wait for an agent to visit the node. Furthermore, agent-based approaches do not perform well in large scale WSNs. Since the number of sensor nodes increases, so do the number of agents.

The authors in [17] proposed a fault-tolerant mechanism for link failure in optical networks since optical fiber cuts is the most frequent failure in metropolitan areas. The proposed approach is based on pre-configured protection cycles (p-cycles) and Double Rings topology with Dual Attachment (DRDA). The network consists of two rings with the same number of nodes. The nodes of the inner ring and the outer one are connected through bidirectional links. The authors formed two p-cycles that cover all nodes with disjoint links. They defined three recovery strategies in order to recover single and multiple link failure. Also, they provided a formal model of the network using Continuous Time Markov Chains, and they verified dependability properties using probabilistic model checking. The proposed work gives useful information for designing dependable optical networks in metropolitan areas. However, this redundant topology may provide poor availability in a case of multiple link failure by putting an excessive load on the other links. Moreover, it may be very expensive and hard to manage or maintain.

3   Bayesian Classifier

In order to detect sensors that are prone to failure, their captured data should be analyzed and classified to know whether it is correct or not. This classification is based on Bayesian classifier.

A Bayesian classifier is a simple probabilistic classifier based on the Bayes theorem, it allows to classify the samples measured on observations [18,19].

Bayes theorem estimates the probability of an event, based on prior knowledge of conditions that might be related to this events [18,19]. Equation 1 represents the Bayes theorem.

$$P(A|B) = \frac{P(B|A)P(A)}{P(B)}, P(B) \neq 0 \tag{1}$$

Where A and B are events.
P(A) and P(B) are the probabilities of observing A and B independently.
$P(B|A)$ a conditional probability, is the probability of B given that A is true.
$P(A|B)$ a conditional probability, is the posterior probability of A given that B is true.

To classify the measured values in the predefined classes using the Bayesian classifier, we will have to define a Bayesian network.

A Bayesian network is an acyclic oriented graph in which nodes present random variables and arcs indicate the dependencies between these variables [7].

Figure 1 represents the structure of a Naive Bayesian Classifier. Node C represents the class c_i and nodes $X = x_1, x_2, ..., x_n$ are the features of the Naive Bayesian Classifier which are independent (i.e. there are strong or naive independence assumptions between the features) [19].

The Bayesian network allows us to calculate the posterior probability $P(C = c_i|X)$ for each possible class c_i by using the Bayes theorem as stated in Eq. 2.

$$P(C = c_i|X) = \frac{P(X|C = c_i)P(C = c_i)}{P(X)} \tag{2}$$

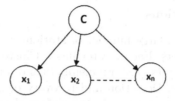

Fig. 1. A Bayesian Network corresponding to a Naive Bayesian Classifier

The probability $P(X|C = c_i)$ is often impractical to compute without imposing independence assumptions between features X. The Naive Bayesian Classifier assumes that each features x_j is conditionally independent of every other feature [19], this yields Eq. 3.

$$P(X|C = c_i) = \prod_{j=1}^{n} P(x_j|C = c_i) \tag{3}$$

Now, we obtain the Eq. 4.

$$P(C = c_i|X) = \frac{\prod_{j=1}^{n} P(x_j|C = c_i)P(C = c_i)}{P(X)} \tag{4}$$

The target class c_i, output by the Bayesian classifier, can be inferred using the concept of Maximum A Posteriori (MAP) [19] as indicated in Eq. 5.

$$c_{MAP} = argmax_{c_i \in C}P(C = c_i|X) = argmax_{c_i \in C} \frac{\prod_{j=1}^{n} P(x_j|C = c_i)P(C = c_i)}{P(X)} \tag{5}$$

Since the denominator is constant and identical for all classes, it does not affect the maximization. After deleting it, we got the Eq. 6.

$$c_{MAP} = argmax_{c_i \in C}P(C = c_i|X) = argmax_{c_i \in C} \prod_{j=1}^{n} P(x_j|C = c_i)P(C = c_i) \tag{6}$$

According to Eq. 6, we can classify the measured data given various observations.

4   Reliability of Captured Data (RCD)

Our contribution is based on Bayesian classifier where every sensor node checks its own sensed data based on his last reading and current readings of their two nearest neighbors. Then, it detects if the captured data is correct or not. Finally, it recovers faulty data by adjusting this latter to avoid transmission of erroneous values. Thus, it ensures the reliability of data before decision-making process.

4.1 System Assumptions

Our network consists of a large number of stationary and homogeneous sensor nodes organized in clusters. Each cluster has a Cluster Head (CH) that aggregates data transmitted by its cluster's members and forwards the collected data to the Base Station (BS). Detection and recovery of inaccurate sensed data is distributed since every sensor nodes in the cluster checks its current data using the Bayesian Classifier. We also assume that we do not have malicious attacks in the network.

4.2 Mechanism Design

In our system model, the sensed data can belong to the interval values $I = [a, b]$. We divide this range into many classes that are mutually exclusive and exhaustive. According to Sect. 3, the Bayesian Network of our mechanism is represented in Fig. 2.

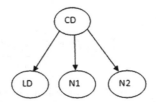

Fig. 2. RCD Bayesian Network

CD: the current sensed data that belongs to a class c_i
LD: the last sensed data that belongs to a class c_i
N1: the current data of the first neighbor that belongs to a class c_i
N2: the current data of the second neighbor that belongs to a class c_i
These variables ensure spatial-temporal correlation in the network.

The RCD's algorithm is composed of three phases:

Learning Phase. This phase is executed after the deployment of the network. We assume that it is free from faults.

The learning phase aims to calculate the parameters of the classifier which are: the probability of each class $P(CD \in c_i)$ and conditional probabilities $P(LD|CD)$, $P(N1|CD)$ and $P(N2|CD)$.

Algorithm 1 explains the steps of the learning phase executed by sensor nodes.

Every sensor node sense 'P' samples and gets the samples (readings) of its two neighbors over a period of time. Then, it calculates the classes' probabilities and the conditional probabilities. The complexity of the Algorithm 1 is quadratic since the running time of the first instruction is proportional to the number of

---

Algorithm 1. Learning phase

Input : Predefined classes
Output : The parameters of the classifier
BEGIN
Defines the two nearest neighbors.
Gets current data of the two neighbors.
Calculates the classes' probabilities and the conditional probabilities.
END

---

the cluster's members while the second one has constant running time. The last instruction has two running times; one is proportional to the number of classes for calculating the classes' probabilities, and the other is proportional to the square of the number of classes for calculating the conditional probabilities.

Inference Phase. In this phase, sensor node detects faulty data by calculating the posterior probability of c_i using Bayes theorem. Then, it deduces the most probable class that its sensed value should belong to. The most probable class is calculated using the MAP (See Eq. 5).

Each sensor node calculates the posterior probability of c_i according to Eq. 6. We replace x_j by three observations; the last sensed data (LD), the current data of the first neighbor (N1), and the current data of the first neighbor (N2). Hence, we obtain the Eq. 7.

$$c_{MAP} = argmax_{c_i \in I} P(LD|c_i)P(N1|c_i)P(N2|c_i)P(c_i) \qquad (7)$$

After calculating the posterior probability of each predefined or learned class c_i, the sensor node compares the class that has the highest probability with the class of the captured value. If they are identical, the sensor considers that the captured value is correct. Otherwise, it identifies it as a faulty one.

The steps of the inference phase are detailed in Algorithm 2. The time complexity of the inference algorithm is linear since the first instruction has constant running time. The second one has a running time proportional to the number of classes, and the running time of if-else statements is constant in the two possibilities (i.e. if sequence and else sequence).

Recovery Phase. In this phase, sensor nodes adjust faulty data by approximate the data value to the average of the most probable class (the result of the classifier) (See Algorithm 3). The recovery phase aims to connect the regions isolated by faulty nodes. The algorithm of the recovery phase has constant complexity.

We can conclude that the time complexity of our RCD's algorithm is quadratic.

Algorithm 2. Inference phase

Input : CD, LD, N1 and N2
Output : The most probable class
BEGIN
Exchange with the two neighbors.
Calculate the posterior probability of each class using Eq. 7.
if (CD ∈ the most probable class) **then**
    Captured data is correct.
else
    Captured data is incorrect.
    Recovery phase.
end if
END

Algorithm 3. Recovery phase

Input : Posterior probability of all classes
Output : Adjusted data
BEGIN
CD = the average of the most probable class (the class with the highest posterior probability)
END

---

5   Simulation Results

In order to evaluate our contribution, we have conducted several series of simulation using Network Simulator 2 (NS2). The simulation results are compared to FDS detailed in Sect. 2. To illustrate the performance of our proposal, we injected some faulty nodes in the network.

5.1   Simulation Parameters

The simulation parameters are summarized in Table 1.

Table 1. Simulation parameters

Parameter	Value
Network size	$100\,\text{m} * 100\,\text{m}$
Number of sensor nodes	100
Initial energy	$2\,\text{J}$
Number of clusters	4
Round duration	$20\,\text{s}$
Slot duration	$0,02\,\text{s}$

5.2   Performance Metrics

For comparison purposes, we have considered the following performance metrics.

Fault Detection Accuracy (FDA) is the ratio of the number of faulty sensor nodes detected to the total number of faulty sensor nodes. Ideally, this ratio should be equal to 1.

False Alarm Rate (FAR) is defined as the ratio of the number of non-faulty nodes diagnosed as faulty to the total number of non-faulty nodes. The FAR should be very small to achieve a reliable detection.

5.3   Results and Discussions

As we mentioned above, to evaluate the performance of our proposal, we injected 15 faulty nodes in the network. Figure 3 shows the fault detection accuracy in both of FDS and RCD.

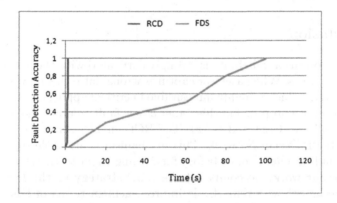

Fig. 3. Fault Detection Accuracy in FDS and RCD with 15 faulty nodes

According to Fig. 3, we notice that the curve of our protocol evolves very quickly to reach 1 during 1.5 s. This is due to the distribution of the detection which is performed by nodes before data transmission. The data transmission is performed in rounds. Sensor nodes transmit data during its time slot of the round. Unlike the RCD protocol, the FDS reaches the value 1 after 100 s because the detection of faulty sensed data is performed in two-levels: sensor node's level and CH's level. We conclude that both of RCD and FDS reach the ratio 1 which is the ideal case. However the fault detection accuracy using the RCD algorithm is faster than when using the FDS one.

Figure 4 represents the false alarm rate for both FDS and RCD.

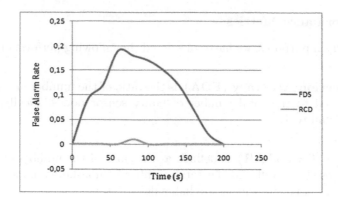

Fig. 4. False Alarm Rate in FDS and RCD with 15 faulty nodes

According to Fig. 4, we notice that the maximum rate of false alarm in the FDS is 0,18 at 60 s, then, it decreased to 0 at 200 s. While in the RCD protocol, it reaches 1 at 80 s. This proves that our proposal can achieve an accurate and reliable detection.

6   Conclusion

In IoT, the application reliability is strongly correlated with the accuracy of data captured by sensors. When a sensor captures wrong data, this may lead to wrong decision-making, which is highly undesirable in critical applications. To overcome such a problem, we proposed in this paper a new fault-tolerant mechanism called RCD (Reliability of Captured Data). The RCD enables sensors to detect faulty data using the Bayesian Network. This mechanism is also able to recover these inaccurate and unreliable data before forwarding them to the BS. To show the efficiency of our work, we compared the RCD strategy to the FDS one using NS2. The results show a remarkable improvement in terms of false alarm rate and fault detection accuracy.

As a future perspective of our current work, we aim to integrate safety into data communication of IoT applications.

References

1. Li, S., Xu, L.D., Zhao, S.: The Internet of Things: a survey. Inf. Syst. Front. **17**(2), 243–259 (2015)
2. Silva, I., Leandro, R., Macedo, D., Guedes, L.A.: A dependability evaluation tool for the Internet of Things. Comput. Electr. Eng. **39**(7), 2005–2018 (2013)
3. Lee, I., Lee, K.: The Internet of Things (IoT): applications, investments, and challenges for enterprises. Bus. Horiz. **58**(4), 431–440 (2015)
4. Cauchi, N., Hoque, K.A., Abate, A., Stoelinga, M.: Efficient probabilistic model checking of smart building maintenance using fault maintenance trees. arXiv:1801.04263 (2018)

5. Kakamanshadi, G., Gupta, S., Singh, S.: A survey on fault tolerance techniques in wireless sensor networks. In: International Conference on Green Computing and Internet of Things, pp. 168–173 (2015)
6. Raj, S.A.A., Ramalakshmi, K., Priyadharsini, C.: A survey on classification of fault tolerance techniques available in wireless sensor network. Int. J. Eng. Res. Technol. (IJERT) **3**(1), 688–691 (2014)
7. Ben-Gal, I.: Bayesian networks. In: Ruggeri, F., Faltin, F., Kenett, R. (eds.) Encyclopedia of Statistics in Quality and Reliability. Wiley, Hoboken (2008)
8. Zhang, Y., Meratnia, N., Havinga, P.: Outlier detection techniques for wireless sensor networks: a survey. IEEE Comm. Surv. Tutor. **12**(2), 159–170 (2010)
9. Ripley, B.D.: Pattern Recognition and Neural Networks. Cambridge University Press, Cambridge (1996)
10. Ghahramani, Z.: An introduction to Hidden Markov Models and Bayesian Networks. Int. J. Pattern Recogn. Artif. Intell. **15**(1), 9–42 (2001)
11. Ayadi, A., Ghorbel, O., Obeida, A.M., Abida, M.: Outlier detection approaches for wireless sensor networks: a survey. Comput. Netw. **129**(1), 319–333 (2017)
12. Yuvaraja, M., Sabrigiriraj, M.: Fault detection and recovery scheme for routing and lifetime enhancement in WSN. Wirel. Netw. **23**(1), 267–277 (2017)
13. Titouna, C., Aliouat, M., Gueroui, M.: FDS: Fault Detection Scheme for wireless sensor networks. Wirel. Pers. Commun. **86**(2), 549–562 (2016)
14. Titouna, C., Gueroui, M., Aliouat, M., Ari, A.A.A., Amine, A.: Distributed fault-tolerant algorithm for wireless sensor network. Int. J. Commun. Netw. Inf. Secur. (IJCNIS) **9**(2), 241–246 (2017)
15. Yuan, H., Zhao, X., Yu, L.: A distributed Bayesian algorithm for data fault detection in wireless sensor networks. In: International Conference on Information Networking, pp. 63–68 (2015)
16. Sutagundar, A.V., Bennur, V.S., Bhanu, K.N.A.A.M.: Agent based fault tolerance in wireless sensor networks. In: International Conference on Inventive Computation Technologies, pp. 1–6 (2016)
17. Siddique, U., Hoque, K.A., Johnson, T.T.: Formal specification and dependability analysis of optical communication networks. In: 2017 Design, Automation and Test in Europe Conference and Exhibition (DATE), pp. 1564–1569 (2017)
18. Zhang, H.: The optimality of Naive Bayes. In: Seventeenth Florida Artificial Intelligence Research Society Conference (2004)
19. Murphy, K.P.: Machine Learning: A Probabilistic Perspective. MIT Press, Cambridge (2012)

Towards an Extensible Context Model for Mobile User in Smart Cities

Boudjemaa Boudaa[1]([⊠])[iD], Slimane Hammoudi[2],
and Sidi Mohamed Benslimane[3]

[1] Département d'Informatique, Université Ibn Khaldoun, Tiaret, Algeria
boudjemaa.boudaa@univ-tiaret.dz
[2] MODESTE, ESEO, Angers, France
slimane.hammoudi@eseo.fr
[3] École Supérieure en Informatique, Sidi Bel-Abbés, Algeria
s.benslimane@esi-sba.dz

Abstract. In smarts cities environments, a recommender system (RS) has for goal to recommend relevant services to the user who is sometimes mobile. Thus, to be able to provide accurate personalized recommendations, the RS should be aware to the user's context (preferences, location, activities, environment, ...), thereby, it should be Context-Aware Recommender System (CARS, for short). Therefore, the context modeling becomes crucial for developing CARSs. Although there is a lack of context models in the RS literature, several ones have been proposed in pervasive computing field. Nevertheless, most of them are dedicated for closed spaces and should be reviewed to be more suitable for open intelligent environments such as smart cities. This paper aims to propose an extensible ontology-based context model for representing contextual information within a smart city. The proposed context model would subsequently allow to design and develop CARSs.

1 Introduction

A recommender system [1] in smart cities [2,3] aims to propose a relevant service to a given user at a given instant, according to her profile, geolocation and surrounding environment. The real challenge here is to bring the services of a city closer to its citizens. The service in this context can be information about an event, a product to buy, a movie to watch, a web page to consult or a place to see, even a road to avoid. In order to be able to provide accurate personalized recommendations, the RS should be aware to the user's context, it should be Context-Aware Recommender System (CARS) [4]. The context includes information about profile, preferences, mobilities, location of a user, even her emotions [5], and information about the surrounding environment (e.g., weather condition) in which the user evolves. Consequently, the context modeling [6] becomes crucial for developing CARSs.

While a lack of context models is found in the literature of recommender systems, several ones have been proposed in pervasive computing field. However, most of them are dedicated for closed spaces and should be reviewed to be more suitable for CARSs in open intelligent environments such as smart cities.

By exploring the most popular context models of pervasive computing [7–9], this paper aims to propose a generic context model for representing context information taking into account the user's mobility information within a smart city. This model is based on ontologies by benefiting from their semantic expressiveness and reasoning capabilities which together enhance the prediction and accuracy to provide services recommendations.

Also, by using the ontologies for modeling, this context model can fulfill to some requirements such as genericity and flexibility as mentioned in [10].

After this introduction, some theoretical background regarding the scope of the paper are given in Sect. 2. Section 3 presents the proposed ontology-based context model for mobile user in a smart city that is instantiated for a tourism example, and related works are discussed in Sect. 4. Finally, Sect. 5 concludes this paper and indicates some future works.

2 Fundamental Concepts

In order to be familiar with the concepts used in the present work, this section gives a theoretical background regarding the scope of the paper.

2.1 Context and Context-Awareness

Context. The context designates all information elements that can influence the understanding of a particular situation. Despite its wide use in different research areas namely in Ubiquitous/Pervasive Computing, Ambient Intelligence and Intelligent Environments, we find a variety of its definition in the literature [11] that does not reach to a consensus among them [7]. However, we highlight that the definition proposed in [12] is the most acknowledged one, considering it as *"any information that can be used to characterize the situation of an entity"*, where *"an entity can be a person, place, or object that is considered relevant to the interaction between a user and an application, including the user and application themselves"*. The authors give a general definition that can be used in a wide range of context-aware applications. To refine their definition, they identify four categories of context that they feel are more practically important than others. These are location, identity (user), activity (state) and time [13].

In our scope of context-aware recommender systems, we believe that this definition is quite sufficient to identify the context change about a user and her various mobilities in a smart city compared to the application that remains unchangeable in most cases. For instance, imagine that a tourist takes the tramway (activity) to reach a tourist place (location), and that the trip will take a duration of 15 min (time). The recommender system in this case may suggest watching a video according to the user's preferences or about this touristic place (identity, location) whose duration is slightly less than 15 min.

Context-Awareness. The characteristic "context-aware" is generally used to describe any type of system that benefits from using context [7]. Nowadays, computing applications operate in a variety of new settings; for example, embedded in cars or wearable devices. They use information about their context to respond and adapt to changes in the computing environment. These kind of applications are, in short, increasingly context-aware (sometimes called "smart" or "intelligent") like as Google Now applications (https://www.google.co.uk/landing/now/).

In the literature, there have been numerous attempts to define context-awareness. In [13], the authors also defined a system as context-aware if *"it uses context to provide relevant information and/or services to the user, where relevancy depends on the user's task"*. In our previous example (Subsect. 2.1), the RS will be aware on the context information about user's preferences to recommend her a suitable set of videos.

2.2 Smart City

Owing to the fast and continual development of new technologies (e.g., Internet of Things) and infrastructures (e.g., smart administration, smart mobility, smart economy, ...) that have been (and/or will be) applied under the smart city concept, it is difficult to precise a unified definition of a smart city. Recent few years witnessed several evolutionist descriptions about smart cities according to these developed technologies/infrastructures. In [2], *"a Smart City can be viewed as an urban innovation and transformation that aims to harness physical infrastructures, Information and Communication Technologies (ICT), knowledge resources, and social infrastructures for economic regeneration, social cohesion, better city administration, and infrastructure management"*. The authors in [3] present the Smart City of today as *"a city planning/urban development methodology heavily relying on ICT to gather necessary input and make optimal engineering and planning decisions"*.

Briefly, a smart city is a city that uses information and communication technologies (ICT) to improve the quality of urban services or reduce its costs.

The most distinguishing feature of the smart city concept is the welfare of city residents whom are not taking any active role in the development and daily management of the city and remain passive [3]. The main goal of smart cities is the transformation of life and work of city inhabitants for the best and make them (life and work) easier.

In this perspective, context-aware recommender systems find their place in the smart city as intelligent applications to facilitate the professional and social lives of the user.

2.3 Context-Aware Recommender System

Recommender systems research [1] typically explores and develops techniques and applications for recommending various products or services or other items

to individual users based on the knowledge of users' tastes and preferences as
well as users' past activities (such as previous purchases).

Context-aware recommender systems (CARS) [4,5] are an important subclass
of recommender systems that take into account the context in which an item
(e.g., service) will be consumed or experienced. In context-aware recommendation research, a number of contextual features have been identified as important
in different recommendation applications [5]: such as *companion* in the movie and
tourism domains, *time* and *mood* in the music domain, and *weather* or *season*
in the travel domain. The context allow CARS to be more accurate in providing recommendations services using the prediction task into a multidimensional
rating function — *R: Users × Items × Contexts → Ratings* [4].

For example, a traditional RS can give us a list of recommended movies
to watch without taking into account any appropriate place/day or hour, but
CARS can give this list if time and location permit it (e.g., at cinema/weekend
or during a trip in train). Here, the context suggests the best time/location to
watch the preferred movies.

3   Context Modeling for User-Centered Smart Cities

The context model is a centerpiece for the design of a context-aware recommender systems in mobile and intelligent environments [14]. In the literature of
context modeling [8,9], we find many modeling techniques of context (key-value,
graphical, object-oriented, logic-based, ontology-based) [7], although all techniques have their strong points and drawbacks, ontologies are the most widely
adopted approaches [8]. They [15] (i) represent an extensible large-scale formalism to model the context in which it is possible to add context elements
(concepts, relationship, axioms, ...), delete or modify others in a flexible way,
(ii) offer more expressiveness and semantic richness to describe complex context
information with a consensual conceptualization, (iii) are fitted with inference
capabilities enabling reasoning about the modeled context information in order
to deduce other ones, (iv) make the possibility to share and/or reuse context
among different information sources in intelligent environment, and (v) have
fairly sophisticated tools available.

Once the context is modeled, it is possible to reason about the currently
captured/sensed contextual information and to deduce other new knowledge.
The reasoning techniques can be divided into [16]: supervised learning, unsupervised learning, rules, fuzzy logic, ontological, probabilistic. For this paper, we
will based on ontologies for context modeling and SWRL rules for ontological
reasoning.

3.1   Context Modeling Requirements for Intelligent Environments

In pervasive computing, several context models have been proposed. Nevertheless, most of them can not be more suitable for the CARS development, especially, in smart cities. The latter are fitted abundantly by renew and smart

technologies/infrastructures that enabling profound changes in the user's social life mode. For CARS in a smart city, the resident has not almost any role and remains a passive actor in the whole smart city process which allows her to receive pushed services recommendation.

In order to propose a generic context model, the authors in [10] specify some requirements that can be adopted for this work as follow:

- be sufficiently general to be used by different user centered mobile applications in smart cities (such as, recommender systems).
- be sufficiently specific to cover the main contextual entities proposed in the state of the art of context-aware mobile applications (such as, context-aware recommender systems).
- be sufficiently flexible to allow an extension and to take into account new entities specific to a given application domain in smart cities.

Also, we can specify the context more precisely in particular for recommendation services in smart cities. Therefore, the two following remarks are considered in our context model:

1. The user should not need to interact explicitly with the application to receive recommendations. Pushed recommendations could be very relevant for mobile users and the system could decide to recommend a user a given service because of her current context.
2. The notion of context is defined in a very general manner and would benefit from being more precise. We argue that *profile, Activity* and the *environment* are three main entities of the context that should be precised, specifically for context-aware recommender systems.

3.2   Ontology-Based Context Model

Considering the previous requirements, Fig. 1 depicts our ontology-based context model represented with Ontology Definition Meta-model (ODM) specification [17] (for Model-Driven Engineering purpose).

Extending two works [14,15], it is structured on three sub-ontologies at three different abstraction levels: generic, domain and application ontologies. They cover four main context dimensions that we believe contribute to develop context-aware recommender systems:

- *User* (on which the CARS is centered): A person that has a profile and preferences. A user evolves in an environment, and her state can be mobile or static. The profile and preferences are strongly attached to the user and contains the information that describes her. For example, the goal of a tourist searching for a restaurant is to have dinner. A profile/preferences in this case can give information concerning the tourist's culinary preferences or whereabouts.
- *MobileDevice:* It is the mobile computing system used by the user to invoke or receive services, and to capture contextual information from the environment. The device can obtain information concerning its type (e.g., tablet, laptop, smart-phone), the application, and the network.

- *Activity:* The activity in which the user is involved may be a key to decide which service is relevant to her. It can be of professional, social or leisure type (e.g., working, meeting, visiting, playing, and so on.). However, recognizing users' activities is still a tough task.
- *Environment:* It contains all the information that describes the surroundings of the user and its device, and that can be relevant for the recommender application. It includes different categories of information such as: spatial context information (e.g., Location, city, destination, speed), temporal context information (e.g., Time, date, season), Climatic context information (e.g., Temperature, type of weather).

The relationships between the *User* Entity and the others are respectively: *uses-Device, involvedIn* and *evolvedIn*. The dimensions are regrouped in the Generic ontology as top level abstraction entities.

The Domain ontology, on the other hand, allows covering all main contextual entities proposed in the literature of context modeling [7–9], and the Application ontology is to specify the context for a confined application space.

For a Tourism example, this Application ontology contains a set of specific context classes such as *TouristProfile* and *TouristPreferences* for the *User* entity (i.e., the tourist), and *TouristLocation* as well as the *TouristPlace* related to the *Environment* entity. In practice, a CARS can recommend to a tourist to visit a place that it is closer to her location and according to his preferences.

In this proposed context model, the ontological attributes (OWLDataType-Property) is responsible to contain context information and can be modeled using also the OUP (Ontology UML Profile) notation proposed with ODM specification [17].

The Domain and Application ontologies are sufficiently flexible to enable storing all context entities involved in a contextual situation and adding each new entity that could be introduced as context information. In addition, other context information can be derived by using reasoning capabilities of ontologies to complete describing contextual situations.

3.3 Context Reasoning with SWRL

The context sources allowing individualizing described ontologies can be provided in two different ways: sensing or inferring methods. Sensing is used directly for basic and low-level context; whereas, inferring is used to provide a high-level context from low-level one using rule-based inference languages [8].

Semantic Web Rule Language (SWRL) [18] allows to express reasoning rules that can be executed through a rule-based engine such as Java Expert System Shell (JESS).

As an example, if the *"Language"* of a tourist is only *"Arabic"* (as low-level context information sensed from her profile), we can infer that *"Arab Countries"* (as high-level context information) is among her "Favorite tourist places", that is a useful information for Tourism Recommender Systems. This is expressed by

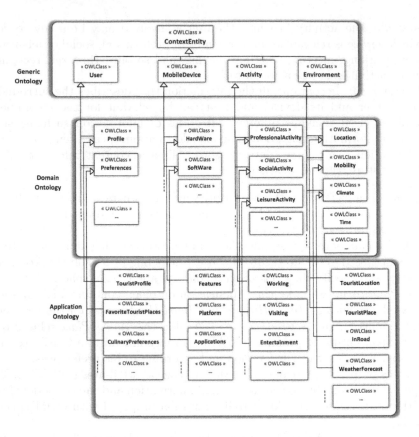

Fig. 1. An ontology-based Context Model

the next SWRL rule: Tourist(?t) **and** hasLanguage(?t, "Arabic") **implies** hasFavoriteTouristPlaces(?t, "ArabCountries").

In the literature, the low-level and high-level context are also called atomic and composite context, respectively. The relationship between them can be represented by the UML composite design pattern (see Fig. 2), in which, high-level context can be compose from others context information (whether low or high levels context).

Thereby, the low-level context can be provided by both types of context sources (Local or Remote). The first is for local provisioning of context information that is available at client (*MobileDevice*) or server (*Recommender Application*) sides. For example, the location of a tourist is captured in *LocationTourist*, which is used to locate the tourist. The second source type is for extra context information that can be captured by a third-party (e.g., a Web service to provide the weather forecast for a tourist place to visit, ...).

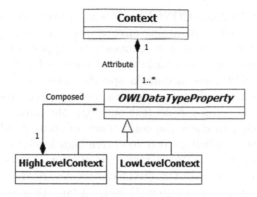

Fig. 2. Low and high levels context

4 Related Work

To the our best knowledge and except [14] which proposed a generic context model, there have not ontology-based context models for context-aware recommender systems (CARS), In contrast, several ones have been proposed in pervasive computing [7–9]. Nevertheless, most of them can not represent all context information available in a smart city and needed for its processes. In this paper, we present the most popular of these models by showing the pros and cons of each one of them regarding our scope of CARS in smart cities.

Chen et al. have developed CoBra-Ont [19], an OWL ontology which is a collection of ontologies for describing places, agents, events and their associated properties in an home area intelligent environments. This work is registered in their Context Broker Architecture (CoBrA) that provides knowledge sharing, context reasoning, and privacy protection supports for pervasive context-aware systems.

Based on the previous ontology, the same authors are presented another context ontology in [20]. The ontology of Standard Ontology for Ubiquitous and Pervasive Applications (SOUPA) provides knowledge sharing, reasoning on context and interoperability in a pervasive and ubiquitous environment. SOUPA deals with more areas of pervasive computing than CoBra-Ont and map many of its concepts to concepts of existing common ontologies like friend-of-friend (FOAF)[1] ontology to enable interoperability by using ontology mapping constructs OWL standard (owl: equivalentClass and owl: equivalentProperty). It is constituted of two sets of ontologies called SOUPA-Core and SOUPA-Extensions. SOUPA-Core defines a generic vocabulary which is universal for different pervasive computing applications (e.g., person, agent, policy, time, space), while SOUPA-Extensions define additional ontologies to support specific kind of applications and provide examples for extending future ontologies (e.g., home, office, attraction).

---

[1] http://www.foaf-project.org/.

The CONtext ONtology (CONON) [21] is a context OWL ontology for modeling context in pervasive computing environment and supporting logic based context reasoning. Wang et al. have proposed an upper context ontology that captures general concepts about basic context such as location, activity, person or computational entity, and a domain specific ontologies for detaining these general concepts and their features in each sub-domain covered in a hierarchical manner. Also and based on this context ontology, the authors have studied the use of logic reasoning to check the consistency of context information, and to reason over low-level, explicit context to derive high-level, implicit context. The notions of upper ontology and domain-specific ontologies are discusses in similar work [22] for presenting Service-Oriented Context-Aware Middleware (SOCAM).

In their project called Context-Driven Adaptation of Mobile Services (CoDAMoS)[2] and suiting some requirements of mobile computing, Preunveneers et al., have presented an adaptable and extensible context ontology for creating context-aware computing infrastructures, ranging from small embedded devices to high-end service platforms [23]. It is defined around four modeling entities Users, Environment, Platforms and Services. This ontology has been designed with the aim of solving many challenges, such as application adaptation, automatic code generation, code mobility, and generation of device-specific user interfaces for allow interoperability in an Ambient Intelligence environment.

Strang et al. [24] describe a context modeling approach using ontologies as a formal foundation. They introduce their Aspect-Scale-Context (ASC) model and show how it is related to other models. A Context Ontology Language (CoOL) is derived from the model, which is used to enable context-awareness and contextual interoperability during service discovery and execution in a distributed architecture.

All proposed context models are specific for closed environment (e.g., a smart home), and can not be applied for wide intelligent spaces. While ours is extensible to new appeared context information in opened environments such as smart cities.

5   Conclusion and Future Work

In this paper, we have proposed an extensible context model based on ontologies and benefiting from their expressiveness and reasoning capabilities. It combine three levels of ontologies, Generic, Domain ontologies for popular and shared concepts used by context-aware applications developers and Application ontology for a specific description of context (e.g., Tourism domain) in wide space such as smart city.

In near future, we aim to extend much more our context model in order to be sufficiently generic and can be reused to build several context-aware recommendation systems for the different smart city applications, and to implement it taking into account how to create new context information attributes at run-time

---

[2] http://distrinet.cs.kuleuven.be/projects/CoDAMoS/.

and integrate them to this ontological model in order to capture any new context entities could be appeared over time. Once, this is achieved, the design and the development of context-aware recommender system (CARS) can be undertook.

References

1. Bobadilla, J., Ortega, F., Hernando, A., GutiéRrez, A.: Recommender systems survey. Knowl. Based Syst. **46**, 109–132 (2013)
2. Curry, E., Dustdar, S., Sheng, Q.Z., Sheth, A.: Smart cities: enabling services and applications. J. Internet Serv. Appl. **7**(1), 6 (2016)
3. Dustdar, S., Nastić, S., Šćekić, O.: Introduction to smart cities and a vision of cyber-human cities. Smart Cities, pp. 3–15. Springer, Cham (2017). https://doi.org/10.1007/978-3-319-60030-7_1
4. Adomavicius, G., Tuzhilin, A.: Context-aware recommender systems. In: Ricci, F., Rokach, L., Shapira, B. (eds.) Recommender Systems Handbook, pp. 191–226. Springer, Boston (2015). https://doi.org/10.1007/978-1-4899-7637-6_6
5. Zheng, Y., Mobasher, B., Burke, R.: Emotions in context-aware recommender systems. In: Tkalčič, M., De Carolis, B., de Gemmis, M., Odić, A., Košir, A. (eds.) Emotions and Personality in Personalized Services. HIS, pp. 311–326. Springer, Cham (2016). https://doi.org/10.1007/978-3-319-31413-6_15
6. Sheng, Q.Z., Benatallah, B.: ContextUML: a UML-based modeling language for model-driven development of context-aware web services. In: Proceedings of 2005 International Conference on Mobile Business (ICMB 2005), Sydney, Australia, pp. 206–212, 11–13 July 2005
7. Alegre, U., Augusto, J.C., Clark, T.: Engineering context-aware systems and applications. J. Syst. Softw. **117**(C), 55–83 (2016)
8. Bettini, C., Brdiczka, O., Henricksen, K., Indulska, J., Nicklas, D., Ranganathan, A., Riboni, D.: A survey of context modelling and reasoning techniques. Pervasive Mob. Comput. **6**, 161–180 (2010)
9. Ye, J., Coyle, L., Dobson, S., Nixon, P.: Ontology-based models in pervasive computing systems. Knowl. Eng. Rev. **22**, 315–347 (2007)
10. Hammoudi, S., Monfort, V., Camp, O.: Model driven development of user-centred context aware services. IJSSC **5**(2), 100–114 (2015)
11. Bazire, M., Brézillon, P.: Understanding context before using it. In: Dey, A., Kokinov, B., Leake, D., Turner, R. (eds.) CONTEXT 2005. LNCS (LNAI), vol. 3554, pp. 29–40. Springer, Heidelberg (2005). https://doi.org/10.1007/11508373_3
12. Dey, A.K., Abowd, G.D.: Towards a better understanding of context and context-awareness. Technical report GIT-GVU-99-22, Institute of Technology, Georgia, June 1999
13. Dey, A.K., Abowd, G.D., Salber, D.: A conceptual framework and a toolkit for supporting the rapid prototyping of context-aware applications. Hum. Comput. Interact. J. **16**(2), 97–166 (2001)
14. Nicolas, G.N., Amghar, T., Camp, O., Hammoudi, S.: A framework for context-aware service recommendation for mobile users: a focus on mobility in smart cities. From Data to Decision (2017)
15. Boudaa, B., Hammoudi, S., Mebarki, L.A., Bouguessa, A., Chikh, M.A.: An aspect-oriented model-driven approach for building adaptable context-aware service-based applications. Sci. Comput. Program. **136**, 17–42 (2017)

16. Perera, C., Zaslavsky, A., Christen, P., Georgakopoulos, D.: Context aware computing for the internet of things: a survey. IEEE Commun. Surv. Tutorials **16**(1), 414–454 (2014)
17. Gasevic, D., Djuric, D., Devedzic, V.: Model Driven Engineering and Ontology Development, 2nd edn. Springer, Heidelberg (2009)
18. Horrocks, I., Patel-Schneider, P.F., Boley, H., Tabet, S., Grosof, B., Dean, M.: SWRL: a semantic web rule language combining OWL and RuleML. Technical report, World Wide Web Consortium, May 2004
19. Chen, H., Finin, T., Joshi, A.: An ontology for context-aware pervasive computing environments. Knowl. Eng. Rev. **18**, 197–207 (2003). Special Issue on Ontologies for Distributed Systems
20. Chen, H., Perich, F., Finin, T., Joshi, A.: SOUPA: standard ontology for ubiquitous and pervasive applications. In: International Conference on Mobile and Ubiquitous Systems: Networking and Services, pp. 258–267 (2004)
21. Wang, X.H., Zhang, D.Q., Gu, T., Pung, H.K.: Ontology based context modeling and reasoning using OWL. In: IEEE International Conference on Pervasive Computing and Communications Workshops, p. 18 (2004)
22. Gu, T., Wang, X.H., Pung, H.K., Zhang, D.Q.: An ontology-based context model in intelligent environments. In: Proceedings of Communication Networks and Distributed Systems Modeling and Simulation Conference, pp. 270–275 (2004)
23. Preuveneers, D., Van den Bergh, J., Wagelaar, D., Georges, A., Rigole, P., Clerckx, T., Berbers, Y., Coninx, K., Jonckers, V., De Bosschere, K.: Towards an extensible context ontology for ambient intelligence. In: Markopoulos, P., Eggen, B., Aarts, E., Crowley, J.L. (eds.) EUSAI 2004. LNCS, vol. 3295, pp. 148–159. Springer, Heidelberg (2004). https://doi.org/10.1007/978-3-540-30473-9_15
24. Strang, T., Linnhoff-Popien, C., Frank, K.: CoOL: a context ontology language to enable contextual interoperability. In: Stefani, J.-B., Demeure, I., Hagimont, D. (eds.) DAIS 2003. LNCS, vol. 2893, pp. 236–247. Springer, Heidelberg (2003). https://doi.org/10.1007/978-3-540-40010-3_21

Combining Proactive and Reactive Approaches in Smart Services for the Web of Things

Nawel Sekkal[1]([⊠]) [iD], Sidi Mohamed Benslimane[1] [iD],
Michael Mrissa[2] [iD], and Boudjemaa Boudaa[1] [iD]

[1] LabRi Laboratory, Ecole Superieure en Informatique, BP 73, El Wiam City,
Sidi Bel Abbes, Algeria
n_iles@mail.univ-tlemcen.dz, s.benslimane@esi-sba.dz,
boudjemaa.boudaa@univ-tiaret.dz
[2] LIUPPA, University Pau and Pays Adour, BP 1155, 64013 Pau Cedex, France
michael.mrissa@univ-pau.fr

Abstract. The Web of Things (WoT), facilitates the interconnection of different types of real-world objects, integrating them into the virtual world and ensuring their interoperability through Web services. However, it remains a challenge to automate the tasks connected objects need to deal with. In this paper, we focus on the development of smart web services that automate service tasks, autonomously adapt to context changes in the object's environment, and to users' preferences. In this article, we propose a software framework for smart services that relies on a reactive and proactive approach to deal with context and its temporal aspects. Smart web services developed according to these principles can react to current situations and proactively anticipate an unforeseen situation in order to take the right decision.

1 Introduction

In recent years, the development of the Web has been characterized by the increased use of connected devices, such as sensors, actuators and mobiles devices. These devices enable a direct integration of an information source depending on the client application context. As such, the WoT (Web of Things) interconnects any sort of physical object deployed in the real world (surveillance devices, medical sensors, temperature sensors, intelligent sensors, etc.), to the Web in order to promote global interoperability.

In the perspective of contributing in the emergence and development of the WoT, existing Web standards are used such as: (a) Uniform Resource Identifier (URI), (b) Representational State Transfer (REST), (c) Hypertext Transfer Protocol (HTTP) [1]. Furthermore, the usage of Web Semantics, such as Resource Description Framework (RDF) [2] and Linked Data (LD) [3] is necessary in order to: (a) provide a semantic value to the data generated by heterogeneous sources, (b) enable contextual interpretation, (c) ease the management of the growing volume of data and (d) promote interoperability at the semantic level.

In order to benefit from the vision of WoT, we adopt the notion of "Smart Services" developed in [4–6], that not only permit remote access to resources and their embedded

© IFIP International Federation for Information Processing 2018
Published by Springer International Publishing AG 2018. All Rights Reserved
A. Amine et al. (Eds.): CIIA 2018, IFIP AICT 522, pp. 509–520, 2018.
https://doi.org/10.1007/978-3-319-89743-1_44

functions, but offer intelligent services as well as adaptation to the application context. As a result, it provides the user with means to carry out its tasks automatically and in an autonomous manner. For instance: patient diagnostic, itinerary planning, adjusting home devices settings (thermostat, air-conditioning, radiator, etc.).

In this paper, we take interest in smart Web services and their diverse functionalities (acquiring, reasoning, adaptation rules depending on context changes). The latter enables the target system to make an autonomous decision without user intervention. Our approach to context adaptation relies on context and time Web. The temporal aspect is expressed with a hybrid proactive-reactive method [27]. The proactive part is interesting in the sense that we can predict the behavior of the system in the future and act before the situation happens. The reactive part is useful when the system needs to make a real-time decision. The rest of this paper is set out as follows: In Sect. 2, we describe the problem statement and introduce a motivating scenario in the smart home domain. Section 3 contains background information about WoT, context and smart Web services. The related works are presented in Sect. 4. In Sect. 5, we presented Framework for context aware proactive-reactive service. Finally, Sect. 6 outlines the conclusions and future work.

2   Problem Statement and Motivating Scenario

The problem we address in this paper is the following:

- How can smarts web services make decisions autonomously about the behavior of the system, or the temporal aspect is important in its actions?
- How can the smart web services adapt to the current situation of the user and to the change of the context of the different objects of the system of objects with a dynamic link?

Currently, there is a lack of work for the smart Web service to incorporate reactive and proactive behavior. The smart service must make decisions in advance, taking into account the events that will occur in advance and depending on the context. It must also adapt autonomously to changes in the context. Our research is motivated by these challenges to provide high-level information such as: detecting situations in the environment [10]; making predictions of situations in the future and reacting in advance, and determining the policy of actions to consume the appropriate services to adapt the environment in advance of the envisaged situation.

In rest of this section, we present a scenario in the field of smart home in order to illustrate the challenge and highlight our contribution. Our scenario involves a family that waters their plants on per-need basis and depending on water requirements. However, with the spring holydays (rain probability and invariable weather), they must leave their home vacant. The question is, how can they insure the daily watering tasks for their plants when they leave on holidays?

To solve this problem, the family installs an intelligent watering system that will monitor the garden remotely through a smart phone. This system enables surveillance in an autonomous way of the pertinent context in real-time (lighting, temperature, humidity, etc.) and decides when and how to act with an appropriate water dose depending on the plant type. Furthermore, a wireless sensor network is placed to collect and transmit the aforementioned parameters. This application takes into account weather conditions and forecast. To perform this task, a GPS API is used to determine the location of the family home. We consider that the software system can control the existing watering system. Consequently, the watering task is triggered automatically by sending notification messages to the user smart phone and react in anomaly cases.

To provide a smart watering service, the following questions arise:

- How will the watering system take into consideration the different types of plants and their preferences in dosage and frequency?
- When and how will the watering system be triggered, taking into account the plant environment and the weather forecast?
- How to adapt the watering system in case of unforeseen situation (unexpected adversaries, other water source, etc.)?

An implementation of the scenario is necessary to validate our approach.

3   Background

Smart services are the result of the combination of multiple technologies: The Web of objects, semantic Web, the context -ware systems, Web services and linked data.

3.1   The Web of Things

An intelligent object is any object capable of processing information and act on its environment in an autonomous way. One of its principle characteristics is its capacity to communicate with other objects. It can be represented in a software/hardware form having autonomous capacity, to communicate and cooperate in addition to reasoning features. Moreover, each object [26]:

- Is identifiable in a unique manner, (barcode, RFID, IP address, etc.);
- A processing capacity in order to control and manage the object;
- Is able to store data and its priorities;
- Is equipped with a network interface card to interact with other objects [8, 26].

The objects use communication protocols such as 6LowPan, Bluetooth, Low Energy, ZigBee, etc. [14, 15].

In the course of the recent years, the Internet of Things has evolved at an accelerating speed, connecting numerous heterogeneous objects (sensors, actuators, smart phones, applications, home devices, etc.) over the Internet. Due to multiple communication protocols, it was difficult to connect objects to each other's. Therefore, on top of the Internet, the Web of things appeared to support the application that operates over this infrastructure, linked with the technologies and relying of the Web, enabling the users to interact with remote services (simple and complex). A physical object is seen as a set of services accessible over the Web and for which the environment defines its context. The principle advantage of the WoT is the usage of the Web norms/protocols like Hypertext Transfer Protocol (HTTP) and URIs to promote interoperability [7].

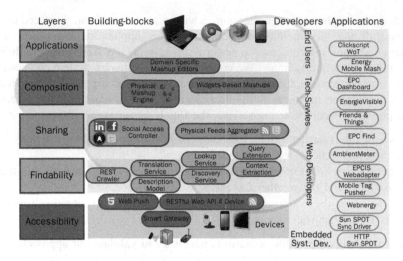

Fig. 1. Architecture of the WoT [8]

An architecture founded on resources, was proposed in [8] in the field of WoT (Fig. 1). It enables the integration of objects through services. This proposition is elaborated in the form of a layered architecture, structured around five layers:

Accessibility layer: deals with the integration of the object in the Web;
Findability layer: enables searching and locating the pertinent services in the WoT;
Sharing layer: Manage the access to objects (social networks);
Composition layer: enables the composition of services and introduces the notion of physical mashup;
Application layer;

For the conception of the WoT, we apply the same tools, techniques, models and languages used in Web applications. On the other hand, the applications concern objects with limited resources, thus other techniques must be adopted such as: AJAX (Asynchronous JavaScript and XML), mashups techniques (physical-environment and virtual- environment), Event Driven approach [8]. The development of WoT requires the extension of the existing Web so that the real world objects can be integrated in it, either in a direct or indirect manner. Consequently, the usage of Web services is necessary in order to exploit the data and functionalities of the physical objects.

Recently, the development of Web services and their APIs turned away from traditional services that use SOAP and WSDL. Instead, the REST architectural style has been recognized as a good engineering practice. In the WoT, it is more interesting to use REST which is conceived in a resource oriented architectural model in addition to the HTTP protocol for a better representation of resources and coordinated Client/Server communication. RESTful services are based on representative state transfer [9], an architectural model that considers that each physical object is addressable/identifiable through an URI. They are defined by the following four concepts: (a) identification of a resource with a URI, (b) the definition of a uniform interface using HTTP standard verbs (GET, POST, PUT and DELETE), (c) the usage of hypertext links, (d) communication through HTTP using formats such as JSON, YAML or XML. A comparative table between Ws-* and REST services is presented in [9].

3.2   Defining Context

The term context in "smart services" can refer to any kind of piece of information about an object, its placement, its modifications and also to the identity of users. Additionally, it can include computing environments, physical environment and even user's environments (social situation). The context is defined in [17] depending on three aspects: (a) usage context, (b) physical context, (c) information context. A subsequent study extended this definition by adding the time context and social context [3]. Hence, a physical object is characterized by:

Physical context: describes the state of the object that contains the acquired information through sensors placed in the device or its running software components. They can be dynamic (such as placement, activity, relation with others, etc.) or static (personal information, preferences, etc.).

Usage context: represents the information that concerns the user, they can be dynamic or static. Dynamic information usually changes over time and requires periodic monitoring. On the other hand, static information is rarely modified such as the user habit, identity and personal preferences.

Environmental context: are all information related to the environment of the user such as the ambient lighting, temperature, humidity, etc.

Social context: is external information that concerns the preferences of the users, their public profiles and their behavior in friend group. The set of information enables the system to recognize the user in its social context.

There are context modelling techniques, such as the techniques of central context management based on the P2P model, as well as others using XML based ontology (i.g. in Context project). In the field of WoT, the Web services can become more intelligent through the use of context. Meaning that the service is aware of the events, the physical situations of the target objects or application. Thus, it can respond proactively and intelligently. Furthermore, they can share their contexts with other services in order to respond to users' requests.

3.3   Smart Web Services

The set of technologies associated to Web service enables remote software applications to communicate through the Web. This technology is based on the principles of service oriented computing and uses the languages of the Web for data description. Its development has progressed with the emergence of semantic Web services which represent a promising path in order to automate Web services tasks (search, invocation, composition). Another approach is to extend API semantics with the principles of "linked data" [11]. This technology is useful to provide access for the resource to its set of related data, to integrate and use them through RDF links which makes the Web an interconnected space. Currently, another interest is taken towards social networks by the use of AJAX technologies [13].

All these approaches aim to introduce new functionalities to Web services and make them smart and autonomous. With the technological advances in sensors and their growing availability in intelligent objects, context is becoming an essential source of information from which we can understand the different situations and adapt specific services. These services are equipped with intelligent capacities; they are called "smart services". They are available today on smart phones, highways, smart televisions, smart cities and health care devices, with the objective to provide customized functionalities without the need of direct users' intervention. In addition to the context, we recommend a proactive approach that allows a smart web service to meet the future needs of the user. We propose in this approach to proactively recommend a service that can respond to users or devices invoked.

A smart service is defined in [5] as a software service that helps the users in their daily lives activities, with an important productivity, an improved quality of services and an efficient communication between the users and objects. In [6], a smart service is defined as a general model that enables the integration of any functions of an ontology of a field where one or multiple Web services are exploited automatically in order to satisfy the specific need of an object. In [16] a smart Web service "SmartWS" is defined as a Web API that conform to standards (HTTP, URI), that uses/produces semantic data (RDF) and encapsulates a logical decision in an autonomous fashion. By comparison with the traditional Web service, a smart Web service is characterized by: (a) autonomy, (b) automatic behavior, (c) intelligence, (d) communication capacity, (e) customization, (f) surveillance and control capacities of objects.

4   Related Work

Numerous works have been conducted in the field of WoT. In [13] the works in this field are classified into five categories: pre-processing and storage of data, data analysis, services management, security and confidentiality, network and communication. A significant amount of work has been done for WoT modelling in order to represent real world objects in the Web such as: TinyRest, WebPlug, AutoWeb, SpitFire, SOCRADES, Avatar, etc. [5]. Generally, these models are founded on LoWPAN and CoAP for the integration of objects in the Web as well as the REST/WS-* services in order to insure the interaction between intelligent objects. Nevertheless, very little work has been made on WoT smart services modelling. In [7, 16] two intelligent applications are shown: "Energy Visualizer" for energy consumption surveillance and control of household devices, and "Ambient Meter" in order to obtain energy consumption of machines. A model of smart services (SSF) based on context adaptation is presented in [5] with a central system scenario. The authors base their work on [17] to define context. A centralized architecture is proposed and the processing is based on pre-requisite knowledge that enables to adapt the offered services to better respond to user requirements. The work in question represents a first step towards intelligent services. Unfortunately, it is concentrated only on the notion of context adaptation and does not address heterogeneous data without explaining how sensors/actuators are registered in the platform or how they communicate.

In [6] an example of intelligent service for plant watering based on context is presented. In order to control the dosage and frequency of water spray, the service must detect the dryness of the object, the knowledge on the field of plants, plant watering "preferences", and use the weather prediction service. An intelligent interface was developed based on Thing-REST model. Subsequently, an extension for the afore-mentioned work by replacing Web Service RESTful with CoAP is presented in [9], introducing the semantic aspect in the event detection service. A new tendency to develop Web services based on micro-services also recently emerged [17].

While the aforementioned contributions were based on the notion of context, others were proposed that are focused on semantic Web. For instance, a personalized mete-orological semantic service as well as an improvement of functionalities of a service for disaster management is presented in [18]. In [19], the authors describe an application in the field of social networks that is oriented towards integrating objects in the daily lives of people. Both applications use the WoT technology as well as the semantic, smart services. Nevertheless, there was no extension of the intelligence notion in these works. The model of smart Web service "SmartWS" in a form of a Web API introduced in [16] enables the automatic adjustment of settings and consequently, adapts to the context where the event is triggered. It is based on semantic technologies in order to describe input and output as well as linked data in order to provide access to all related resources. In [21] a testing model was proposed to evaluate and validate intelligent applications as well as QoS depending on user requirements, with the help of the Quality Model (QM) of the International Organization for Standardization (ISO). In [17, 20] a new avatar model in the Web of objects discovers the capacities of objects and exposes them as functionalities based on semantic techniques as part of the ASAWoO project [17]. In the same project, a meta-model of context [22, 23] capable of responding to different adaptation questions for a given WoT application is proposed.

Table 1 is a summary of the works cited in the state of the art section. The different studies aim to introduce context of objects with a semantic description in Web services in order to make them "Smart". In the pursuit of smart service development, the discussed works are based on context adaptation with semantic description of data while taking decisions in an autonomous way and in real time (reactive approach).

Our proposition is founded on the generic model of smart Web service. It is not based only on context, but we have added temporal aspects as well in order to predict the future behavior of the system. This approach is original because it combines reactive and proactive decision-making, seeking to adapt automatically to the environment according to the time-evolving situations.

Table 1. Summary of the state of the art on smart web services

Approach	Context	Semantic	Architecture	Adapt	Protocol	Data format	Linked data	Protocol network	Application
Guinard [20]	X	X	Web architecture (SUN SPOT)	X	Rest	JSON	X	–	Ambient meter, energy visualizer
He et al. [6]	✓	✓	Resource oriented ROA	✓	Thing-rest	XML	X	HTTPs	Plant watering system
Ramaswamy et al. [9]	✓	✓	Multi-layer	✓	Rest	XML	X	6LowPAN/COAP	Plant watering System
Lee et al. (SSF) [5]	✓	X	Multi-agent	✓	Reasoning algorithm (HMM, KNN)	XML	X		Smart home
Lee et al. WISE [18]	X	✓	M2M/cloud	X	–	RDF	X	–	Meteorological service
Beltran et al. [19]	X	✓	Multi-layer	X	REST	JSON/LD	X	–	Social field
Maleshkova et al. (SmartWS) [4]	✓	✓	Multi-layer	✓	REST	XML/JSON	✓	HTTPs	Medical diagnostic
Zeiner [17]	X	X	Multi-service	✓	REST	JSON/XML	X	HTTPs	/
Gyrad et al. [25]	X	✓	M3	X	–	RDF	✓		Mobile application
Mrissa et al. [17, 18]	X	✓	Multi-agent	X	REST	XML	X		Temperature regulator
Terdjimi et al. [24]	✓	X	Multi agent	X	/	Contextual data	X	–	Agriculture field

5 Proposed Architecture

Figure 2 gives an overview of a multi-layered architecture that represents the development of our smart web service that is context-based and proactive-reactive.

It provides the following functionalities:

- Data acquisition in order to define the context of objects and user;
- Identifying the current situation of the user in order to understand the user requested services;
- Selecting the appropriate service and adapt it in a proactive and reactive way in order to obtain the appropriate decision;
- Adapting the service in an autonomous way depending on the changes of the context;

The framework involves different types of objects that must collaborate. Each object is considered as a resource. Based on the REST architectural style, any resource represented with a sensor, is addressed by a unique identifier of standard format via Uniform Resource Locators (URL) using the Hypertext Transfer Protocol (HTTP) and its methods (e.g. GET, POST, PUT and DELETE) to access them.

Furthermore, we use the ontologies and domain knowledge that are needed to unify and understand the context of things and their preferences (e.g., dose, frequency, or properties) and user preferences. We think that our proactive-reactive approach can

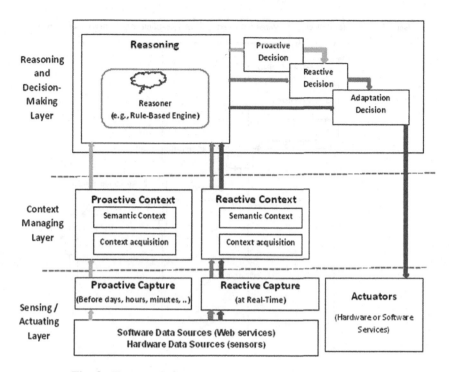

Fig. 2. Framework for context aware proactive-reactive service

offer an added value to Web service mechanisms that adapt to context and usage preferences, in order to make it a smart Web service. For the development of our smart web service, we present a generic model in terms of architectural view made of multiple layer types.

"Sensing actuating layer": includes all data sources: hardware data sources, such devices, wearables, sensors, algorithms, and software components, such as web services. The capture of the data can be either reactive, i.e. real-time, or otherwise predict the data as it will be in the future (e.g., weather forecasts), this may help us make the right decision and avoid unwanted events.

"Context managing layer": enables to comprehend the situation and define the corresponding service. It is composed by the following sub components: a Proactive context and a Reactive context that are necessary for the smart web service in order to make the decision autonomously and adapts dynamically.

For the acquisition context, each resource is identifiable with a unique URI and HTTP, we use the REST model and its methods (GET, PUT, DELETE, POST) to gather the state of resources represented by a sensor. The semantic context enables to unify the context of the objects using the context ontology in order to facilitate interfacing between Web services.

"Reasoning and decision-making layer": depending on the acquired context, this component deduces the corresponding situation based on the reasoning algorithms that are of three types: Sequential reasoning algorithms, relational algorithms, decision tree.

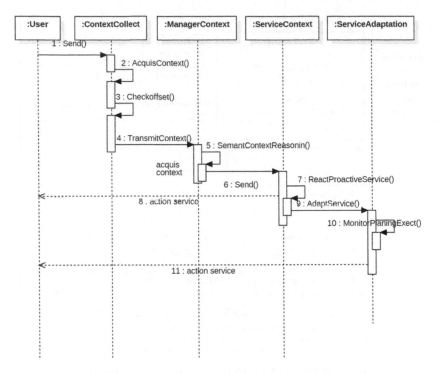

Fig. 3. Sequence diagram for the proactive-reactive framework.

The reactive decision takes the decision to execute a service depending on the situation in real-time, according to the context reactive and the proactive decision is based on the proactive context to recommend a course of action and the appropriate service in the near future. This mechanism can be implemented based on Bayesian Networks to estimate the probability of a situation to expire in the future. Particularly, the final decision taken by our approach is founded on the decisions of the reactive and proactive mechanisms, since the temporal aspect is crucial to respond to the intelligent system requirement. The decision adaptation adapts and customizes the associated service depending on the changes of context. For example, following unexpected changes or services errors. As a result, our system must adapt and reconsider the decision. We can use a loop such as MAPE-K that contains the state indicators: monitoring, analysis, planning, and execution. It is necessary to implement our scenario to validate this approach. Figure 3 shows a sequence diagram for the proposed proactive-reactive framework:

6 Conclusion

A major challenge of the WoT is to allow applications to adapt to their environment. In this article, we discuss the notion of smart web services that plays an important role in our society. We have established a state of the art in this field, and we have proposed a framework of smart web service solution for domain of web of things, based a proactive and reactive approach. The proactive part is interesting in the sense that we can predict the behavior of the system in the future and act before the situation happens. The reactive part is useful when the system needs to make a real-time decision. This approach is supported by semantic technology and the concept of the context, to enable the smart web service to make the most effective decision in a way. Autonomous and adapt it according to the change of context.

Possible directions for future research include:

(i)   To propose a context meta-model in Wot that allows transforming raw context data, into semantically annotated contextual information, and then solving easily adaptation requests. The model can be represented based on UML diagram of formal declarations,

(ii) An approach for handling reactive and proactive behaviors using different techniques: dynamic Bayesian network, Markov chain, fuzzy logic ...,

(iii) Improve the proposed architecture in relation to the proposed model,

(iv) Extend the use case of watering by introducing a scenario that supports proactive action (drought, abundance of rain, failure of irrigation equipment, etc.), and testing with the architecture proposed over a real world automated environment.

References

1. Klyne, G., Carroll, J.J.: Resource description framework (RDF): concepts and abstract syntax. In: W3C Recommendation (2004)

2. Heath, T., Bizer, C.: Linked data: evolving the web into a global data space. Synth. Lect. Antic Web: Theory Technol. **1**(1), 1–136 (2011)

3. Garriga, M., Mateos, C., Flores, A., Cechich, A., Zunino, A.: RESTful service composition at a glance: a survey. J. Netw. Comput. Appl. **60**, 32–53 (2015)

4. Maleshkova, M., Philipp, P., Sure-Vetter, Y., Studer, R.: Smart web services (SmartWS) – the future of services on the web. IPSI BgD Trans. Adv. Res. **12**(1), 15–26 (2016)

5. Lee, J.Y., Kim, M.K., La, H.J., Kim, S.D.: A software framework for enabling smart services. In: SOCA (2012)

6. He, J., Zhang, Y., Huang, G.: A smart web service based on the context of things. J. ACM Trans. Internet Technol. (TOIT) **11**(3), 13 (2012)

7. Guinard, D., Trifa, V., Mattern, F., Wilde, E.: From the internet of things to the web of things: resource oriented architecture and best practices. In: Uckelmann, D., Harrison, M., Michahelles, F. (eds.) Architecting the Internet of Things, pp. 97–129. Springer, Heidelberg (2011). https://doi.org/10.1007/978-3-642-19157-2_5

8. Guinard, D.: A web of things application architecture - integrating the real-world into the web. These doctorat (2011)

9. Ramaswamy, P., Vinob, C., Elias, S., Manoj, S.S.: A REST based design for web of things in smart environments. In: 2nd IEEE International Conference on Parallel, Distributed and Grid Computing (2012)

10. Khalfi, M.F., Benslimane, S.M.: Proactive approach for service discovery using web service for devices on pervasive computing. In: ICCASA 2014 Proceedings, Dubai, United Arab Emirates (2014)

11. Jamont, J.P.: Démarche, modèles et outils multi-agents pour l'ingénierie des collectifs cyber-physiques (2016)
12. Terdjimi, M., Médini, L., Mrissa, M., Le Sommer, N.: An avatar-based adaptation workflow for the web of things. In: 25th IEEE International Conference on Enabling Technologies: Infrastructure for Collaborative Enterprises (2016)
13. Liu, W., Li, X., Huang, D.: A survey on context awareness. In: Proceeding of the International Conference on Computer Science and Service System (CSSS 2011), pp. 144–147. IEEE (2011)
14. Mathew, S.S., Atif, Y., Sheng, Q., Maamar, Z.: The web of things - challenges and enabling technologies. In: Bessis, N., Xhafa, F., Varvarigou, D., Hill, R., Li, M. (eds.) Internet of Things and Inter-Cooperative Computational Technologies for Collective Intelligence, vol. 460, pp. 1–23. Springer, Berlin (2013). https://doi.org/10.1007/978-3-642-34952-2_1
15. Zeng, D., Guo, S., Cheng, Z.: The web of things: a survey. J. Commun. 6(6), 424–438 (2011)
16. Guinard, D., Trifa, V., Wilde, E.: A resource oriented architecture for the web of things. In: Proceedings of the Internet of Things Conference (2010)
17. Schilit, B.N., Theimer, M.: Disseminating active map information to mobile hosts. IEEE Netw. 8, 22–32 (2002)
18. Lee, J.W., Ki, Y.W., Kwon, S.: WISE: an applying of semantic IoT platform for weather information service engine. In: The International Semantic Web Conference (2014)
19. Beltran, V., Ortiz, A.M., Hussein, D.: A semantic service creation platform for Social IoT. In: The Meeting of the WF, IoT (2014)
20. Mashal, I., Alsaryrah, O., Chung, T., Yang, C., Kuo, W., Agrawal, D.: Choices for interaction with things on Internet and underlying issues (2014)
21. Zeiner, H., Goller, M., Juan, V., Salmhofer, F., Haas, W.: SeCoS: web of things platform based on a microservices architecture and support of time-awareness. e & I Elektrotech. Inftech. 133(3), 158–162 (2016)
22. Mrissa, M., Médini, L., Jamont, J.: Semantic discovery and invocation of functionalities for the web of things. In: IEEE WETICE, pp. 281–286 (2014)
23. Mrissa, M., Médini, L., Jamont, J., Sommer, N.L., Laplace, J.: An avatar architecture for the web of things. IEEE Internet Comput. 19(2), 30–38 (2015)
24. Terdjimi, M., Médini, L., Mrissa, M., Le Sommer, N.: An avatar-based adaptation workflow for the web of things. In: 25th IEEE International Conference on Enabling Technologies (2016)
25. Gyrard, A., et al.: Assisting IoT projects and developers in designing interoperable semantic web of things applications. In: IEEE International Conference on Data Science and Data Intensive Systems (DSDIS) (2016)
26. Mattern, F., Floerkemeier, C.: From the internet of computers to the internet of things. In: Sachs, K., Petrov, I., Guerrero, P. (eds.) From Active Data Management to Event-Based Systems and More. LNCS, vol. 6462, pp. 242–259. Springer, Heidelberg (2010). https://doi.org/10.1007/978-3-642-17226-7_15
27. Krupitzera, C., Rotha, F.M., VanSyckela, S., Schieleb, G., Beckera, C.: A survey on engineering approaches for self-adaptive systems. Pervasive Mob. Comput. 17, 184–206 (2015)

Multi-agent System Based Service Composition in the Internet of Things

Samir Berrani$^{(\boxtimes)}$, Ali Yachir, Badis Djamaa, and Mohamed Aissani

Artificial Intelligence Laboratory, Military Polytechnic School (EMP),
PO BOX 17, 16111 Bordj-El-Bahri, Algiers, Algeria
samir.berrani@yahoo.fr, a_yachir@yahoo.fr, badis.djamaa@gmail.com,
maissani@gmail.com

Abstract. Service composition is seen as the key issue to create innovative, efficient, flexible and dynamic applications on the Internet of things (IoT). Accordingly, we propose, in this paper, an approach for IoT service composition based on multi-agent system where several agents are engaged to satisfy the user request. This approach is designed using SysML and implemented using Netlogo platform. The use-cases scenarios and extensive tests show clearly the interest, the feasibility and the suitability of the multi-agent system for service composition.

Keywords: Internet of things · Service composition
Multi-agent system · Semantic web

1 Introduction

As a natural continuity of Ubiquitous Computing (UbiComp) and Ambient Intelligence (AmI), the Internet of Things (IoT) envisions a future Internet architecture integrating both physical and cyber worlds by combining sensing and actuating with digital services. IoT is the future all-IP ecosystem where smart objects (or IP networked things) connected to the Internet, exchange their data and capabilities as services, just like today's millions of Web services. The challenging question, however, is not only how to make these smart objects communicate over the Internet using open and standardized protocols, but also how their provided services can be exchanged and aggregated efficiently to create novel and dynamic IoT applications. This research problem is known as service composition.

Service composition is one of the core principles of the Service Oriented Computing (SOC) paradigm. It aims to reuse several existing component services (or atomic services) to create more complex services (or composite services) by joining them in creative ways [1]. The idea, when applied to IoT, promises to provide value-added services that none of IoT smart objects could provide individually. In fact, service composition allows the aggregation of smart object services to

A. Amine et al. (Eds.): CIIA 2018, IFIP AICT 522, pp. 521–532, 2018.
https://doi.org/10.1007/978-3-319-89743-1_45

meet complex requirements from various application domains. It can be used to create innovative IoT applications in an efficient, flexible and dynamic manner. A robust service composition mechanism also makes it possible to support applications in a dynamic network environment, which is the key to foster the development of IoT applications [2].

In this paper, an IoT application is created, using IoT service composition, to control and/or observe specific targets defined by users. To do so, appropriate IoT devices hosting services are deployed on/in the defined targets. In addition, we propose a subdivision of IoT services and client requests into three types, namely: actuating, processing and trigger. Accordingly, appropriate services involved in the composition process are selected using the user request target. To perform the service composition process, we propose an approach based on a multi-agent system where several agents (targets, devices, services, requests and composer) are engaged in collaboration to meet the same goal: satisfying the user (client) request.

The remainder of this paper is organized as follows. Section 2 gives a background and discuss some existing work in IoT service composition. Section 3 presents the proposed semantic model describing IoT services, messages and user requests. This is followed by an explanation of the proposed multi-agent system for IoT service composition in Sect. 4. Section 5 describes the proposed approach implementation and some use cases scenarios. Finally, Sect. 6 gives conclusions and ideas for future work.

2 Related Works

Service composition issues in classic and Pervasive Computing have been largely discussed in literature, and several solutions have been also proposed. A user-centric service composition approach [3] assists users in composing and selecting their applications by analyzing dependencies among published application templates, collaboration templates, workflows, and services. In [4], a Service Composition Planner (SCP) is proposed to generate all the feasible composite services from the user request and the set of available services. In [5], the problem of automatic service composition is addressed using a graph search-based algorithm and two different pre-processing techniques to handle the defined Same-Intension Different-Extension (SIDE) services. A Graph plan based approach is proposed in [6] for automatic service composition using branch structures in the resulting composite services to ensure their correct execution. In [7], the authors address the issue of selecting and composing web services via a genetic algorithm and give a transaction and QoS-aware selection approach. A service composition is proposed in [8] using two main phases: off-line phase and on-line phase. In the off-line phase, a global graph that links all the available abstract services is generated using a rule-based technique. In the on-line phase, a sub-graph is extracted spontaneously from the global graph according to the occurred and detected events in the environment.

Besides, the service composition approaches based on multi-agent systems are investigated regarding the similarity between services and agents. In [9], authors have tackled the dynamicity issues of ambient intelligence environment by using service-oriented architecture, semantic web services, and multi-agent systems. This approach empowers the dynamic service composition which allows the development of the AmI applications that improve autonomous reconfiguration to face the unstable and uncertain nature of AmI context and to maintain the user activities available. An automatic P2P semantic web service composition approach is presented in [10]. It is based on multi-agent systems to carry out web service composition. It uses the multi-agent System Engineering (MaSE) methodology to define the whole life cycle of the studied system including analyzing, designing and developing the expected multi-agent system. In [11], the authors propose a decentralized multi-agent model for service composition. This approach of service composition is based on local service interactions in order to form agent coalitions to create new non-existing functionalities. The proposed decentralized multi-agent model of service composition is implemented using the Java Agent DEvelopment Framework (JADE) and the web services integration gateway. In [12], the proposed service composition approach is based on software agents. It aims to reduce the complexity of composition service global networks which are too complex to be dealt with. The proposed service compositor is based on agent coordination to calculate the composite service that fulfills the user request. This system elaborates an activity plan and controls its execution in order to verify if the user requirements are satisfied. The control strategy improves the time response of the user request and decreases the network traffic by compacting the overhead of exchanged messages.

In the previous works, IoT application based services are partially described without including its actors, such as targets, devices, services, and client requests. Moreover, relationships between such actors are not taken into account. In addition, a subdivision of services and requests into sub-categories to reduce the composition process complexity is not considered.

To overcome the aforementioned shortcomings of existing works, we propose a new model based multi-agent system for service composition in IoT. It represents the first developed part of our global solution for the applications based on IoT services. Our model takes into consideration services and client requests classification (actuating, processing and trigger) in order to enhance the composition process performances. We implement targets, devices, services and requests as agents. The request agent initiates the composition process through the composer agent which selects the appropriate service agents based on the target specified by request agent

3   Service and Request Description Model

Most of IoT service systems consider that semantic or knowledge model is a basic foundation that provides indispensable means to define communication rules. The exchanged data are considered structured messages which can be requests or responses.

3.1 Context Messages

In IoT service ecosystem, we suppose that a message is a basic entity used in communication between services. It represents an observed context state, treated data or a command at a specified time. We assume that the IoT message contains functionally dependent context parameters.

3.2 Service Description

In this work, we adopt the service definition given in [8], in which concrete and abstract services are introduced.

Concrete Services. They implement a physical or logical function which provides context data or applies an effect on the environment. A concrete service has a real implementation that can be invoked through a specific input message in order to produce an output message.

Abstract Services. It is a class of concrete services that are functionally equivalent. Besides, they are based on input/output data and category similitude criteria. So, two concrete services are considered functionally equivalent, if and only if, they have the same input and output parameters, and reference an identical category.

3.3 User Request Description

In the system engineering discipline, any system is described through its parameters which represent pertinent means to observe the system behavior during its execution. We consider IoT services as parameters which allow us to study the evolution of the application targets. We suppose that the client does not need to know the system's architecture or the available services. Simply, he defines his request using the reference ontology. The reasoner must well understand the user's request through the semantic interpretation and calculates its relative response which can be a basic service or a composite one. We have distinguished three main types of requests as mentioned above. The trigger request concerns the commands to control actuators through actuating services or to ask for raw context data in periodic, event-aware or on-demand manner. The processing request is dedicated to carry out context data in order to calculate another treated data or decisions (command). Finally, the actuating request is used to apply directly an effect on/in specific targets.

4   Multi-agent System Based Service Composition Model

4.1 Adequacy of Multi-agent System for Service Composition

Multi-agent systems are constituted from a well-defined environment and various agents. The environment is formed by a set of passive and active objects that can be manipulated by agents. The latter are endowed with communication ability, some own resources and a satisfaction function which represents their individual tendencies or individual objectives. In addition, they can perceive the environment in a limited way, reproduce and die. Moreover, agent-agents and/or agent-environment interactions allow coordination and cooperation to achieve a global objective in a very close way to reality. For these reasons, the multi-agent approach is largely used to study phenomena and systems from several domains.

In addition, an agent is an autonomous entity that operates on an environment with an ability to communicate with other agents, receives and transmits messages, in order to improve coordination, cooperation and avoid deadlock and starvation when they are in need to use some critical resources. These mechanisms allow efficient collaboration to achieve global tasks. Comparing the IoT service systems, we can distinguish some similarities, namely: between agents and abstract services on the plans of structure, behavior, and communication way, also the collaboration to achieve a global goal by agents and the service composition process to satisfy the user's request. Based on these points, we believe that multi-agent systems constitute a faithful model for IoT service systems. In the following section, we introduce a description of a model based on the multi-agent system.

4.2   Functional Description of the Proposed Multi-agent System

The proposed multi-agent system calculates a solution for a user's request. The relative responses can be simple or composite services. In this section, we define a reasoner model based on the multi-agent system. According to Fig. 1 in which is depicted the use case of the proposed reasoner system, we remark that this system calculates response for all kinds of requests aforementioned. The calculated responses are based on the list of available services provided by the service registry. Services involved in the composition process are selected according to the user's request subject.

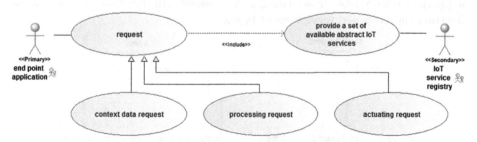

Fig. 1. Multi-agent system use case diagram

The reasoner under construction has to provide a simple or a composite service that satisfies the user's requirements. The calculation of the user's request response must be in a finite time. The developed system must avoid the starvation and the deadlock of the user's requests.

4.3   Structural Description of the Proposed Multi-agent System

As shown in diagram: A of Fig. 2, the proposed reasoner contains a request agent, service agents, and link agents. The reasoner agent plays the role of an observer agent which handles the entire agent world including environment and different types of existing agents.

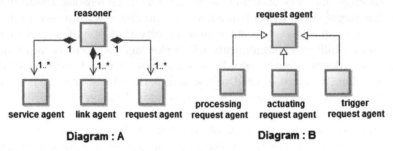

Diagram : A   Diagram : B

Fig. 2. Reasoner and request agent block definition diagram

The request agent represents the user's request. It is described by an output message and a category. The output message denotes the expected user result and the category is introduced to simplify the request complexity. We have distinguished three types of categories. Each one among these categories will be satisfied by a special composition agent. Diagram: B of Fig. 2 illustrates the user request types.

The service agents symbolize the services of IoT systems. The service agent is featured with an input message, an output message, a category, a number of contained concrete services (it is optional, this parameter is used in a case of abstract service) and an identifier. The service agent can be an actuating service agent that applies directly an effect on specific targets, a processing service agent that provides treated data or a command based on raw context data, or a trigger service agent which collects context data from observed targets or provides commands for actuating service agents. In diagram: A of Fig. 3, we illustrate the proposed service agent types.

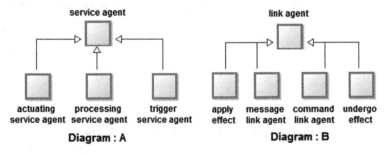

Diagram : A   Diagram : B

Fig. 3. Service and link agent block definition diagram

The link agents denote the possible links between all types of agents including service and request ones. Two agents are connected by a link agent, if and only if the output message type of one of them is similar to the input message type of the second one. On the other side, they can be seen as a communication channel which joins two service agents. In diagram: B of Fig. 3, we present the existing link agent types, namely a message and a command which are exchanged between service agents. However, apply an effect link agent is a link between a specific target and a service agent. In this case, we represent the physical influence of actuating service agents on a specific target. Furthermore, we have proposed an undergo link agent type to represent the influence of a specific target on trigger service agents.

4.4   Dynamic Description of the Proposed Multi-agents System

The proposed reasoner based on the multi-agent system has mainly three behavior patterns. The composition process is launched by the user's request which is described by a subject, an expected result, and a category. According to this latter, the reasoner selects the appropriate services and the relevant composition process to calculate a composite agent that fulfills the user's request. As we have mentioned, a trigger request is answered by a trigger agent, a processing request is satisfied by a processing agent and an actuating request is responded by an actuating agent.

Figure 4 shows the dynamic behavior of the proposed reasoner based on the multi-agent system. Each state depicts which kind of agents is involved in the composition process. In addition to the request agent, three states can be distinguished, namely: trigger, processing and actuating. Each transition is featured by a condition that must be validated to cross it. Meeting such condition means also that the aggregation between their agents is allowed. So, we can deduce that transitions between states cited above depict link agents.

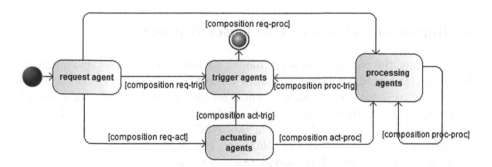

Fig. 4. Reasoner agent state machine diagram

In Fig. 5, we have described the reasoner process to calculate a composite agent that fulfills the user's request. First of all, the reasoner loads the user's request and according to its category, the reasoner creates the appropriate agents. Thereafter, the reasoner checks, if there are mappings between the request expected result and the output message of the other agents. These latter are selected according to the category of the request agent. If the result set is empty, it does mean, that no solution is found, otherwise, the process of composition continues. After each composition round, the reasoner checks, if there is a composite agent that satisfies the request agent. In other words, the reasoner verifies, if there is a connection between the trigger agents and the set of the composite agent under construction.

The processing-processing composition has two major problems. The first one concerns the direct composition loop. For example, let's suppose two agents "S1" and "S2" described respectively by "msg1" and "msg2" as inputs and "msg2" and "msg1" as outputs. The category of both agents is "processing". We remark that it is possible to create an infinite composite using "S1" and "S2" in an alternative way such as "S1-S2-S1-S2-, etc.".

On the other side, the second problem is related to the indirect composition loop. It does mean that we have a cycle of a sub-composite agent which can be repeated in an infinite way. To illustrate this case let's assume three agents "S1", "S2" and "S3" described respectively by "msg1", "msg2" and "msg3" as inputs and "msg2", "msg3" and "msg1" as outputs. We observe that the elementary composite is "(S1-S2-S3)". Therefore, we can notice the possibility to get as composition result an infinite composite "(S1-S2-S3)-(S1-S2-S3)-(S1-S2-S3)-, etc.", which are considered a negative composition.

To solve these problems, we have supposed that in a composite agent, an agent must be used at most once time. Taking into account this hypothesis, we have introduced a mechanism that detects whether the reasoner is developing a negative (infinite) composition or not. In processing-processing composition, the reasoner checks, if there are new composites after each composition cycle. If it is the case, the reasoner continues the development, otherwise, the composition process will be stopped in order to avoid negative construction.

5   Implementation and Use Cases Scenarios

To implement the proposed reasoner model, we have used the Netlogo language. As known, the programming task is an interpretation and a translation of a model that includes functional, structural and behavioral aspects, in a specific language. In this work, we have implemented the trigger, processing and actuating agents as turtle agents. The link agents symbolize the communication channels between the system's agents. However, the reasoner agent is an observer agent, so it can control all the agent world. Our proposed system is based on coordination and this task is ensured by the observer agent perfectly.

Based on the state machine diagram (Fig. 4) and the activity diagram (Fig. 5) which describe the behavior of the multi-agent system, we have elaborated its implementation. We have specified for each agent of the system its own properties and methods. For the main one, the observer plays the role of the reasoner, which manages the coordination between all types of agents in order to look for an agent composite that satisfies the user's request agent.

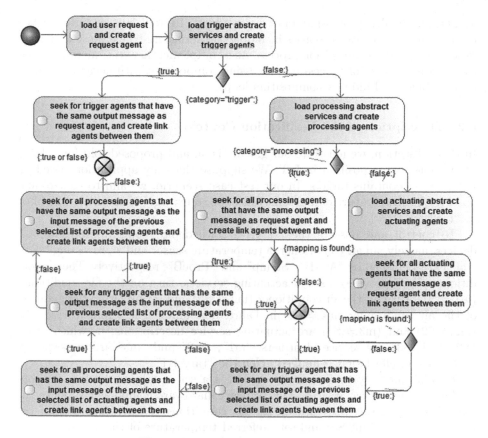

Fig. 5. Reasoner agent activity diagram

5.1 Verification of the Proposed Multi-agent System: Unit Tests

During the development of the reasoner system, we have introduced unit tests to avoid code mistakes. Indeed, we have tested the composer program for each possible scenario, namely: trigger, processing and actuating requests.

In a case of trigger request, we have elaborated scenarios for negative and positive responses. In the former, it does mean no connections between request and triggers agents. However in the latter, it signifies that at least one of the trigger agents which satisfied the request agent was found.

Concerning the second case, the processing request, we have prepared negative tests including non existing mapping between request and processing agents. Thereafter, we have tested the reasoner with non existing mapping between processing and trigger agents. The remaining tests concern the positive responses, in which the composite agent that satisfies the request was calculated.

The last verification case concerns the actuating request. In the first step, we have tested the developed reasoner in a non existing mapping between request and actuating agents, actuating and processing agents, processing and processing agents, and finally, processing and trigger agents. In addition, we have verified the reasoner in positive responses including the actuating-trigger and actuating-processing-trigger composition process. As a complementary test, we have tested the reasoner in case of processing-processing composition with negative calculation (a direct and indirect composition loop).

5.2   Description of the Verification Context

In this subsection, we illustrate our global vision and proposed solution related to the service composition in IoT. We suppose that any application based on IoT services contains targets. In our test case scenario, we aim to control and observe the temperature of a specific office. So, our target is an office, identified as "target-1".

To control and observe the "target-1", we deploy some appropriate IoT devices, namely: an air conditioner, a temperature sensor and a processing board which are identified by "AC-1", "Stemp" and "ProcB", respectively. The air conditioner "AC" provides a set of actuating services, turn on the air conditioner (m1,#,S1), turn off the air conditioner (m2,#,S2) and set preferred temperature of the air conditioner (m4,#,S3) in Fahrenheit units. The triples (m1,#,S1), (m2,#,S2) and (m3,#,S3) are actuating service descriptions, in which "S1", "S2" and "S3" are service identifiers, "m1", "m2" and "m3" are messages, # represents an effect that is applied in/on the "target-1".

In addition, a virtual controller provides a set of trigger services for controlling the actuating services. This software/hardware component supplies the following trigger services: turn on the air conditioner (#,m1,S4), turn off the air conditioner (#,m2,S5) and set preferred temperature of the air conditioner (#,m4,S6) in Fahrenheit units. These trigger services can be called directly by user requests. The triples (#,m1,S4), (#,m2,S5) and (#,m3,S6) are trigger service descriptions, in which "S4", "S5" and "S6" are service identifiers, # represents the "target-1" effects/orders,"m1", "m2" and "m3" are trigger messages.

To complete this example, we suppose that "Stemp" publishes a trigger service (#,m4,S7) that provides the ambient temperature of the office "target-1" in Fahrenheit units. The triple (#,m4,S7) is a trigger service description, in which # represent an effect applied by observed context, target-1, on sensor device, "m4" is a trigger message that represents measurements of the observed phenomena and "S7" is the service identifier.

Finally, the processing board "ProcB" provides a processing service that converts temperature values form Fahrenheit to Celsius (m4,m3,S8).

5.3   Request Examples and Obtained Results

In this section, we introduce some elementary test cases in order to illustrate the process patterns and the execution of the developed reasoner. To do it, we have prepared three types of test case scenarios, namely: trigger, processing and actuating requests.

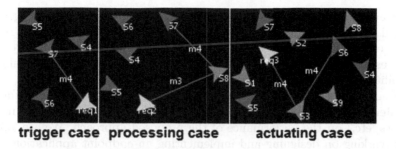

trigger case   processing case   actuating case

Fig. 6. The responses of reasoner agent to the trigger, processing and actuating requests

Trigger Request: We assume the following trigger request: "Give us the ambient temperature of the office in Fahrenheit units". The translation of this request is as follows: ("target-1",m4,req1). The "target-1" represents the observed target (subject/office), the message "m4" depicts the temperature measurements in Fahrenheit unit and "req1" is the identifier of the request. Figure 6, trigger case, represents the reasoner's response to the request defined above. We remark that "req1" is satisfied by trigger service (S7) which provides the current temperature of the office in Fahrenheit units.

Processing Request: We suppose the following processing request: "Give us the ambient temperature of the office in Celsius units". The translation of this request is as follows: ("target-1",m3,req2). The "target-1" represents the observed target (subject/office), the message "m3" depicts the temperature measurements in Celsius unit and "req2" is the identifier of the request. Figure 6, processing case, represents the reasoner's response to the request defined above. We remark that "req2" is satisfied by the processing service (S8) which converts to Celsius units the current temperature of the office provided by trigger service "S7" given in Fahrenheit units.

Actuating Request: We suppose the following actuating request: "Set the preferred temperature of the air conditioner to 77 Fahrenheit". The translation of this request is as follows: ("target-1",m4,req3). The agent "target-1" represents the target (subject/office), the message "m4" depicts the temperature value in Fahrenheit units and "req3" is the identifier of the request. Figure 6, actuating case, represents the reasoner's response to the request defined above. We remark that "req3" is satisfied by a composite agent "S3"+"S6". The agent "S6" represents a trigger service that provides the command to set temperature of the air conditioner in Fahrenheit unit. The agent "S3" is an actuating service that applied the received order sent by the agent "S6".

6   Conclusion

In this work, we have presented a new multi-agent system based service composition approach. Such an approach is based on four main agents namely: reasoner (observer) agent, service agent, request agent and link agent. These agents are engaged in coordination to achieve the same goal of satisfying the user's request. We have described the role of each agent and their collaborative process. The use-cases scenarios and extensive tests show clearly the interest, feasibility and suitability of the multi-agent system for service composition. As future work, it is still necessary to conduct further empirical evaluations to study the impact of the different pre-configured parameters (number of concrete services, number of requests, etc.) on the performance of the proposed approach. We are also currently working on designing and implementing an endpoint application and an automatic concrete services classification approach.

References

1. Lemos, A.L., Daniel, F., Benatallah, B.: Web service composition: a survey of techniques and tools. ACM Comput. Surv. **48**(3), 33:1–33:41 (2015)
2. Chen, I., Guo, J., Bao, F.: Trust management for SOA-based iot and its application to service composition. IEEE Trans. Serv. Comput. **9**, 482–495 (2016)
3. Tsai, W.T., Zhong, P., Bai, X., Elston, J.: Dependence-guided service composition for user-centric SOA. IEEE Syst. **8**(3), 889–899 (2014)
4. Wang, P.W., Ding, Z.J., Jiang, C.J., Zhou, M.C.: Automated web service composition supporting conditional branch structures. Enterp. Inf. Syst. **8**(1), 121–146 (2014)
5. Wang, P.W., Ding, Z.J., Jiang, C.J., Zhou, M.C.: Constraint-aware approach to web service composition. IEEE Trans. Syst. Man Cybern. Syst. **44**(6), 770–784 (2014)
6. Wang, P.W., Ding, Z.J., Jiang, C.J., Zhou, M.C., Zheng, Y.W.: Automatic web service composition based on uncertainty execution effects. IEEE Trans. Serv. Comput. **9**, 551–565 (2015)
7. Ding, Z.J., Liu, J.J., Sun, Y.Q., Jiang, C.J., Zhou, M.C.: A transaction and QoS-aware service selection approach based on genetic algorithm. IEEE Trans. Syst. Man Cybern. Syst. **45**(7), 1035–1046 (2015)
8. Yachir, A., Amirat, Y., Chibani, A., Badache, N.: Towards an event-aware approach for ubiquitous computing based on automatic service composition and selection. Ann. Telecommun. **67**(7), 341–353 (2012)
9. Vallée, M., Ramparany, F., Vercouter, L.: A multi-agent system for dynamic service composition in ambient intelligence environments. In: The 3rd International Conference on Pervasive Computing (PERVASIVE 2005) (2005)
10. Paikari, E., Livani, E., Moshirpour, M., Far, B.H., Ruhe, G.: Multi-agent system for semantic web service composition. In: Xiong, H., Lee, W.B. (eds.) KSEM 2011. LNCS (LNAI), vol. 7091, pp. 305–317. Springer, Heidelberg (2011). https://doi.org/10.1007/978-3-642-25975-3_27
11. Papadopoulos, P., Tianfield, H., Moffat, D., Barrie, P.: Decentralized multi-agent service composition. Multiagent Grid Syst. **9**(1), 45–100 (2013)
12. Bennajeh, A., Hachicha, H.: Web service composition based on a multi-agent system. In: Silhavy, R., Senkerik, R., Oplatkova, Z.K., Prokopova, Z., Silhavy, P. (eds.) Software Engineering in Intelligent Systems. AISC, vol. 349, pp. 295–305. Springer, Cham (2015). https://doi.org/10.1007/978-3-319-18473-9_29

Crowdsourced Collaborative Decision Making in Crisis Management: Application to Desert Locust Survey and Control

Mohammed Benali(✉), Abdessamed Réda Ghomari,
and Leila Zemmouchi-Ghomari

LMCS (Laboratoire des Méthodes de Conception des Systèmes),
ESI (Ecole nationale Supérieure d'Informatique), Algiers, Algeria
{m_benali,a_ghomari,l_zemmouchi}@esi.dz

Abstract. Dealing with crisis situations involves significant collaborative decision-making to recover lives and preserve properties. In order to reach timely and appropriate decisions, crisis response organizations need to rapidly obtain an accurate situation awareness of the crisis context. A process by which they need to gather, access, and exchange near real-time information about the events circumstances throughout the entire crisis life cycle. Such activities imply a huge participation of collaborating organizations and goes even beyond their internal borders to reach the entire crisis-stricken community. Recent crisis and disaster situations have demonstrated the crucial role citizens can and must play in responding to such events. The growing development of advanced technologies has open the door for large public to be a key factor in making decisions and conducting guided response actions. We propose in this paper a comprehensive approach that integrates crisis crowdsourcing tasks and techniques to crisis decision-making activities. To demonstrate its relevance, we present a study of crisis scenario based on the context of Desert Locust Plague Survey and Control in the Algerian National Institute of Plant Protection (INPV).

Keywords: Crowdsourcing · Collaborative decision-making
Crisis management · Case study · Desert locust survey and control

1 Introduction

The dynamically changing environment of crisis events creates highly unpredictable and complex situations to be handled, with human lives and properties at stake. In such contexts, crisis response organizations look for an accurate situational awareness by retrieving and exchanging real-time contextual information and knowledge, share a collective response action plan for purpose of making

© IFIP International Federation for Information Processing 2018
Published by Springer International Publishing AG 2018. All Rights Reserved
A. Amine et al. (Eds.): CIIA 2018, IFIP AICT 522, pp. 533–545, 2018.
https://doi.org/10.1007/978-3-319-89743-1_46

efficient and well-informed decisions. Crisis management activities imply taking complex and collaborative decisions, involving a huge and diverse information and knowledge owned and distributed among a variety of contributing actors from different areas of interests and expertise [1]. Moreover, real crisis situations have shown that response activities exceed the capabilities of crisis response organizations staff members and resources and have compelled them to open their borders and bring up together the whole affected community including unskilled citizens. Moreover, studies have concluded that unexpected problems such as crisis and disaster cases involve giving the decision authority to those closest to the situation in either location or knowledge [2]. In fact, the growing development of collaborative technologies have shifted citizens role and participation during crisis events from being just victims watching and sharing posts about the event consequences using social media and social networks, to crisis responders involved in a more proactive contribution by performing specific tasks related to crisis decision-making. According to [3] crowdsourcing is a dynamic form of cooperative work intended for a broad range of citizens engaged in semi-autonomous tasks related to information management issues.

The main benefits of implementing crowdsourcing mechanism lie in its ability to promote civic involvement and connect it with organizations by facilitating active and collaborative problem solving and decision-making [4]. However, a remaining challenge lies on finding new ways of understanding, conceptualizing and defining specific design methodologies for aligning crisis crowd-tasking activities and techniques with respondent organizations decision-making processes. In addition, there is still more studies of real crisis scenarios to be conducted to analyze and examine the role that ordinary citizens may play in the different steps of decision-making activities in the whole crisis life cycle [5]. To address this challenge, we propose a comprehensive crowdsourced collaborative decision-making approach for the crisis management domain. To this end, we adopt the well-known decisional model IDC of Simon [6] composed of the three following phases: Intelligence (information gathering and sharing for problem identification and recognition), Design (generating alternative courses of action), Choice (evaluating, prioritizing, and selecting the best course of action). For each step of Simons decisional process, we specify the main crisis crowdsourcing tasks which the crowds are called to perform, and the appropriate technique for outsourcing those tasks. Regarding the collaborative nature of the crisis decision-making process, we use the BPMN4Social modeling notation [7], which is a social extension of Business Process Model and Notation (BPMN), that aims to define a specific notation for describing Social Business Process behaviors, to which we add some crowdsourcing related specifications.

In order to demonstrate the applicability of our approach in a practical scenario, we carry out a study of the Desert Locust Plague in the Algerian National Institute of Plant Protection (INPV). The mission of the INPV is to ensure sustained survey and control of desert locust situation across the entire Algerian territory with 2.4 million km^2. To assess the current Desert Locust situation, plan further survey and control operations, and request external assistance, a variety of data and information are required. This information is provided by a set of collaborating organizations such as survey teams of INPV, the Food and Agricultural Organization (FAO), regional governmental organizations, and local humanitarian organizations. Given its limited staff members and resources compared to the vast area to be monitored, the INPV have to rely on additional information coming from less experienced people such as agricultural extension agents and scouts as well as from non-locust experts such as travelers, truck drivers, farmers, government agents, villagers and nomads. These features make the Desert Locust Plague a relevant case for the analysis of the challenges and opportunities that arise when integrating crowdsourcing processes to the participatory decision-making process of the collaborating stakeholders.

The reminder of this paper is structured as follows. Section 2 reviews the main related work addressing the crowdsourcing applicability to decision-making process. Section 3 presents the fundamental concepts of the proposed approach. Section 4 presents the evaluation of the approach by its application to the desert locust plague case. Finally, Sect. 5 concludes by summarizing the main contributions achieved in this research and makes recommendations for future work.

2   Related Work

Crowdsourcing have proven to be a promising field of research, giving the wide-ranging studies in which it has been addressed by scholars and practitioners over the last decade. In this work, we focus on the alignment of crowdsourcing process with organizational decision-making process in a crisis management context. Bonabeau [8] in his work provides a general framework to assess crowdsourcing application in business problem solving and decision-making. Considering that the decision-making is divided on two phases: the generation of potential alternatives and the evaluation of them, he discussed for each phase the main behavioral biases (social interference, availability, anchoring, and stimulation) that adversely influence the process, as well as the proper collective intelligence approaches (outreach, additive aggregation, and self-organization) to handle them. The crowdsourcing framework examine the key implementation issues such as control (loss of control, unpredictability, liability, etc.), diversity versus expertise, engagement (incentives to motivate people's participation), and policy (policies to control participants' behaviors). Another study we could mention is that carried out by [9]. In this study, a simulation experiment of the Citizens Emergency Response Portal System (CERPS) have been conducted, to examine the hypothesis that citizen participation in crisis response decision-making can improve the outcome of a crisis. The findings of this study have shown that citizen involvement provided additional first-hand situation awareness, such as the locations of fires or pictures of suspicious packages. The crowdsourced information helped crisis managers to make better response decisions, understand the population needs, and decide when and how to provide information to the public. Closer to our work, Chiu et al. [10] propose a crowdsourcing based decision-making framework. The framework specifies for each phase of the IDC model of

Simon the main tasks to be crowdsourced. At the intelligence phase, the crowds are implicated in knowledge discovery activities, making predictions, and providing opinions. The crowds in the design phase, are called for idea and proposals generation. At the choice phase, the main crowds' concern is the evaluation and assessment of the proposed ideas in the previous phase. Another relevant study in this topic is that conducted by [11]. They propose a crowdsourcing-based framework to build decision support systems based on the crowdsourced data. The proposed system provides on-the-fly decision support, and is based on two aspects: constructing the knowledge base in run-time using crowds' contributions and providing decision support using the collected knowledge. Another related study we mention is the proposal of [12] who propose an interactive crowdsourcing based decision support framework for post disaster situation awareness and decision support. The framework gathers situational information using SMS from the crowd present at the disaster site, and summarizes such responses to have situational awareness and appropriate decision-making.

Most of the prior works are oriented towards addressing the main issues and challenges facing organizations when incorporating crowdsourcing into their decision-making processes. In addition, most of the existing literature addresses the applicability of crowdsourcing to decision-making for general business problems and lacks more research that examines in particular the crisis management area. To the best of our knowledge current literature fails to establish a comprehensive approach to support the design of processes to integrate crowdsourcing into the different phases of crisis decision-making, which represents the main concern of this study.

3   Crowdsourced Collaborative Decision-Making Approach in Crisis Situations

In this section, we delineate the way crowdsourcing processes can be incorporated into the decision-making process within a crisis management context. The crisis crowdsourcing approach is built upon cooperation between citizens and crisis response organizations (CROs), in each stage of the decision-making process at the various crisis management phases. For each phase of Simon's decisional model (IDC), citizens are engaged either by providing information about the crisis or as crowd workers performing specific tasks related to information processing and management issues. Thus, the proposed approach will allow response organizations to identify and express their information needs, be aware of the crisis evolving context, and strengthen their decision-making capabilities by providing relevant recommendations based on volunteer citizens' evaluations and feedbacks. As depicted in Fig. 1, the decisional process relies on a two-way flow of information, where crisis stakeholders may act both as consumers and as providers of information. From one hand, citizens provide near real-time information about the event to display an accurate situational awareness for the affected population and the involved response organizations. Respondent organizations, on the other hand, share their plans and resources to both citizens and

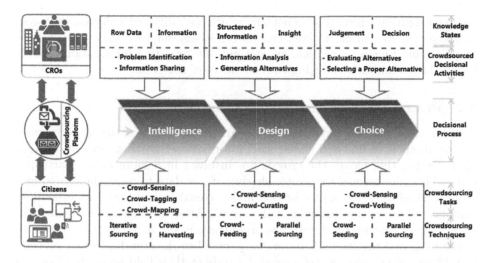

Fig. 1. Overview of the proposed approach

the others crisis management organizations to share a clear activity awareness with the crisis management community and proceed to distributed planning and execution of response plans.

3.1 Phase I: Information Sharing and Problem Identification

In this phase, the first concern of the respondent organizations is to rapidly obtain a clear on the ground picture of the affected zone. Maintaining an updated situation awareness of the crisis context requires gathering information from sensor systems in conjunction with crowdsourced data from citizens reporting observations and direct eyewitness accounts. In fact, local population tends to immediately gather and share geolocated data and information related to the crisis effects. They contribute to the information chain either passively by providing raw data using their phones, tablets, and social networking technologies or actively by reporting more established and controlled information using a dedicated crowdsourcing platform. Whereas this crowd-sensing activity represents a key pillar for response decisions, it may generate information overload problems, due to the great volume of the collected data and information. To overcome this challenge, crowd-tagging comes as an efficient means by which crowds can be involved in categorizing and classifying of the provided data based on their local knowledge and the perception of their own environment. Results of the tagging process are then presented and visualized to enhance the awareness of both response actors and the public. Response organizations, in turn, and based on the information harvested automatically from social networks or gathered via the platform, proceed to information categorization and classification using their internal expertise. In this regard, situational awareness ontologies are relevant tools used to avoid ambiguity and confusion of terms and share a common language with the different stakeholders, which have different cultures, protocols and systems. At the meanwhile, crowds iteratively improve these information mapping and tagging activities, by continuously reporting information according to the dynamically changing circumstances of the event.

3.2   Phase II: Generating Possible Courses of Action

At this stage of the decisional process, the primary objective of the crisis response organizations is to find actionable crisis information that translates into action plan. During this phase, the crowds are involved in a more active participation, which goes beyond the simple sharing of raw data to more elaborated and structured information. A process that makes crowd-tasking activities even more complicated for certain member of the crowd. After identifying the main tasks to be assigned to the crowd, response organizations and based on the complexity of the tasks and the available resources, proceed to the crowd-selection process. In addition, they share georeferenced data about their field teams positions as well as their associated resources, which enhance activity awareness of the different collaborating organizations, facilitate the distributed response planning, and guide the crowds participations. In this regard, mapping and tagging activities can facilitate and improve the crowds understanding of the different tasks to be performed by giving them a global picture of the response plan. Once crowds are selected, the platform accordingly, generates the necessary instances of tasks and automatically assign them to each selected member. The crowds and based on the available information, proceed to the crowd-curating activity. They are consequently involved in various information processing and management tasks such as information filtering and synthesizing to support respondent actors with practical insights and guidances for the development of potential alternative courses of actions.

3.3   Phase III: Evaluating and Selecting a Proper Alternative

In this phase, the objective behind engaging the crowds is to get evaluations, opinions, and suggestions to prioritize response actions and improve alternatives created in the previous phase. To this end, response organizations share the proposed response plan through the platform with the different response stakeholders. The shared plan provides relevant basis for the other involved organizations to adjust their priorities and ensure an optimal use of the available response resources. An overview of the response plan can be presented to citizens using the crowd-mapping technique to enhance their awareness and gain their confidence. Evaluations are then collected, analyzed, and reviewed by the internal experts of the response organizations. While ranking activity provides explicit information, some evaluation can include complex responses containing implicit information, which require more clarifications or additional work to be requested from the crowd. Thus, response organizations can ask for additional details and explanations about a specific contribution. Crowdsourced data or tasks performed by the crowd are processed and reviewed and then shared back to the crowd to get more adjustments and feedbacks. In this context, the collection process can be improved using the parallel sourcing technique, where participants evaluate the same alternative independently of other's evaluations, which reduces the probability for participants to be influenced and guided by previous works.

4   Case Study

The desert locust plague is considered as one of the major threats to food security worldwide due to its devastating effects on crops and pastures, which potentially results in disturbing socio-economic consequences. Locust outbreaks have occurred on all continents except Antarctica and they can harm the livelihood and well-being of 10% of the worlds population [13]. In fact, the desert locust outbreak of 2003–2005 affected 8 million people, in over 20 countries mostly sub-Saharan Africa, with an estimated 80 to 100% of crops lost [14]. Besides the financial losses, the large amounts of chemical insecticides commonly used for control operations can result in serious environmental damage [14]. In Algeria, the last desert locust invasion, in 2004–2005, has required an urgent mobilization of financial resources estimated at USD 120 million to treat 4.5 million hectares which contains mainly hopper bands and swarms [15]. To cope with such a calamity, application of an effective prevention and control strategy is necessary to protect the agricultural production. In this regard, the National Institute of Plant Protection[1] have been charged to take over this critical task, which aims to organize and conduct during the recession period the monitoring and control operations against the Desert Locust, develop and coordinate during the invasion period a response plan for control operations.

4.1   Desert Locust Survey and Control Process at the INPV

The National Institute of Plant Protection carries out a sustained survey and control of desert locust situation all over the Algerian territory with 2.4 million km^2 and especially in the southern Saharan zone. Usually, monitoring and control operations are undertaken by qualified Locust Field Officers who send field reports to the national unit headquarter. Received reports are then corrected, stored, and visualized by the Locust Information Officer (LIO) using RAMSES (Reconnaissance and Monitoring System of the Environment of Schistocerca) a centralized system set up by the desert locust information service (DLIS) of FAO. In addition, the LIO receives information from the DLIS of the FAO which consists of a bulletin that contains a summary of the current situation at the locust-affected countries and a forecast of future developments. Despite the amount and the variety of collected data and information, experienced locust officers at the INPV assume that just 50% of information about the locust situation are detected by survey operations for a given area, due to

---

[1] http://www.inpv.edu.dz/.

the limited capacity of the INPV Staff regarding the vast area to be monitored and the accessibility of locust habitat. To strengthen its locust situational awareness, the INPV rely on additional information coming from less experienced people such as agricultural extension agents and scouts as well as from non-locust experts such as travelers, truck drivers, farmers, government agents, villagers and nomads. After verifying these information, the LIO incorporate meteorological and SPOT-vegetation data, assess the current situation and forecasts future developments. In order to have a global view of the locust situation, the LIO send a weekly (in the recession period) or monthly (in the invasion case) report to the DLIS of FAO which incorporate it to its locust bulletin. Citizens are also informed about the locust situation and the targeted areas of control campaigns. Based on the analysis performed by the LIO, the administrative board decides when and where to conduct survey operations, what are the priorities of control campaigns, evaluate the effectiveness of control operations, and determine where further control is required.

4.2   Crowdsourcing the Decisional Process of the INPV

Information Sharing and Problem Identification

A better understanding of the locust situation means better assessment of the current situation, which leads to accurate forecasts of future developments and make the basis of all locust response decisions. The LIO receives locust information from locust monitoring teams, which contain a detailed report of the surveyed area. Locust situation reports are also provided by the DLIS of FAO, which represents a global situational image of the locust affected countries. However, experienced locust field officers of the INPV suppose that these information cover only 50% of the overall locust situation. Especially in unsurveyed areas, non-locust experts such as villagers, nomads, travelers, truck drivers, and government agents play a crucial role in monitoring and reporting information about habitat conditions and locust presence. In addition to directly using the crowdsourcing platform, posts and comments on locust situation can be automatically collected incorporated from social networks to the platform using dedicated web-based sensors. Four key data are required: ecology, rainfall, locusts and control. As highlighted in Fig. 2, locals and based on their knowledge of their own environment, provide ecological data which comprise a description of the habitat nature (such as valleys, plains, pastures), density and the greenness of the vegetation, and soil moisture. In this regard, SPOT-vegetation data are provided by the Algerian Space Agency (ASA) to allow a guided citizens participation. They also provide meteorological data, which consists of Rainfall (to identify vegetation and breeding areas), temperature (to estimate the development rate of eggs and hoppers), wind and atmospheric pressure (to evaluate the possibility of adult development and their migrations direction). These data are compared with confirmed data from National Office of Meteorology (NOM). In addition, citizens can provide critical locust data such as locust maturity, density, appearance, phase, and behavior. To this end, mapping tools should be incorporated to

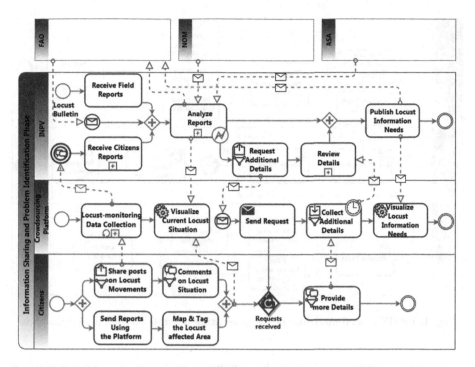

Fig. 2. BPMN4Social modeling of the information sharing and problem identification phase

the crowdsourcing platform to enable geo-locating and geo-tagging of the locust affected area. The LIO analyze the received reports and compare the analysis results with historical data to understand the dynamics of locust populations. Crowds can be asked to report particular pieces of information such as locust colors, pictures of the observed location, etc. Based on the analysis results, he evaluates the current locust situation and forecast future development.

Generating Possible Courses of Action

After identifying the main information to be collected, the INPV evaluate its internal staff capabilities and its own resources regarding the extent of the locust situation. Thereafter, the LIO identifies the main tasks to be crowdsourced based on situational reports analysis and collected contributions as depicted in Fig. 3. In this regard, profile analysis techniques may be of great value in selecting crowd members and assigning tasks. Usually, the previous phase involves a large number of contributors providing near-real time eye-witnesses in form of raw data. The huge amount of crowdsourced data demand considerable work, time, and concerted efforts to extract actionable information used for locust forecasts and response planning. The main objective of the crowd is to reduce irrelevance and redundancy of information in received reports, analyze and synthesize posts

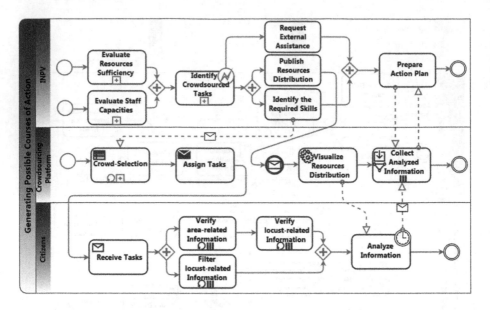

Fig. 3. BPMN4Social modeling of the generating possible courses of actions phase

and comments from social networks. Furthermore, the crowdsourcing platform should allow selected citizens to fill out pre-prepared forms such as selecting images of ecological conditions (plain, dunes, crops), rainfall (light, moderate, heavy), locust maturity (hopper instars, fledgling, immature, maturing, mature adult), locust behavior (isolated, scattered, copulating, laying, flying), and its size or density (very small, small, medium, large, very large). Contributions are then visualized to get feedbacks and improvements. A synthesized report is then sent to the INPV decision makers in order to prioritize the response operations, and elaborate a coordinated plan for exploitation and distribution of resources.

Selecting the Appropriate Alternative

The main preoccupation of the INPV in this phase is to generate a shared and distributed locust response action plan. To this end, the INPV locust experts prepare and share a draft action plan with selected crowds and the other concerned organizations such as the Agricultural Services Branch, General Direction of Forests, Civil Protection Services, and Local Communities. Using mapping features of the platform, the INPV in conjunction with the other involved organizations display geo-tag locations of risk areas, confirmed locust-related information, potential targets for intervention, means and control methods (mechanical control, baiting, dusting, and spraying), route itineraries, and the estimated time for arrival of the locust survey and control teams. For their part, citizens directly engage in crowd-voting activities, which involve commenting, ranking and evaluating of the proposed draft action plan. They can give feedbacks on

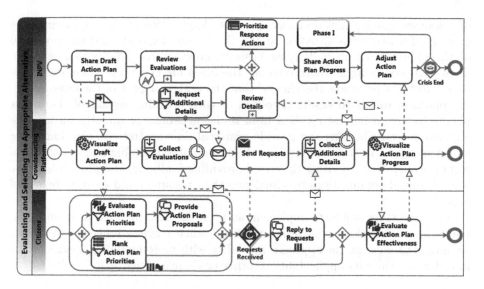

Fig. 4. BPMN4Social modeling of evaluating and selecting the appropriate alternative phase

particular points or alternatives and provide evaluation and action plan proposals. Evaluations are collected, analyzed, and reviewed by locust response teams, who may request additional details on citizens proposals such as confirmation about route itineraries, habitat conditions, and locust behavior. After reviewing citizens analysis, desert locust control commission prioritize response actions, prepare and implement the operational action plan. On the other hand, citizens continuously evaluate the effectiveness of the action plan progress and send feedbacks. According the changing circumstances of the locust situation and citizens insights and reactions to the response plan, locust response teams adjust the later as shown in Fig. 4.

5 Conclusion

Handling crisis events implies collaborative, distributed and complex decision-making activities involving diverse sources of spatiotemporal information and knowledge, distributed among stakeholders from different organizations, agencies and even ordinary citizens. The increasing development of collaborative technologies has conducted crisis decision-making towards a more civic-driven process. Unfortunately, available literature lacks studies and design methodologies intended to align and leverage the emerging power of crowdsourcing to enhance organizational decision-making process especially in crisis situations. The contribution outlined in this paper is a comprehensive approach for crisis collaborative decision-making, which identifies the type of required information and data, and the main crowd-tasking activities and techniques for each phase of response organizations decisional process. Application of the approach in the context of a real crisis situation at the INPV has demonstrated the potential of the approach as an underlying basis to crowdsource information management and processing needs throughout the crisis life cycle. Furthermore, application to this case study also showed that the central aspect of the crowdsourcing platform role is to ensure detection of early warning signals of the locust plague by crowd-harvesting data from social networks or directly incorporating citizens contributions and provide near-real time spatial visualization of the locust situation. In addition, it allows the coordinated planning and evaluation of response operations by either the crowd providing feedbacks and proposals or the other involved organizations by visualizing their resources and sharing their field expertise.

Our next steps will be to implement and test the described platform on a simulated locust recession and invasion event to evaluate its feasibility. Furthermore, we intend to increase the approach flexibility with more modeling specifications in order to ensure its applicability to other crisis situations.

References

1. Poblet, M., García-Cuesta, E., Casanovas, P.: Crowdsourcing roles, methods and tools for data-intensive disaster management. Inf. Syst. Front., 1–17 (2017)
2. Turoff, M., White, C., Plotnick, L., Hiltz, S.R.: Dynamic emergency response management for large scale decision making in extreme events. In: Proceedings of the 5th International ISCRAM Conference, ISCRAM, Washington, DC, USA, pp. 462–470 (2008)
3. Liu, S.B.: Crisis crowdsourcing framework: designing strategic configurations of crowdsourcing for the emergency management domain. Comput. Support. Coop. Work (CSCW) **23**(4–6), 389–443 (2014)
4. Brabham, D.C., Ribisl, K.M., Kirchner, T.R., Bernhardt, J.M.: Crowdsourcing applications for public health. Am. J. Prev. Med. **46**(2), 179–187 (2014)
5. Benali, M., Ghomari, A.R.: Towards a crowdsourcing-based approach to enhance decision making in collaborative crisis management. In: Proceedings of the 14th International ISCRAM Conference, ISCRAM, Albi, France, pp. 554–563 (2017)
6. Simon, H.A.: Administrative Behavior, vol. 3. Cambridge University Press, Cambridge (1976)
7. Yahya, F., Boukadi, K., Maamar, Z., Abdallah, H.B.: Enhancing business processes with web 2.0 features. In: 2015 12th International Joint Conference on e-Business and Telecommunications (ICETE), vol. 2, pp. 183–190. IEEE (2015)
8. Bonabeau, E.: Decisions 2.0: the power of collective intelligence. MIT Sloan Manag. Rev. **50**(2), 45 (2009)
9. Laskey, K.B.: Crowdsourced decision support for emergency responders. Technical report, George Mason Univ., FairFax, VA, Center for Excellence in Command Control Communications Computers-Intelligence (2013)
10. Chiu, C.M., Liang, T.P., Turban, E.: What can crowdsourcing do for decision support? Decis. Support Syst. **65**, 40–49 (2014)
11. Hosio, S., Goncalves, J., Anagnostopoulos, T., Kostakos, V.: Leveraging wisdom of the crowd for decision support. In: Proceedings of the 30th International BCS Human Computer Interaction Conference: Fusion! p. 38. BCS Learning & Development Ltd. (2016)

12. Basu, M., Bandyopadhyay, S., Ghosh, S.: Post disaster situation awareness and decision support through interactive crowdsourcing. Procedia Eng. **159**, 167–173 (2016)
13. Latchininsky, V.: Locusts. In: Breed, M.D., Moore, J. (eds.) Encyclopedia of Animal Behavior, vol. 2, pp. 288–297. Academic Press, Oxford (2010)
14. Brader, L., Djibo, H., Faye, F., Ghaout, S., Lazar, M., Luzietoso, P., Babah, M.O.: Towards a more effective response to desert locusts and their impacts on food security, livelihoods and poverty. In: Multilateral Evaluation of the 2003–2005 Desert Locust Campaign. Food and Agriculture Organisation, Rome (2006)
15. Lazar, M.: La dynamique des populations du criquet pèlerin (Schistocerca gregaria, Forsk. 1775) dans ses aires grégarigènes du sud algérien. Apport des données historiques et satellitaires pour améliorer la prévision des pullulations. Ph.D. thesis, ENSA (2015)

Pattern Recognition and Image Processing

Combined and Weighted Features for Robust Multispectral Face Recognition

Nadir Kamel Benamara[1]([✉]), Ehlem Zigh[2], Tarik Boudghene Stambouli[1], and Mokhtar Keche[1]

[1] Laboratory Signals and Images (LSI), University of Sciences and Technology of Oran Mohamed Boudiaf (USTO-MB), Oran, Algeria
nadirkamel.benamara@univ-usto.dz
[2] Laboratory of Research in Applied ICT (LARATIC), National Institute of Telecommunications and ICT of Oran (INTTIC), Oran, Algeria

Abstract. Face recognition has been very popular in recent years, for its advantages such as acceptance by the wide public and the price of cameras, which became more accessible. The majority of the current facial biometric systems use the visible spectrum, which suffers from some limitations, such as sensitivity to light changing, pose and facial expressions. The infrared spectrum is more relevant to facial biometric, for its advantages such as robustness to illumination change. In this paper, we propose two multispectral face recognition approaches that use both the visible and infrared spectra. We tested the new approaches with Uniform Local Binary Pattern (uLBP) as a local descriptor and Zernike Moments as a global descriptor on IRIS Thermal/Visible and CSIST Lab 2 databases. The experimental results clearly demonstrate the effectiveness of our multispectral face recognition system compared to a system that uses a single spectrum.

Keywords: Multispectral face recognition · Infrared
Uniform Local Binary Pattern · Zernike moments · Feature fusion

1 Introduction

Face recognition is a routine task for humans to identify persons in their everyday lives. Since 1960, automatic face recognition become an area that interest more and more researches in computer vision and biometric technologies [1]. Their applications are useful mainly for security purposes, such as access control, authentication systems and crime investigations. Compared to the other biometric modalities (fingerprint, iris and palmprint), face recognition has advantages of the ease of capturing subject samples without interacting with the person to identify and it is accepted by the wide public.

© IFIP International Federation for Information Processing 2018
Published by Springer International Publishing AG 2018. All Rights Reserved
A. Amine et al. (Eds.): CIIA 2018, IFIP AICT 522, pp. 549–560, 2018.
https://doi.org/10.1007/978-3-319-89743-1_47

Face recognition, using visible spectrum (from 0.38 μm to 0.78 μm), has been a major interest of researchers in recent years, with the evolution of cameras, which have become cheaper and more sophisticated. Several algorithms have been developed in this field, like eigenfaces [2], ficherfaces [3] and elastic bunch graph matching [4], but the visible images are vulnerable to light changing, poses, facial expressions and also disguise and faking. To overcome these limitations, multispectral face recognition has grown in interest for the advantages that offers, such as more discriminative features than those given by other spectra.

Infrared spectrum is divided into 2 parts, the active infrared, with Near Infra-Red (NIR) (0.74 μm–1 μm) and short-wave IR (1 μm–3 μm), and the passive infrared or thermal infrared, with middle-wave IR (3 μm–5 μm) and the long-wave IR (8 μm–12 μm). The infrared does not suffer from the limitations of the visible spectrum, mainly the light changing, However, these spectra face other limitations and challenges, due to face expressions and outdoor applications for NIR, dark face images caused by skin moisture absorption of infrared wavelength above 1.45 μm in SWIR, glasses opacity and body metabolism (fever, sporting activity) that change the thermal image for MWIR and LWIR (thermal infrared).

Based on the advantages and disadvantages of each spectrum, it is possible to design a multispectral facial recognition biometric system with a better recognition rate than a system with a single spectrum modality.

To this effect, we propose, in this article, a face recognition method that combines the infrared and visible spectrum. The features from each spectrum are merged to obtain a more discriminative information.

This paper is organized as follows: In Sect. 2, we present some related work. Then, in Sect. 3, the two proposed multispectral face recognition approaches are described. In Sect. 4, we talk about the used benchmark databases for our experiments. Following, in Sect. 5, we present and discuss the obtained results. Finally, we draw some conclusions in Sect. 6.

2   Related Works

Many research works have been carried out, in order to ensure robust solutions for face recognition problems, by combining visible and infrared spectra. We can cite for example Kong et al. [5] who proposed a multiscale fusion of visible and thermal face images. They detected and replaced the glasses opacity in thermal infrared with an eye template, to improve the recognition performance. Buddharaju et al. [6] proposed a multispectral system based on score fusion between results of eigenspace matching on visible faces and those obtained by the physiology-based face recognition method in thermal infrared spectrum, introduced by [7–9]. Bhowmik et al. [10] presented the effect of infrared spectrum on the enhancement of recognition rate when it is fused with visible spectrum. Hermosilla et al. [11], proposed a multispectral face recognition system based on fusion of visible and thermal descriptors using a genetic algorithm. More recently, Guo et al. [12] proposed a deep network with an adaptive score fusion strategy for visible - near infrared face recognition.

3   Proposed Approaches

The proposed multispectral face recognition system comprises four stages: data base preprocessing, features extraction, features fusion and finally classification. Our contribution in this paper is related to the third stage, in which two features vectors are merged into a single characteristic vector containing the visible and thermal infrared information.

Two fusion approaches, named the features weighted average and the combined feature vectors, were tested. The flowcharts of these approaches are presented in Figs. 1 and 2, respectively.

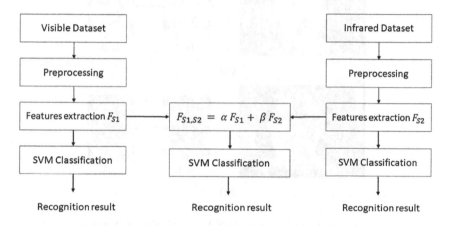

Fig. 1. Flowchart of the features weighted average approach

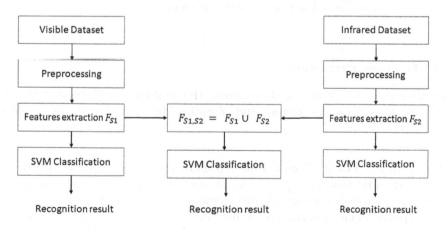

Fig. 2. Flowchart of the combined feature vectors approach

3.1   Database Preprocessing

The databases images are generally taken with a background, which it is useless for facial recognition. Therefore, the images must be cropped to keep only need the face of the person. Also, since the color information is not used, the cropped images should be converted from the color space to the grayscale one, as shown below in Fig. 3.

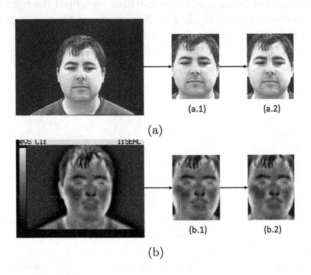

Fig. 3. Face preprocessing applied on an IRIS Thermal/Visible image. (a) Visible image (a.1) Visible cut image (a.2) Visible grayscale cut image. (b) Thermal image (b.1) Thermal cut image (b.2) Thermal grayscale cut image

3.2   Features Extraction

We have used two different descriptors: the uniform Local Binary Pattern (uLBP), as a local descriptor, and Zernike Moments (ZMs), as a global descriptor.

The Uniform Local Binary Pattern Feature. Local Binary Pattern was used for the first time in [13], for texture analysis. It proved its effectiveness in many image analysis applications, like biomedical, motion and biometric applications. Applied in face recognition for first time by Ahonen [14], the advantages of LBP are: the invariance to monotonic gray level changes or in otherwise illumination changing, the powerfulness for textural descriptions and the computational efficiency.

This local descriptor consists of the distribution (histogram) of the pixels, according to their neighborhoods. The neighborhood of a central pixel $P_0(x_c, y_c)$

is characterized by the pair (P, R), where P is the set of points (pixels) located around P_0, inside the circle of radius R. The coordinates of point P_i are given by:

$$P_i = (x_c + R\cos(\frac{2\pi i}{P}), y_c - R\sin(\frac{2\pi i}{P})) \tag{1}$$

As illustrated in Fig. 4, the value of the LBP pixel, for a neighborhood of P = 8 pixels, is calculated by the following formula:

$$I_{LBP}(P_0) = \sum_{i=1}^{8} P_i' * 2^{i-1} \tag{2}$$

where

$$P_i' = \begin{cases} 1 & \text{for } I(P_i) \geq I(P_0) \\ 0 & \text{for } I(P_i) < I(P_0) \end{cases} \tag{3}$$

where $I(P)$ denotes the intensity, $i.e.$ the gray level of pixel P.

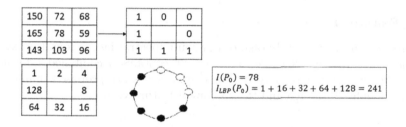

Fig. 4. LBP Calculation $(P = 8, R = 1)$, for a circular binary pattern representation.

Uniform Local Binary Pattern was proposed in [15], where the smaller non-uniformity measure is described as the less likely pattern that undergoes unwanted changes, as rotation, the non-uniformity measure represents the number of transition in the circular bitwise LBP representation, for example: 00000100 and 11111000 have non-uniformity measure of 2, 0 and 255 (00000000 and 11111111) have a measure of 0. Other patterns have at least a non uniformity of 4. In [15], they selected nine uniform pattern that have non uniformity measure of at most 2 which are: 00000000, 00000001, 00000011, 00000111, 00001111, 00011111, 00111111, 01111111 and 11111111, these patterns and their circular rotated versions correspond to a subset of 58 patterns, from the original 256 patterns LBP set. The remaining patterns are accumulated in the 59th bin.

The resulting histogram has 59 bins could be expressed as a feature vector that describes the face image.

The Zernike Moments Feature. Zernike moments are a set of orthogonal polynomials that describes the whole image. Defined as a global descriptor, they have very interesting properties [16,17], like orthogonality, which means less information redundancy, rotation invariance and high accuracy for detailed shapes. Zernike moments are calculated by:

$$Z_{n,m} = \frac{n+1}{\pi} \sum_{r \leq 1} \sum_{\theta \leq 2\pi} I(r,\theta).[V_{n,m}(r,\theta)]^* \tag{4}$$

where $I(r,\theta)$ is the representation of image in polar coordinates, $V_{n,m}(r,\theta)$ represents an orthogonal radial basis function, on which the image is projected, defined by:

$$V_{n,m}(r,\theta) = R_{n,m}(r).e^{jm\theta} \tag{5}$$

$R_{n,m}(r)$ is equal to:

$$R_{n,m}(r) = \begin{cases} \sum_{s=0}^{\frac{n-m}{2}} \frac{(-1)^s (n-s)!}{s![\frac{1}{2}(n+m)-s]![\frac{1}{2}(n-m)-s]!}.r^{n-2s} & \text{for } (n-m) \text{ even} \\ 0 & \text{for } (n-m) \text{ odd} \end{cases} \tag{6}$$

and $n = 0,1,2,3...$, $m = 0,1,2,3...$ and $m \leq n$.

3.3 Features Fusion

Features fusion is the third stage of our multispectral face recognition system. The aim of this stage is to get a data that contains a combined information from different spectrums. Two different approaches are proposed, namely: the features weighted average and the combined feature vectors.

The Features Weighted Average Approach. In this approach, a new feature vector is obtained by a linear combination of the feature vectors, obtained from two different spectrums, the visible and the infrared ones:

$$F_{S1,S2} = \alpha F_{S1} + \beta F_{S2} \tag{7}$$

where F is a feature vector, $S1$ and $S2$ are two different spectra and α, β are the weights of each spectrum, with $\alpha + \beta = 1$.

The Combined Feature Vectors Approach. In the combined feature vector approach, the features vectors, from the visible and infrared spectrums, are concatenated to form a new feature vector:

$$F_{S1,S2} = F_{S1} \cup F_{S2} \tag{8}$$

3.4 Classification

The role of a classification algorithm is to assign a class to each input feature vector.

We have chosen, as a classifier, the well-known powerful one-versus-all multiclass Support Vector Machine (SVM) classifier, with a linear kernel.

4   The Used Face Databases

Two databases have been used to evaluate the performance of the proposed multispectral face recognition methods, the CSIST database and IRIS Thermal/Visible face database (Fig. 5).

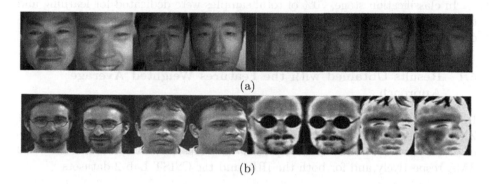

(a)

(b)

Fig. 5. Database samples: (a) CSIST Lab 2 (b) IRIS database

4.1   CSIST Database

CSIST Database was built by Harbin Institute of Technology Shenzhen Graduate School [18]. It contains two subsets of two different spectrum: Near Infrared & Visible spectrum. The first subset named Lab1 contains 1000 face images at a resolution of 100×80, 500 images for each spectrum, of 50 different subjects (10 images for each subject), The second subset, Lab2, contains 2000 face images at a resolution of 200×200, 1000 images for each spectrum, of 50 different subjects (20 images for each subject).

For our experiments, we have chosen the second subset Lab2 because the first subset, Lab1, does not contain an illumination change in its face images. The step of preprocessing will not be performed on CSIST Database, because face images in this data set are already cropped by its founder.

4.2   IRIS Thermal/Visible Face Dataset

IRIS Thermal/visible face dataset [19] is a public database comprising 4228 pairs 320×240 pixel visible and thermal face images of 30 individuals, with different poses, variable illumination conditions and different facial expressions.

In our experiment, we have selected a sub-set of 954 images, 477 images from each spectrum: front view, 2 different poses (right and left orientation), with presence and absence of light, and 3 different facial expressions, with different right and left orientations too.

5   Results and Discussion

To evaluate our proposed approaches, we have used two different descriptors: Uniform Local Binary Pattern, as a local descriptor, and Zernike Moments, as a global descriptor. The evaluation was performed on two different datasets, IRIS Thermal/Visible Database and CSIST Lab 2, as described above.

In classification stage, 70% of total samples were dedicated for learning and the remaining 30% for testing. All experiments were carried out with the same computer configuration which is Intel i3 5010u 2.10 Ghz with 4 GB RAM.

5.1   Results Obtained with the Features Weighted Average Approach

In order to get the features weighted average, we calculate features (uLBP or ZMs) from each spectrum and fuse them, according to Eq. (7), for different values of α. The classification results are presented in Figs. 6a and b, for the uLBP and ZMs, respectively and for both the IRIS and the CSIST Lab 2 datasets.

Results Obtained with the uLBP Local Features. For the IRIS Dataset, Fig. 6a clearly shows that the features weighted average method provides a recognition rate that is significantly higher than those given by using the single spectrums. The highest 88.8% recognition rate was obtained with $\alpha = 0.8$. This rate is to be compared to the 82.6 % and 81.9% rates, obtained using the single thermal and visible spectra, respectively. We notice that the lowest performance was obtained with $\alpha = 0.3$, *i.e.* a heavy weighting in favor of thermal data. This is because third of the used subset contains face images with glasses, which rises the glasses opacity problem that the thermal spectrum suffers from.

The results obtained with CSIST Lab 2 Dataset confirm that combining the visible spectrum and the NIR spectrum features, using the features weighted average approach, gives better results than using these features separately. The highest recognition rate of 87%, obtained with $\alpha = 0.5$, is to be compared to the rates 81% and 74.7%, obtained, respectively, with the near infrared and the visible spectrums features.

Results Obtained with the ZMs Global Descriptor. For the second evaluation of the features weighted average approach, the Zernike moments with a polynomial degree of order 10 were used, as a global descriptor. The classification results are presented in Fig. 6b.

With the IRIS Dataset, using the features weighted average approach, gives, for $\alpha = 0.5$, a 88.9% recognition rate that is better than the 83.3% rate, obtained with the thermal spectrum features, and slightly better than the 88.2% rate, obtained using the visible spectrum features only.

With the CSIST Lab 2 Dataset, the highest recognition rate, obtained by combining the features extracted from the visible and the near infrared spectra, using the features weighted average approach, with $\alpha = 0.4$, is 83.7%. It is higher

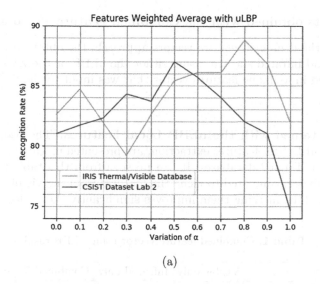

(a)

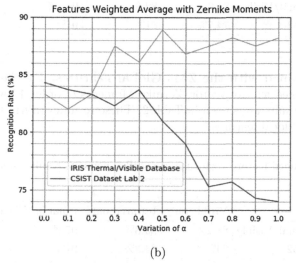

(b)

Fig. 6. Features weighted average results: (a) uLBP 2 (b) ZMs $n = 10$

than the 74% rate, obtained with the visible spectrum features, and slightly lower than the 84.3% rate, obtained by using the near infrared features alone.

The optimal value of α depends essentially on the conditions in which the database has been captured, for that reason we notice the difference between IRIS database and CSIST Lab2 dataset regarding the optimal weighting α and β.

5.2   Results obtained with the Combined Feature Vectors Approach

To evaluate the combined feature vectors approach, the obtained features from the visible and infrared spectra, using either the uLBP or the ZMs descriptors, were gathered in a unique features vector that was input to the SVM for classification.

Results Obtained with the uLBP Local Features. The results obtained by using the uLBP, for features extraction, are presented in Table 1, for both the IRIS and CSIST Lab2 databases. It can be seen from this table that for both of these databases, compared to using the features from the visible and infrared spectra separately, merging them improves significantly the performance.

Table 1. Combined feature vector using uLBP results

Database	Visible only	Infrared only	Combined feature vector
IRIS Thermal/Visible	81,9%	82.6%	91,7%
CSIST Lab2	74,7%	81%	94%

Results Obtained with the ZMs Global Features. For the second evaluation of combined feature vector, Zernike Moments of order 10 were used. The classification results are shown in Table 2. These results confirm those obtained with the uLBP local features.

Table 2. Combined feature vector using Zernike moments results (polynomial degree $n = 10$)

Database	Visible only	Infrared only	Combined feature vector
IRIS Thermal/Visible	88.2%	83.3%	97.2%
CSIST Lab2	74%	84.3%	91%

5.3   Discussion

We have noticed that the combined feature vectors proposed approach gives better results than the features weighted average approach for both databases, in terms of recognition rate, however it has a longer training phase as shown in Table 3. The reason of this result is that the first features fusion approach combines the totality of the visible and invisible vectors, whereas the second one takes just a percentage from each spectrum.

Concerning the near infrared-visible database, the local descriptor has a better recognition rate than the global one for both features fusion approaches. This can be explained by the robustness of the local descriptors, in comparison to the

Table 3. The computational learning time for the two proposed approaches

Approach	IRIS Thermal/Visible		CSIST Lab2	
	uLBP	Zernike moments	uLBP	Zernike moments
Features weighted average	31.3 ms	15.6 ms	78.1 ms	62.5 ms
Combined feature vectors	62.5 ms	31.3 ms	125 ms	109.4 ms

global descriptors, to variations, like illumination changes, which is the case for the CSIST Lab 2 database that contains an important light changing in face images.

Regarding the IRIS thermal/visible database, the global descriptor has a better performance than the local one, for both proposed approaches. This result is due to the rotation invariance property of the Zernike moments, which performs well for the slightly rotated faces, present in this database.

To summarize, our proposed approaches perform well in presence of illumination changes and slight variations of poses (rotated faces), which could be ideal for building a robust multispectral face recognition system.

6 Conclusion

In this paper, a features weighted average and a combined feature vectors approaches were proposed and applied for multispectral face recognition. The features from both the visible and the infrared spectra were extracted by using either the local uLBP descriptor or the global Zernike moments descriptor. These features were combined, by using one of the above mentioned approaches, and then input to a SVM classifier. The results obtained by fusing the features from the visible and invisible spectra were compared to those obtained by using the features from these spectra singly. The comparison shows that fusing the features from the visible and infrared spectra improves the performance. It also shows that the fusion by the combined feature vectors approach is better than the fusion by features weighted average approach.

For future work, a more elaborated fusion method and an other SVM kernels will be applied for a further improvement of classification rate.

References

1. Li, S.Z., Jain, A.K. (eds.): Handbook of Face Recognition, 2nd edn. Springer, London (2011). https://doi.org/10.1007/978-0-85729-932-1
2. Turk, M., Pentland, A.: Eigenfaces for recognition. J. Cogn. Neurosci. **3**(1), 71–86 (1991)
3. Belhumeur, P., Hespanha, J., Kriegman, D.: Eigenfaces vs. fisherfaces: recognition using class specific linear projection. IEEE Trans. Pattern Anal. Mach. Intell. **19**(7), 711–720 (1997)

4. Wiskott, L., Fellous, J.M., Kuiger, N., von der Malsburg, C.: Face recognition by elastic bunch graph matching. IEEE Trans. Pattern Anal. Mach. Intell. **19**(7), 775–779 (1997)
5. Kong, S.G., Heo, J., Boughorbel, F., Zheng, Y., Abidi, B.R., Koschan, A., Yi, M., Abidi, M.A.: Multiscale fusion of visible and thermal IR images for illumination-invariant face recognition. Int. J. Comput. Vis. **71**(2), 215–233 (2007)
6. Buddharaju, P., Pavlidis, I.: Multispectral face recognition: fusion of visual imagery with physiological information. In: Hammoud, R.I., Abidi, B.R., Abidi, M.A. (eds.) Face Biometrics for Personal Identification, pp. 91–108. Springer, Heidelberg (2007). https://doi.org/10.1007/978-3-540-49346-4_7
7. Buddharaju, P., Pavlidis, I., Tsiamyrtzis, P.: Physiology-based face recognition, pp. 354–359. IEEE (2005)
8. Buddharaju, P., Pavlidis, I., Tsiamyrtzis, P.: Pose-invariant physiological face recognition in the thermal infrared spectrum, pp. 53–53. IEEE (2006)
9. Buddharaju, P., Pavlidis, I.T., Tsiamyrtzis, P., Bazakos, M.: Physiology-based face recognition in the thermal infrared spectrum. IEEE Trans. Pattern Anal. Mach. Intell. **29**(4), 613–626 (2007)
10. Kanti Bhowmik, M., Saha, K., Majumder, S., Majumder, G., Saha, A., Nath, A., Bhattacharjee, D., Basu, D.K., Nasipuri, M.: Thermal infrared face recognition a biometric identification technique for robust security system. In: Corcoran, P., (ed.) Reviews, Refinements and New Ideas in Face Recognition. InTech, July 2011
11. Hermosilla, G., Gallardo, F., Farias, G., Martin, C.: Fusion of visible and thermal descriptors using genetic algorithms for face recognition systems. Sensors **15**(8), 17944–17962 (2015)
12. Guo, K., Wu, S., Xu, Y.: Face recognition using both visible light image and near-infrared image and a deep network. CAAI Trans. Intell. Technol. **2**(1), 39–47 (2017)
13. Ojala, T., Pietikinen, M., Harwood, D.: A comparative study of texture measures with classification based on featured distributions. Pattern Recogn. **29**(1), 51–59 (1996)
14. Ahonen, T., Hadid, A., Pietikäinen, M.: Face recognition with local binary patterns. In: Pajdla, T., Matas, J. (eds.) ECCV 2004. LNCS, vol. 3021, pp. 469–481. Springer, Heidelberg (2004). https://doi.org/10.1007/978-3-540-24670-1_36
15. Topi, M., Timo, O., Matti, P., Maricor, S.: Robust texture classification by subsets of local binary patterns, vol. 3, pp. 935–938. IEEE Computer Society (2000)
16. Teh, C.H., Chin, R.: On image analysis by the methods of moments. IEEE Trans. Pattern Anal. Mach. Intell. **10**(4), 496–513 (1988)
17. Singh, C., Mittal, N., Walia, E.: Face recognition using Zernike and complex Zernike moment features. Pattern Recogn. Image Anal. **21**(1), 71–81 (2011)
18. Xu, Y.: Bimodal biometrics based on a representation and recognition approach. Opt. Eng. **50**(3), 037202 (2011)
19. OTCBVS: IRIS Thermal/Visible Face Database. http://vcipl-okstate.org/pbvs/bench/. Accessed 17 Feb 2017

Conjugate Gradient Method for Brain Magnetic Resonance Images Segmentation

EL-Hachemi Guerrout[1(✉)] , Samy Ait-Aoudia[1] , Dominique Michelucci[2] , and Ramdane Mahiou[1]

[1] Ecole nationale Supérieure en Informatique, Laboratoire LMCS, Oued-Smar, Algiers, Algeria
{e_guerrout,s_ait_aoudia,r_mahiou}@esi.dz
[2] Université de Bourgogne, Laboratoire LE2I, Dijon, France
dominique.michelucci@u-bourgogne.fr

Abstract. Image segmentation is the process of partitioning the image into regions of interest in order to provide a meaningful representation of information. Nowadays, segmentation has become a necessity in many practical medical imaging methods as locating tumors and diseases. Hidden Markov Random Field model is one of several techniques used in image segmentation. It provides an elegant way to model the segmentation process. This modeling leads to the minimization of an objective function. Conjugate Gradient algorithm (CG) is one of the best known optimization techniques. This paper proposes the use of the nonlinear Conjugate Gradient algorithm (CG) for image segmentation, in combination with the Hidden Markov Random Field modelization. Since derivatives are not available for this expression, finite differences are used in the CG algorithm to approximate the first derivative. The approach is evaluated using a number of publicly available images, where ground truth is known. The Dice Coefficient is used as an objective criterion to measure the quality of segmentation. The results show that the proposed CG approach compares favorably with other variants of Hidden Markov Random Field segmentation algorithms.

Keywords: Brain image segmentation
Hidden Markov Random Field · The Conjugate Gradient algorithm
Dice Coefficient metric

1 Introduction

Automatic segmentation of medical images has become a crucial task due to the huge amount of data produced by imaging devices. Many popular tools as FSL [42] and Freesurfer [11] are dedicated to this aim. There are several techniques to achieve the segmentation. We can broadly classify them into thresholding methods [21,28,43], clustering methods [7,31,39], edge detection

© IFIP International Federation for Information Processing 2018
Published by Springer International Publishing AG 2018. All Rights Reserved
A. Amine et al. (Eds.): CIIA 2018, IFIP AICT 522, pp. 561–572, 2018.
https://doi.org/10.1007/978-3-319-89743-1_48

methods [5,30,35], region-growing methods [22,34], watersheds methods [3,24], model-based methods [6,20,25,38] and Hidden Markov Random Field methods [1,14,14–19,29,42]. Threshold-based methods are the simplest ones that require only one pass through the pixels. They begin with the creation of an image histogram. Then, thresholds are used to separate the different image classes. For example, to segment an image into two classes, foreground and background, one threshold is necessary. The disadvantage of threshold-based techniques is the sensitivity to noise. Region-based methods assemble neighboring pixels of the image in non-overlapping regions according to some homogeneity criterion (gray level, color, texture, shape and model). We distinguish two categories, region-growing methods and split-merge methods. They are effective when the neighboring pixels within one region have similar characteristics. In model-based segmentation, a model is built for a specific anatomic structure by incorporating prior information on shape, location and orientation. The presence of noise degrades the segmentation quality. Therefore, noise removal phase is generally an essential prior. Hidden Markov Random Field (HMRF) [12] provides an elegant way to model the segmentation problem. It is based on the MAP (Maximum A Posteriori) criterion [40]. MAP estimation leads to the minimization of an objective function [37]. Therefore, optimization techniques are necessary to compute a solution. Conjugate Gradient algorithm [26,33,36] is one of the most popular optimization methods.

This paper presents an unsupervised segmentation method based on the combination of Hidden Markov Random Field model and Conjugate Gradient algorithm. This method referred to as HMRF-CG, does not require preprocessing, feature extraction, training and learning. Brain MR image segmentation has attracted a particular attention in medical imaging. Thus, our tests focus on BrainWeb[1] [8] and IBSR[2] images where the ground truth is known. Segmentation quality is evaluated using Dice Coefficient (DC) [9] criterion. DC measures how much the segmentation result is close to the ground truth. This paper is organized as follows. We begin by introducing the concept of Hidden Markov Field in Sect. 2. Section 3 is devoted to the well known Conjugate Gradient algorithm. Section 4 is dedicated to the experimental results and Sect. 5 concludes the paper.

2   Hidden Markov Random Field (HMRF)

Let $S = \{s_1, s_2, \ldots, s_M\}$ be the sites, pixels or positions set. Both image to segment and segmented image are formed of M sites. Each site $s \in S$ has a neighborhood set $V_s(S)$ (see an example in Fig. 1).

A neighborhood system $V(S)$ has the following properties:

$$\begin{cases} \forall s \in S, s \notin V_s(S) \\ \forall \{s, t\} \in S, s \in V_t(S) \Leftrightarrow t \in V_s(S) \end{cases} \tag{1}$$

---

[1] http://www.bic.mni.mcgill.ca/brainweb/.
[2] https://www.nitrc.org/projects/ibsr.

s_1	s_2			
		n		
	n	s	n	
		n		
			s_M	

Fig. 1. An example of a lattice S, n is the set of sites neighboring s.

A r-order neighborhood system $V^r(S)$ is defined by the following formula:

$$V_s^r(S) = \{t \in S \mid \text{distance}(s,t)^2 \leq r \wedge s \neq t\} \tag{2}$$

where distance(s,t) is the Euclidean distance between pixels s and t. This distance depends only on the pixel position *i.e.*, it is not related to the pixel value (see examples in Fig. 2). For volumetric data sets, as slices acquired by scanners, a 3D neighborhood system is used.

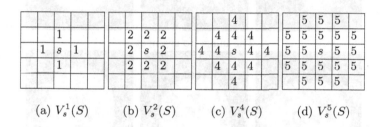

| (a) $V_s^1(S)$ | (b) $V_s^2(S)$ | (c) $V_s^4(S)$ | (d) $V_s^5(S)$ |

Fig. 2. First, second, fourth and fifth order neighborhood system of the site s.

A clique c is a subset of S where all sites are neighbors to each other. For a non single-site clique, we have:

$$\forall \{s,t\} \in c, s \neq t \Rightarrow (t \in V_s(S) \wedge s \in V_t(S)) \tag{3}$$

A p-order clique noted C_p contains p sites i.e. p is the cardinal of the clique (see an example in Fig. 3).

$V_s^2(S)$

a	b	c	
d	s	e	
f	g	h	

$C_1 : \{s\}$
$C_2 : \{s,a\}, \{s,b\}, \{s,c\}, \{s,d\}, \{s,e\}, \{s,f\}, \{s,g\}, \{s,h\}$
$C_3 : \{s,a,b\}, \{s,b,c\}, \{s,c,e\}, \{s,e,h\}, \{s,h,g\}, \{s,g,f\},$
$\qquad \{s,f,d\}, \{s,d,a\}, \{s,d,b\}, \{s,b,e\}, \{s,e,g\}, \{s,g,d\}$
$C_4 : \{s,d,a,b\}, \{s,b,c,e\}, \{s,e,h,g\}, \{s,g,f,d\}$

Fig. 3. Cliques associated to the second order neighborhood system for the site s.

Let $y = (y_1, y_2, \ldots, y_M)$ be the pixels values of the image to segment and $x = (x_1, x_2, \ldots, x_M)$ be the pixels classes of the segmented image. y_i and x_i are respectively pixel value and class of the site s_i. The image to segment y and the segmented image x are seen respectively as a realization of Markov Random families $Y = (Y_1, Y_2, \ldots, Y_M)$ and $X = (X_1, X_2, \ldots, X_M)$. The families of Random variables $\{Y_s\}_{s \in S}$ and $\{X_s\}_{s \in S}$ take their values respectively in the gray level space $E_y = \{0, \ldots, 255\}$ and the discrete space $E_x = \{1, \ldots, K\}$. K is the number of classes or homogeneous regions in the image. Configurations set of the image to segment y and the segmented image are respectively $\Omega_y = E_y^M$ and $\Omega_x = E_x^M$. Figure 4 shows an example of segmentation into three classes.

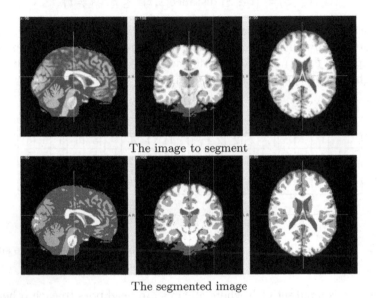

The image to segment

The segmented image

Fig. 4. An example of segmentation using FSL tool.

The segmentation of the image y consists of looking for a realization x of X. HMRF models this problem by maximizing the probability $P[X = x \mid Y = y]$.

$$x^* = \underset{x \in \Omega_x}{\arg\max} \{P[X = x \mid Y = y]\} \tag{4}$$

$$\begin{cases} P[X = x | Y = y] = A \exp(-\Psi(x, y)) \\ \Psi(x, y) = \sum_{s \in S} \left[\ln(\sigma_{x_s}) + \frac{(y_s - \mu_{x_s})^2}{2\sigma_{x_s}^2} \right] + \frac{B}{T} \sum_{c_2 = \{s,t\}} (1 - 2\delta(x_s, x_t)) \\ A \text{ is a positive constant} \end{cases}$$

where B is a constant, T is a control parameter called temperature, δ is a Kronecker's delta and μ_{x_s}, σ_{x_s} are respectively the mean and standard deviation of the class x_s. When $B > 0$, the most likely segmentation corresponds to the constitution of large homogeneous regions. The size of these regions is controlled by the B value.

Maximizing the probability $P[X = x \mid Y = y]$ is equivalent to minimizing the function $\Psi(x, y)$.

$$x^* = \arg\min_{x \in \Omega_x} \{\Psi(x, y)\} \tag{5}$$

The computation of the exact segmentation x^* is practically impossible [12]. Therefore optimization techniques are necessary to compute an approximate solution $\hat{x}$.

Let $\mu = (\mu_1, \ldots, \mu_j, \ldots, \mu_K)$ be the means and $\sigma = (\sigma_1, \ldots, \sigma_j, \ldots, \sigma_K)$ be the standard deviations of the K classes in the segmented image $x = (x_1, \ldots, x_s, \ldots, x_M)$ i.e.,

$$\begin{cases} \mu_j = \frac{1}{|S_j|} \sum_{s \in S_j} y_s \\ \sigma_j = \sqrt{\frac{1}{|S_j|} \sum_{s \in S_j} (y_s - \mu_j)^2} \\ S_j = \{s \mid x_s = j\} \end{cases} \tag{6}$$

In our approach, we will minimize $\Psi(\mu)$ defined below instead of minimizing $\Psi(x, y)$. We can always compute x through μ by classifying y_s into the nearest mean μ_j i.e., $x_s = j$ if the nearest mean to y_s is μ_j. Thus instead of looking for x^*, we look for μ^*. The configuration set of μ is $\Omega_\mu = [0 \ldots 255]^K$.

$$\begin{cases} \mu^* = \arg_{\mu \in \Omega_\mu} \min \{\Psi(\mu)\} \\ \Psi(\mu) = \sum_{j=1}^{K} f(\mu_j) \\ f(\mu_j) = \sum_{s \in S_j} [\ln(\sigma_j) + \frac{(y_s - \mu_j)^2}{2\sigma_j^2}] + \frac{B}{T} \sum_{c_2 = \{s,t\}} (1 - 2\delta(x_s, x_t)) \end{cases} \tag{7}$$

where S_j, μ_j and σ_j are defined in the Eq. (6).

To apply unconstrained optimization techniques, we redefine the function $\Psi(\mu)$ for $\mu \in \mathbb{R}^K$ instead of $\mu \in [0 \ldots 255]^K$ as recommended by [4]. Therefore, the new function $\Psi(\mu)$ becomes as follows:

$$\Psi(\mu) = \sum_{j=1}^{K} F(\mu_j) \quad \text{where } \mu_j \in \mathbb{R} \tag{8}$$

$$F(\mu_j) = \begin{cases} f(0) - u_j * 10^3 & \text{if } \mu_j < 0 \\ f(\mu_j) & \text{if } \mu_j \in [0 \ldots 255] \\ f(255) + (u_j - 255) * 10^3 & \text{if } \mu_j > 255 \end{cases} \tag{9}$$

3   Hidden Markov Random Field and Conjugate Gradient algorithm (HMRF-CG)

To solve the minimization problem expressed in Eq. 7, we used the nonlinear conjugate gradient method. This latter generalizes the conjugate gradient method to nonlinear optimization. The summary of the algorithm is set out below.

Let μ^0 be the initial point and $d^0 = -\Psi'(\mu^0)$ be the first direction search.

Calculate the step size α^k that minimizes $\varphi_k(\alpha)$. It is found by ensuring that the gradient is orthogonal to the search direction d^k.

$$\varphi_k(\alpha) = \Psi(\mu^k + \alpha d^k) \tag{10}$$

At the iteration $k+1$, calculate μ^{k+1} as follows:

$$\mu^{k+1} = \mu^k + \alpha^k d^k \tag{11}$$

Calculate the residual or the steepest direction:

$$r^{k+1} = -\Psi'(\mu^{k+1}) \tag{12}$$

Calculate the search direction d^{k+1} as follows:

$$d^{k+1} = r^{k+1} + \beta^{k+1} d^k \tag{13}$$

In conjugate gradient method there are many variants to compute β^{k+1}, for example:

1. The Fletcher-Reeves conjugate gradient method:

$$\beta^{k+1} = \frac{\left(r^{k+1}\right)^T r^{k+1}}{\left(r^k\right)^T r^k} \tag{14}$$

2. The Polak-Ribière conjugate gradient method:

$$\beta^{k+1} = \max\left\{ \frac{\left(r^{k+1}\right)^T \left(r^{k+1} - r^k\right)}{\left(r^k\right)^T r^k}, 0 \right\} \tag{15}$$

To use conjugate gradient algorithm, we need the first derivative $\Psi'(\mu) = (\Delta_1, \ldots, \Delta_i, \ldots, \Delta_K)$. Since no mathematical expression is available, it is approximated with finite differences [10]. In our tests, we have used a centered difference approximation to compute the first derivative as follows:

$$\Delta_i = \frac{\Psi(\mu_1, \ldots, \mu_i + \varepsilon, \ldots, \mu_n) - \Psi(\mu_1, \ldots, \mu_i - \varepsilon, \ldots, \mu_n)}{2\varepsilon} \tag{16}$$

The good approximation of the first derivative relies on the choice of the value of the parameter ε. Through the tests conducted, we have selected 0.01 as the best value. In practice, the application is implemented in the cross-platform Qt creator (C++) under Linux system. We have used the GNU Scientific Library implementation of Polak-Ribière conjugate gradient method [13, 32]. The HMRF-CG method is summarized hereafter.

Input:
y the image to segment, K the number of classes, B the constant parameter of HMRF, T the control parameter of HMRF, μ^0 the initial point, ε the parameter used by the first derivative.

Initialization:
we define s as the minimizer structure of gsl_multimin_fdfminimizer type and we initialize it by: K the size problem, Ψ the function to minimize using the Eq. 8, Ψ' the first derivative of the function to minimize using the Eq. 16, μ^0 the start point, gsl_multimin_fdfminimizer_conjugate_pr (Polak-Ribière conjugate gradient) the minimizer function.

Iterations:
we perform one iteration to update the state of the minimizer using the function gsl_multimin_fdfminimizer_iterate(s) and after that, we test s for convergence.

The stopping criterion:
in our case, the minimization procedure should stop when the norm of the gradient ($\|\Psi'\|$) is less than 10^{-3}.

Output:
an approximation $\hat{\mu}$ of $\mu^* \in \mathbb{R}^n$, $\hat{x}$ the segmented image using $\hat{\mu}$.

4   Experimental Results

In this section, we show the effectiveness of HMRF-CG method. To this end, we will make a comparison with some methods that are: improved k-means and MRF-ACO-Gossiping [41]. Next, we will show, the robustness of HMRF-CG method against noise, by doing a comparison with FAST FSL(FMRIBs Automated Segmentation Tool) and LGMM (Local Gaussian Mixture Model)[23]. To perform a fair and meaningful comparison, we have used a metric known as Dice Coefficient [9]. Morey et al. [27] used interchangeably Dice coefficient and Percentage volume overlap. This metric is usable only when the ground truth segmentation is known (see Sect. 4.1). The image sets and related parameters are described in Sect. 4.2. Finally, Sect. 4.3 is devoted to the yielded results.

4.1   Dice Coefficient Metric

Dice Coefficient (DC) measures how much the result is close to the ground truth. Let the resulting class be $\hat{A}$ and its ground truth be A^*. Dice Coefficient is given by the following formula:

$$DC = \frac{2|\hat{A} \cap A^*|}{|\hat{A} \cup A^*|} \tag{17}$$

4.2   The Image Sets and Related Parameters

To evaluate the quality of segmentation, we use four volumetric (3D) MR images, one obtained from IBSR (real image) and the others from BrainWeb (simulated images). Three components were considered: GM (Grey Matter), WM (White Matter) and CSF (Cerebro Spinal Fluid). IBSR image has the following characteristics: dimension is $256 \times 256 \times 63$, with voxel $= 1 \times 3 \times 1$ mm and T1-weighted modality. The three BrainWeb image sets BrainWeb1, BrainWeb2 and Brain-Web3 have the following characteristics: dimensions are $181 \times 217 \times 181$, with voxels $= 1 \times 1 \times 1\,mm$ and T1-weighted modality. They have different levels of noise and intensity non-uniformity that are respectively: (0%,0%), (3%,20%) and (5%,20%). In this paper we have retained a subset of slices, which are cited in [41]. The IBSR slices retained are: 1-24/18, 1-24/20, 1-24/24, 1-24/26, 1-24/30, 1-24/32 and 1-24/34. The BrainWeb slices retained are: 85, 88, 90, 95, 97, 100, 104, 106, 110, 121 and 130.

Table 1 defines some parameters necessary to execute HMRF-CG method.

Table 1. Related parameters to images used in our tests.

Image	Constant B	Temperature T	Initial point μ^0
IBSR	1	10	(1, 5, 140, 190)
BrainWeb1	1	10	(4, 45, 110, 149)
BrainWeb2	1	4	(4, 45, 110, 150)
BrainWeb3	1	2	(0, 45, 110, 150)

Table 2. Mean DC values (the best results are given in bold type).

Methods	Dice Coefficient			
	GM	WM	CSF	Mean
K-means	0.500	0.607	0.06	0.390
MRF-ACO-Gossiping	0.778	0.827	0.262	0.623
HMRF-CG	**0.859**	**0.855**	**0.381**	**0.698**

Table 3. Mean DC values (the best results are in bold type).

Tissue	Method	Dice Coefficient		
		BrainWeb1	BrainWeb2	BrainWeb3
GM	HMRF-CG	**0.970**	**0.945**	**0.921**
	LGMM	0.697	0.905	0.912
	FSL FAST	0.727	0.737	0.735
WM	HMRF-CG	**0.990**	**0.971**	**0.954**
	LGMM	0.667	0.940	0.951
	FSL FAST	0.877	0.862	0.860
CSF	HMRF-CG	**0.961**	**0.942**	**0.926**
	LGMM	0.751	0.897	0.893
	FSL FAST	0.635	0.647	0.643

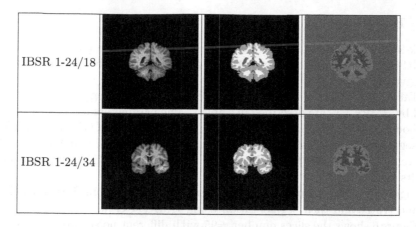

Fig. 5. (a) Slices number, (b) slices to segment from IBSR, (c) ground truth slices, (d) segmented slices using HMRF-CG.

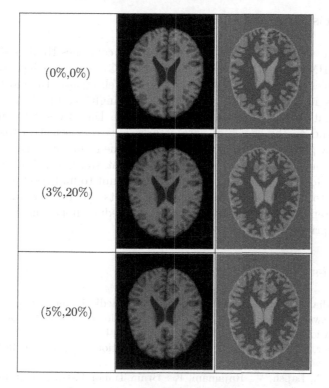

Fig. 6. The slices number #95 of BrainWeb images. (a) Noise and intensity non-uniformity, (b) slices to segment from BrainWeb images, (c) segmented slices using HMRF-CG.

4.3 Results

Table 2 shows the mean DC values using IBSR image. The parameters used by HMRF-CG are described in Table 1. The parameters used by the other methods are given in [41].

Table 3 shows the mean DC values using BrainWeb images. The parameters used by HMRF-CG are described in Table 1. The parameters used by the LGMM method are given in LGMM [23]. The implementation of LGMM is built upon the segmentation method [2] of SPM 8 (Statistical Parametric Mapping[3]), which is a well known software for MRI analysis. As reported by [23], LGMM has better results than SPM 8.

Figure 5 shows a sample of slices to segment obtained from IBSR image, their ground truths and their segmentation using HMRF-CG method.

Figure 6 shows the slices number #95 with different noise and intensity non-uniformity from BrainWeb images and their segmentation using HMRF-CG.

5   Discussion and Conclusion

In this paper, we have described a method which combines Hidden Markov Random Field (HMRF) and Conjugate Gradient (GC). The tests have been carried out on samples obtained from IBSR and BrainWeb images, the most commonly used images in the field. For a fair and meaningful comparison of methods, the segmentation quality is measured using the Dice Coefficient metric. The results depend on the choice of parameters. This very sensitive task has been conducted by performing numerous tests. From the results obtained, the HMRF-GC method outperforms the methods tested that are: LGMM, Classical MRF, MRF-ACO-Gossiping and MRF-ACO. Tests permit to find good parameters for HMRF-CG to achieve good segmentation results. To further improve performances a preprocessing step can be added to reduce noise and inhomogeneity using appropriate filters.

References

1. Ait-Aoudia, S., Guerrout, E.H., Mahiou, R.: Medical image segmentation using particle swarm optimization. In: 18th International Conference on Information Visualisation (IV) 2014, pp. 287–291. IEEE (2014)
2. Ashburner, J., Friston, K.J.: Unified segmentation. Neuroimage **26**(3), 839–851 (2005)
3. Benson, C., Lajish, V., Rajamani, K.: Brain Tumor Extraction from MRI Brain Images Using Marker Based Watershed Algorithm, pp. 318–323 (2015)
4. Boyd, S., Vandenberghe, L.: Convex optimization. Cambridge University Press, Cambridge (2004)
5. Canny, J.: A computational approach to edge detection. IEEE Trans. Pattern Anal. Mach. Intell. **6**, 679–698 (1986)

---

[3] http://www.fil.ion.ucl.ac.uk/spm/software/spm8/.

6. Chan, T.F., Vese, L., et al.: Active contours without edges. IEEE Trans. Image Process. **10**(2), 266–277 (2001)
7. Chuang, K.S., Tzeng, H.L., Chen, S., Wu, J., Chen, T.J.: Fuzzy c-means clustering with spatial information for image segmentation. Comput. Med. Imaging Graph. **30**(1), 9–15 (2006)
8. Cocosco, C.A., Kollokian, V., Kwan, R.K.S., Pike, G.B., Evans, A.C.: BrainWeb: Online Interface to a 3D MRI Simulated Brain Database (1997)
9. Dice, L.R.: Measures of the amount of ecologic association between species. Ecology **26**(3), 297–302 (1945)
10. Eberly, D.: Derivative Approximation by Finite Differences. Magic Software, Inc (2003)
11. Fischl, B.: Freesurfer. Neuroimage **62**(2), 774–781 (2012)
12. Geman, S., Geman, D.: Stochastic relaxation, Gibbs distributions, and the Bayesian restoration of images. IEEE Trans. Pattern Anal. Mach. Intell. **6**, 721–741 (1984)
13. Grippo, L., Lucidi, S.: A globally convergent version of the Polak-Ribière conjugate gradient method. Math. Program. **78**(3), 375–391 (1997)
14. Guerrout, E.H., Ait-Aoudia, S., Michelucci, D., Mahiou, R.: Hidden Markov random field model and BFGS algorithm for brain image segmentation. In: Proceedings of the Mediterranean Conference on Pattern Recognition and Artificial Intelligence, pp. 7–11. ACM (2016)
15. Guerrout, E.H., Ait-Aoudia, S., Michelucci, D., Mahiou, R.: Hidden Markov random field model and Broyden-Fletcher-Goldfarb-Shanno algorithm for brain image segmentation. J. Exp. Theor. Artif. Intell. 1–13 (2017)
16. Guerrout, E.H., Mahiou, R., Ait-Aoudia, S.: Medical image segmentation on a cluster of PCs using Markov random fields. Int. J. New Comput. Architectures Appl. (IJNCAA) **3**(1), 35–44 (2013)
17. Guerrout, E.H., Mahiou, R., Ait-Aoudia, S.: Medical image segmentation using hidden Markov random field a distributed approach. In: The Third International Conference on Digital Information Processing and Communications. The Society of Digital Information and Wireless Communication, pp. 423–430 (2013)
18. Guerrout, E.H., Mahiou, R., Ait-Aoudia, S.: Hidden Markov random fields and swarm particles: a winning combination in image segmentation. IERI Procedia **10**, 19–24 (2014)
19. Held, K., Kops, E.R., Krause, B.J., Wells III, W.M., Kikinis, R., Muller-Gartner, H.W.: Markov random field segmentation of brain MR images. IEEE Trans. Med. Imaging **16**(6), 878–886 (1997)
20. Ho, S., Bullitt, L., Gerig, G.: Level-Set Evolution with Region Competition: Automatic 3-D Segmentation of Brain Tumors, vol. 1, pp. 532–535 (2002)
21. Kumar, S., et al.: Skull Stripping and Automatic Segmentation of Brain MRI Using Seed Growth and Threshold Techniques, pp. 422–426 (2007)
22. Lin, G.C., Wang, W.J., Kang, C.C., Wang, C.M.: Multispectral MR images segmentation based on fuzzy knowledge and modified seeded region growing. Magn. Reson. Imaging **30**(2), 230–246 (2012)
23. Liu, J., Zhang, H.: Image segmentation using a local GMM in a variational framework. J. Math. Imag. Vis. **46**(2), 161–176 (2013)
24. Masoumi, H., Behrad, A., Pourmina, M.A., Roosta, A.: Automatic liver segmentation in MRI images using an iterative watershed algorithm and artificial neural network. Biomed. Sig. Process. Control **7**(5), 429–437 (2012)
25. McInerney, T., Terzopoulos, D.: Deformable models in medical image analysis: a survey. Med. Image Anal. **1**(2), 91–108 (1996)

26. Møller, M.F.: A scaled conjugate gradient algorithm for fast supervised learning. Neural Netw. **6**(4), 525–533 (1993)
27. Morey, R.A., Petty, C.M., Xu, Y., Hayes, J.P., Wagner, H.R., Lewis, D.V., LaBar, K.S., Styner, M., McCarthy, G.: A comparison of automated segmentation and manual tracing for quantifying hippocampal and amygdala volumes. Neuroimage **45**(3), 855–866 (2009)
28. Natarajan, P., Krishnan, N., Kenkre, N.S., Nancy, S., Singh, B.P.: Tumor Detection Using Threshold Operation in MRI Brain Images, pp. 1–4 (2012)
29. Panjwani, D.K., Healey, G.: Markov random field models for unsupervised segmentation of textured color images. IEEE Trans. Pattern Anal. Mach. Intell. **17**(10), 939–954 (1995)
30. Perona, P., Malik, J.: Scale-space and edge detection using anisotropic diffusion. IEEE Trans. Pattern Anal. Mach. Intell. **12**(7), 629–639 (1990)
31. Pham, D.L., Xu, C., Prince, J.L.: Current methods in medical image segmentation. Ann. Rev. Biomed. Eng. **2**(1), 315–337 (2000)
32. Polak, E., Ribière, G.: Note sur la convergence de méthodes de directions conjuguées. Revue française d'informatique et de recherche opérationnelle, série rouge **3**(1), 35–43 (1969)
33. Powell, M.J.D.: Restart procedures for the conjugate gradient method. Math. Program. **12**(1), 241–254 (1977)
34. Roura, E., Oliver, A., Cabezas, M., Vilanova, J.C., Rovira, À., Ramió-Torrentà, L., Lladó, X.: MARGA: multispectral adaptive region growing algorithm for brain extraction on axial MRI. Comput. Methods Programs Biomed. **113**(2), 655–673 (2014)
35. Senthilkumaran, N., Rajesh, R.: Edge detection techniques for image segmentation-a survey of soft computing approaches. Int. J. Recent Trends Eng. **1**(2), 250–254 (2009)
36. Shewchuk, J.R.: An introduction to the conjugate gradient method without the agonizing pain. Lecture available on internet (1994)
37. Szeliski, R., Zabih, R., Scharstein, D., Veksler, O., Kolmogorov, V., Agarwala, A., Tappen, M., Rother, C.: A comparative study of energy minimization methods for Markov random fields with smoothness-based priors. IEEE Trans. Pattern Anal. Mach. Intell. **30**(6), 1068–1080 (2008)
38. Wang, L., Shi, F., Li, G., Gao, Y., Lin, W., Gilmore, J.H., Shen, D.: Segmentation of neonatal brain MR images using patch-driven level sets. NeuroImage **84**, 141–158 (2014)
39. Wu, Z., Leahy, R.: An optimal graph theoretic approach to data clustering: theory and its application to image segmentation. IEEE Trans. Pattern Anal. Mach. Intell. **15**(11), 1101–1113 (1993)
40. Wyatt, P.P., Noble, J.A.: MAP MRF joint segmentation and registration of medical images. Med. Image Anal. **7**(4), 539–552 (2003)
41. Yousefi, S., Azmi, R., Zahedi, M.: Brain tissue segmentation in MR images based on a hybrid of MRF and social algorithms. Med. Image Anal. **16**(4), 840–848 (2012)
42. Zhang, Y., Brady, M., Smith, S.: Segmentation of brain MR images through a hidden Markov random field model and the expectation-maximization algorithm. IEEE Trans. Med. Imaging **20**(1), 45–57 (2001)
43. Zhao, M., Lin, H.Y., Yang, C.H., Hsu, C.Y., Pan, J.S., Lin, M.J.: Automatic threshold level set model applied on MRI image segmentation of brain tissue. Appl. Math. **9**(4), 1971–1980 (2015)

Facial Expressions Recognition: Development and Application to HMI

Zekhnine Chérifa$^{(\boxtimes)}$ ⓘ and Berrached Nasr Eddine$^{(\boxtimes)}$

University of Science and Technology of Oran Mohamed Boudiaf,
BP. 1505, El M'naouer, 31 000 Oran, Algeria
elnfi@yahoo.fr, laresi.usto.2015@gmail.com

Abstract. We present in this paper, a facial expressions recognition system to command a mobile robot (Pionner-3DX). The proposed system mainly consists of two modules: facial expression recognition and robot command. The first module aims to recognize the facial expressions like happiness, sadness, surprise, anger, fear, disgust and neutral using Gradient Vector Flow (GVF) snake to find ROI (Region Of Interest like: mouth, eyes, eyebrow) segmentation from FEEDTUM database (video file). While the second module, analyses the segmented ROI to recognize with Euclidian distance calculation (compatible with the MPEG-4 description of the six universal emotions) and Time Delay Neural Network classifier. Finally, the recognized facial expressions were used as control commands for the mobile robot displacement (forward; backward; turn left; turn right) in ROS (Robot Operating System).

Keywords: Facial expression recognition · GVF snake
MPEG-4 description · Time Delay Neural Network · Mobile robot
Robot Operating System

1 Introduction

Automatic analysis of facial expressions constitutes an important tool for research in human machine interaction (HMI). Autonomous robot control systems are complex systems that consist of a sensor, a decision-making control system and a motor drive system.

The sensor can be either a visual system, a speech system, or a manual control system [1]. In [2, 3], numerous systems are proposed to control a robot or a wheelchair using head or face movement. Such systems involve body movement and are not suitable for people with extreme physical disabilities where head or face movement is difficult. Speech controlled systems [4] are also not suitable for people with speech disability. Thus, current research has been focused on design of systems, which can be a good solution to these problems. The best alternative is to design a system where command is derived from recognizing the user's facial expressions like happiness, sadness, surprise, anger and neutral.

Recently, many works were about the automatic localization of the face and its characteristic features. Indeed, the objective is to develop some interactive systems capable to analyze and to interpret the user's behavior. Several works have been made

A. Amine et al. (Eds.): CIIA 2018, IFIP AICT 522, pp. 573–584, 2018.
https://doi.org/10.1007/978-3-319-89743-1_49

in this sense. The methods described in [5, 6] use models of facial movements, and those of [7, 8] the methods of classification by neural networks.

We already developed a FER (Facial Expression Recognition) system in the static case (JAFFE database) with MLP neural network, trained by the backpropagation of gradient algorithm. We used data extracted by a local method to model the facial expressions. Thus introducing the geometric coordinates of 19 points characterizing the ROI manually as made in [9]. These extracted data are normalized, so that the facial expressions remain invariant for either the change of scale or slant of the individual head. The performances of our recognition system, have been established on 4 facial expressions "neutral, joy, surprise, fear", among the seven universal expressions, because there is no very big difference (in geometric side) between neutral and fear expressions, and between joy and surprise expressions [10].

We have developed another FER system in the dynamic case (FEDTUM database [11]) with Time Delay Neural Network classifier, trained by the backpropagation of gradient algorithm. In this system, we use GVF snake and the Euclidian distance calculation (compatible with the MPEG-4 description of the six universal emotions) manually and we interpret this analysis in articular movements of an arm manipulator robot "Mentor 5-dof" in the setting development of HMI application [12].

In this article, we present our FER system (the dynamic case) with some improvements in the modeling, the analysis and the interpretation of the facial expressions from image sequences, to command the mobile robot displacement (forward; backward; turn left; turn right) in ROS (Robot Operating System).

2   Facial Expression Modeling

Features extraction is the crucial and complex part in any shapes recognition system. It often uses the results of the statistics theory, of evaluation, to get a transformation from the representation space toward the space of interpretation. The major problem of this part can be seen as the resolution of a two sub-problems:

- What measures must be made?
- What features from row data to use as input?

In our case, input data are features of facial expressions modeling frames. The facial expression is measured by the temporary non-rigid distortion (0.25–5 s) of the facial features (eyebrows, eyes, mouth). We have used the Gradient Vector Flow snake algorithm to extract the deformable and non-rigid measurement. However, this algorithm presents some limitations: it is only used for binary images, selects manually the initial snake and extracts only one geometric object per images. We propose to modify the GVF algorithm in a way to overcome those limitations. Our method is described in Sect. 2.1.

The FEEDTUM database was chosen because it is of the most used databases in facial expression recognition. It consists of elicited spontaneous emotions of 18 subjects within the MPEG-4 emotion-set plus added neutrality.

Kass et al. [13] have introduced the active contours or snakes. Since, numerous variants of these deformable models have been studied for multiple applications.

Their utility was particularly well illustrated in medical imagery, but also in electronic surveillance domain, and spatio-temporal tracking in video [14].

The GVF method has been developed to propose a solution to some limitations of the approach as the initialization of the snake and its convergence toward the concave regions [15].

2.1   GVF Field

The GVF method proceeds in two stages to calculate the GVF field:

- Calculate the gradient of image
- Calculate the gradient vector flow

The classical snake model proposed by Kass et al., defines the active contour as a parametric curve, $r(s) = (x(s); y(s))$, that moves in the spatial domain until the energy functional in Eq. 1 reaches its minimum value.

$$E_{snake} = \int (E_{int}(r(s)) + E_{ext}(r(s))) ds \tag{1}$$

Eint and *Eext* represent the internal and external energy, respectively. The internal energy enforces smoothness along the contour. A common internal energy function is defined as follows:

$$E_{int}(r(s)) = \left(\alpha |r'(s)|^2 + \beta |r''(s)|^2 \right) \tag{2}$$

Where α and β are weighting parameters, r' and r'' are the first and second derivative of $r(s)$ with respect to s. The first term, also known as tension energy, prevents the snake to remain attracted to isolated points. The second term, known as bending energy, prevents the contour of developing sharp angles. Constraints based on more complex shape models, such as Fourier descriptors.

The external energy is derived from the image, so that the snake will be attracted to features of interest. Given a gray level image $I(x; y)$, a common external energy is defined as:

$$E_{ext}(I(x,y)) = -|\nabla(G_\sigma(x,y) * I(x,y))|^2 \tag{3}$$

Where ∇ is the gradient operator, $G_\sigma(x,y)$ a 2D Gaussian kernel with standard deviation σ and where $*$ is the convolution operator. Minimizing the energy function of Eq. 1 results in solving the following associated Euler-Lagrange equations:

$$\alpha \frac{d^2 x(s)}{ds^2} - \beta \frac{d^4 x(s)}{ds^4} - \frac{\partial E_{ext}}{\partial x} = 0 \tag{4a}$$

$$\alpha \frac{d^2 y(s)}{ds^2} - \beta \frac{d^4 y(s)}{ds^4} - \frac{\partial E_{ext}}{\partial y} = 0 \tag{4b}$$

This can be seen as a force balance equation:

$$F_{int} + F_{ext} = 0 \tag{5}$$

These equations can be solved using gradient descent by considering $r(s)$ as a function of time, i.e. $r(s; t)$. The partial derivative of r with respect to t is then

$$\frac{dx(s,t)}{dt} = \alpha \frac{d^2x(s,t)}{ds^2} - \beta \frac{d^4x(s,t)}{ds^4} - \frac{\partial E_{ext}}{\partial x} \tag{6a}$$

$$\frac{dy(s,t)}{dt} = \alpha \frac{d^2y(s,t)}{ds^2} - \beta \frac{d^4y(s,t)}{ds^4} - \frac{\partial E_{ext}}{\partial y} \tag{6b}$$

When the snake stabilizes, i.e. when an optimum is found, the terms $\frac{dx(s,t)}{dt}$ and $\frac{dy(s,t)}{dt}$ vanish.

2.2   Gradient Vector Flow

The external force field defined in the previous section requires a good initialization [15], close to the object boundary, in order to segment the object. This limitation is caused by the nature of the external force field, whose vectors point towards the object only in the proximity of the object's boundary. As we move away from the boundary the external fields rapidly become zero, therefore reducing the possibilities that a contour located in such regions will converge correctly. To overcome this problem, Xu and Prince [16] proposed another external force field $v(x; y) = (u(x; y); v(x; y))$. This vector field minimizes the following energy functional:

$$E_{GVF}(u, v) = \iint \mu \left(\frac{du^2}{dx} + \frac{du^2}{dy} + \frac{dv^2}{dx} + \frac{dv^2}{dy} \right) + |\nabla f|^2 |v - \nabla f|^2 dxdy \tag{7}$$

Where μ is a nonnegative parameter expressing the degree of smoothness of the field v and where f is an edge map, e.g. $|\nabla I|$. The first term in Eq. 7 keeps the field v smooth, whereas the second term forces the field v to resemble the original edge force in the neighborhood of edges. This new external force is called gradient vector flow (GVF) field. The GVF-field can be found by solving the following associated Euler-Lagrange equations:

$$\mu \nabla^2 u - \left(u - \frac{\partial f}{\partial x} \right) \left(\frac{\partial f^2}{\partial x} + \frac{\partial f^2}{\partial y} \right) = 0 \tag{8a}$$

$$\mu \nabla^2 v - \left(u - \frac{\partial f}{\partial x} \right) \left(\frac{\partial f^2}{\partial x} + \frac{\partial f^2}{\partial y} \right) = 0 \tag{8b}$$

Where ∇^2 is the Laplacian operator.

2.3   Semi-automatic Initialization of GVF Snake

We have automated the initialization procedure of the snake by a geometric shape (ellipsis) that corresponds in its shape to the region of interest (the eyes), in the goal to give solution for the number of initial snake points problem, (see Fig. 1).

$$x = x_0 + a\cos(t) \tag{9}$$

$$y = y_0 + b\sin(t) \tag{10}$$

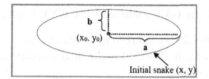

Fig. 1. Geometric shape of the initial snake.

Where (x_0, y_0) defines the region of interest center coordinates, (x, y) are the initial snake coordinates, a and b are respectively the width and the height of the ellipsis, t represents the step displacement of the snake.

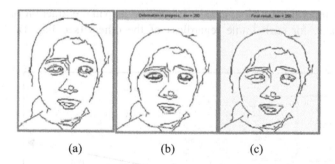

(a)   (b)   (c)

Fig. 2. (a) Automatic snake initialization. (b) Snake progression. (c) Final snake.

We modify the initial snake so that is initialized automatically (see Fig. 2a), with the GVF snake parameters: $\alpha = 0.2$, $\beta = 0$, $\gamma = 1$, $\tau = 2$, *iterMax* = 5.

The snake could converge toward the zone of concavity during 250 iterations, from a snake automatically initialized without any user intervention.

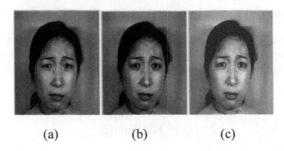

(a)   (b)   (c)

Fig. 3. (a) Superposition Image. (b) Automatic snake initialization. (c) Final snake.

2.4   Superposition of the GVF Field and the GVF Snake

The second idea that proved very interesting to achieve, is to superpose the GVF snake and the GVF field as shown in (Fig. 3a) in the goal to understand why the snake chose to converge toward this result.

We notice then, that the snake of (Fig. 3c) converged well toward the ROI in the GVF sense because it perfectly integrates the zone of the potential biggest GVF.

We achieved two strategies of attributes extraction by GVF snake from video frames using FEEDTUM database. These strategies are spatio-temporal extraction and temporal extraction.

2.5   Spatio-Temporal Extraction of the Attributes

The spatio-temporal extraction, consist in making a spatial extraction of eyes contour by GVF snake, for each frame regardless of the others (see Fig. 4). The result of

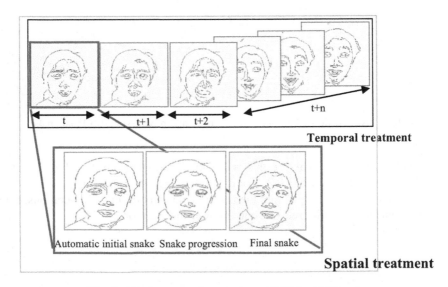

Fig. 4. Spatio-temporal extraction of ROI by GVF snake.

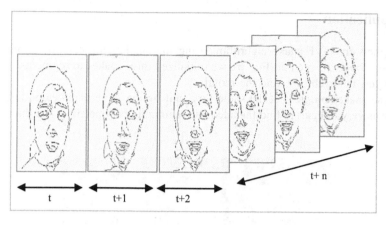

Fig. 5. Temporal extraction of ROI by GVF snake.

temporal extraction is defined by the set of results of all frames sequence, that we created while gathering pictures of three expressions (neutral, disgust, joy) by the software Ulead Medium Studio Pro 7.0, in order to test the performances of the GVF snake.

2.6   Temporal Extraction of the Attributes

This extraction, consist in initializing snake of the first frame, then the resulting contour of frame t is used as an initialization of the contour of frame $t + 1$ (see Fig. 5). The GVF force should range as far as the ROI can move between two subsequent frames as:

$$(Initial\ snake)|_{t=0} = Ellipsis\ shape.$$

$$(Initial\ snake)|_{t=t+1} = (final\ snake)|_{t=t}.$$

Fig. 6. Extraction of the ROI of FEEDTUM database by GVF snake.

3 Application

We applied our GVF snake algorithm, using the spatio-temporal extraction strategy described above. As seen in (Fig. 6), we initialized three snakes to segment the regions of the eyes, the eyebrows, and the mouth.

These snakes progress at the same time and by the same parameters. Let us note that it was very difficult to find the adequate regulating parameters of the three snakes.

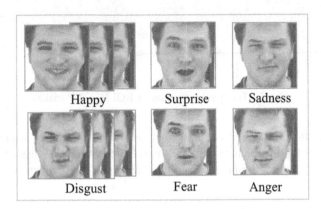

Happy	Surprise	Sadness
Disgust	Fear	Anger

Fig. 7. Skeletons of the six facial expressions extracted by the GVF snake (FEEDTUM database).

3.1 Modeling the ROI by MPEG-4 Description

We exploited the present information in the skeletons of the sequences of segmented pictures by the GVF snake (Fig. 7).

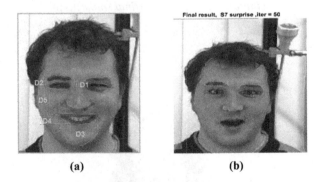

(a)   (b)

Fig. 8. (a) The facial expression model by MPEG-4 description. (b) Automatic extraction of points of interest from GVF snake results.

We took the work in [17] as a basis, according to the description of the MPEG-4 norm of the six universal expressions (Expressions → translation of these descriptions) (see Fig. 8a). We extract automatically points of interest from segmented ROI to calculate the 5 Euclidian distances (see Fig. 8b).

The neuter is represented by five distances takes as reference. For the rest of the expressions their description will be a combination of comparison of their distances in relation to those of the neutral expression. That, every facial expression (in our case, video sequence = 4 frames) is represented by four normalized features vectors.

We have applied this normalization procedure:

The change of axis (O, X, Y) linked to image, to (O', Uh, Uv) linked to face is established as follows:

$$x_{o'} = \frac{x_{og} + x_{od}}{2} \tag{11}$$

$$y_{o'} = \frac{y_{og} + y_{od}}{2} \tag{12}$$

Where $x_{o'}$ and $y_{o'}$ are the coordinates of new origin. $x_{od}, y_{od}, x_{og}, y_{og}$ are respectively coordinates of right and left eyes (see Fig. 9).

$$Uh = \sqrt{\left[x_{od} - x_{og}\right]^2 + \left[y_{od} - y_{og}\right]^2} \tag{13}$$

$$Uv = \sqrt{\left[x_{nez} - x_{o'}\right]^2 + \left[y_{nez} - y_{o'}\right]^2} \tag{14}$$

After having made a change of landmark, we can calculate the coordinate of the points of interest by taking:

$$X = \frac{x - x_{o'}}{U_h/2} \tag{15}$$

$$Y = \frac{y - y_{o'}}{U_v} \tag{16}$$

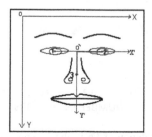

Fig. 9. Normalization with changing landmark.

With, x, y are coordinates of points of interest in old landmark (o, x, y). The facial expressions remain invariant for either the change of scale or slant of the individual head.

3.2 Analysis of Distances Features

To recognize the facial expressions from video sequences, time delay neural network classifier is used with the 5 feature distances in inputs layer, and the 7 expressions in output layer trained by the backpropagation gradient algorithm (with 90% of good recognition). Once the facial expressions are recognized, the corresponding control signals are sent to a real mobile robot (Pionner-3DX) using serial port via the Robot Operating System (establishing communication) which makes the robot move forward, backward, left, right.

The control signals are a sequence of the distance features that define a scenario displacement of robot corresponding to recognized facial expressions like:

Move forward:
IF   D5   (surprise) $\gg$ D5   (fear) <=> D5   (surprise) $\gg$ T$_1$ = 1,991034 => Move forward.
Move backward:
IF D5 (joy) $\ll$ D5 (disgust) <=> D5 (joy) $\ll$ T$_2$ = 1,77888 => Move backward.
Turn right:
IF D1 (fear) $\gg$ D1 (neutral) <=> D1 (fear) $\gg$ T$_3$ = 1,711864 => Turn right.

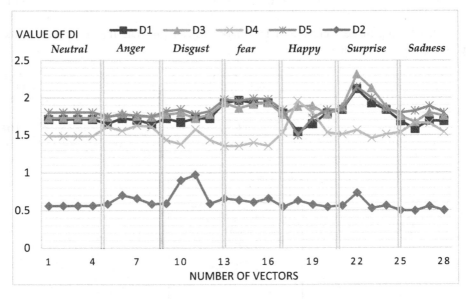

Fig. 10. Evolution of Di in the 7 expressions 'Anton'.

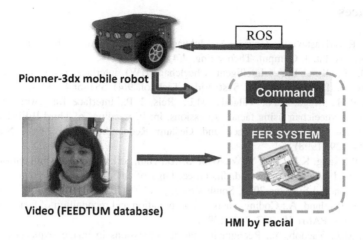

Fig. 11. Mobile robot command by the facial expression recognition system.

Turn left:

IF D2 (disgust) $\gg$ D2 (neutral) <=> D2 (disgust) $\gg T_4 = 0,561078$ => Turn left.

Where T_i defines thresholds values deduced from analyzing the features distances evolution in all facial expressions recognized of an individual to the database (see Fig. 10).

An example of scenarios that takes place in the Intelligent Systems Research Laboratory (LARESI) is seen in (Fig. 11): forward -> forward -> turn right -> forward > turn left -> forward, corresponding to: D5, D5, D1, D5, D2, D5, D5. We could check that, the robot reacts to all these commands successfully with 100% accuracy.

4 Conclusion

In this paper, the command of mobile robot using facial expressions, is the second application that we achieved in the setting of human machine interaction, intended especially to people having a reduced mobility (motor handicap) and that have only face movements to express their intents. Moreover, we conceived a dynamic facial expression model using active contour segmentation (GVF snake) by following a temporal extraction of the regions of interests. Then we got the 5 distance features while applying the MPEG-4 norm on the result of extraction automatically. These distance features are sent as signal command via ROS to a mobile robot. The results are very encouraging to achieve this application in real world from live video, like intelligent wheelchairs, human computer interaction and security systems.

References

1. Mittal, R., Srivastava, P., George, A., Mukherjee, A.: Autonomous robot control using facial expressions. Int. J. Comput. Theory Eng. **4**(4), 631 (2012)
2. Ju, D.J.S., Kim, E.Y.: Intelligent wheelchair (IW) interface using face and mouth recognition. IEEE Trans. Circ. Syst. Video Technol. **9**(4), 551–564 (1999)
3. Faria, P.M., Braga, R.A., Valgode, M.E., Reis, L.P.: Interface framework to drive an intelligent wheelchair using facial expressions. In: Proceedings of Third IEEE International Conference on Automatic Face and Gesture Recognition (FG 1998), Nara, Japan, pp. 124–129 (1998)
4. Shim, B., Kang, K.K., Lee, W.W., Won, S.J.B., Han, S.H.: An intelligent control of mobile robot based on voice command. In: Proceedings of International Conference on Control, Automation and Systems 2010, South Korea, p. 2107 (2010)
5. Irfan, E., Pentland, A.: Coding, analysis interpretation and recognition of facial expressions. IEEE Trans. PAMI **19**, 757–763 (1997)
6. Black, M.J., Yacoob, Y.: Recognizing facial expressions in image sequences using local parametrized models of image motion. Int. J. Comput. Vis. **25**(1), 23–48 (1997)
7. Tian, Y., Kanade, T., Cohn, J.: Recognition actions units for facial expression analysis. IEEE Trans. PAMI **23**(2), 97–115 (2001)
8. Lanitis, A., Taylor, C.J., Coots, T.F.: Automatic interpretation and coding of faces images using flexible models. IEEE Trans. PAMI **19**(7), 743–756 (1997)
9. Shinza, Y., Saito, Y., Kenmochi, Y., Kotani, K.: Facial expression analysis by integrating information of feature-point positions and grey levels of facial images. In: IEEE ICIP, Vancouver, Canada (2000). **2**(5), 99–110 (2016)
10. Zekhnine, C., Berrached, N.E.: Realization of facial expressions recognition system by MLP neural network. In: Fifth Conference on Electric Engineering, Polytechnic Military School, Bordj El Bahri, Algiers, Algeria (2005)
11. wallhoff@ei.tum.de. http://www.mmk.ei.tum.de/~waf
12. Zekhnine, C., Berrached, N.E.: Command of an arm manipulator robot by facial expression. In: Second International Conference on Image and Signal Processing and Their Applications. University Mohamed Khider of Biskra, Algeria (2010)
13. Kass, M., Witkin, A., Terzopoulos, D.: Snakes: active contour models. IJCV **1**(4), 321–331 (1988)
14. Cohen, I.: Modèles Déformables 2-D et 3-D: Application à la Segmentation d'Images Médicales (1992)
15. Vylder, J.D., Ochoa, D., Wilfried, P., Chaerle, L., Van Der Straeten, D.: Tracking multiple objects using intensity-GVF snakes. In: EUSIPCO (2009)
16. Xu, C., Pince, J.L.: Generalized gradient vector flow external forces for active contours. Sig. Process. **71**, 131–139 (1998)
17. Hammal, Z., Caplier, A., Rombaut, M.: Classification d'expressions faciales par la théorie de l'évidence. LIS INPG, Grenoble (2004)

SSD and Histogram for Person Re-identification System

Abdullah Salem Baquhaizel$^{(\boxtimes)}$ ⓘ, Safia Kholkhal, Belal Alshaqaqi,
and Mokhtar Keche

Laboratoire Signaux et Images, Département d'électronique,
Université des sciences et de la Technologie d'Oran Mohamed Boudiaf, USTO-MB,
El M'naouer., BP 1505, 31000 Oran, Algeria
abdullah.baquhaizel@univ-usto.dz, abdullah.baquhaizel@gmail.com

Abstract. In this paper, we give the design and implementation of a system for person re-identification in a camera network, based on the appearance. This system seeks to construct an online database that contains the history of every person that enters the field of view of the cameras. This system is qualified to associate an identifier to each detected person, which keeps this identifier in the same camera and in other cameras even if he or she disappears and then appears again. Our system comprises a moving objects detection step that is implemented using the Mixture of Gaussians method and a proposed difference method, to improve the detection results. It also comprises a tracking step that is implemented using the sum of squared differences algorithm. The re-identification stage is realized using two steps: the intersection of tracking and detection for the temporal association, the histogram for comparison. The global system was tested on a real data set collected by three cameras. The experimental results show that our approach gives very satisfactory results.

Keywords: Person re-identification · MoG · SSD · Histogram
Intersection

1 Introduction

In recent years, video surveillance has grown more and more. This resulted in an increase of cameras installed in different places (private or public), making their exploitation and monitoring very difficult for human being. That is why much research has been done to create intelligent vision systems that can help the human being, in interpreting scenes and reacting with alarms in case of any anomaly. Currently there are several types of video surveillance systems (access control in sensitive locations, people recognition, control of traffic congestion, ...etc.).

ⓒ IFIP International Federation for Information Processing 2018
Published by Springer International Publishing AG 2018. All Rights Reserved
A. Amine et al. (Eds.): CIIA 2018, IFIP AICT 522, pp. 585–596, 2018.
https://doi.org/10.1007/978-3-319-89743-1_50

In this paper we are interested in the problem of person re-identification in a camera network. Re-identification in computer vision systems aims to follow a person, associate an identifier to him, and store it in a database. If the person leaves the scene then reappears in the field of view of any camera, it will be assigned the same identifier. In a crowded and uncontrolled environment observed by cameras from unknown distances, person re-identification relying upon conventional biometrics, such as face recognition, is neither feasible nor reliable, due to insufficiently constrained conditions and insufficient image details for extracting robust biometrics [17]. Instead, visual features based on the appearance of people, determined by their clothing and objects carried or associated with them, can be exploited more reliably for re-identification.

The remainder of this paper is organized as follows: Sect. 2 presents some related works from the literature. Section 3 descibes in details each block of the proposed system. The experimental results and their discussion are presented in Sect. 4. Finally, some conclusions are drawn in Sect. 5.

2   Related Works

In the literature, the approaches of re-identification can be grouped in several classes, according to several criteria [12]:

1. The number of images per person:
   This class comprises two families. The first family is the family of mono-sample methods, where the signature of a person is extracted from a single image as in [1,3,6,15,16,24] . The second family is the family of multi-sample methods, where multiple images are used to calculate the signature of a person as in [4,5,7,11,14,19,21].
2. The type of representation:
   The first family in this class is the family of global approaches, where the whole information in the image is exploited for calculating the person's signature, as in [1,2,13]. The second family is that of local approaches, which represent the image by several feature vectors, each vector describes a region or a locally detected point, such as in [5,8,9].
3. The existence of a set of images mapped a priori:
   This class includes supervised approaches like in [2,3,14] and unsupervised approaches as in [10,23].

A very nice survey of people re-identification approaches is presented in [22]. They are therein grouped as a multidimensional taxonomy according to camera setting, sample set cardinality, signature, adoption of a body model, machine learning techniques and application scenario.

3   Description of the Proposed System

In this section, we describe the different stages of the proposed system, for person re-identification in non-overlapping camera network. These stages are: person detection, their localization and verification, their tracking and their re-identification. The overall flowchart of the proposed system is shown in Fig. 1.

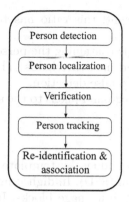

Fig. 1. The flowchart of the proposed system

3.1 Person Detection

This initial stage is accomplished by combining the Mixture of Gaussians (MoG) method [20] and the difference method. The MoG is one of the most used and successful methods in surveillance systems, because it is adaptive, and can handle multimodal backgrounds.

In the difference method, we first take the difference between two successive images in grayscale $I_{g(t)}$ and $I_{g(t-1)}$, as in Eq. (1), and then we compare the resulting difference image I_{diff} to a threshold to detect pixels in movement.

$$I_{diff} = I_{g(t)} - I_{g(t-1)} \tag{1}$$

The hybrid of detections resulting from the MoG and difference methods is performed using the logical OR operation.

After detecting moving objects we fill the holes [18]. The holes of a binary image correspond to the set of its regional minima, which are not connected to the image border.

3.2 Person Localization

The localization of the detected person is done using the labeling technique. This technique consists in separating the areas in the mask obtained from the detection step. We associate with each area an integer value (label) by using an 8-connected neighborhood, then we calculate some proprieties for each area, e.g. x and y coordinates, height, width and sum of foreground pixels.

3.3 Verification

To eliminate false detections, we propose a verification phase. To be validated each detected person has to verify the following three conditions:

- The ratio of width to height: this ratio has to lie between min and max thresholds.
- The surface of the rectangle containing the person (surface = height × width) has to lie between min and max thresholds. This is to eliminate very small and very big objects due to false detection.
- The ratio of the sum of foreground pixels to the surface also has to be limited.

3.4   Person Tracking

The person tracking process is done by template matching using the Sum of Squared Differences Algorithm (SSD). In digital image processing, the SSD is a measure of the similarity between image blocks. It is calculated by taking the square of the difference between each pixel in the original block X (a portion from the current frame) and the corresponding pixel in the Y block being used for comparison (Model from previous detection).

These differences are summed to create a simple metric of block similarity as in   Eq. (2), zero means that the two blocks are identical. We sweep all the positions in the frame, then the block with the smallest metric is the tracked block.

The SSD value for two blocks X and Y calculated by:

$$SSD = \sum_{i=1}^{M} \sum_{j=1}^{N} (X(i,j) - Y(i,j))^2 \tag{2}$$

For a given Y model, the most similar block X is the one that minimizes the SSD.

3.5   Re-identification and Association

Following the stages of detection, localization, verification, and tracking, we have the stage of re-identification and online construction of database DB containing the history of each person that appeared in the view field of the cameras. Figure 2 presents a detailed flowchart of this stage.

This stage deals with the moving objects obtained from the detection and tracking stages, which are called 'found person'.

First, we calculate the intersection between the found persons resulting from the detection and tracking, the intersection $(A \cap B)$ of two rectangles A and B is the rectangle that contains all elements of A that also belong to B.

Then we test if the found persons resulted from detection only, tracking only or from both. If found person comes from intersection or tracking only, we update the database with the identifier of tracked person.

On the other hand, if that found person comes from detection only, then we calculate its histogram. An image histogram is a type of histogram that acts as a graphical representation of the tonal distribution in a digital image. It plots the number of pixels for each tonal value. The histogram of the found person is

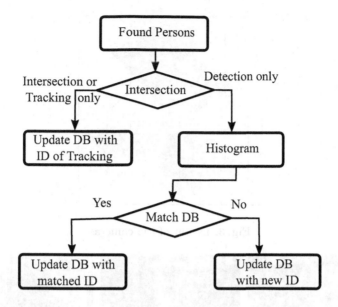

Fig. 2. Detailed flowchart of re-identification and online construction of database

compared to the histograms of identified persons stored in the database. If there is a match, then we update the database by associating that person with the matched identifier, otherwise, we consider this person as a new one and assign to it a new identifier that is added to the database.

4   Experimental Results

In this section, we present the material and the database used, the experimental results, and their discussion.

4.1   System development environment

The material that was used for the development of our application is:

1. A laptop with:
    - Processor: Intel core i7 4702MQ CPU @ 2.20 GHz 2.20 GHz.
    - RAM memory: 8.00 Go.
    - Operating system: Windows 8.1, 64-bits
    - Hard Drive: 1 TB.
2. Digital video recorder DVR.
3. Camera with characteristics:
    - 1/3 Sony HR CCD
    - 420 TV lines
    - 0.2 Lux
    - Adjustable Focal between (3 mm and 8 mm).

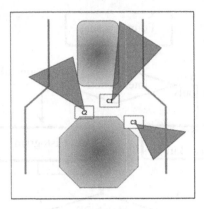

Fig. 3. Layout of the cameras

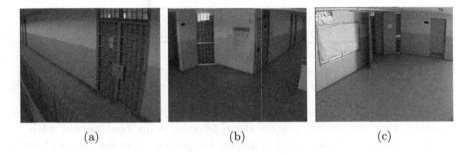

(a)   (b)   (c)

Fig. 4. Fields of view of the 3 cameras: (a) Camera 1, (b) Camera 2, (c) Camera 3

To test our system we build our own database, composed of sequences of images recorded on the third floor of the Department of Electronics at USTO university. Three cameras, set to a height of $(2.30\,\text{m})$ and with an angle of (-30), were used to take these images. Each sequence contains from one to three people who walk in the fields of view of the three cameras. The cameras were placed as shown in the layout presented in Fig. 3.

To fulfil the condition of a non-overlapping camera network, the database was realized so that a person lies in the field of view of only one camera, at a given instant. Figure 4 shows the fields of view of the three cameras.

4.2 Experimental Results, and Discussion

In this section, we will present and discuss the results of each step of the proposed system. The Mixture of Gaussian gives us raw results of detection from each camera, after having defined suitable settings according to some criteria, like: indoor or outdoor environment, people movement speed and lighting changes. Figure 5(b) presents an example of these results. To improve these raw results, we combine them with the results of the difference method (Fig. 5(c)), which allows

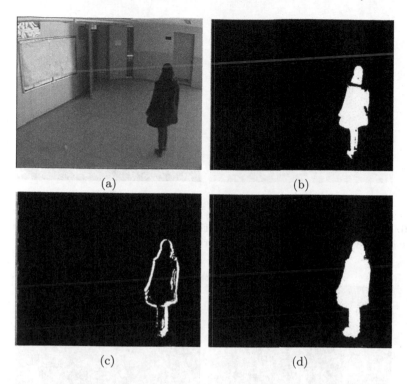

Fig. 5. Example of person detection. (a) Original image, (b) Results of detection by MoG, (c) Results of detection by difference, (d) Results of the holes filling of the OR between b and c

for the detection of the edges of moving objects, then we proceed to a holes filling of the resulting image to obtain better results as illustrated in Fig. 5(d).

In Fig. 6, we present the localization and verification results. After the localization by the labeling technique, we apply the verification procedure to each person. In Fig. 6(a), only the persons that verify the validation conditions are kept (the person in green rectangle), the others in red are ignored. Figure 6(b) presents the detection results.

The tracking step is run in parallel with the detection step and it is realized by the SSD. To accelerate its execution we decided to apply it only to a limited region of interest, instead of searching in the whole frame. This region is determined by the coordinates of the model to track. The obtained results of tracking are satisfactory. Figure 7 shows the tracking results, Fig. 7(a) is a detection in frame 159 and Fig. 7(b) is its tracking in frame 222.

The re-identification stage is realized using two techniques, the intersection of detection and tracking for the temporal association, and the histogram for comparison. In Fig. 8, we present the multiplication of the detected person with its mask to extract the silhouette only. Then, we calculate histograms of Red, Green, Blue channels and grayscales as shown in Fig. 9. We use the histogram of

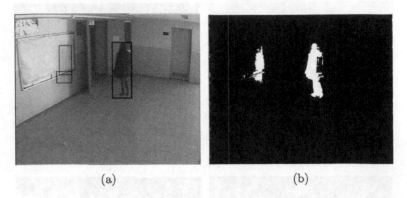

Fig. 6. Example of localization and verification, (a) Localization and verification results on the original image, (b) Results of detection

Fig. 7. Tracking results, (a) Detection in frame 159, (b) The tracking of detection of frame 159 in frame 222

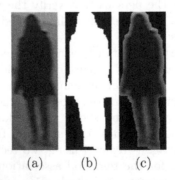

Fig. 8. The multiplication of detected person with its mask, (a) Detected person, (b) Mask of detected person and (c) Results of multiplication

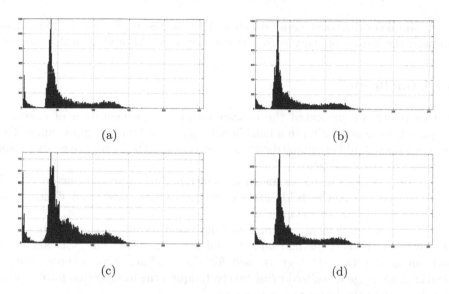

Fig. 9. Different histograms of the silhouette, (a), (b), (c) and (d) are the histograms of Red, Green, Blue channels and grayscales respectively (Color figure online)

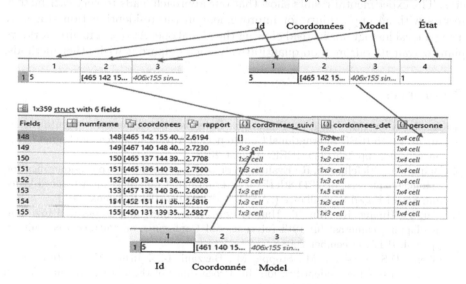

Fig. 10. A sample of the constructed database

the silhouette to avoid the effect of the background. These different histograms are used for comparison with the models stored in the database. If there is a match, we associate the matched identifier to the actual person, otherwise, we consider that the actual person is new and add it to the database with a new identifier.

A sample of the constructed database is presented in Fig. 10. This database contains the history of every person that enters the field of view of the cameras.

5 Conclusion

In this paper, we presented the conception and implementation of a system for person re-identification in a camera network, based on the appearance. This system aims to build an online database that contains the history of every person captured by the cameras.

This system is able to assign an identifier to each detected person, that it keeps everywhere in the fields of view of the cameras and even if he or she disappears and then appears again.

Our system implements an improved detection technique that combines the Mixture of Gaussians method and the difference method. The SSD algorithm with an acceleration strategy is used for the tracking step, whereas the re-identification stage is realized using two techniques: the intersection for temporal association and the histogram comparison.

The global system was tested on a real data set collected by three cameras. The experimental results show that our approach leads to very satisfactory results with an opportunity for improvement in the re-identification stage, by using a local histograms instead of using the global one. Also as a future work, we plan to evaluate our method quantitatively and compare it with other methods.

References

1. An, L., Kafai, M., Yang, S., Bhanu, B.: Reference-based person re-identification. In: 2013 10th IEEE International Conference on Advanced Video and Signal Based Surveillance, pp. 244–249. IEEE, August 2013
2. Bauml, M., Stiefelhagen, R.: Evaluation of local features for person re-identification in image sequences. In: 2011 8th IEEE International Conference on Advanced Video and Signal Based Surveillance (AVSS), pp. 291–296. IEEE August 2011
3. Cai, Y., Huang, K., Tan, T.: Human appearance matching across multiple non-overlapping cameras. In: 19th International Conference on Pattern Recognition, pp. 1–4. IEEE December 2008
4. Cheng, D.S., Cristani, M., Stoppa, M., Bazzani, L., Murino, V.: Custom pictorial structures for re-identification. In: Proceedings of the British Machine Vision Conference, pp. 68.1–68.11. British Machine Vision Association (2011)
5. de Oliveira, I.O., De Souza Pio, J.L.: People reidentification in a camera network. In: 2009 Eighth IEEE International Conference on Dependable, Autonomic and Secure Computing, pp. 461–466. IEEE, December 2009
6. Dikmen, M., Akbas, E., Huang, T.S., Ahuja, N.: Pedestrian recognition with a learned metric. In: Kimmel, R., Klette, R., Sugimoto, A. (eds.) ACCV 2010. LNCS, vol. 6495, pp. 501–512. Springer, Heidelberg (2011). https://doi.org/10.1007/978-3-642-19282-1_40

7. Farenzena, M., Bazzani, L., Perina, A., Murino, V., Cristani, M.: Person re-identification by symmetry-driven accumulation of local features. In: 2010 IEEE Computer Society Conference on Computer Vision and Pattern Recognition, pp. 2360–2367. IEEE, June 2010
8. Gheissari, N., Sebastian, T.B., Hartley, R.: Person reidentification using spatiotemporal appearance. In: 2006 IEEE Computer Society Conference on Computer Vision and Pattern Recognition - Volume 2 (CVPR 2006), vol. 2, pp. 1528–1535. IEEE (2006)
9. Hamdoun, O., Moutarde, F., Stanciulescu, B., Steux, B.: Person re-identification in multi-camera system by signature based on interest point descriptors collected on short video sequences. In: 2008 Second ACM/IEEE International Conference on Distributed Smart Cameras, pp. 1–6. IEEE, September 2008
10. Hirzer, M., Roth, P.M., Bischof, H.: Person re-identification by efficient impostor-based metric learning. In: 2012 IEEE Ninth International Conference on Advanced Video and Signal-Based Surveillance, pp. 203–208. IEEE, September 2012
11. Huang, C.-H., Wu, Y.-T., Shih, M.-Y.: Unsupervised pedestrian re-identification for loitering detection. In: Wada, T., Huang, F., Lin, S. (eds.) PSIVT 2009. LNCS, vol. 5414, pp. 771–783. Springer, Heidelberg (2009). https://doi.org/10.1007/978-3-540-92957-4_67
12. Khedher, M.I.: Ré-identification de personnes à partir des séquences vidéo. Dissertations, Institut National des Télécommunications, Paris (2014)
13. Ijiri, Y., Lao, S.: Human re-identification through distance metric learning based on jensen-shannon kernel. In: VISAPP: - International Conference on Computer Vision Theory and Applications, pp. 603–612 (2012)
14. Jungling, K., Arens, M.: View-invariant person re-identification with an implicit shape model. In: 2011 8th IEEE International Conference on Advanced Video and Signal Based Surveillance (AVSS), pp. 197–202. IEEE, August 2011
15. Park, U., Jain, A.K., Kitahara, I., Kogure, K., Hagita, N.: ViSE: visual search engine using multiple networked cameras. In: 18th International Conference on Pattern Recognition (ICPR 2006), pp. 1204–1207. IEEE (2006)
16. Schwartz, W.R., Davis, L.S.: Learning discriminative appearance-based models using partial least squares. In: 2009 XXII Brazilian Symposium on Computer Graphics and Image Processing, pp. 322–329. IEEE, October 2009
17. Gong, S., Cristani, M., Loy, C.C., Hospedales, T.M.: The re-identification challenge. In: Gong, S., Cristani, M., Yan, S., Loy, C.C. (eds.) Person Re-Identification. ACVPR, pp. 1–20. Springer, London (2014). https://doi.org/10.1007/978-1-4471-6296-4_1
18. Soille, P.: Morphological Image Analysis. Springer, Heidelberg (2004). https://doi.org/10.1007/978-3-662-05088-0
19. Souded, M.: People Detection, Tracking and Re-identification Through a Video Camera Network. Ph.D. thesis, Signal and Image processing, Institut National de Recherche en Informatique et en Automatique, Universite de Nice - Sophia Antipolis Ecole, Nice (2013)
20. Stauffer, C., Grimson, W.E.L.: Adaptive background mixture models for real-time tracking. In: IEEE Computer Society Conference on Computer Vision and Pattern Recognition, vol. 2, pp. 246–252. IEEE Computer Society (1999)
21. Cong, D.-N.T., Achard, C., Khoudour, L.: People re-identification by classification of silhouettes based on sparse representation. In: 2010 2nd International Conference on Image Processing Theory, Tools and Applications, pp. 60–65. IEEE, July 2010
22. Vezzani, R., Baltieri, D., Cucchiara, R.: People reidentification in surveillance and forensics: a Survey. ACM Comput. Surv. 46(2), 1–37 (2013)

23. Wang, T., Gong, S., Zhu, X., Wang, S.: Person re-identification by video ranking. In: Fleet, D., Pajdla, T., Schiele, B., Tuytelaars, T. (eds.) ECCV 2014. LNCS, vol. 8692, pp. 688–703. Springer, Cham (2014). https://doi.org/10.1007/978-3-319-10593-2_45
24. Wang, X., Doretto, G., Sebastian, T., Rittscher, J., Tu, P.: Shape and appearance context modeling. In: 2007 IEEE 11th International Conference on Computer Vision, pp. 1–8. IEEE (2007)

Automatic Recognition of Plant Leaves Using Parallel Combination of Classifiers

Lamis Hamrouni[1,2(✉)], Ramla Bensaci[1,2],
Mohammed Lamine Kherfi[1,3], Belal Khaldi[1,2], and Oussama Aiadi[1,2]

[1] Department of Computer Science and Information Technologies,
University of Kasdi Merbah Ouargla, 30 000 Ouargla, Algeria
lamis0215@gmail.com, ramla.bensaci@gmail.com,
{khaldi.belal,aiadi.oussama}@univ-ouargla.dz
[2] LAGE Laboratory, University of Kasdi Merbah Ouargla,
30 000 Ouargla, Algeria
[3] LAMIA Laboratory, Université du Québec à Trois-Rivières,
3351, boul. des Forges, C.P. 500, Trois-Rivières, Canada

Abstract. Because they are exploited in many fields such as medicine, agriculture, chemistry and others, plants are of fundamental importance to life on earth. Before it can be used, a plant need to firstly be identified and categorized. However, a manual identification task requires time, and it is not an easy task to do. This is because some plants look visually similar to the human eye, whereas some others may be unknown to it. Therefore, there has been an increasing interest in developing a system that automatically fulfils such tasks fast and accurate. In this paper, we propose an automatic plant classification system based on a parallel combination technique of multiple classifiers. We have considered using three widely known classifiers namely Naïve Bayes (NB), K-Nearest Neighbour (KNN) and Support Vector Machine (SVM). Our system has been evaluated using the well-known Flavia dataset. It has shown a better performance than those obtained using only one classifier.

Keywords: Morphological features · Parallel classifiers · Leaf classification
Plant leaves · Image recognition

1 Introduction

Plants play an important role in our life, without them there will be no existence of the earth's ecology. They are widely exploited in our life such as in food, breath, health, and even in industry fields such as medicine, economic agriculture and so on. There are millions of plants species, some of them are subject to the danger of extinction [1]. Therefore, there is an urgent need for identifying plant species.

Traditionally, botanists classify plants using molecular biology and cellular features of leaves. Nevertheless, this task is very tedious, requires time and need the presence of expertise which is not available in all times. Additionally, an expert on one species or family may be unfamiliar with another. Subsequently, fast and accurate automatic plant identification system is highly needed.

© IFIP International Federation for Information Processing 2018
Published by Springer International Publishing AG 2018. All Rights Reserved
A. Amine et al. (Eds.): CIIA 2018, IFIPAICT 522, pp. 597–606, 2018.
https://doi.org/10.1007/978-3-319-89743-1_51

Plants are usually classified using their leaves, stems, fruits or flowers. Leaves seem to be the most suitable parts that can be used to identify a plant. This is due to their availability in all seasons. In addition, leaves flatness makes it easy to be represented by the computer in some 2D.

In recent years, plant identification techniques have become a hot topic of research [2]. Authors in [3] have exploited the visual features of leaves in combination with Random Forest (RF) and Linear Discriminant Analysis (LDA) to classify and identify 30 plant species. In [4] a similar technique, but using Artificial Neural Networks (ANN) this time, has been opted for to identify 12 plant species. Silva et al. [5] have focused on developing a system that can automatically identify medicinal plants such as herbs, shrubs and trees. As a classifier, they use ANN then Support Vector Machine (SVM).

Reader should notice that the former works use only one classifier in their systems. Thus, some other works have tried to improve them by combining and using more than one classifiers in the system. In [6] as instance, authors suggest using a serial combination of two SVMs. Their main idea was to devote one classifier for color features, and the other one for both shape and texture features. Their evaluation has been carried out on a dataset that contains six diseases classes. The system reports an 87.7% accuracy. In [7], authors have used a parallel combination of two classifiers namely, ANN and SVM. The first one was devoted to texture, color and shape features, whereas, then the second one uses shape and texture features. The evaluation has been carried out on a dataset that is composed of six diseases classes and they reported 91.46% accuracy.

In this paper, we introduce a system that parallel combines three classifier namely SVM, NB and KNN. As features, we consider extracting shape features (i.e., morphological features) from the leaves. The evaluation of our system has been accomplished using the well-known Flavia dataset. More details about the proposed system will be given in the next section.

The rest of the paper is organized as follow: Sect. 2 present the architecture of our system and then discuss the used morphological features. In Sect. 3, we conduct experimentations on the proposed system and report results. Finally, we draw some conclusions.

2   Proposed System

In order to achieve a better performance, our system consists in a parallel combination of three classifiers namely SVM, NB and KNN. These classifiers are trained using a set of morphological features that we extract from the leaves. In Fig. 1, we illustrate a general scheme that resumes the different main stages of our system.

2.1   Preprocessing

The preprocessing stage is, generally, responsible for applying a set of treatment (e.g., noise reduction, rotation, transformation, etc.) on the image before employing it for features extraction. In our work, we firstly converted the original color images to

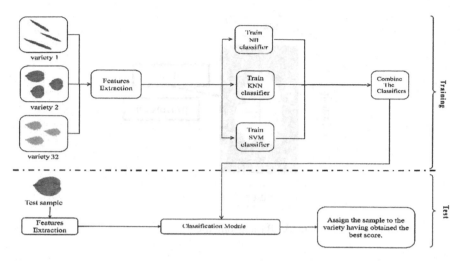

Fig. 1. Architecture of the proposed system.

gray-level then to binary images. Thereafter, a smoothing filter is applied to these binary images to reduce the noise. The steps involved in pre-processing are illustrated in Fig. 2.

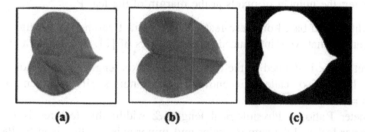

Fig. 2. Preprocessing stage (a) Image in RGB color space, (b) Gray-level image, (c) Binary image. (Color figure online)

2.2 Features Extraction

This stage aims to transform the objects into a vector of numeric values (i.e., feature vector). There are many types of features that can be extracted from an image, such as shape [8], texture [9], and color [10] features.

In this stage and after having the original image transformed into binary, we extract a set of shape features that describe the morphology of a leaf. Morphological features are obtained by extracting the basic geometrical properties [11] of the leaf such as: diameter, area, perimeter, major and minor. Figure 3 shows an example of some geometrical features extracted from a leaf image.

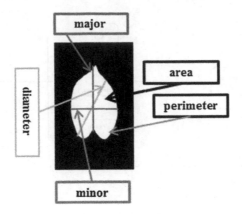

Fig. 3. Basic geometrical features

In our method, we extract the following five morphological features:

1. Diameter: is the longest distance between two points of the leaf contour D.
2. Area: is the number of pixels that constitute the area of the leaf.
3. Major axis length: is the distance between two terminal points orthogonal to minor axis length LP.
4. Minor axis length: is the longest distance orthogonal to major axis length WD.
5. Perimeter: the number of pixels at the margin of the leaf P.

In addition and based on these features, we extract another set of digital morphological features that were introduced by the authors of [11, 12], which are:

6. Aspect Ratio: is defined as the ratio of major axis length LP to minor axis length WP. It is also called Eccentricity or Slimness. It is given by Aspect Ratio = LP/WP.
7. Perimeter Ratio of Physiological length & width: this features is the ratio of perimeter leaf and the sum of major and minor axis length, given by PRPW = P/(LP + WP).
8. Perimeter Ratio of Diameter: it is the ratio of perimeter to the diameter, given by PRD = P/D.
9. Rectangularity: The similarity between the leaf and a rectangle, given by R = (Lp * Wp)/A.
10. Narrow Factor: the ratio of the diameter D and length Lp (i.e., NF = D/Lp).
11. Circularity: The ratio involving the area A of the leaf and the square of its perimeter P, given by $C = 4\pi A/P2$.
12. Solidity: The ratio between *A* the area of the leaf and *Ach* the area of a convex hull, given by S = A/Ach.

Figure 4 shows some of morphological features.

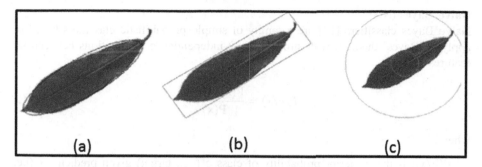

Fig. 4. Morphological features. (a) Form ellipse, (b) Rectangularity, (c) Circularity

2.3   Classification

Classification or categorization is, generally, the process in which images are recognized, differentiated, and understood. The classifier needs to, firstly, be subjected to a set of labeled data (i.e., training set). Then, test samples will be provided to the classifier in order to recognize them.

In our work, a portion of the feature vectors, which are extracted in the previous stage, will be used to train our system that consists in three types of classifiers: KNN [11, 12], NB [13] and SVM [14, 15]. These three classifiers are combined in parallel. The parallel approach allows the different classifiers to operate independently of each other. The results of each classifier are then merged in order to obtain a higher recognition rate.

K-Nearest Neighbor (KNN)
KNN [11] is a simple yet very effective classification method. For a given test sample s, KNN mainly consists in determining the k closest training example to this s. It then states the class C that has the max membership degree to s as a class of s. the similarity could be extracted using the next formula (1).

$$d(x_r, x_s) = \left[\sum_{i=1}^{P} c_i(x_{ri}, x_s i) \right]^{\frac{1}{2}} \tag{1}$$

Support Vector Machine (SVM)
SVM is a supervised classifier that has a great effectiveness especially with high dimensional data [14]. Formally, SVM constructs a hyperplane (alt. hyperplanes) that has the highest distance to the nearest training-data point of any class [15].

Naive Bayes (NB)
Naive Bayes classifiers [13] are a family of simple probabilistic classifiers based on applying Bayes' theorem with strong (naive) independence assumptions between the features [13].

$$P(c/x) = \frac{P(x/c)P(c)}{P(x)} \tag{2}$$

where:

- $P(c/x)$: is the posterior probability of class c (i.e., target) given predictor x (i.e., attributes).
- $P(c)$: is the prior probability of class.
- $P(x/c)$: is the likelihood, which is the probability of predictor given a class.
- $P(x)$: is the prior probability of predictor.

2.4   Combining Classifiers

The main aim of combining classifiers is to improve the accuracy. Several works have been proposed in this context [6]. Classifier combination schemes could be roughly categorized into three main approaches, namely: sequential, parallel and hybrid combination.

2.4.1   Sequential Combination
The sequential combination consists in placing one classifier after the other. It simpler words, the outcome of one classifier will be the input of another. Such a cascade structure helps to improve the decision taken from the previous step by including reliable samplers or excluding unreliable ones [6].

2.4.2   Parallel Combination
In parallel combination, the different classifiers operate independently of each other. Then, the obtained results are fused together, by some method, to produce a final decision [7].

2.4.3   Hybrid Combination
The hybrid combination scheme takes the advantages of the two previous schemes (i.e., sequential and parallel combination) in order to reach a more reliable decision. It illustrates the two aspects of the combination which are: in on the one hand reducing all possible classes and in the other hand finding the consensus between all classifiers, in order to reach a final decision [19].

In our system, we combine the classifiers in parallel way and we opted for voting-based methods [16]. In such module, each classifier provides a certain number of votes (i.e., potential classes for a given image). The final decision is then made based on these votes using the following formula:

$$E(x) = \begin{cases} c_i si \sum_i e(i) = max_{c_i \in \{1,...,M\}} \sum_j e(j) \geq \alpha K \\ else \ reject \end{cases} \tag{3}$$

where K is the number of the combined classifiers, α is a threshold that represents the needed number of votes for the same class to be relevant.

Voting-based method could be categorized into two main categories, the simple majority vote and the weighted vote.

In simple majority vote [17], each classifier votes for one class to be relevant to an input image. The final decision will, then, be made regarding the number of votes for each class. The relevant class is the one with the heights number of votes. Although this method is simple and effective, it suffers from the problem of rejection if all classes have the same votes number [18].

In the other hand, the weighted vote method associates each classifier with some coefficient (i.e., weight) that indicates the importance of the corresponding classifier in the combination. Selecting weights for different classifiers is a critical process and highly affects the quality of the results.

To take advantage of both the former voting methods, we suggest combining them in one module. In our used module, the system tries, firstly, majority vote method. If a conflict accrues using majority vote, then, the system opts for weighted vote method. The following algorithm resumes these two steps.

```
If (DSVM==DKNN==DNb) Then
        FD ← DSVM+DKNN+DNB
Else
        FD ← α DNB + β DKNN+Ω DSVM
Endif
```

3   Experimental Results

The evaluation of the used approach has been carried out on Flavia dataset [4]. Flavia contains 1907 images that are categorized into 32 classes. Figure 5 shows representative simples from this dataset.

In our experiments, the dataset has been divided into two subsets. The First one consists of 1284 images (70%) used for training, whereas the rest 623 images (30%) have been used to test the system. We opted for Accuracy to be used as an evaluation metric of our system. It is given by the following formula.

$$Accuracy \ (\%) = \frac{Nc}{Nt} * 100 \tag{4}$$

where Nc is the number of correctly classified images and Nt is the total number test.

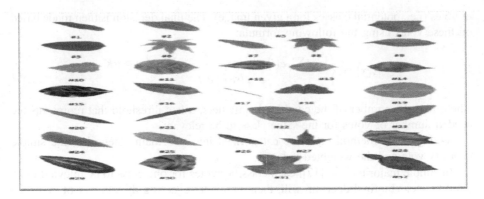

Fig. 5. Representative samples from Flavia dataset.

Table 1. Accuracy results using each classifier separately.

Classifier	Average accuracy%
NB	72,231
KNN	65,810
SVM	59,390

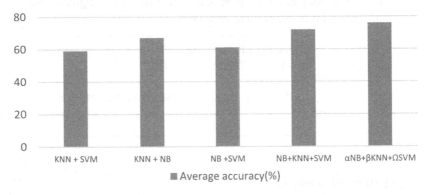

Fig. 6. Comparing the accuracy yielded by different classifiers.

In order to prove the performance of our method, we have firstly classify images using independently each classifier. Table 1 shows the obtained results.

From the Table 1 we can see that the best result has been yielded by naive-bayes (72%). Because It makes use of all the features contained in the data, and analyses them individually as though they are equally important and independent of each other. Additionally, we can see that SVM has not performs well compared to the others (59%). This low performance can be attributed to the low-dimensionality of the used data [15].

After evaluating each classifier separately, we will, next, evaluate the combinations of different classifiers. The obtained results are shown in Fig. 6.

As Fig. 6 shows, combining classifiers does not improve results but rather decrease them in most cases. Because of suffering from low-dimensionality of data, SVM degrade the results of the different combinations by declaring conflicts in classification. Such issue could be resolved by assigning weights to the different classifiers. Thus and according to Table 1, the best classifiers has been associated with higher weights, combined then evaluated (i.e., $\alpha = 2, \beta = 2, \Omega = 1$). As it is illustrated in Fig. 6, combining these three classifiers with their corresponding weighs has yielded better results (76%).

4 Conclusion

In this paper, we have proposed an automatic plant classification system. Our system is based on a parallel combination of three classifiers namely KNN, NB and SVM with two combining modules namely sample majority vote and weighted vote. These classifiers are firstly trained with a set of morphological features that describes the shape of the leaf. Our system has been evaluated using the well-known Flavia dataset. After evaluating each classifier separately, we have found out that NB is the best classifier among all. We, then evaluate the different possible combinations of classifiers. Results indicates that SVM negatively effects the results because of the low-dimensionality of data especially by using simple majority vote module. To solve this issue, we have opted for weighted vote with the following weights 2, 2 and 1 that corresponds respectively to NB, KNN and SVM. By associating the lowest weight to SVM, our system reduces its negative effect to the voting process. The former combination has yielded best results. Future works should consider using more powerful classifiers such as neural networks or discriminate analysis.

References

1. Murphy, G.E., Romanuk, T.N.: A meta-analysis of declines in local species richness from human disturbances. J. Ecol. Evol. 4(1), 91–103 (2014)
2. Tharwat, A., Gaber, T., Hassanien, A.E.: One-dimensional vs. two-dimensional based features plant identification approach. J. Appl. Logic 24, 15–31 (2017)
3. Anami, B.S., Suvarna, S.N., Govardhan, A.: A combined color, texture and edge features based approach for identification and classification of Indian medicinal. Int. J. Comput. Appl. 6(12), 45–51 (2010)
4. Wu, S.G., Bao, F.S., Xu, E.Y., Wang, Y.X., Chang, Y.F.: A leaf recognition algorithm for plant classification using probabilistic neural network. In: IEEE International Symposium on Signal Processing and Information Technology, Giza, pp. 11–16 (2007)
5. Silva, P.F.B., Marçal, A.R.S., da Silva, R.M.A.: Evaluation of features for leaf discrimination. In: Kamel, M., Campilho, A. (eds.) ICIAR 2013. LNCS, vol. 7950, pp. 197–204. Springer, Heidelberg (2013). https://doi.org/10.1007/978-3-642-39094-4_23
6. Rahman, A.F.R., Fairhurst, M.C., Es-saady, Y., El Massi, I., El Yassa, M., Mammass, D., Benazoun, A.: Automatic recognition of plant leaves diseases based on serial combination of two SVM classifiers. In: International Conference on Electrical and Information Technologies (ICEIT), Morocco, pp. 1–13 (2016)

7. El Massi, I., Es-saady, Y., El Yassa, M., Mammass, D., Benazoun, A.: Automatic recognition of the damages and symptoms on plant leaves using parallel combination of two classifiers. In: 13th International Conference Computer Graphics, Imaging and Visualization, Morocco, pp. 131–136. IEEE (2016)

8. Asrani, K., Jain, R.: Contour based retrieval for plant species. Int. J. Image Graph. Sig. Process. **5**(9), 29–35 (2013)

9. Casanova, D., de Mesquita Sa Junior, J.J., Bruno, O.M.: Plant leaf identification using Gabor wavelets. Int. J. Imaging Syst. Technol. **19**(3), 236–243 (2009)

10. Kadir, A., Nugroho, L.E., Susanto, A., Santosa, P.I.: Leaf classification using shape, color, and texture features. Comput. Vis. Pattern Recogn. (2013)

11. Harish, B.S., Hedge, A., Venkatesh, O., Spoorthy, D.G.: Classification of plant leaves using Morphological features and Zernike moments. In: International Conference Advances in Computing, Communications and Informatics (ICACCI), India, pp. 1827–1831 (2013)

12. Buttrey, S.E., Karo, C.: Using k-nearest-neighbor classification in the leaves of a tree. J. Comput. Stat. Data Anal. **40**(1), 27–37 (2002)

13. Russell, S., Norvig, P.: Artificial Intelligence: A Modern Approach, 3rd edn. Prentice Hall, Englewood Cliffs (1995)

14. Vapnik, V.: The Nature of Statistical Learning Theory, 2nd edn. Springer, New York (1995). https://doi.org/10.1007/978-1-4757-2440-0

15. Chapelle, O., Haffner, P., Vapnik, V.N.: Support vector machines for histogram based image classification. Trans. Neural Netw. **5**(10), 1055–1064 (1999)

16. Ho, T.K., Hull, J.J., Srihari, S.N.: Decision combination in multiple classifier systems. IEEE Trans. Pattern Anal. Mach. Intell. **16**(1), 66–75 (1994)

17. Bahler, D., Navarro, L.: Methods for combining heterogeneous sets of classifiers. In: 17th National Conference on Artificial Intelligence (AAAI), Austin, USA (2000)

18. Rahman, A.F.R., Fairhurst, M.C.: Multiple classifier decision combination strategies for character recognition: a review. Int. J. Doc. Anal. Recogn. (IJDAR) **5**, 166–194 (2003)

19. El Massi, I., Es-saady, Y., El Yassa, M., Mammass, D., Benazoun, A.: A hybrid combination of multiple SVM classifiers for automatic recognition of the damages and symptoms on plant leaves. In: Mansouri, A., Nouboud, F., Chalifour, A., Mammass, D., Meunier, J., ElMoataz, A. (eds.) ICISP 2016. LNCS, vol. 9680, pp. 40–50. Springer, Cham (2016). https://doi.org/10.1007/978-3-319-33618-3_5

Semantic Web Services

Enhancing Content Based Filtering Using Web of Data

Hanane Zitouni[1]([⊠]), Souham Meshoul[1], and Kamel Taouche[2]

[1] Department of Computer Science, Abdelhamid Mehri University,
Constantine, Algeria
{hanane.zitouni,souham.meshoul}@univ-constantine2.dz
[2] Claude Bernard Lyon 1 University, Lyon, France
kamel.taouche@liris.cnrs.fr

Abstract. Recommender systems are very useful to help access to relevant information on the web and to customize search. Content based filtering (CBF) is an alternative among others used to design recommender systems by exploiting items' contents. Basically, they recommend items based on a comparison between the content of items and user profile. Usually, the content of an item is represented as a set of descriptors or terms; typically the words that occur in text documents. The user profile is represented by the same terms and built up by analysing the content of items he used before. However current CBF recommender systems are mostly devoted to deal with textual resources and cannot be used in their current form to handle the variety of data published on the web especially unstructured data. Another challenge for the existing CBF methods is the issue of new user for whom the system cannot draw any inference due to the lack of information about the user. This paper describes an approach to CBF that aims to deal with these problems on which CBF systems perform poorly. The basic feature of the proposed approach is to incorporate linked data cloud into the information filtering process using a semantic space vector model, and FOAF vocabulary, which is used to define a new distance measure between users, based on their FOAF profiles. We report on some experiments and very promising results of the proposed approach.

Keywords: Content based filtering · RDF · Linked data · FOAF vocabulary
Web of data · SPARQL

1 Introduction

Nowadays, the explosive growth of data published on the web in all different fields such as e-learning, social networks, e-commerce among many others is not slowing down soon according to recent studies [1, 2]. The expanding data universe makes it difficult to get benefit from the web content. Furthermore, predicting user responses to options for recommendation purpose becomes an enormous challenge for an extensive class of web applications. Recommending a resource is usually achieved through information filtering. There exist two major approaches to information filtering [1]: Collaborative filtering and Content-based filtering. A Collaborative Filtering (CF) system chooses items based on the correlation between people with similar

© IFIP International Federation for Information Processing 2018
Published by Springer International Publishing AG 2018. All Rights Reserved
A. Amine et al. (Eds.): CIIA 2018, IFIP AICT 522, pp. 609–621, 2018.
https://doi.org/10.1007/978-3-319-89743-1_52

preferences, while a content-Based Filtering system (CBF) selects items based on the correlation between the content of the items and the user preferences.

Despite the demonstrated effectiveness of CBF technology in many cases, some drawbacks make it inappropriate in its current form for other cases. Indeed, CBF requires analyzing the content of a document which is computationally expensive and even impossible to perform on multimedia items which do not contain descriptive text [3]. Furthermore, CBF presents difficulties to handle the new user problem where no preference is available. At the beginning, a new user does not have any preference value. Therefore, it is very hard to issue any recommendation to him.

In this paper, we propose solving these issues by enhancing CBF systems using semantic derived from the *Web of Data*. In this latter, the World Wide Web is viewed as a global database by creating links between data which known as *Linked Data*. When these linked data enable describing people, they are called FOAF (*Friend of A Friend*). The proposed approach is based on Vector Space Modeling of CBF [4], and enhanced by a semantic level extracted from the web of data leading to a new model that we refer to as *Semantic Vector Space Model* (SVSM).

Following this introduction, CBF based on Vector Space Model is described in Sect. 2. Section 3 presents key features of the web of data. In Sect. 4, a review of some related works that propose recommender systems in web of data context is given. In Sect. 5, we describe the proposed approach SCBF and we report on the conducted experimental study and obtained results. Finally, conclusion and future work are given.

2   Content Based Filtering (CBF)

Information filtering deals with the delivery of information that would be interesting and useful to a user given his profile and preferences. An information filtering system assists users by filtering the data source and deliver relevant information to them. When the delivered information comes in the form of suggestions such information filtering system is called a recommender system. A CBF technique, also referred to as cognitive filtering [1], recommends items based on a correlation between the content of the items and a user profile. The content of each item is represented as a set of descriptors or terms, classically the words that occur in a document. The user profile is represented by a set of terms built up by analyzing the content of items seen by the user. Typically, a content based filtering system selects relevant items based on the correlation between the content of the items and the user's preferences.

One of the most important approaches is *Vector Space Model (VSM)* or term vector model [5]. In the vector space model, a document D *(item)* is represented as an m-dimensional vector, where each dimension corresponds to a distinct term [6]. The term frequency (tf) is a numerical statistic that measures the importance a term would have with regard to a document in a collection or corpus:

$$\mathrm{tf}_{vi} = \frac{n_{vi}}{N} \tag{1}$$

Where, n_{vi} is the number of times term t_i appears in a vector v; it models the taste of user and N is the total number of terms in the vector v.

To measure the extent to which documents contain a given term t_i we need to calculate the inverse document frequency (*idf*).

$$idf_i = \log\left(\frac{D}{n_j}\right) \tag{2}$$

Where, D is the total number of documents, n_j is the number of documents d_j containing term t_i.

From *tf* and *idf* we can calculate the *weight* (W) or *tfidf*. This latter is a concept that can be used to create a profile of an item for example a document or an object... etc.

$$W_i = tf_{vi} * idf_i \tag{3}$$

A content-based filtering system selects relevant items based on the correlation between the content of the items and the user's preferences [3]. However this technique suffers too from some disadvantages such as: it requires analyzing the content of the document which is expensive and even impossible to perform on multimedia [7] and the problem of new user or no preferences problem. At the beginning, a new user does not have any preference values; this makes it impossible to give him any recommendation. To address these problems, we propose to enhance CBF using the *Web of Data*.

3 Web of Data

Typically, a data set published in the web contains knowledge about a particular domain, like books, music, encyclopedic data and companies to name just few. If these data sets were interconnected i.e. linked to each other, this makes the World Wide Web a global database termed by Tim Berners Lee as *Web of Data*.

The most important concepts related to the web of data are: *Linked Open Data (LOD), Friend of A Friend (FOAF) vocabulary*, and *Resource Description Frame work (RDF)*.

3.1 Linked Open Data

The term Linked Open Data refers to a set of best practices for publishing and connecting structured data on the web using international standards of the World Wide Web Consortium.[1] LOD cloud is considered as a network or collection of data silos.

The diagram of Fig. 1 is maintained by Richard Cyganiak, and Anja Jentzsch (http://lod-cloud.net/).

---

[1] http://www.w3.org/standards/.

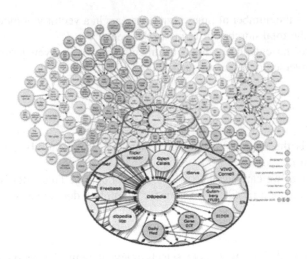

Fig. 1. Liked data cloud diagram [8].

The core of this diagram is *DBpedia*[2] which is a community effort to extract structured information from Wikipedia and to make this information available on the Web.

3.2 FOAF Vocabulary

The FOAF project began as an "experimental linked information project." Dan Brickley and Libby Miller are responsible for its inception, and EddDumbill and Leigh Dodds (http://www.foaf-project.org) notably contributed to its success. FOAF enables to describe people, their interests, their achievements, their activities, and their relationship with other people [8]. In Table 1 below all FOAF classes and proprieties are presented.

3.3 Resource Description Framework (RDF)

Resource Description Framework or in short RDF provides a common data model for Linked Data [8] and is particularly suited for representing data on the Web. Linked Data uses RDF as its data model and represents it in one of several syntaxes. There is also a standard query language called *SPARQL*. A single RDF statement describes two things and a relationship between them. Technically, this is called an *Entity-Attribute-Value (EAV)* data model.

---

[2] http://www.wiki.dbpedia.org/.

Table 1. Classes and proprieties of FOAF vocabulary.

Classes		
Agent	Document	Group
Image	LabelProprety	OnlineAccount
OnlineChatAccount	OnlineEcommerceAccount	OnlineGamingAccount
Organization	Person	PersonalProfileDocument
Project		
Proprieties		
Account	AccountName	AccountServiceHomepage
Age	AimChatID	Based near
Birthday	CurrentProject	Depiction
Depicts	DnaChechsum	FamilyName
First name	FirstName	Focus
Funded	Geekcode	Gender
GivenName	Givenname	HoldsAccount
Homepage	IcqChatID	Interest
IsPrimaryTopicOF	JabberID	Knows
LastName	Logo	Made
Maker	Mbox	Mbox sha1sum
Member	MembershipClass	MsnChatID
MyersBriggs	Name	Nick
Openid	Page	PastProject
Phone	Plan	PrimaryTopic
Publications	SchoolHopage	Sha1
SkypeID	Status	Surname
Theme	Thumbnail	Tipjar
Title	Tobic	Topic interst
Weblog	WorkInfoHomepage	WorkPlaceHomepage
YahooChatID		

4 Related Work

Few recommender systems based on web of data have been developed till date. The following Table 2 reviews some recent approaches. It provides a short description of the methods and indicates the web of data concepts used.

From Table 2, we can observe that most proposed approaches are dedicated to a specific domain example movies or music and use either FOAF vocabulary or linked data cloud. Almost half of these methods are based on Collaborative filtering.

Table 2. Recent recommender systems enhanced by web of data.

Systems	Technique	Recommendation domain	Using FOAF vocabulary	Using linked open data	Description
Foafing the music [9]	CBF	Music	Yes	No	It used RDF data. In order to provide its recommendations it crawled data from a large number of web sites
Mining recommendations [10]	CF	Unspecified	No	No	Uses the new metric to compare the recommendations that were generated from the public web
Music-related recommender systems [11]	CBF	Music	No	Yes	A Content-based recommendation is discussed, which uses not only meta-data but also the audio signal of the music for recommendations
Linked data recommendation [12]	CF	Music	No	Yes	It presents how to exploit the benefits of the LOD community effort to build recommender systems. By providing public, collaboratively created and semantically structured data
SPrank [13]	CF	Unspecified	No	Yes	Semantic path-based ranking SPrank a novel hybrid recommendation algorithm able to compute top-N item recommendations from implicit feedback exploiting the information available in the so called web of data
LOD for content-based recommender systems [14]	CBF	Movies	No	Yes	It implemented a content-based RS that leverages on the data available within Linked Open Data DBpedia and LinkedMDB datasets in order to recommend movies to the end users
Movie recommender system [15]	Hybrid	Movies	No	No	Investigate the use of folksonomies to generate tag-clouds that can be used to build better user profiles to enhance the movie recommendation. They use an ontology to integrate both IMDB and Netflix data
Linked data datasets for same as interlinking using recommendation techniques [16]	CF	Unspecified	No	Yes	In this work they to treat

Our work is motivated by the fact that combination of FOAF vocabulary and linked data cloud would have the potential to further improve the ability of CBF to achieve suitable recommendations. Using the FOAF vocabulary helps in solving the problem of new user and the extracted linked data from the cloud provide a semantic description of non-structured items.

5   Proposed Semantic Content Based Filtering (SCBF)

CBF selects items based on the correlation between the content of the items and the user's preferences. As aforementioned, the problem with CBF is that it requires analyzing the content of the items which is expensive or impossible with multimedia items. To solve this issue along with the new user problem, we describe in this section how the *Web of Data* technologies could be used to enhance CBF systems. We refer to the proposed web of data based variant of CBF as Semantic Content Based Filtering (SCBF). In SCBF, we suggest integration of the following technologies:

- **FOAF Vocabulary:** if new user is connected, his FOAF description will be compared with the other users' FOAF descriptions. The comparison is based on the proposed formula:

$$D_{FOAF}(u, v) = 1 + \log\left(\frac{1+K}{P}\right) \tag{4}$$

Where, $D_{FOAF}(u, v)$ is the FOAF distance between users u and v, $K = L + S$ with S is number of the similar FOAF proprieties between users u and v, L is number of links between u and v and P stands for the total number of FOAF proprieties describing target user u.

Following some important properties for the *class person* [8]:

- *Based near* - A location that something is based near, for some broadly human notion of near (The *based near* relationship relates two "spatial things").
- *Age* - The age in years of some person.
- *Gender* - The gender of this person (typically but not necessarily 'male' or 'female').
- *Title* - Title (Mr, Mrs, Ms, Dr. etc.).
- *Knows* - A person known by this person (indicating some level of reciprocated interaction between the parties).
- *dMaker* - An agent that made this thing.
- *Member* - Indicates a member of a Group.
- *Interest* - A page about a topic of interest to this person.
- *Topic_interest* - A thing of interest to this person.

- ***Linked Data Cloud***

  The vector space model is a representation often used for *text items* In this model, an item *i* is represented as an m-dimensional vector, where each dimension corresponds to a distinct term. However, this technique is too limited with unstructured and even with semi-structured items.

  To fix this problem, we propose in SFBC to enhance the *m* dimensional vector by *n* other textual or semantic attributes extracted from the linked data cloud. Therefore, the representation of the item will include $(m + n)$ attributes and expressed of a $(m + n)$- dimensional vector that we refer to as *Semantic Vector Space Model (SVSM)*. The example below brings more explanation about the proposed SVSM.

  In the dataset Movielens[3], the movie *"No escape"* is represented by the following textual attributes:

Id	Title	Realise date	Genre
1416	No escape	1994-01-01	Action, science fiction

On the same movie and using DBpedia, we can extract other information such as those given in the following Table 3.

Table 3. Textual and semantic attributes describing the movie "no escape".

Textual attributes			Semantic attributes		
Budget	Director	Country	Language	Subject	Image size
2.0E7	Martin campbell	USA	English	Prison films	250

For that we propose a new version of the *tf* denoted by $\widetilde{tf}$ defined as follows:

$$\widetilde{tf}(v, i) = \frac{NS_{vi}}{T} \tag{5}$$

Where, NS_{vi} is the number of times triplet t_i appears in the semantic segment of the vector v and T is the total number of triplets in the semantic segment of the vector v.

$$\widetilde{idf_i} = \log\left(\frac{Tt}{n_j}\right) \tag{6}$$

Where, Tt is the total number of triplet and n_j is the number of documents d_j where triplet $t_i \in d_j$.

---

[3] https://movielens.org/.

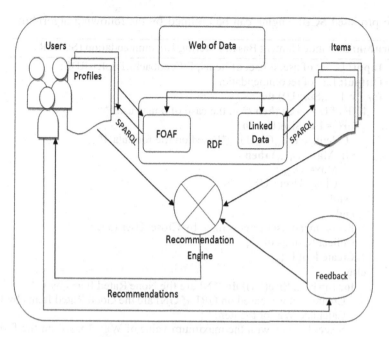

Fig. 2. General architecture of SCBF.

Therefore, the semantic weight is given by:

$$\widetilde{W_i} = \widetilde{tf_{v,i}} * \widetilde{idf_i} \tag{7}$$

And the global weight Wg for the item is defined as follows:

$$Wg_i = Wi * \widetilde{W_i} \tag{8}$$

Based on the above description, the proposed SCBF approach suggests the following architecture of recommender systems shown on Fig. 2.

In the case of new user (the feedback is empty), his D_{FOAF} is calculated using other users, just after we recommend the set of items liked by the user who has the maximum D_{FOAF}.

The Space of attributes that describe the items is enhanced by semantic and textual attributes extracted from linked data cloud, which gives further descriptions of the items.

The proposed SCBF engine can be outlined by the following algorithm.

Algorithm: Semantic Content Based Filtering Recommendation (SCBFR)

Input: U: set of users; I:set of items; F:Feedback; Triplet-RDF;
Output: List of recommendation;

1:   **for** (i=1,...,Size(U)) **do**
2:       **if** (F of U_i = = ∅) **then** /* in the case of a new user */
3:         **for**(j=1 to Size(U)) **do**
4:           Calculate $D_{FOAF}(U_i,U_j)$; /*based on the formula 4*/
5:           **if** (Max< D_{FOAF}) **then**
6:             Max= D_{FOAF};
7:             Close_User= U_j;
8:           **end**
9:         **end**
10:        Recommend list of items liked by Close_User to U_i;
11:        Collect Ratings of U_i;
12:        Create F of U_i;
13:      **else**
14:        **for** (n=1,..., Size(NI)) **do** /*NI are the None Rated Items by U_i*/
15:          Create vector v based on GRI /*GNI are the Good Rated Items by U_i*/
16:          Calculate Wg for the item I_n;
17:          Select k items with the maximum value of Wg; /*based on the formulas
18:          6,7,8*/
19:          Collect Ratings of U_i;
20:          Update F of U_i;
21:        **end**
22:      **end**
     end

---

6   Experiments

In the dataset Movielens[4], all movies are characterized by the following attributes: *Id*, *Title*, *Realize date*, and *Genre*, using following SPARQL query based on the federation (released by FedX) [17], between DBpedia and Linked Movie DataBase (LDMDB[5]). We can extract more information about these movies like: *Director*, *Country*, *Actor*, and *Abstract*. The common attribute between Movielens and the federated query is the movies titles. Following is the SPARQL query that extract more information about Movielens movies:

---

[4] https://movielens.org/.
[5] http://www.linkedmdb.org/.

```
SELECT * WHERE {?film
<http://data.linkedmdb.org/resource/movie/filmid> ?uri.

    ?film <http://purl.org/dc/terms/title> ?Title .

    ?film
<http://data.linkedmdb.org/resource/movie/actor> ?cast .

  ?cast
<http://data.linkedmdb.org/resource/movie/actor_name>
?Actor .   ?film
<http://data.linkedmdb.org/resource/movie/country> ?Coun-
tryID .

    ?CountryID
<http://data.linkedmdb.org/resource/movie/country_name>
?Country .?film <http://purl.org/dc/terms/date> ?Year .

    ?film
<http://data.linkedmdb.org/resource/movie/director> ?di-
rectorID .

    ?directorID
<http://data.linkedmdb.org/resource/movie/director_name>
?Director.
            ?film  <http://www.w3.org/2002/07/owl#sameAs>   ?x.
        ?x  <http://dbpedia.org/ontology/director>   ?direc-
        tor.?x <http://dbpedia.org/ontology/abstract>   ?ab-
        stact.}
```

To measure the effectiveness of our approach, we calculated the Mean Absolute Error MAE, and Root Mean Square Error (RMSE) using the following formulas:

$$MAE = \frac{\sum_{u,i}|P_{u,i} - n_{u,i}|}{n} \qquad (9)$$

$$RMSE = \sqrt{\frac{1}{n}\sum_{u,i} P_{u,i} - n_{u,i}^2} \qquad (10)$$

Where $n_{u,i}$ is the note given by the user u on item I, $p_{u,i}$ is the predicted note, n is the total number of predicted notes.

The value MAE and RMSE of SBCF are compared with other values of state of the art techniques described in [18]. The results are shown on Table 4 where we can observe that the proposed approach offers the minimum error value (Fig. 3).

Table 4. Comparative results.

	CBF	CF_U	CF_I	SCBF
MAE	0.83	0.86	0.83	0.46
RMSE	1.04	1.04	1.09	0.49

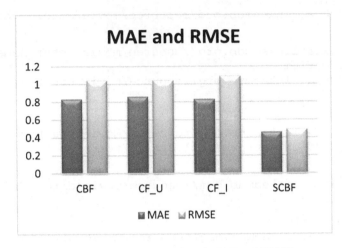

Fig. 3. Experimental results of MAE and RMSE.

7   Conclusion

In this work, we described a new approach to content based recommendation using web of data which is mainly supported by some of intelligent technologies namely: FOAF vocabulary and Linked Data Cloud. We were faced with a challenge to use the technique of CBF while reducing the impact of new user issue and the difficulty of analyzing unstructured items. Promising preliminary results have been obtained. As future work, our plan is to test and evaluate the proposed approach with other metrics like recall and precision, and apply new user problem solution to Collaborative Filtering (CF) algorithm to reduce the impact related to cold start issues.

References

1. Burke, R.: Hybrid recommender systems: survey and experiments. User Model. User-Adap. Inter. **12**, 331–370 (2002)
2. Zitouni, H., Nouali, O., Meshoul, S.: Toward a new recommender system based on multi-criteria hybrid information filtering. In: Amine, A., Bellatreche, L., Elberrichi, Z., Neuhold, Erich J., Wrembel, R. (eds.) CIIA 2015. IAICT, vol. 456, pp. 328–339. Springer, Cham (2015). https://doi.org/10.1007/978-3-319-19578-0_27
3. Shoval, P., Maidel, V., Shapira, B.: An ontology-content-based filtering method. Int. J. Inf. Theor. Appl. **15**, 303–314 (2008)

4. Raghavan, V.V., Wong, S.K.M.: A critical analysis of vector space model for information retrieval. J. Am. Soc. Inf. Sci. **37**(5), 279–287 (1986)
5. Salton, G., Wong, A., Yang, C.S.: A vector space model for automatic indexing. Commun. ACM **18**(11), 613–620 (1975)
6. Karen, S.J.: A statistical interpretation of term specificity and its application in retrieval. J. Doc. **28**(1), 11–21 (1972)
7. Dai, H., Mobasher, B.: Using ontologies to discover domain-level web usage profiles. In: Proceedings of the Second Semantic Web Mining Workshop at PKDD 2001, Helsinki, Finland (2001)
8. Wood, D., Zaidman, M., Ruth, L., Hausenblas, M.: Linked Data. Manning Publications, New York (2014)
9. Celma, O., Serra, X.: FOAFing the music: bridging the semantic gap in music recommendation. Web Semant.: Sci. Serv. Agents World Wide Web **6**, 250–256 (2008)
10. Shani, G., Chickering, M., Meek, C.: Mining recommendations from the web. In: ACM Conference on Recommender Systems, pp. 35–42. ACM, New York (2008)
11. Passant, A., Raimond, Y.: Combining social music and semantic web for music-related recommender systems. In: Social Data on the Web Workshop (2008)
12. Passant, A., Heitmann, B., Hayes, C.: Using linked data to build recommender systems. In: Proceedings of RecSys, New York, USA (2009)
13. Ostuni, V.C., Di Noia, T., Di Sciascio, E., Mirizzi, R.: Top-n recommendations from implicit feedback leveraging linked open data. In: Proceedings of the 7th ACM Conference on Recommender Systems, pp. 85–92. ACM (2013)
14. Mirizzi, R., Di Noia, T., Ostuni, V.C., Ragone, A.: Linked Open Data for Content-Based Recommender Systems (2012)
15. Szomszor, M., Cattuto, C., Alani, H., O'Hara, K., Baldassarri, A., Loreto, V., Servedio, V. D.: Folksonomies, the Semantic Web, and Movie Recommendation (2007)
16. Liu, H., Wang, T., Tang, J., Ning, H., Wei, D., Xie, S., Liu, P.: Identifying linked data datasets for sameAs interlinking using recommendation techniques. In: Cui, B., Zhang, N., Xu, J., Lian, X., Liu, D. (eds.) WAIM 2016. LNCS, vol. 9658, pp. 298–309. Springer, Cham (2016). https://doi.org/10.1007/978-3-319-39937-9_23
17. Schwarte, A., Haase, P., Hose, K., Schenkel, R., Schmidt, M.: FedX: optimization techniques for federated query processing on linked data. In: Aroyo, L., Welty, C., Alani, H., Taylor, J., Bernstein, A., Kagal, L., Noy, N., Blomqvist, E. (eds.) ISWC 2011. LNCS, vol. 7031, pp. 601–616. Springer, Heidelberg (2011). https://doi.org/10.1007/978-3-642-25073-6_38
18. Haddad, M.R., Baazaoui, H., Ziou, D., Ghézala, H.B.: Un modèle de recommandation contextuel pour la prédiction des intérêts des consommateurs sur le Web. In: IC2015 (2015)

Selecting Web Service Compositions Under Uncertain QoS

Remaci Zeyneb Yasmina$^{(\boxtimes)}$, Hadjila Fethallah$^{(\boxtimes)}$,
and Didi Fedoua$^{(\boxtimes)}$

LRIT Laboratory, Computer Sciences Department, UABT University – Tlemcen,
Tlemcen, Algeria
yasmina_rmc@hotmail.fr,
{f_hadjila, f_didi}@mail.univ-tlemcen.dz

Abstract. The uncertain QoS management is gaining a lot of interest in the service oriented computing area. In this work, we propose a framework that allows to select the TopK compositions of services that best meet the user's requirements. This framework not only handles the user's global constraints but it also takes into account the fluctuating nature of the QoS informations. More specifically we present two algorithms that ensure the aforementioned purposes. The first one ranks the services of each abstract class according to the probabilistic dominance heuristic. The second one explores the compositions search space by leveraging the backtracking search. The experimental evaluation shows that the proposed heuristic is more effective than the ranking based on average QoS.

Keywords: Service oriented architecture · Web service selection
Uncertain quality of service · Backtracking search · Combinatorial optimization

1 Introduction

Over the last decade, the web services have been increasingly published and deployed over the web. Consequently, many providers offer the same functionality, (i.e. the same interface/behavior) but they differ according to their non-functional attributes (or quality of service such as response time, cost, reputation, availability…). In this context, the user has to leverage the QoS to select the best advertised services that meet his/her requirements. On the other hand, we also observe that the service QoS is generally fluctuating and non-deterministic. This is mainly due to the environment circumstances (i.e. the price of a service depends to the season; the response time/the throughput depend to the network load…). As a result, our selection/aggregation models should take into account these fluctuations. Additionally, we notice that a complex user's request is generally fulfilled with a composition of services rather than a single component. This means that the optimization/selection algorithms should not only handle the non-deterministic QoS, but also the global optimization aspects (i.e. global constraints, aggregated QoS,…). In the example cited in Table 1, we assume a user's request that consists in invoking two types of services: a currency service and a purchase order. Each service is characterized by two criteria: the cost (denoted C in Table 1), and the latency (denoted L in Table 1). The selected combination must have a

© IFIP International Federation for Information Processing 2018
Published by Springer International Publishing AG 2018. All Rights Reserved
A. Amine et al. (Eds.): CIIA 2018, IFIP AICT 522, pp. 622–634, 2018.
https://doi.org/10.1007/978-3-319-89743-1_53

Table 1. Normalized QoS of service instances

Currency conversion						Purchase order					
X			Y			S			T		
	C	L		C	L		C	L		C	L
X₁	0.3	0.3	Y1	0.2	0.1	S1	0.5	0.2	T1	0.3	0.5
X₂	0.2	0.5	Y2	0.3	0.4	S2	0.5	0.3	T2	0.2	0.5
X₃	0.6	0.3	Y3	0.3	0.6	S3	0.7	0.3	T3	0.4	0.7
X₄	0.5	0.6	Y4	0.6	0.6	S4	0.8	0.6	T4	0.6	0.4

global cost (the QoS sum of the composition services) less or equal than 0.8 (according to a given unit such as $) and a global (aggregated) latency less or equal than 0.9 (according to a given unit such Seconds).

Furthermore, the Table 1, also shows the QoS variation (see the instances lines such as X1, ..., X4) of each service. By considering the elements of the currency class (i.e. the services X and Y that have same functionality), we notice that the comparison of their performances is not always self-evident. More specifically, if we use the mean QoS as comparison mechanism, this can create a misleading result. Simply speaking, the mean QoS of X is (0.4, 0.42), likewise the mean QoS of Y is (0.35, 0.42), and therefore Y is better than X (i.e. $Y \gg X$), but if we consider all the instances (i.e. the QoS variations), then we observe that Y dominate X in 37% of the cases and X is dominate Y in 43% of the cases, furthermore the QoS instances of X have a reduced variance in comparison with those of Y. Consequently our initial ordering may be erroneous. To tackle these difficulties, we should use an ordering scheme that takes into account all the sampled QoS (and not the aggregated values). In addition, any proposed service selection system should differentiate between feasible compositions and non-feasible compositions. For example if we consider the median QoS of each component service as the representative value, then the composition $c = <Y, S>$, is not feasible because:

MedianCost(Y) + MedianCost(S) = 0.3 + 0.6 > 0.8. (The first global constraint is violated).

On the other hand, the composition $c' = <X, T>$ is feasible since:

MedianCost(X) + MedianCost(T) = 0.4 + 0.35 ≤ 0.8, and
MedianLatency(X) + MedianLatency(T) = 0.4 + 0.5 ≤ 0.9.

By analyzing the literature approaches, we notice that the majority of the service composition works don't handle the non-deterministic QoS aspects and global constraints, at the same time. To deal with this situation, we propose in this paper a general framework that selects the Top-k compositions while managing the following requirements:

- The user's needs (global constraints, QoS optimization, number of services classes, control flow).
- The QoS fluctuations of web services over time.

Since the number of services per class might still be extremely high, it would be preferable to reduce the computational cost of the selection process. This aim can be

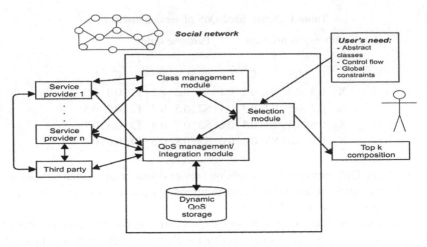

Fig. 1. Service selection framework

ensured by introducing heuristics that select the best services of each class. It is worth mentioning that the complexity of this issue is known to be NP-Hard [1, 2].

In summary, our main contribution referred to as the "selection module" (see Fig. 1) can be described as follows:

1. Firstly, we rank the services of each class, according to the probabilistic dominance relationship shown in formula (5), and we retain only the Top-K services having the highest scores. This step aims to reduce the search space.
2. To rank the service compositions, we leverage an objective function based on the median QoS of the components of the composite solution (see formula (1)).
3. We explore the search space constituted of the first K services of each class (see step I), by implementing a backtracking search (inspired from the constraint satisfaction problems) and we return the Top-K optimal compositions.

The reminder of the paper is organized as follows: Sect. 2 demonstrates a literature review on the QoS aware service selection issue. Section 3 specifies the problem, in Sect. 4 we show the proposed framework as well as the selection algorithms, and finally we present in Sect. 6 our conclusions and perspectives.

2 State of the Art

The service composition and selection has drawn a lot of attention during the past decade, The existing works either focus on global selection with deterministic/non-deterministic QoS [1–3, 9, 10, 12, 16] or local service selection with non-deterministic QoS [4, 14, 18].

The service selection with uncertain QoS is gaining a lot of interest in the service oriented computing area. Existing works such as [11, 14] leverage the dominance probability relationship to extract the most dominant services from a predefined dataset.

In nutshell, the work presented in [11] extends the traditional concept of skyline [5] to cover the uncertain data (i.e. notion of probabilistic skylines). To get the probabilistic skylines, the authors extract the services that have at least a percentage p to not be dominated by another component. As mentioned in [14], the P-skyline prefers noisy services to the detriment of consistent services. In [14] the authors propose a new concept called P-dominant skyline, which is less sensitive to noisy (inconsistent) services, in addition it is more suitable for including good services. Furthermore the authors leverage an R-tree [6] structure in order to efficiently extract the p-dominants services. In [4], the authors leverage the possibility theory in order to compute the dominants services. The possibility theory is preferred, when the probability distributions of QoS criteria is unknown or cannot be computed. As a results, the QoS attribute are modeled as possibility distribution. The authors also present two novel concepts: the possibility based skyline and the necessity based skyline, in addition they provide a mechanism to control the size of the skylines set. In [13], the authors propose an approach for computing the top k dominant compositions without taking into account the global constraints. The authors handle the QoS uncertainty by proposing the concept of dominance ability (which is based on the dominance probability). In [17, 18] the authors present a set of formulas for estimating the uncertain QoS (mainly the execution time) of a composite service. To this end, they model each QoS metric of a component service as a probability distribution; in addition the composite service is represented as a graph that leverages several basic patterns (Sequential, Parallel, conditional, and Loop).

In the area of deterministic service selection, we can review a lot of approaches that handle the QoS as a non-varying phenomenon. In [3], the scholars handle both the functional aspects (inputs/outputs) and the non-functional aspects (QoS/global constraints), they propose an optimization framework based on the harmony search meta-heuristic. In [15, 16], the authors aim to avoid the user's implication (which usually assigns a set of numerical weights to the criteria) by focusing on skylines compositions (denoted C-SKY). In addition, the authors present a set of heuristics in order to accelerate the computation of C-SKY. To this end, they sort the skylines of each abstract class according to a predefined objective function (that sums all the QoS criteria), thereafter they explore the compositions space by scanning at first, the top services of each class. In [9] the authors leverage the harmony search meta-heuristic to get the near optimal compositions; the results can be further improved by tuning the meta-heuristic parameters. In [2] the authors address this issue by taking into account multiple control flow. Their main idea consists in extracting the skylines of each abstract class, thereafter; the authors create a hierarchical clustering of each skyline's set by leveraging the K-means Algorithm. Finally they explore the compositions space by combining the clusters heads and checking the global constraints fulfillment.

3 Problem Formalization

In this section, we will formalize the problem of Top-k dominant compositions under uncertain QoS. In what follows, we will assume a set of hypothesis and notations in order to simplify the problem specification:

- All the QOS attributes are positive (i.e. all the positive attributes need to be maximized).
- The composition model is sequential.
- The QoS criteria of a composition are aggregated according to the sum function (such as reputation), if there are multiplicative criteria, then we replace them with their log value and we treat them as additive criteria. The other types of QoS criteria are not handled in this paper.
- n: is the number of abstract classes.
- Cl_1, Cl_2, ..., Cl_n: are the set of abstract classes.
- m: is the number of services per classes.
- r: is the number of QoS attributes.
- l: is the number of service instances (i.e. the number of QoS realizations or the sample size).
- QoS_{piju}: is the value of the pth QoS attribute related to the uth instance of the service $S_i \in Cl_j$.
- $AVGQoS_{pij}$: the average QoS computed over all the instances of $S_i \in Cl_j$.
- b_1, b_2, ..., b_r: are the user's global constraints (i.e. the limits which need to be met by the QoS of the composition).
- W_1, ..., w_r: are the weight of the QoS criteria, the default value of each w_p is $1/r$.
- k: the size of the returned list (of compositions).
- The overall utility of a service composition $c = (x_1, x_2, ..., x_n)$ is computed as follows:

$$U'(c) = \sum_{p=1}^{r} w_p * ((MedianQ'_p(c) - Qmin'(p))/(Qmax'(p) - Qmin'(p)) \tag{1}$$

- x_1 (resp x_2, ..., x_n): represents the id of the selected service related to Cl_1 (resp Cl_2, ..., Cl_n)

$$Qmin'(p) == \sum_{j=1}^{n} Qmin(j, p), \tag{2}$$

(the first normalization constant), were $Qmin(j, p) = MIN_{Si \in Clj, u \in \{1,...l\}} (QoS_{piju})$

$$Qmax'(p) == \sum_{j=1}^{n} Qmax(j, p), \tag{3}$$

(the second normalization constant), were $Qmax(j, p) = MAX_{Si \in Clj, u \in \{1,...l\}} (QoS_{piju})$

$$MedianQ'_p(c) = \sum_{j=1}^{n} Median_{u \in \{1,...l\}} (QoS_{pxjju}) \tag{4}$$

Since a component service S_{xi} is characterized by several QoS realizations, we choose the median QoS Value to evaluate its performance; consequently the composition c is also evaluated according to the median performance (see formula 1). We have chosen the median aggregation for the QoS realizations, because it is less sensitive to the variations and the outliers of the sample. In addition, we use the formula (5) to compare the compositions according to their degree of satisfying the global constraints. Roughly speaking, a composition c is ranked above another composition c' if the score of c with respect to (5) is higher than the score of c' with respect to (5). If c ties with c', then we order them according to formula 1 (which is also termed fitness or function U'(.)), the higher the score of U' the better the rank. Formally: c is ranked above c' iff:

$$\begin{cases} 1/r. \sum_{p=1}^{r} \Pr(\text{MedianQ'}p(c)) > 1/r. \sum_{p=1}^{r} \Pr(\text{MedianQ'}p(c')) \textbf{ or} \\ 1/r. \sum_{p=1}^{r} \Pr(\text{MedianQ'}p(c)) > 1/r. \sum_{p=1}^{r} \Pr(\text{MedianQ'}p(c')) \text{ and } U'(c) \geq U'(c') \end{cases}$$

We also notice that the computational cost of formula (1) is $O(r.n)$, (we assume that median values of the services are already computed), likewise the computational cost of formula (5) is $O(r.n)$. In summary, our main objective is to select the Top-K compositions, $C_1, \ldots, C_k$ which:

- Maximize the chance of satisfying the global constraints:

$$1/r. \sum_{p=1}^{r} \Pr(\text{MedianQ'}_p(c_y)) \geq b_p), \text{ where } y \in \{1, \ldots, k\}. \tag{5}$$

- Maximize the function U'(.).

4 Proposed Approach

In this section, we present our selection framework (shown in Fig. 1). It is constituted of three main modules:

The Class Management Module: Its purpose is to assign each service to a given abstract class (which represents the main functionality of the service such as: hotel booking, currency conversion, maps services...), the module also updates the classes.

The QoS Management and Integration Module: It allows to store the fluctuating QoS of each service, the QoS data can be drawn from: the service provider itself (ex: the cost), the social networks (ex: the reputations) the third parties (ex: the latency...).

The Selection Module: Its main goal is to provide the Top K dominants service compositions for each user's request. This module is constituted of two algorithms (Algorithm 1: service ranking and Algorithm 2: backtracking search). The Algorithm 1 aims to reduce the search space of Algorithm 2, thereafter the backtracking search is executed in order to give the final compositions.

The service ranking sorts the services of each abstract class through the use of the probabilistic dominance relationship. We notice that the dominance relationship and its variants are widely used in the preference queries [4, 8] as well as the service discovery [7]. Simply speaking, we compare the QoS of each pair of services S_i, $S_{i'}$ with respect to the probabilistic dominance, thereafter we increment the ranking score of the wining service. The more the score is high the better the rank. The probabilistic dominance between two services $S_{i'}$ and S_i measures the average fraction of the instances of S_i that are weakly dominated by an instance of $S_{i'}$. It is given as follows:

$$\text{prob-dom}(S_{i'}, S_i) = 1/1 \sum_{u'=1}^{1} \text{individual-prob-dom}(u', i', i) \tag{6}$$

and individual-prob-dom(u', i', i)

$$= (|\{(QoS_{1iju}, \ldots, QoS_{riju})/(QoS_{1i'ju}, \ldots, QoS_{ri'ju'}) \gg (QoS_{1iju}, \ldots, QoS_{riju})\}|/1),$$
$$u \in \{1, \ldots 1\}$$

The relation $\gg$ denotes the weak dominance relationship, it is defined as follows:

Let X and Y be two vectors of R^r
$X \gg Y$ iff for each dimension $i \in \{1, \ldots, r\}$: $X(i) \geq Y(i)$
We assume that $\geq$ denotes "better than".

The pseudocode of Algorithm 1 is given below:

Algorithm 1: ServiceRanking

Input: $Cl_1, \ldots Cl_n$

Output: $RankedCl_1, \ldots RankedCl_n$

1. For i=1 to n Do RankedCl$_i$=<>;

2. For i=1 to n Do Begin

2.1 For y=1 to m Do score(y)=0;

2.2 For j=1 to m Do Begin

2.2.1 For j'=1 to m Do Begin

2.2.1.1 If (j!=j') Then Begin

2.2.1.1.1 If (probdom(($S_j, S_{j'}$)$\geq$ probdom(($S_{j'}, S_j$)) Then
        Begin

2.2.1.1.1.1 score(j)=0 score(j)+1;
            End

        End

        End

    End

2.3 RankedCl$_i$=decreasing-sort(Cl$_i$)
End

3. return <RankedCl$_1, \ldots$ RankedCl$_n$>

The explanation, of Algorithm 1 is given as follows:

In line 1, we initialize the ranked Class $RankedCl_i$ (with an empty structure).
In line 2.1, we initialize the ranking score of each service of the current class i.
In lines 2.2 up to 2.2.1.1.1, we compare each pair of services $(S_j, S_{j'})$ of the same class i, through the use of the probabilistic dominance formula (6).
In line 2.2.1.1.1.1, we update the score of S_j if it wins the test.
In line 2.3, we sort the elements of RankedCli according to the scores updated in 2.2.1.1.1.1.
We return the ranked classes in line 3.

It is worth noting that, the overall complexity of Algorithm 1 is $O(nm^2.r.l^2 + n.$ $mlogm)$, and the complexity of formula (6) is $O(r.l^2)$. The pseudo-code of Algorithm 2 is given below (we notice that the symbol $\Leftrightarrow$ denotes an empty structure):

Algorithm 2: BacktrackingSearch

Input: $RankedCl_1,\ldots,RankedCl_n$

$b_1,b_2,..b_r$: global constraints

k: size of the results set,   t: the minimum % of the preserved constraints.

Output: TopKCompositions

1. TopKCompositions $=\Leftrightarrow$

2. For i=1 to k^n Do Begin

2.1 c=GetNextComposition($RankedCl_1,\ldots,RankedCL_n$);

2.2 degree= $1/r.\sum_{p=1}^{r}$ $Pr(MedianQ'_p(c)) \geq b_p)$

2.3 If (degree) $\geq$ t) Then Begin

2.3.1   If better (c, TopKCompositions) Then Begin

2.3.1.1   Update (c, TopKCompositions)

            End

       End

End

3. Return (TopKCompositions)

The explanation is given as follows:
In line 2, we explore all the possible compositions. In line 2.1, we get the current composition c. In line 2.2, we compute the fraction of satisfied global constrained. In line 2.3, we check that the fraction of the preserved global constraints is above the threshold. In line 2.3.1, we compare c with the existing "TopKComposition" elements through the use of formulas (5) and (1) (see Sect. 3 for more details about the ordering of compositions). In line 2.3.1.1, we update the result TopKCompositions if c is better

than an existing composition. We return the final result in line 3. It is worth noting that, the overall complexity of Algorithm 2 is $O(k^n(k.r.n + n + k\log k))$.

5   Experiments

In this section we analyze the performance of our framework in terms of execution time and optimality. To this end, we conduct a set of experiments, with several configurations of parameters (see Table 2). The experiments were conducted on a machine having an Intel I3 core 2.53 GHz processor, 4 GB RAM, and running Windows 7. The Figs. 1 up to 5 are related to Algorithm 2; however the Fig. 6 is related to Algorithm 1.

Table 2. Parameters and examined values

Parameters	Values
Number of tasks (n)	10, 15, 20
Number of services (m)	100 to 1100
Number of QOS criteria (r)	2 to 10
Instances (l)	100 to 400
size of the result (k)	2, 6, 10

The Fig. 2 shows the exponential growth of the execution time with respect to n. If $k = 2$, then the execution time is acceptable for all values of n. however when $k = 6$, the time overhead is not tolerable for $n \geq 10$.

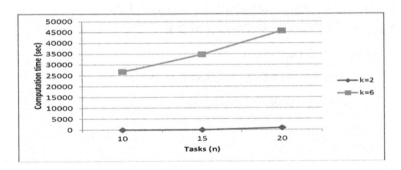

Fig. 2. CPU time versus n ($r = 3$, $m = 50$, $l = 10$)

The Fig. 3 shows the impact of r over the execution time. We observe that all values of r, are tolerable for $k = 2$ and $k = 6$, however for $k = 10$ and $r \geq 4$ the execution time will be inacceptable (more than 10 min). The same observation is made for Fig. 4, for $k = 2$ and $k = 6$, the execution time is tolerable, however for $k = 10$ and $l \geq 200$ the computational is not acceptable.

As depicted in Fig. 5, the global search (Algorithm 2) is not very sensitive to m, this is mainly due to the fact that the generation of compositions depends on the number of filtered services i.e. K. In what follows, we compare the effectiveness/efficiency of Algorithm 1 DSR (Dominance service ranking) with respect to the ranking based on

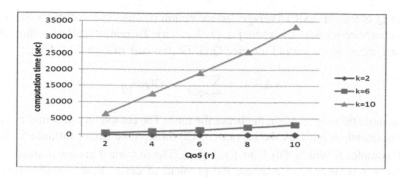

Fig. 3. CPU time versus r (n = 5, m = 200, 1 = 100)

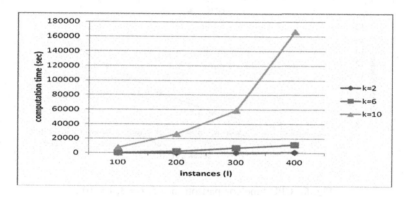

Fig. 4. CPU time versus 1 (n = 5, r = 3, m = 200).

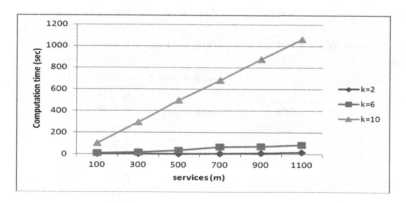

Fig. 5. CPU time versus m (n = 5, r = 3, 1 = 10).

average QoS termed ASR (Average service ranking). The latter computes the average QoS for each service $S_i \in Cl_j$ where $j \in \{1, 2, \ldots, n\}$. Thereafter ASR sorts the elements of Cl_j according to the sum of average QoS, i.e. the rank of each $S_i \in Cl_j$ is:

$$rank(i, j) = \sum_{p=1}^{r} AVGQoS_{pij}. \tag{7}$$

The more the score is high, the better the rank. The complexity of formula 6 is O(r. l^2), consequently if we rank the services of Cl_j through the use of formula 6, then the overall complexity will be $O(r.l^2.m + m \log m)$. The formula 7 is chosen instead of the dominance relationship, to alleviate the problem of curse dimensionality (i.e., with large r, the probability that a service s dominates another service s' is very weak).

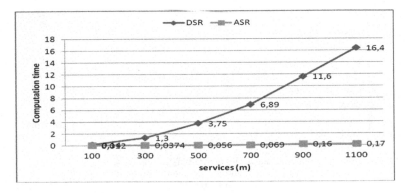

Fig. 6. CPU time comparison (n = 5, r = 3, l = 10).

Table 3. Performance comparison between DSR and ASR

Parameters	Solutions	Algorithms	Fitness	Respected constraints (%)
n = 5, r = 3, m = 200, l = 300	TOP 2 solutions	DSR + Algo2	0.4063	100
			0.4053	100
		ASR + Algo2	0.3823	100
			0.3813	100
	TOP 6 solutions	DSR + Algo2	0.4075	100
			0.4073	100
			0.407	100
			0.4068	100
			0.4065	100
			0.4064	100
		ASR + Algo2	0.3997	100
			0.3992	100
			0.3989	100
			0.3984	100
			0.3983	100
			0.3982	100

As shown in Fig. 6, the ASR approach is better than DSR in terms of execution time. This is due to the fact that ASR is principally based on formula (7) which is only $O(r.l)$, in addition ASR doesn't depend on m, however the DSR algorithm is based on formula 6 (which is $O(r.l^2)$), and depends on m.

According to Table 3, we observe that the percentage of respected global constraints is the same for both approaches (ASR and DSR). We also notice a slight fitness superiority (i.e. the function U') of DSR with respect to ASR. This observation is valid for all values of K.

6 Conclusion

In this work, we have investigated the problem of service selection under uncertain QoS. Our approach consists of two steps: the first one sorts the uncertain services according to the probabilistic dominance relationship, and the second one explores the search space by using a backtracking algorithm. The effectiveness/efficiency of the approach is confirmed with a set of experiments.

For future work, we will consider alternative sorting relationships (such as the dominance related to the necessity/possibility distributions). In addition we will adapt this framework to the selection of cloud services.

References

1. Alrifai, M., Risse, T.: Selecting skyline services for QoS-based web service composition. In: Proceedings of WWW 2010, 26–30 April 2010, Raleigh, NC, USA (2010)
2. Alrifai, M., Risse, T., Nejdl, W.: A hybrid approach for efficient Web service composition with end-to-end QoS constraints. ACM Trans. Web (TWEB) 6(2), 7 (2012)
3. Bekkouche, A., Benslimane, S.M., Huchard, M., Tibermacine, C., Hadjila, F., Merzoug, M.: QoS-aware optimal and automated semantic web service composition with user's constraints. Serv. Oriented Comput. Appl. 11(2), 183–201 (2017)
4. Benouaret, K., Benslimane, D., Hadjali, A.: Selecting skyline web services from uncertain QoS. In: 2012 IEEE 9th International Conference on Services Computing (SCC), pp. 523–530 (2012)
5. Borzsony, S., Kossmann, D., Stocker, K.: The skyline operator. In: Proceedings of 17th International Conference on Data Engineering, pp. 421–430. IEEE (2001)
6. Guttman, A.: R-trees: a dynamic index structure for spatial searching. vol. 14, no. 2, pp. 47–57. ACM (1984)
7. Fethallah, H., Amine, B., Amel, H.: Hybrid web service discovery based on fuzzy condorcet aggregation. In: Morzy, T., Valduriez, P., Bellatreche, L. (eds.) ADBIS 2015. LNCS, vol. 9282, pp. 415–427. Springer, Cham (2015). https://doi.org/10.1007/978-3-319-23135-8_28
8. Hamiche, M., Drias, H., Allel, H.: A strong-dominance-based approach for refining the skyline. In: 12th International Symposium on Programming and Systems (ISPS), pp. 1–8. IEEE (2015)
9. Merzoug, M., Chikh, M.A., Hadjila, F.: QoS-aware web service selection based on harmony search. In: 4th International Symposium IEEE-ISKO-Maghreb: Concepts and Tools for Knowledge Management, Algiers, Algeria, pp. 1–6 (2014)

10. Minjung, K., Byungkook, O., Jooik, J., Kyong-Ho, L.: Outlier-robust web service selection based on a probabilistic QoS model. Int. J. Web Grid Serv. **12**(2), 162–181 (2016)
11. Pei, J., Jiang, B., Lin, X., Yuan,Y.: Probabilistic skylines on uncertain data. In: VLDB Endowment, Vienna, Austria (2007)
12. Rosenberg, F., Müller, M.B., Leitner, P., Michlmayr, A., Bouguettaya, A., Dustdar, D.: Metaheuristic optimization of large-scale QoS-aware service compositions. In IEEE International Conference on Services Computing (SCC 2010), pp. 97–104. IEEE (2010)
13. Wen, S., Tang, C., Li, Q., Chiu, D.K.W., Liu, A., Han, X.: Probabilistic top-K dominating services composition with uncertain QoS. Serv. Oriented Comput. Appl. **8**(1), 91–103 (2014)
14. Yu, Q., Bouguettaya, A.: Computing service skyline from uncertain QoWS. IEEE Trans. Serv. Comput. **3**(1), 16–29 (2010)
15. Yu, Q., Bouguettaya, A.: Computing service skylines over sets of services. In: IEEE International Conference on Web Services (ICWS), pp. 481–488 (2010)
16. Yu, Q., Bouguettaya, A.: Efficient service skyline computation for composite service selection. IEEE Trans. Knowl. Data Eng. **25**(4), 776–789 (2013)
17. Zheng, H., Yang, J., Zhao, W., Bouguettaya, A.: QoS analysis for web service compositions based on probabilistic QoS. In: Kappel, G., Maamar, Z., Motahari-Nezhad, H.R. (eds.) ICSOC 2011. LNCS, vol. 7084, pp. 47–61. Springer, Heidelberg (2011). https://doi.org/10.1007/978-3-642-25535-9_4
18. Zheng, H., Zhao, W., Yang, J., Bouguettaya, A.: QoS analysis for web service compositions with complex structures. IEEE Trans. Serv. Comput. **6**(3), 373–386 (2013)

Unified-Processing of Flexible Division Dealing with Positive and Negative Preferences

Noussaiba Benadjimi(✉) and Walid Hidouci

Laboratoire de la Communication dans les Systèmes Informatiques, Ecole Nationale Supérieure d'Informatique, BP 68M, 16309 Oued-Smar, Alger, Algeria
{an_benadjimi,hidouci}@esi.dz
http://www.esi.dz

Abstract. Nowadays, current trends of universal quantification-based queries are been oriented towards flexible ones (tolerant queries and-or those involving preferences). In this paper, we are interested in universal quantification-like queries dealing with both positive or negative preferences (requirements or prohibitions), considered separately or simultaneously. We have emphasised the improvement of the proposed operator, by designing new variants of the classical Hash-Division algorithm, presented in [1], for dealing with our context. The parallel implementation is also presented, and the issue of answers ranking is dealt with. Computational experiments are carried out in both sequential and parallel versions. They shows the relevance of our approach and demonstrate that the new operator outperforms the conventional one with respect to performance (the gain exceeds a ratio of 40).

Keywords: Universal quantification-like queries · Relational division
Relational anti-division · Preferences · Tolerant division · Hash-division

1 Introduction

Relational operators including universal quantification are an interesting type of queries. They are very useful for many applications, especially in business intelligence applications and in recommendation systems [2]. In relational algebra, universal quantification-like queries are the most complex operators. That is why a lot of research focuses on their implementation, algorithms and optimisation [3]. Universal quantification-like queries are, often, about **division** or **anti-division** operators. The division searches elements associated with all members of a set of requirements, while the anti-division aims to find all elements that are associated with none of the members of a set of prohibitions [4]. In this paper, we are concerned with some relevant issues related to the improvement of queries combining both of required and forbidden associations.

© IFIP International Federation for Information Processing 2018
Published by Springer International Publishing AG 2018. All Rights Reserved
A. Amine et al. (Eds.): CIIA 2018, IFIP AICT 522, pp. 635–647, 2018.
https://doi.org/10.1007/978-3-319-89743-1_54

1.1   The Division and Anti-division Operators

Relational division is used when an element that satisfies a whole set of requirements is sought for. Whereas, the anti-division operator is used to select elements
that exclude any association with a set of prohibitions [6].

In relational algebra, the division (resp. anti-division) of relation $r(X,Y)$,
called **'dividend'**; by relation $s(Y)$, called **'divisor'**; is a new relation $q(X)$,
called **'quotient'** that includes some parts of $Projection(r,X)$ satisfying the
following condition: x is in $q(X)$ if and only if x is in $Project(r, X)$ and for **all**
(resp. **none**) y in $s(Y)$, $r(X,Y)$ contains (resp. doesn't contain) $< x,y >$ [6].
X and Y are two compatible sets of attributes. More formally, the relational
division is characterised by Eq. 1, and the anti-division by Eq. 2 :

$$Div(r, s, X, Y) = \{x \in projection(r, X) \mid \forall y, (y \in s) \Rightarrow (\langle x, y \rangle \in r)\} \quad (1)$$

$$Anti - Div(r, s, X, Y) = \{x \in projection(r, X) \mid \forall y, (y \in s) \Rightarrow (\langle y, s \rangle \notin r)\} \quad (2)$$

Example 1: Consider a distribution company of some products. In its com-
mercial activity, the company wants to select its most valued customers
(buyers). Customers ranking is based on some categories of products. Let
Customer_Order $(\#customer, \#product, \#order\_state)$, **Critical_Product**
$(\#product, \#order\_state)$ and **Golden_Product** $(\#product, \#order\_state)$ be
three crisp relations as sketched in Fig. 1.

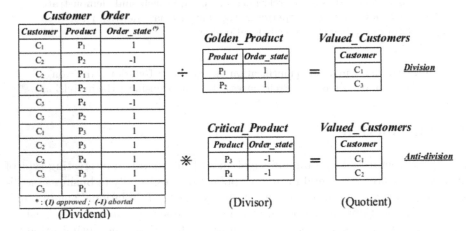

Fig. 1. Division query: *"Which customers have made an approved order for
each golden products?"*; Anti-division query: *"Which customers have not made
an aborted order for any of the critical products?"*

In the figure above, C_1 and C_3 are the resulting quotients of the division
because they have made an approved order for all golden products. Whereas, for
the anti-division, C_1 and C_2 are the valid quotients, since both of them have not
made an aborted order of any critical product.

1.2   Current Trends

Both relational division and anti-division often provide an empty answer. This is a widely studied problem in the last two decades [7]. **Flexible operators** (tolerant operators and operators dealing with user's preferences), is the most desirable technique to solve this problem and improve the DBMS answer quality [8], especially for recommender systems [9]. Flexible division (anti-division) consists in the weakening of the quantifier 'all' ('none') used in the classical operator [6,10].

1.3   Related Work and Motivation

Two main areas of research on division and anti-division can be identified. The first concerns the improvement of those operators, while the second area investigate them in a flexible context.

In literature, several studies have been focused on how to efficiently implement the division, including those surveyed in [1–3] in the relational model, and [5] in the object-oriented model. Indeed, the approach proposed in [1] and called **'Hash-Division'**, has proven through the experimental results to be better than the traditional algorithms in processing time in most cases. Further, there are only as far as we know, the work of *Bosc et al.* for the relational anti-division [6,11]. Nonetheless, their implementation is based upon the SQL query derivation and is far from being optimal.

In the flexible area, some authors have suggested new operators for relational division [10,12] and anti-division [6,11,13], which are tailored for the flexible context. However, the performance aspect has not been adequately dealt with. Besides, some extended variants of the hash-division algorithm have been discussed in our earlier work [14] to tailor with some forms of the flexible division and division with preferences. However, to the best of our knowledge, the only experimentations done for the anti-division are those presented in [6,11]. Although, their implementation is based on the nested loops algorithm which is far from being acceptable. Moreover, queries evaluation are performed with a reduced size of data (dividend and divisor). This does not fit reality, especially for analysis treatments on extra-large databases. In addition, authors in [15] have suggested a way for combining the division and the anti-division operators. However, neither implementation nor experimentations are presented in the paper.

1.4   Main Contributions in This Paper

This paper is carried out as a continuation of our previous work detailed in [14], which is proven to be an efficient processing of the flexible division. Hence, extended variant will be proposed in this work to cover additional forms of the universal-quantification based queries.

In fact, the main purpose of our work is to design a unified processing to handle queries involving requirements and prohibitions simultaneously, with a single operator. Such queries allow users to express several kinds of their preferences, which is very useful in information systems especially in artificial intelligence.

We also address the performance enhancement of the new operator drawing to the Hash-Division strategy as used in our previous work [14].

Example 2: Let's take relations in the previous example. Thanks to the mixing query, customers can be evaluated through the following query: *"Find customers who have made an **approved order for all golden products** and they haven't made **any aborted order for the critical products**?"*.

Here, C_3 is no longer a valid quotient because he has made an aborted order of one of the critical products (P_2). Idem for C_2, he hasn't made an approved order for all golden product. Thus, we can conclude that the customers can be better distinguished through the mixed query. In addition, a unified (single operator) and fast processing of such queries will improve them even more. This is the backdrop behind our work. Hereafter we summarise our contributions:

- Investigate performance enhancement of the flexible queries involving both of division and anti-division, essentially for very large volumes of data.
- Investigate the parallel implementation feasibility for the extended approach.

We consider in this work the flexible division and anti-division over crisp databases exclusively. Fuzzy relations will be studied in future work.

1.5   Outline of the Paper

The remainder of this paper is organised as follows. In Sect. 2, we present the classical Hash-Division algorithm. Section 3 gives an overview of the flexible division and the flexible anti-division. In Sect. 4, our contribution is presented together with analytics and discussion of the experimental results obtained. Section 5 introduces a parallel implementation of the proposed operator. Finally, Sect. 6 concludes the paper and suggests directions for future work.

2   Review of Hash-Division Algorithm

In this section, we give a brief description of the hash-division algorithm (**HD**) (see [1] for further details). It uses two hash tables, in order to avoid the exhaustive comparison, used in the traditional algorithms. The first table is for the divisor and the second for the quotient. Thanks to these two structures, both dividend and divisor relations are scanned exactly once, that makes the division operator faster. Hash-Division algorithm is proceeding in three stage:

Stage 01: Building the Hash-Divisor Table: during the scan of the divisor table, we insert all divisor tuples into buckets in the hash-divisor table. Each entry in this table, is stored together with an integer called **divisor number** *'Num_div'*. Num_div is initialized to 0 and it is incremented whenever a new insertion in the hash-divisor table occurs.

Stage 02: Building the Hash-Quotient Table: during the scan of the dividend; for each row that corresponds to one of the divisors, stored in the hash-divisor table, we insert a quotient candidate into hash buckets in the hash-quotient table. Together with each inserted candidate, a bitmap is kept with one bit for each divisor. All bits are initialized to 0, and updated to 1 whenever a match with the corresponding divisor occurred.

Stage 03 (end): Building the Result: in this last stage, we select from the constructed hash-quotient table all quotient candidates whose bitmaps contain only ones as valid quotients.

3   Review of Flexible Division and Flexible Anti-division

Flexible (or tolerant) division and anti-division were essentially proposed in order to avoid the empty result problem, which may occur mostly whenever we use **'for all'** or **'for none'** quantifiers [6,10]. There are a plethora of suggestions, in literature, showing that original relational division (anti-division) can be extended to different types of flexible queries. We are interested in this work on the following forms of flexible operators: *(i)* Exception-based tolerant division, *(ii)* Exception-based tolerant anti-division.

3.1   Principle

This category is based on exceptions into the requirements set for the division or the prohibitions set for the anti-division (divisor). The principle is to weak the quantifier *'all'* (resp. *'none'*) to the fuzzy quantifier *'almost all'* (resp. *'almost none'*) to express tolerant division (resp. anti-division) [6,10,12]. Thus, depending on the desired level of relaxation, some elements, in the divisor set, are allowed to be not associated (resp. associated) with the quotient in the dividend relation.

3.2   Modelling

In fact, a maximum number of exceptions is allowed to be ignored. Satisfaction-level SL of a quotient is measured by Eq. 3 for the division and Eq. 4 for the anti-division. A threshold is required for accepted quotients [10,13]. Valid quotients are sorted depending on their satisfaction levels.

$$SL_{Division} = \frac{Number\ of\ divisors\ associated\ with\ the\ candidate}{total\ number\ of\ divisors} \tag{3}$$

$$SL_{Anti-Division} = \frac{Number\ of\ divisors\ not\ associated\ with\ the\ candidate}{total\ number\ of\ divisors} \tag{4}$$

4   Our Proposed Approach for the Mixed Query

This section is devoted to a tolerant universal-quantification queries in which both division and anti-division are considered *simultaneously*. We first give a novel way for combining those two types of associations: required and forbidden associations. Then, the performance of the proposed approach is highlighted.

In fact, we propose to improve the effectiveness of the mixed query by inspiring from the strategy of the hash-division algorithm. We have made various alterations to the structures and the procedures used in the classic algorithm, to deal with the unified mixed operator. Moreover, we describe an adequate technique to better discriminate final quotients, with no additional cost.

It should be noted that our work differs from *Bosc et al.*'s work presented in [15] in our formulating query. All preferences, requirements and prohibitions, are expressed thanks to a single operator. While the key *issue* with the approach presented in [15] is that is based on the decomposition of the mixed operator on several successive relational division and-or anti-division operations, depending on the number of layers, which is a very time-consuming process.

4.1   Strict and Gradual Mixed Query

To deal with the mixed query, the divisor is subdivided into two sets, positive part (requirements) P, and negative part (prohibitions) N.

In the strict version, to be selected as a valid quotient, an element x must be associated with all values in P **and** must be not associated with any value in N. Thereby, P and N must be totally independents. In this strict version, all results are equally ranked. For the gradual mixed query, since some tolerances are allowed in both subsets P and N, results are discriminated depending on their satisfaction levels. Hence, for each accepted quotient we define two sublevel: S_p and S_n stand for the satisfaction level for the positive and the negative part respectively. S_p is computed as in Eq. 3 with respect to the positive part, and S_n is computed as in Eq. 4 regarding the negative part.

4.2   Hash-Mixed Query: An Improvement of the Mixed Query

Here we will describe how we have improved the processing time of the mixed query relying on the Hash-Division like algorithm. Hence, the three altered phases of the hash-mixed query are described hereafter.

The First Stage:
As in the classic algorithm, we store all divisor tuples in a hash table. Whereas for ours, each tuple is stored together with two integers:

- *ind_lyr:* index of the layer, 0 for P and 1 for N. This integer is used to indicate the offset of the divisor tuple inside the bitmap. Bits corresponding to divisors in P are located, in the bitmap, before those belonging in N.

– **num_div_lyr:** the divisor number (*rank*) of the tuple in its layer (*P* or *N*).

The data structure of a divisor tuple in the hash-divisor table is shown in the following Fig. 2.

Divisor value	ind_lyr	num_div_lyr

Fig. 2. Data structure of a hash-divisor tuple

Hence, for each layer, P and N, we keep its own divisors counter. These two counters are initialized to 0 and incremented whenever we insert a new divisor, of the corresponding layer, into the hash-divisor table. Pseudo-code of the hash-divisor table building for the mixed query is given hereafter:

Algorithm 1. Building of the hash-divisor table for the Hash-mixed query

---

num_divisors$_P$ ← 0; num_divisors$_N$ ← 0; /* *initialize the two counters to zero* */
for each tuple t in the divisor relation **do**
    Calculate its hash bucket ($Hdiv$);
    if t belongs in P **then**
        divisor.ind_lyr ← 0;
        divisor.num_div_lyr ← num_divisor$_P$; /*assign the offset to the current divisor*/
        num_divisor$_P$ ++;
    end if
    if t belongs in N **then**
        divisor.ind_lyr ← 1;
        divisor.num_div_lyr ← num_divisor$_N$; /*assign the offset to the current divisor*/
        num_divisor$_N$ ++;
    end if
    Insert the divisor tuple into the corresponding hash bucket ($Hdiv$);
end for

The Second Stage:
In the second stage (*Construction of the Hash-quotient table*) of the hash-mixed query, we have made two major differences from the basic algorithm. The first is how to update the bitmap. Hence, if a divisor matching (*P* or *N*) with the quotient candidate occurs, we set the bit to 1 whose position, in the quotient bitmap[1], is equal to '*offst_lyr+num_div_lyr*' where:

– **num_div_lyr:** the divisor number stored together with the matching divisor.
– **offst_lyr:** is set to 0 if the matching divisor belongs to P, otherwise (belongs to N) it is set to $|P|$ (the cardinality of the positive subset).

---

[1] As in the basic version, the bitmap is initialized with 0 in all their bits.

Therefore, the data structure of the bitmap of candidates is as shown below (Fig. 3):

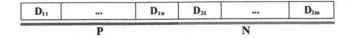

Fig. 3. Data structure of the bitmap for hash-mixed query.

The second difference is that we kept with each quotient candidate counters of ones ($bit = 1$), in its bitmap, for each layer. We called these counters Nb_ones_1 for the layer P and Nb_ones_2 for the layer N. These latter are incremented at each bit switching (0 to 1) in the corresponding layer of the quotient candidate bitmap. Hereafter is a pseudo-code of this stage:

Algorithm 2. Building of the hash-quotient table for the hash-mixed query

---

  for each tuple t in the dividend table **do**
    Calculate the hash bucket $Hdiv$ over the divisor attributes of the tuple t;
    if the divisor is contained in the hash-divisor table in the bucket Hdiv **then**
      layer ← ind_lyr of the matching divisor;
      rank ← num_div_lyr of the matching divisor;
      Calculate the hash bucket $Hqot$ over the quotient attributes of the tuple t;
      if the candidate (quotient value) is already contained in the hash-quotient table
  at the bucket Hqot **then**
        if $rank^{th}$ bit in the $layer^{th}$ part of the candidate bitmap is set to 0 **then**
          Set this bit to 1;
          Nb_ones $_{layer}$++; /*Increment the counter of ones in the corresponding layer*/
        end if
      else {/*quotient candidate does not yet exist*/}
        Insert a new quotient candidate into the hash-quotient table at the bucket
  $Hqot$, with a bitmap where all bits are set to zero;
        Set the $rank^{th}$ bit in the $layer^{th}$ part of the bitmap to 1;
        Nb_ones $_{layer}$ ← 1; /*Initialize the counter of ones to 1*/
        Set the other counter (Nb_ones $_{\overline{layer}}$) to 0;
      end if
    end if
  end for

The Third Stage:
In the third stage, and for the strict version of the mixed query, quotient candidates whose bitmaps contain *only ones* in the positive part **and** *only zeros* in the negative part will be selected as valid quotients. Thereby, all final results will be equally ranked. Besides, for the tolerant version, we identify two manners to consider the satisfaction sub-levels (S_p and S_n) mentionned in Subsect. 4.1 in order to discriminate and rank the accepted quotients:

- **Strong Symmetrical Impact:** both of positive and negative part have the same impact on the result ranking. So the final satisfaction-level S_f is defined as '$S_f=S_p+S_n$'. To sort accepted quotients, we propose to use a mechanism close to that used in our previous work, where we have used an indexed table for this sorting phase. Hence, a final quotient Q, whose satisfaction-level S_f is greater than the threshold chosen by the user, is stored in a bucket of index '$(|P|-Nb\_ones_1)+Nb\_ones_2$' (see Fig. 4.a). Nb_ones_1 and Nb_ones_2 are the two counters stored with the bitmap of the quotient candidate.
- **Positive and Negative Part as Hierarchical Preferences:** here, the positive and the negative part haven't the same impact in results discrimination, one part is more important than the other. Indeed, an indexed table with two levels is used to rank valid quotients. The first level corresponds to the most important part. Each level is subdivided according to the number of exceptions allowed (see Fig. 4.b). The positive bucket ($level_1$ or $level_2$) has the index '$|P|-Nb\_ones_1$', while the negative one is equal to 'Nb_ones_2'.

In such a way, final quotients are automatically sorted in decreasing order according to their satisfaction levels. The cell whose index is 0 points the best quotients (satisfying the whole set of requirements and dissatisfying all prohibitions). Hence, to select the $k-top$ answers, we just need to browse the indexed table from the top (from quotients with the highest satisfaction-level to the lowest ones); until k quotients are found. This sorting technique offers a better discrimination between accepted quotients, while no additional costs is needed.

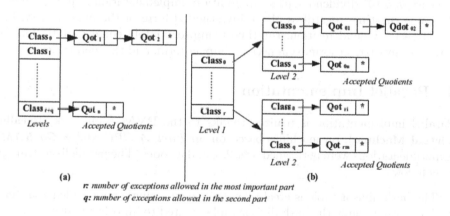

(a)

r: number of exceptions allowed in the most important part
q: number of exceptions allowed in the second part

Fig. 4. Quotient Candidates Discrimination

In the light of the above, it can be said that we have been able to combine two types of associations (positive and negative) in a single operator. The conceived operator is not complex since it does not need to handle each operation (Division and anti-division) separately. Furthermore, it requires no iterations. Hence, thanks to th new unified operator, users can introduce simultaneously requirements and prohibitions in a constructively simple manner.

Table 1. Experimental results for the hash-mixed query algorithms

Size		Classical mixed query	Strict hash-mixed query	Gradual hash-mixed query	
Dividend	Divisor:**P-N**			**SSI**[a]	**PN − HP**[b]
3×10^4	5–5	3.11	0.01	0.02	0.02
5×10^5	5–15	87	0.165	0.173	0.169
3×10^6	30–20	598	3.44	3.42	3.5
5×10^8	50–50	7645	120.3	121.01	123.65

[a] : Strong symmetrical impact.
[b] : Positive and negative part as hierarchical preferences.

4.3 Experimentations

We consider four sizes for the dividend relation: 3.10^4, 5.10^5, 3.10^6, and 5.10^8 tuples, randomly generated[2]. Sizes considered for the divisor relation are: 10, 20, 50, and 100 uniformly distributed over layers P and N. Obtained results are gathered in Table 1. Run-time is measured in seconds.

Table 1 shows the run-times of our variants of the mixed query, comparing with the classic one presented in [15] where several successive classical-divisions are involved. We can notice that our approaches complete performance much faster than the classic one for the four dividend sizes. Indeed, the run-time is improved by several orders of magnitude (the gain factor is greater than **61** in the case of 5.10^8 dividend tuples). In addition, implementation requires roughly the same run-time regardless of the investigated form of the mixed query. For the largest dividend relation: run-time is approximately equal to 120 s for the three variants (strict form, symmetrical impact form, and the hierarchical form).

5 Parallel Implementation

Parallel implementation is realized thanks to the **PVM** framework (**Parallel Virtual Machine**), on machines based on an *Intel i5 CPU* and *8 Go RAM*. Experimentations were performed over 2, 4 and 6 nodes. The parallelism strategy is as follows:

1. The hash-divisor table is created only once on a single node called *master*.
2. The master sends the hash-divisor table created to all other nodes.
3. The dividend table is uniformly partitioned between all nodes.
4. Each node builds its own hash-quotient table. The hash function may be different between the nodes, depending on the memory space of each one.
5. When all sub-tables of the hash-quotient are completely constructed in all nodes, the master collects those sub-tables. Then, it merges all of them in one global hash-quotient table to select valid quotients.

---

[2] In the literature, up to now and as far as we know, the largest set used in the experimentations never exceed a cardinality of 3.10^4 tuples in the dividend relation.

The pseudo-code of the last step (point 5), in the master, is given hereafter:

Algorithm 3. Parallel implementation of the mixed query.

---

for each sub-hash-quotient table received from the slave nodes **do**
  for each quotient-candidate in the sub-hash-quotient table **do**
    Compute the hash bucket (***Hqot***) using the master hash function, over the quotient value of the candidate;
    if the candidate (quotient value) is already contained in the hash-quotient table, constructed in the master, at the bucket *Hqot* **then**
      Update the bitmap of the candidate by calculating the result of the **binary OR** operator between the bitmap in the master and that received from the node;
    else
      Insert a new quotient candidate into the hash-quotient table of the master at the bucket *Hqot*, with a bitmap equal to that received from the node;
    end if
  end for
end for

As well, Fig. 5 illustrates the speed-up behaviour of the parallel algorithm of the hash-mixed query over 2, 4 and 6 nodes.

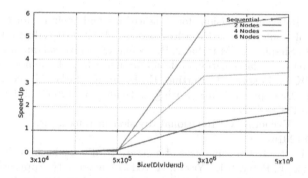

Fig. 5. Speed-up for parallel algorithm of the hash-mixed query.

Through the results obtained from the parallel implementation of the hash-mixed query and illustrated in the figure above, we observed a ***linear effect*** on speed-up in the case of large dividend ($\geq 3.10^6$). However, an additional cost[3], but still negligible, for a relatively small size of the dividend ($\leq 5.10^5$) occurs.

In summary, first results of the hash-mixed query presented in this paper are encouraging. The proposed approach has been successful in processing this complex forms of the universal-quantification based queries effectively. Although, there is still a need for multiple implementations in real SGBD, to firmly validate the hash mixed approach proposed.

---

[3] The additional cost comes from the fact that the communication time between nodes is more expensive than the run-time of algorithms in each node.

6   Conclusion and Perspectives

We have presented in this paper a unified operator to deal with universal quantification based queries involving positive and negative preferences (desired and forbidden associations) simultaneously. Our new technique is then improved relying on the hash-division algorithm. Moreover, the issue of answers ranking is dealt with. We have conducted some experiments particularly for large-sized relations, and compare execution time with the original approaches (nested loop algorithms) proposed in the literature. As expected, the performance got is very interesting. We have been able to improve the response time of some queries by several orders of magnitude. We presented also a parallel version of the mixed query, where we have obtained a near-linear speed-up, especially for large tables. We are currently designed new forms of complex queries where more than two layers, several kinds of preferences, and several connectors come to play. Furthermore, there is still a need for multiple implementations in real SGBD, to firmly validate the hash mixed approach proposed. It will also be exciting to look at other parallelism strategies which take into account the data skew issue that causes deteriorations in performance.

References

1. Graefe, G.: Relational division: four algorithms and their performance. In: Fifth International Conference on Data Engineering, 1989, Proceedings. IEEE (1989)
2. Rantzau, R., Shapiro, L., Mitschang, B., Wang, Q.: Universal quantification in relational databases: a classification of data and algorithms. In: Jensen, C.S., Šaltenis, S., Jeffery, K.G., Pokorny, J., Bertino, E., Böhn, K., Jarke, M. (eds.) EDBT 2002. LNCS, vol. 2287, pp. 445–463. Springer, Heidelberg (2002). https://doi.org/10.1007/3-540-45876-X_29
3. Vaverka, O., Vychodil, V.: Relational division in rank-aware databases. Inf. Sci. **366**, 48–69 (2016)
4. Bosc, P., Pivert, O.: On some uses of a stratified divisor in an ordinal framework. In: Kacprzyk, J., Petry, F.E., Yazici, A. (eds.) Uncertainty Approaches for Spatial Data Modeling and Processing. SCI, vol. 271, pp. 133–154. Springer, Heidelberg (2010). https://doi.org/10.1007/978-3-642-10663-7_9
5. Marin, N., Molina, C., Pons, O., et al.: Semantically-driven flexible division in fuzzy object oriented models. In: IFSA/EUSFLAT Conference, pp. 1039–1044 (2009)
6. Bosc, P., Pivert, O., Soufflet, O.: Strict and tolerant antidivision queries with ordinal layered preferences. Int. J. Approximate Reasoning **52**(1), 38–48 (2011)
7. Bosc, P., Hadjali, A., Pivert, O.: Empty versus overabundant answers to flexible relational queries. Fuzzy Sets Syst. **159**(12), 1450–1467 (2008)
8. Zadrożny, S., Kacprzyk, J.: Bipolarity in database querying: various aspects and interpretations. In: Pivert, O., Zadrożny, S. (eds.) Flexible Approaches in Data, Information and Knowledge Management. SCI, pp. 71–91. Springer, Cham (2014). https://doi.org/10.1007/978-3-319-00954-4_4
9. Pigozzi, G., Tsoukias, A., Viappiani, P.: Preferences in artificial intelligence. Ann. Math. Artif. Intell. **77**(3–4), 361–401 (2016)
10. Bosc, P., Pivert, O., Soufflet, O.: On three classes of division queries involving ordinal preferences. J. Intell. Inf. Syst. **37**(3), 315–331 (2011)

11. Bosc, P., Pivert, O., Soufflet, O.: Anti-division queries with ordinal layered preferences. In: Sossai, C., Chemello, G. (eds.) ECSQARU 2009. LNCS (LNAI), vol. 5590, pp. 769–780. Springer, Heidelberg (2009). https://doi.org/10.1007/978-3-642-02906-6_66

12. Tamani, N., Lietard, L., Rocacher, D.: Bipolarity and the relational division. In: The Joint 7th Conference of the European Society for Fuzzy Logic and Technology (EUSFLAT 2011) and Rencontres Francophones sur la Logique Floue et ses Applications (LFA 2011), pp. 424–430 (2011)

13. Bosc, P., Pivert, O.: A family of tolerant antidivision operators for database fuzzy querying. In: Greco, S., Lukasiewicz, T. (eds.) SUM 2008. LNCS (LNAI), vol. 5291, pp. 92–105. Springer, Heidelberg (2008). https://doi.org/10.1007/978-3-540-87993-0_9

14. Benadjmi, N., Hidouci, K.W.: New variants of hash-division algorithm for tolerant and stratified division. In: Christiansen, H., Jaudoin, H., Chountas, P., Andreasen, T., Legind Larsen, H. (eds.) FQAS 2017. LNCS (LNAI), vol. 10333, pp. 99–111. Springer, Cham (2017). https://doi.org/10.1007/978-3-319-59692-1_9

15. Bosc, P., Pivert, O.: Queries mixing positive and negative associations and their weakening. In: Fuzzy Information Processing Society (NAFIPS), 2010 Annual Meeting of the North American, pp. 1–6. IEEE (2010)

Implementing a Semantic Approach for Events Correlation in SIEM Systems

Tayeb Kenaza[(✉)] [iD], Abdelkarim Machou, and Abdelghani Dekkiche

Ecole militaire polytechnique, BP 17 BEB, 16043 Alger, Algeria
ken.tayeb@gmail.com

Abstract. Efficient reasoning in intrusion detection needs to manipulate different information provided by several analyzers in order to build a reliable overview of the underlying monitored system trough a central security information and event management system (SIEM). SIEM provides many functions to take benefit of collected data, such as Normalization, Aggregation, Alerting, Archiving, Forensic analysis, Dashboards, etc. The most relevant function is Correlation, when we can get a precise and quick picture about threats and attacks in real time. Since information provided by SIEM is in general structured and can be given in XML, we propose in this paper to use an ontological representation based on Description Logics (DLs) which is a powerful tool for knowledge representation and reasoning. Indeed, Ontology provides a comprehensive environment to represent any kind of information in intrusion detection. Moreover, basing on DLs and rules, Ontology is able to ensure a decidable reasoning. Basing on the proposed ontology, an alert correlation prototype is implemented and two attack scenarios are carried out to show the usefulness of the semantic approach.

Keywords: SEIM · Monitoring · Intrusion detection
Alert correlation · Description logic · Rules based reasoning
Ontology · OWL

1 Introduction

Information systems security requires the deployment of a rigorous security policy with several security mechanisms and tools. We generally start with prevention systems such as authentication where the goal is to prove the identity of users, access control where the goal is to define rights of users on data, and firewalls where the role is to control the access to the information system towards the outside world.

However, these mechanisms are not sufficient to fully protect systems against malicious attacks. Indeed, computer systems often exhibit vulnerabilities, which allow attackers to bypass preventive mechanisms. In addition, many security tools focus on the protection against external attacks, while attacks can be also

A. Amine et al. (Eds.): CIIA 2018, IFIP AICT 522, pp. 648–659, 2018.
https://doi.org/10.1007/978-3-319-89743-1_55

internal. For example, client side attacks are a very common nowadays. Therefore, intrusion detection is necessary as a second layer of security after deploying prevention systems. Unfortunately, Intrusion Detection is still imperfect for two reasons. First, intrusion detection systems (IDSs) generate a very large number of low-level alerts, where most of them are false positive; i.e, alerts generated in the absence of attacks. And second, IDSs suffer from false negative which is the absence of alerts in the presence of attacks.

In order to overcome these problems, a promising approach is the so-called cooperative intrusion detection [4, 20], which allows various intrusion detection tools to cooperate. In addition to IDS, other analyzers can be considered such as network and vulnerability scanners in order to correlate alerts by considering contextual information. This can be done by including for example topology and cartography. In fact, nowadays all security tools have to cooperate using a central security information and event management system (SIEM). A SIEM provide many functions to take benefit of the collected data, such as Normalization, Aggregation, Alerting, Archiving, Forensic analysis, Dashboards, etc. The most relevant function is Correlation, when we can get a precise and quick picture about the threats and attacks in real time. However, most of proprietary SIEM use its own data representation and its own correlation techniques which are not always favorable to share knowledge and to do custom reasoning.

In such situation, the use of common and extensible formalism to describe information in intrusion detection is a major concern. This information is generally structured and encoded in XML. For example, this is the case of alerts in IDMEF (for Intrusion Detection Message Exchange Format) and TAXII (Trusted Automated eXchange of Indicator Information) as well as the vulnerabilities in OVAL (Open Vulnerability and Assessment Language) and STIX (Structured Threat Information eXpression). However, information encoded here in XML is limited to a syntactic representation basing in different taxonomies. Consequently, in the absence of a semantic approach correlating this information is a fastidious task. Indeed, it is more interesting to move from taxonomies to ontology specification languages [9, 12], which are able to simultaneously serve as recognition, reporting and correlation languages.

Several existing knowledge representation models can be used in SIEM such as [1, 2, 7, 8]. In this paper, our contribution can be seen as an enhancement of existing representations by regrouping a large amount of information into a single ontology. This will offer a comprehensive and extensible knowledge representation which can be used in many event correlation systems.

On an other hand, given that tools used in SIEM are not totally reliable, usually conflicts appear between them [15, 19]. For example, one can easily see that IDSs are not fully reliable since they generate many false positives and false negatives. Therefore, it is very important to resolve these conflicts in order to exploit the cooperation. Hence, our second contribution is an ontological reasoning approach to correlate alerts in order to reduce the amount of alerts, especially false positives.

The rest of this paper is organized as follows. In Sect. 2 we briefly recall intrusion detection and some works of knowledge representation proposed in the context of intrusion detection, and then we present the proposed ontology. Section 3 presents an architecture of an alert correlation system based on DLs reasoning. In Sect. 4 experiments are conducted and results are discussed. In Sect. 5, some related works are briefly discussed. Section 6 concludes this paper.

2   Related Works

The automatic correlation of information from different security systems has been a vivid topic of research for over a decade [4,20]. Numerous approaches have been developed for correlating alerts and other log entries to strength the power of intrusion detection systems. Here, we briefly discuss only related works regarding the use of ontology in computer security. Ontology can be used in many field in SIEM, such as to analyze user behavior and system activities, or to identify known attack patterns, or also to analysis abnormal behavior and activity of both systems and users. Notice that semantic approaches have many advantages over existing approaches, mainly two aspects: the formal and extensible knowledge representation capability and the decidable reasoning.

Using ontology in computer security is relatively new. The first research work was done by Undercoffer et al. [16]. They produced an ontology that specify a model of computer attack. Their ontology is based on attack strategies which is categorized according to targeted system components, tools of attacks, consequences of attacks, and location of attackers. They present their model as a target-centric ontology. Since the work of Jeffrey many other ontologies was proposed. In [17], Wang et al. propose an Ontology for Vulnerability Management (OVM) which contains several concepts about vulnerabilities, affected products, consequences and countermeasures, etc. Authors have used their own implementation of their ontology without referring to any languages. In [2], Azevedo et al. propose a domain-ontology with more generic and abstract concepts in the field of computer security, serving as the basis for the construction of other specific security-domain-ontologies called CoreSec. In [5], Gao et al. provide an ontology-based attack model which is used to assess the information system security from attack angle. The proposed ontology consists of five dimensions, which include attack impact, attack vector, attack target, vulnerability and defense.

More recently, many semantic description methods for the security policy has been proposed. In [14], an ontology-based method is presented to solve the problem of the semantic description and verification of a security policy. Onto-ACM (ontology-based access control model), is a semantic analysis model proposed by Choi et al. [3] to address the difference in the permitted access control between service providers and users. More over, in [18] ontologies are used to perform threat analysis and develop defensive strategies for mobile security. Authors have proposed on ontology-based approach that can identify an attack profile in accordance with structural signature of mobile viruses, and also overcome the uncertainty regarding the probability of an attack being successful, thanks to semantic reasoning.

3 Ontological Based Specification and Reasoning for Alert Correlation

3.1 Knowledge Representation in Intrusion Detection

In front of an intrusion detection environment characterized by a very low detection rate, a high rate of false alerts, and a poor granularity of the information provided by alerts, a huge effort has been made by the intrusion detection community for the standardization of threats and attacks representation. The resulted data formalisms (e.g. IDMEF, TAXII, STIX, etc.) has provided a workspace for open communication between security tools and has been largely used in many alert correlation systems [4, 6].

Despite their different approaches, alert correlation systems have to share knowledge about attacks and the context in which they occur. However, many security tools do not care about how they represent their knowledge and how they use it. We think that having a coherent and formal model to represent knowledge is important for any correlation system. M2D2 is among the most important work in this area, it is a relational model that regroup essential information used in correlation, such as alerts, events, nodes, softwares, etc. In 2009, this model was revised by adding new concepts and by regrouping concepts into classes, this new model is called M4D4 [8]. In a recent work [10] proposed by Sadighian et al., authors have designed a set of comprehensive and extensible ontologies, and have implemented fusion and detection algorithms based on OWL-DL and SQWRL in order to allow reducing false positives.

3.2 The Proposed Ontology

Strassner defines the ontology as follows: "An ontology is a formal, explicit specification of a shared, machine-readable vocabulary and meanings, in the form of various entities and relationships between them, to describe knowledge about the contents of one or more related subject domains throughout the life cycle of its existence" [13]. This meaning of ontology is used mostly in the context of knowledge sharing.

IDMEF and M4D4 are among the most important work in terms of knowledge representation in the domain of intrusion detection. However, IDMEF does not contain enough information because it describes just alerts, and M4D4 is proposed in the context of network intrusion detection including contextual information (cartography and topology) and the description of vulnerabilities.

In this section, we propose an ontological conceptualization that combines the representation of IDMEF, M4D4, TAXII and other information sources such as OVAL, STIX and NVD. Generally, we can divide knowledge in intrusion detection into 5 groups [8]: Analyzers, Events and alerts, Attacks and Vulnerabilities, Contextual information, and Users and Attackers. Figure 1 shows the main concepts and relations of the proposed ontology, baptized "ONTO-SIEM".

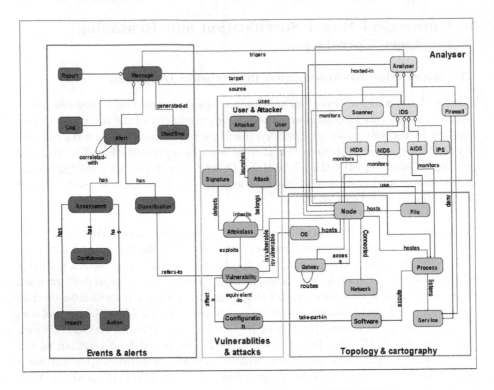

Fig. 1. Main concepts and relations of ONTO-SIEM

4   Ontology Based Event Correlation System

The use ONTO-SIEM is very suitable for event correlation within a SIEM, when many tools have to cooperate and to exchange information. Indeed, we developed a prototype of alert correlation system to show the importance and usefulness of this ontology. The architecture of our system consists of two essential modules: the conversion module that puts reported alerts into the ontology, as well as contextual information (topology and cartography), and the correlation module that allows reasoning about the constructed ontology. Figure 2 summarizes the architecture of the correlation system.

In order to use an ontology within an application, it must be specified in a formal representation. Indeed, a variety of languages exists that are used to represent conceptual models, with varying expressiveness, ease of use and computational complexity. We used OWL, which is a recommendation of The World Wide Web Consortium (W3C), widely used in web semantic. OWL is based on Description Logics. Description Logics are known for their expressiveness and their clearly defined semantics that allow a decidable reasoning.

In this work, we build our ontology using the API Jena (http://jena.apache.org/), and the reasoning is provided by Pellet (http://clarkparsia.com/pellet/) which is a full OWL-DL reasoner.

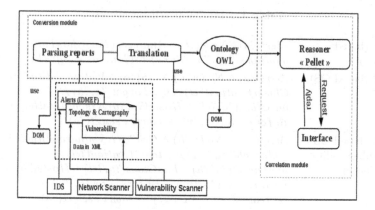

Fig. 2. ONTO-SIEM based alert correlation system architecture.

4.1 Populating the Ontology

To populate our ontology we need to use several tools. Information about hosts and the network topology are given using Nmap (http://nmap.org/). This tool can provide many information such as running hosts and their operating systems, servers listening in these hosts with their corresponding version, and many further information. Information about the vulnerabilities of systems and applications are given using Nessus (http://www.tenable.com/products/nessus/). Information about attacks are given in real time by IDS/IPS, in our system we used Snort (http://www.snort.org/) with a set of VRT and community rules. Notice that it is also possible to insert directly information into ONTO-SIEM by security operators, namely add information about equipments, systems and applications.

4.2 Reasoning with the Ontology

Reasoning is important in ontology because it allows to ensure the quality of ontology. Indeed, through the use of a reasoner, it is possible to test whether concepts are non-contradictory, and also to derive implicit relations.

Filtering Events: In Tables 1 and 2, we present some rules that can be used to filtering pertinent and not pertinent events. Most of theses rules are reused from the Pasagrada framework [10]. Notice that if an event is not classified as pertinent this does not means it is not pertinent. To decide so, the event must satisfy at least one rule from Table 2.

Rule 1 selects events generated by analyzers that can actually monitors the target. For example, an IDS can only detect events that occur in the network to which it is connected. This can be explicitly provided by the relation *monitors* or inferred, for example for NIDS, as follows.

$$monitors \equiv hosted - in \wedge connected \wedge netNodes \qquad (1)$$

Table 1. Filtering pertinent events

Rule-1 (based analyser)	$Alert(?a) \wedge Node(?h) \wedge Analyser(?an) \wedge hasTarget(?a, ?h) \wedge$ $monitors(?an, ?h) \wedge Trigers(?an, ?a) \rightarrow sqwrl : select(?a)$
Rule-2 (based OS)	$Alert(?a) \wedge Node(?h) \wedge OS(?o) \wedge Vulnerability(?v) \wedge$ $Classification(?cl) \wedge hasTarget(?a, ?h) \wedge$ $hasClass(?a, ?cl) \wedge Hosts(?h, ?o) \wedge isvulnerable(?o, ?v) \wedge$ $Refers - to(?cl, ?v) \rightarrow sqwrl : select(?a)$
Rule-3 (based OS with vulnerability equivalence)	$Alert(?a) \wedge Node(?h) \wedge OS(?o) \wedge Vulnerability(?v1) \wedge$ $Vulnerability(?v2) \wedge Classification(?cl) \wedge$ $hasTarget(?a, ?h) \wedge hasClass(?a, ?cl) \wedge Hosts(?h, ?o) \wedge$ $isvulnerable(?o, ?v1) \wedge Refers - to(?cl, ?v2) \wedge$ $equivalent - to(?v1, v2?) \rightarrow sqwrl : select(?a)$
Rule-4 (based application)	$Alert(?a) \wedge Node(?h) \wedge Software(?sof) \wedge$ $Vulnerability(?v) \wedge Classification(?cl) \wedge Process(?p) \wedge$ $Configuration(?conf) \wedge hasTarget(?a, ?h) \wedge$ $hasClass(?a, ?cl) \wedge Hosts(?h, ?p) \wedge Execute(?p, ?sof) \wedge$ $Take - part - of(?sof, ?conf) \wedge Affects(?v, ?conf) \wedge$ $Refers - to(?cl, ?v) \rightarrow sqwrl : select(?a)$
Rule-5 (application with vulnerability equivalence)	$Alert(?a) \wedge Node(?h) \wedge Software(?sof) \wedge$ $Vulnerability(?v1) \wedge Vulnerability(?v1) \wedge$ $Classification(?cl) \wedge Process(?p) \wedge$ $Configuration(?conf) \wedge hasTarget(?a, ?h) \wedge$ $hasClass(?a, ?cl) \wedge Hosts(?h, ?p) \wedge Execute(?p, ?sof) \wedge$ $Take - part - of(?sof, ?conf) \wedge Affects(?v1, ?conf) \wedge$ $Refers - to(?cl, ?v2) \wedge equivalent - to(?v1, v2?) \rightarrow sqwrl :$ $select(?a)$

Table 2. Filtering not pertinent events

Rule-6 (based analyzer)	$Alert(?a) \wedge Node(?h) \wedge Analyser(?an) \wedge hasTarget(?a, ?h) \wedge$ $Trigers(?an, ?a) \wedge NotMonitors(?an, ?h) \rightarrow sqwrl :$ $select(?a)$
Rule-7 (based vulnerability)	$Alert(?a) \wedge Node(?h) \wedge Vulnerability(?v) \wedge$ $Classification(?cl) \wedge hasClass(?a, ?cl) \wedge$ $hasTarget(?a, ?h) \wedge IsNotVulnerable(?h, ?v)$ $\rightarrow sqwrl : select(?a)$

Rules 2 and 4 select events based on the vulnerability of the OS and the Application, respectively. Some tools such as vulnerability scanners can confirm if an OS or an application is vulnerable or not to a given vulnerability. Obviously, this concern only known vulnerabilities, not zero-day vulnerabilities. Rules 4 and 6 are similar to rules 3 and 4, they just consider the equivalence between vulnerabilities reported by several organisms with different names. These tow rules deal with the case when different analyzers (IDS and scanners) refer to the same vulnerability with different names or references.

Rule 6 is the inverse of rule 1, it selects events reported by tools that does not actually monitor the target of the attack. Rules 7 selects events reported for target that is not actually vulnerable to the referred vulnerability. This concern both OS and Software vulnerabilities. The question now is haw to get such information, because traditionally scanners only report affected hosts not protected once. For instance we admit that such information is explicitly given by the relation *IsNotVulnerable*.

Aggregating Events: Here we consider only pertinent event for which we try to group events together in order to generate meta-event. A meta-event represent a summarizing of a single malicious activity that causes multiple elementary events. We distinguish tow types, Host based meta-event and Network-based meta-event. In this latter we can distinguish three sub-classes [10], within certain time interval (Table 3).

1. One-to-One (Rule-8). This can be an attack attempted by a single attacker against a single target, for example a SQL injection.
2. One-to-Many (Rule-9). This can be an attack attempted several time by a single attacker against many targets, for example a network or vulnerabilities scan.
3. May-to-One (Rule-10). This can be an attack attempted by several attackers against a single target, for example a DDoS.

Table 3. Network based events aggregating

Rule-8 (one to one)	$Message(?meg) \wedge Node(?h1) \wedge Node(?h2) \wedge$ $hasSource(?meg, ?h1) \wedge hasTarget(?meg, ?h2) \wedge generated-$ $at(?meg, ?t) \wedge biggerThan(?t, ?t1) \wedge lessThan(?t, ?t2) \rightarrow$ $sqwrl : select(?meg)$
Rule-9 (many to one)	$Message(?meg) \wedge Node(?h1) \wedge Node(?h2) \wedge$ $hasTarget(?meg, ?h1) \wedge generated-at(?meg, ?t) \wedge$ $biggerThan(?t, ?t1) \wedge lessThan(?t, ?t2) \rightarrow oqwrl :$ $select(?meg)^s qwrl : count(?meg)$
Rule-10 (one to many)	$Message(?meg) \wedge Node(?h1) \wedge Node(?h2) \wedge$ $hasSource(?meg, ?h1) \wedge generated-at(?meg, ?t) \wedge$ $biggerThan(?t, ?t1) \wedge lessThan(?t, ?t2) \rightarrow sqwrl :$ $select(?meg)^s qwrl : count(?meg)$

For host-base meta-event, we consider several event' features to decide to group or not events. These features are Node (N), User (U), Process (P), Service (S), and File (F) [10]. Based on these features and in case of a complete availability of data, we can distinguish two main subclasses (Table 4).

1. NUP (Rule-11), when many events have the same node, the same user and the same process.
2. NUF (Rule-12), when many events have the same node, the same user and the same file.

Table 4. Host based events aggregating

Rule-11 (node-user-process)	$Message(?msg) \wedge Node(?h) \wedge User(?u) \wedge Process(?p) \wedge$ $Analyser(?an) \wedge hasTarget(?msg, ?h) \wedge$ $Launches(?u, ?p) \wedge Hosts(?u, ?p) \wedge Moniters(?an, ?p) \wedge$ $Trigers(?an, ?msg) \wedge generated - at(?msg, ?t) \wedge$ $biggerThan(?t, ?t1) \wedge lessThan(?t, ?t2) \rightarrow sqwrl :$ $select(?msg) \wedge sqwrl : count(?msg)$
Rule-12 (node-user-file)	$Message(?msg) \wedge Node(?h) \wedge User(?u) \wedge File(?f) \wedge$ $Analyser(?an) \wedge hasTarget(?msg, ?h) \wedge$ $Launches(?u, ?f) \wedge Hosts(?u, ?f) \wedge Moniters(?an, ?f) \wedge$ $Trigers(?an, ?msg) \wedge generated - at(?msg, ?t) \wedge$ $biggerThan(?t, ?t1) \wedge lessThan(?t, ?t2) \rightarrow sqwrl :$ $select(?msg) \wedge sqwrl : count(?msg)$

5 Experimental Results

ONTO-SIEM is implemented using Protégé which is a powerful editor supporting OWL-Dl, SWRL and other many reasoners such as HermiT, Pollet, etc. Protégé is powerful thanks to many plugins that can be add to it.

To evaluate the proposed rules we have used UNB-ISCX-2012 [11] which is an open Intrusion Detection Evaluation dataset. UNB-ISCX-2012 is an interesting benchmark because it provides a real labeled traffic which contains both attacks and normal activities. Moreover, this benchmark provides a complete capture of the traffic with a set of divers and multi-steps attack scenarios. Table 5 gives a summary about the benchmark and its attack scenarios.

Table 5. UNB-ISCX benchmark description

Day	Scenario description	Size (GB)
Friday	Normal activity	16.1
Saturday	Normal activity	4.22
Sunday	Infiltrating from the inside + normal activity	3,95
Monday	HTTP DDoS + normal activity	6,85
Tuesday	DDoS using an IRC Botnet	23,4
Wednesday	Normal activity	17.6
Thursday	Brute Force SSH + normal activity	12.3

We tested our approach using 2 scenarios, namely "Infiltrating from inside" and "HTTP DDoS". The first scenario consists to obtain access to a host inside the local network, and then the compromised host is used as a pivot to attack computers which are not accessible via the Internet. The second scenario consists of performing a stealthy, low bandwidth denial of service attack without the need to flood the network (for more details about the testbed architecture and the attack scenarios see [11]).

5.1   Populating ONTO-SIEM

The UNB-ISCX-2012 benchmark provides only the raw traffic collected during 7 days, and an xml file containing labeled attacks with their execution periods, but no thing about topology and the cartography of the testbed, as well as the Vulnerabilities. So, we did a had work to manually extract that information from the benchmark. For instance, in Table 6 we show some of used OSs and Softwares. Vulnerabilities are also manually insert into ONTO-SIEM.

Table 6. UNB-ISCX benchmark' OS and Softwares

OS	Softwares
4 Windows xp SP1, Windows xp SP2, Windows 7, Windows Server 2003, Ubuntu 10.04	Acrobat Adobe Reader 8.1, Apache 2.2.9, Bind 9, Postfix, Dovecot, IIS 6, MSSQL Server, OpenSSH 5.3, Vsftp 2.2.2

The raw traffic are analyzed using Prelude-Snort, and then reported alert are translated from xml to ONTO-SIEM. Prelude is an open SIEM which can easily be connected with many analyzers. Prelude' output are in xml, so it will be very simple to translate them to ONTO-SIEM.

5.2   Discussing Results

Concerning the first scenario, snort has reported alerts which are translated into ONTO-SIEM using Prelude (https://www.prelude-siem.org/). Correlation process has given results shown in Table 7. The first filtering level has reduced the amount of alerts by 30% (3307 alerts are removed) which are alerts that refer to no existing hosts, while the second filtering level has reduced about 17% (1340 alerts are removed) of the amount of the remainder alerts. Therefore, after this preliminary filtering stage, more than 43% of alerts are reduced. Only 57% of the initial alerts will be concerned by further correlation processing using the ontology.

Table 7. Filtering alerts of scenario 1

Rules	Not pertinent alerts	
Rule 1	3307/11016	(30.02% of alerts are removed)
Rule 2 to 5	1340/7709	(17.40% of alerts are removed)

The same discussion is given for the second scenario. Table 8 shows correlation process results. The first filtering level has reduced the amount of alerts by 27.95% (1561 alerts are removed) which are alerts that refer to no existing hosts, while the second filtering level has reduced about 10% (410 alerts are removed) of the remainder alerts. Therefore, after this preliminary filtering stage, more than 36% of alerts are reduced. Only 64% of the initial alerts will be concerned by further correlation processing using the ontology.

Table 8. Filtering alerts of scenario 2

Rules	Not pertinent alerts	
Rule 1	1561/5584	(27.95% of alerts are removed)
Rule 2 to 5	410/4023	(10.20% of alerts are removed)

6   Conclusion and Future Work

We proposed in this paper a domain ontology for a cooperative intrusion detection based on several data sources such as IDMEF, TAXII, STIX, M4D4, OVAL, NVD, etc. This ontology is implemented with OWL which is recommended by W3C since 2004 for the representation of ontologies in the Web Semantic. OWL is based on Description Logics which are a decidable fragment of the first order logic and are well suitable to represent structured information.

We have illustrated the usefulness of this ontology through an application in the context of alert correlation. This application allows automatic translation of alerts generated by IDSs to OWL, as well as contextual information generated by network and vulnerability scanners. Furthermore, a set of rules proposed to be inferred over the constructed ontology, these rules aim mainly to remove not pertinent alerts. This is very important to reduce the amount of alerts by analyzing in priority pertinent alerts. Other actions can be performed in the perspective to complete this work. Indeed, the proposed ontology need to be completed by more concepts and relation to allow a more comprehensive correlation rules, and also by using other reasoning mechanisms provided by OWL-Dl such as the verification of consistency and the satisfiability of concepts.

References

1. Abdoli, F., Kahani, M.: Ontology-based distributed intrusion detection system. In: 14th International CSI Computer Conference, pp. 65–70. IEEE (2009)
2. Azevedo, R., Dantas, E., Freitas, F., Rodrigues, C., Almeida, M.J., Veras, W., Santosyi, R.: An autonomic ontology-based multiagent system for intrusion detection in computing environments. Int. J. Infonomics (IJI) **3**, 182–189 (2011)
3. Choi, C., Choi, J., Kim, P.: Ontology-based access control model for security policy reasoning in cloud computing. J. Supercomput. **67**(3), 711–722 (2013)
4. Elshoush, H.T., Osman, I.M.: Alert correlation in collaborative intelligent intrusion detection systems - a survey. Appl. Soft Comput. **11**, 4349–4365 (2011)
5. Gao, J., Zhang, B., Chen, X., Luo, Z.: Ontology-based model of network and computer attacks for security assessment. J. Shanghai Jiaotong Univ. (Sci.) **18**(5), 554–562 (2013)
6. Mirheidari, S.A., Arshad, S., Jalili, R.: Alert correlation algorithms: a survey and taxonomy. In: Wang, G., Ray, I., Feng, D., Rajarajan, M. (eds.) CSS 2013. LNCS, vol. 8300, pp. 183–197. Springer, Cham (2013). https://doi.org/10.1007/978-3-319-03584-0_14

7. More, S., Matthews, M., Joshi, A., Finin, T.: A knowledge-based approach to intrusion detection modeling. In: 2012 IEEE Symposium on Security and Privacy Workshops (SPW), pp. 75–81. IEEE (2012)
8. Morin, B., Mé, L., Debar, H., Ducassé, M.: A logic-based model to support alert correlation in intrusion detection. Inf. Fusion **10**(4), 285–299 (2009)
9. Motik, B., Patel-Schneider, P.F., Parsia, B., Bock, C., Fokoue, A., Haase, P., Hoekstra, R., Horrocks, I., Ruttenberg, A., Sattler, U., et al.: OWL 2 web ontology language: structural specification and functional-style syntax. W3C Recommendation, vol. 27, p. 17 (2009)
10. Sadighian, A., Zargar, S.T., Fernandez, J.M., Lemay, A.: Semantic-based context-aware alert fusion for distributed intrusion detection systems. In: 2013 International Conference on Risks and Security of Internet and Systems (CRiSIS), pp. 1–6, October 2013
11. Shiravi, A., Shiravi, H., Tavallaee, M., Ghorbani, A.A.: Toward developing a systematic approach to generate benchmark datasets for intrusion detection. Comput. Secur. **31**(3), 357–374 (2012)
12. Staab, S., Studer, R.: Handbook on Ontologies. Springer, Heidelberg (2009). https://doi.org/10.1007/978-3-540-92673-3
13. Strassner, J.: Knowledge engineering using ontologies. In: Bergstra, J., Burgess, M. (eds.) Handbook of Network and System Administration, vol. 4. Elsevier, Amsterdam (2008)
14. Tang, C., Wang, L., Tang, S., Qiang, B., Tian, J.: Semantic description and verification of security policy based on ontology. Wuhan Univ. J. Nat. Sci. **19**(5), 385–392 (2014)
15. Tombini, E., Debar, H., Me, L., Ducasse, M.: A serial combination of anomaly and misuse iDSes applied to HTTP traffic. In: 20th Annual Computer Security Applications Conference, pp. 428–437. IEEE (2004)
16. Undercoffer, J., Pinkston, J., Joshi, A., Finin, T.: A target-centric ontology for intrusion detection. In: 18th International Joint Conference on Artificial Intelligence, pp. 9–15 (2004)
17. Wang, J.A., Guo, M.: OVM: an ontology for vulnerability management. In: Proceedings of the 5th Annual Workshop on Cyber Security and Information Intelligence Research: Cyber Security and Information Intelligence Challenges and Strategies, CSIIRW 2009, pp. 34:1–34:4. ACM, New York (2009)
18. Wang, P., Chao, K.-M., Lo, C.-C., Wang, Y.-S.: Using ontologies to perform threat analysis and develop defensive strategies for mobile security. Inf. Technol. Manag. **18**, 1–25 (2015)
19. Yahi, S., Benferhat, S., Kenaza, T.: Conflicts handling in cooperative intrusion detection: a description logic approach. In: 22nd IEEE International Conference on Tools with Artificial Intelligence (ICTAI), vol. 2, pp. 360–362. IEEE (2010)
20. Zhou, C., Leckie, C., Karunasekera, S.: A survey of coordinated attacks and collaborative intrusion detection. Comput. Secur. **29**, 124–140 (2010)

An Improved Collaborative Filtering
Recommendation Algorithm for Big Data

Hafed Zarzour[1($\boxtimes$)], Faiz Maazouzi[2], Mohamed Soltani[1],
and Chaouki Chemam[3]

[1] LIM Research, Department of Computer Science, University of Souk Ahras,
41000 Souk Ahras, Algeria
hafed.zarzour@gmail.com, msoltani@univ-soukahras.dz
[2] Department of Computer Science, University of Souk Ahras, Souk Ahras,
41000 Souk Ahras, Algeria
mazouzi@labged.net
[3] Department of Computer Science, University of El-Tarf,
36000 El-Taref, Algeria
chemam-chaouki@univ-eltarf.dz

Abstract. With the increase of volume, velocity, and variety of big data, the traditional collaborative filtering recommendation algorithm, which recommends the items based on the ratings from those like-minded users, becomes more and more inefficient. In this paper, two varieties of algorithms for collaborative filtering recommendation system are proposed. The first one uses the improved k-means clustering technique while the second one uses the improved k-means clustering technique coupled with Principal Component Analysis as a dimensionality reduction method to enhance the recommendation accuracy for big data. The experimental results show that the proposed algorithms have better recommendation performance than the traditional collaborative filtering recommendation algorithm.

Keywords: Big data · Recommender system
Collaborative filtering recommendation algorithm · K-means
Clustering · PCA

1 Introduction

With the explosive increase in available data on the web and the rapid advances of information technology, big data has become a hot research topic in the field of data mining. Generally, it is commonly used to describe the exponential growth and availability of structured and unstructured data. Nowadays, many governmental and industrial communities become interested in the high potential of this innovative technology. However, it is very difficult for such communities to find relevant contents, recommender systems appear to solve present problems. Recommender system is defined as a decision making strategy for users under complex information platforms [1] in which it can effectively recommend the required information to end-users. Various techniques for developing recommender systems have been proposed, which can use either content-based filtering, collaborative filtering or hybrid methods [2–5]. In particular, the collaborative filtering recommendation algorithm (CFRA) is popular and has been used by many providers and consumers of big data such as: eBay, Amazon and Facebook.

Recently, many researches have reported that applying k-means as clustering technique in collaborative recommender systems can significantly enhance the performance of traditional CFRA [6]. Moreover, it has been proved that using Principal Component Analysis (PCA) as a dimensionality reduction method can significantly improve the clustering techniques [7], therefore, it is necessary to conduct dimensions reeducation before formally conducting clustering tasks. Hence, in this paper, we propose two varieties of algorithms for an effective collaborative filtering recommendation system. The first one uses the improved k-means clustering technique while the second one uses the improved k-means clustering technique coupled with PCA as a dimensionality reduction method to enhance the recommendation accuracy for big data. The experimental results show that the proposed algorithms have better recommendation performance than the traditional collaborative filtering recommendation algorithm.

The rest of this paper is organized as follows: Sect. 2 discusses some related works. Section 3 presents the collaborative filtering recommendation algorithm. Section 4 explains in details the proposed approach. Section 5 describes the experimental results. Finally, Sect. 6 concludes this study and proposes the plans for future work.

2   Related Work

In the recent years, the philosophy of big data attracts great attention from several official organizations including governments, universities, and industries in which the recommender systems are introduced to help them to find what they need via a mechanism that can make prediction depending on different criteria. One of the recommender strategies that can provide several kinds of recommendation is the open source project Apache Mahout [8]. It is primary enables free scalable implementation of machine learning methods [9, 10]. Another free and open source scalable library of recommender system is MyMediaLite [11], which addresses both common rating and item prediction from positive-only feedback. The rating prediction can be a scale of 1 to 5 stars while the item prediction from positive-only implicit feedback can be purchase actions or from clicks. In [12], the authors propose a keyword-aware service recommendation method, named KASR, to indicate users' preferences and generate appropriate recommendations on MapReduce [13] for big data applications. In [14], Lee et al. propose an adaptive recommendation algorithm, ACFSC, that is focused on scalable clustering to solve the problem of scalability by composing neighborhood based on reducing time complexity. They also address the problem of sparsity by making items' and users' feature vectors incrementally learning. CSRS [15] is a customized service recommendation system for Big Data. It uses the MapReduce framework and focuses on service recommendation method to create proper recommendations based on users' preferences. In [16], Zarzour et al. propose a new collaborative filtering recommendation algorithm based on dimensionality reduction and clustering techniques. They use clustering k-means algorithm and Singular Value Decomposition (SVD) to cluster similar users and reduce the dimensionality, respectively. In [17], the authors use k-means algorithm to cluster users according to their interests and then voting algorithm to generate prediction in recommender systems.

3 Collaborative Filtering Recommendation Algorithm

In the field of recommender systems, the collaborative filtering recommendation algorithm (CFRA) is the most successful recommendation method. The behind idea of CFRA is to provide for an active user recommendations or predictions by first looking for users who share the same rating patterns with him and then using the ratings from those like-minded users found to calculate a prediction for him. In other words, CFRA can suggests new similar items or predict the interest of a certain item for an active user based on their previous likings and the preferences of other similar users. More technically, it uses a user-item rating matrix that includes the preferences for items by users for matching users with relevant performances obtained by employing a similarity function between theirs profile to make recommendations or predict the ratings of selected items [18, 19].

$$sim(a, b) = \frac{\sum_{i \in I_{ab}} (r_{ai} - \bar{r}_a) \times (r_{bi} - \bar{r}_b)}{\sqrt{\sum_{i \in I_{ab}} (r_{ai} - \bar{r}_a)^2} \times \sqrt{\sum_{i \in I_{ab}} (r_{bi} - \bar{r}_b)^2}} \tag{1}$$

To compute the similarity between users or items, there are several similarity measure functions. One of the most popular methods is by using Pearson Correlation Coefficient (PCC), which is defined as follows:

$$p_{ti} = \bar{r}_t + \frac{\sum_{u \in U_{nei}} sim(t, u) \times (r_{ui} - \bar{r}_u)}{\sum_{u \in U_{nei}} |sim(t, u)|} \tag{2}$$

Once the similarity is computed, the most N nearest users are selected as a group of similar users called neighborhood and predicted ratings of unrated item can be then computed. The recommendation formula is presented as follow:

The main steps of the collaborative filtering recommendation algorithm (CFRA) are as follows:

Step 1: Input the matrix M[m, n] of user-item rating data, active user, K;

Step 2: Calculate the similarity between users by using Pearson Correlation Coefficient (PCC) and generate the similarity matrix S[m, m];

Step 3: Calculate the similarity between the active user and the clusters;

Step 4: Select the first n similar users of the active user;

Step 5: Calculate the prediction values of active user to every cluster by using the formula (2);

Step 6: Choose the top N items of users as recommendations;

Step 7: Output the recommendations.

4 K-means Based-Collaborative Filtering Algorithm

In this paper, two varieties of algorithms for collaborative filtering recommendation system are proposed. The first one uses directly the k-means clustering technique while the second one uses the k-means clustering technique after performing the PCA method. PCA aims at reducing the dimensions of the big data by extracting the most important information from the data. It can make big data mining more useful and get similar results by the reduction of dimensions [20].

4.1 K-means Algorithm

In data mining, K-means is considered as one of the most widely used method of clustering [21] in which it generates automatically a set of clusters based on a collection of datasets in easiest way. The main aim of k-means is to make the similarity inter-points of the same cluster be high, while the similarity inter-clusters be low. The steps of the algorithm are as follows:

Step 1: Input dataset, clusters number and K;
Step 2: Select randomly initial clustering centers which is the initial value of K;
Step 3: Calculate the distances between centers and objects then assign objects to the most nearest cluster;
Step 4: For each cluster, calculate the average as new partition centers;
Step 5: Use the new partition centers to redistribute points into new clusters;
Step 6: Repeat Steps 4 and 5 until the algorithm converge to a stable partition;
Step 7: Output K clusters.

4.2 CFRA-Km: A Collaborative Filtering Recommendation Algorithm Based on K-means Clustering

The general k-means algorithm is now personalized in order to take into consideration the recommendation requirements as well as the perdition of unknown ratings for a given active user. The specific steps are as follows:

Step 1: Input the matrix M[m, n] of user-item rating data, active user, K;
Step 2: Calculate the similarity between users by using Pearson Correlation Coefficient (PCC) and generate the similarity matrix S[m, m];
Step 3: Use the matrix S[m, m] as dataset and select randomly initial clustering centers which is the initial value of K;
Step 4: Calculate the distances between centers and objects then assign objects to the most nearest cluster;
Step 5: For each cluster, calculate the average as new partition centers;
Step 6: Use the new partition centers to redistribute points into new clusters;
Step 7: Repeat Steps 5 and 6 until the algorithm converge to a stable partition;
Step 8: Calculate the similarity between the active user and the clusters;
Step 9: Select the first n similar clusters of the active user;
Step 10: Calculate the prediction values of active user to every cluster by using the formula (2);

Step 11: Choose the top N items of users as recommendations;
Step 12: Output the recommendations.

4.3   Reducing the Dimension by PCA

One of the purposes of a PCA is the analysis of big data for eliminating noises and finding patterns to reduce the dimensions of the data without loss of relevant information. To do this, it converts a collection of observations of possibly correlated variables into a collection of values of principal components by using a linear transformation called orthogonal transformation. In general, the quantity of the obtained principal components is less than or equal to the quantity of original variables. Therefore, PCA is used as a statistical method to reduce not only the dimension of the user–user ratings matrix but also to reduce the loss of information by employing eigenvalue decomposition of data covariance matrix to obtain principal components of dataset with their weights. The general steps of PCA are as follows:

Step 1: Input the dataset;
Step 2: Normalize the data in the dataset;
Step 3: Calculate the covariance of the corresponding matrix;
Step 4: Calculate the eigenvectors of the covariance matrix;
Step 5: From matrix multiplication, translate the data to be in terms of the principal components.
Step 6: Output principal components.

4.4   CFRA-Km-PCA: A Collaborative Filtering Recommendation Algorithm Based on K-means Clustering and PCA

The first version of our k-means clustering- based collaborative filtering recommendation algorithm does not consider the effect of the dimensions reduction which may significantly influence the prediction results and make them inaccurate. Thus, PCA is applied before conducting the k-means clustering and performing the prediction step to reduce the dimension of the dataset and improve the performance of the prediction results. In other words, the collaborative filtering recommendation algorithm based on K-means clustering and PCA called CFRA-Km-PCA combines the advantages of PCA method with those of k-means clustering technique. The specific steps of CFA-Km-PCA are as follows:

Step 1: Input The matrix $M[m, n]$ of user-item rating data, active user, K;
Step 2: Calculate the similarity between users by using Pearson Correlation Coefficient (PCC) and generate the similarity matrix $S[m, m]$;
Step 3: Normalize the data in the obtained $S[m, m]$;
Step 4: Calculate the covariance of the corresponding matrix;
Step 5: Calculate the eigenvectors of the covariance matrix;
Step 6: From matrix multiplication, translate the data to be in terms of the principal components.
Step 7: Use the obtained principal components matrix as dataset and select randomly initial clustering centers which is the initial value of K;

Step 8: Calculate the distances between centers and objects then assign objects to the most nearest cluster;

Step 9: For each cluster, calculate the average as new partition centers;

Step 10: Use the new partition centers to redistribute points into new clusters;

Step 11: Repeat Steps 5 and 6 until the algorithm converge to a stable partition;

Step 12: Calculate the similarity between the active user and the clusters;

Step 13: Select the first n similar clusters of the active user;

Step 14: Calculate the prediction values of active user to every cluster by using the formula (2);

Step 15: Choose the top N items of users as recommendations;

Step 16: Output the recommendations.

5 Experimentation Results and Evaluation

To evaluate the performance of the k-means clustering-based collaborative filtering recommendation algorithm with and without using PCA compared to traditional collaborative filtering recommendation algorithm, experimentations were conducted on real big data. The experimental dataset was obtained from Netflix [22] which contains over 17,770 movies rated by approximately 480 000 users. In this dataset, there are over 100 million ratings ranging from 1 to 5 stars. A random sample was chosen and 80% of these data were also randomly used for training, and the remaining data were selected to test the performance of the considered algorithms.

In the performance evolution of recommender systems, Mean Absolute Error (MAE) and Root Mean Square Error (RMSE) are the most widely used. Therefore, we used those metrics to evaluate the performance of recommendations in CFRA, CFRA-Km, and CFRA-Km-PCA algorithms.

The formulas of RMSE and MAE are shown as follows, respectively.

$$RMSE = \sqrt{\frac{\sum_{i=1}^{N}(p(i) - q(i))^2}{N}} \tag{3}$$

$$MAE = \frac{\sum_{i=1}^{N}|p(i) - q(i)|}{N} \tag{4}$$

Figure 1 shows the experimental results in terms of RMSE metric for the proposed algorithms. As we can see from the graph, the RMSE results of the proposed CFRA-Km and CFRA-Km-PCA is low in the whole neighbors range compared to that for the CFRA algorithm. More precisely, the CFRA-Km-PCA achieves better results than both other algorithms.

Figure 2 shows the experimental results in terms of MAE metric for the three algorithms. In the same way, we can observe from the graph that the MAE results of the proposed CFRA-Km and CFRA-Km-PCA is low in the whole neighbors range

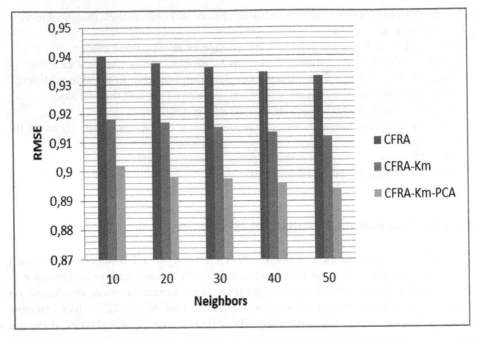

Fig. 1. RMSE results for CFRA, CFRA-Km, and CFRA-Km-PCA.

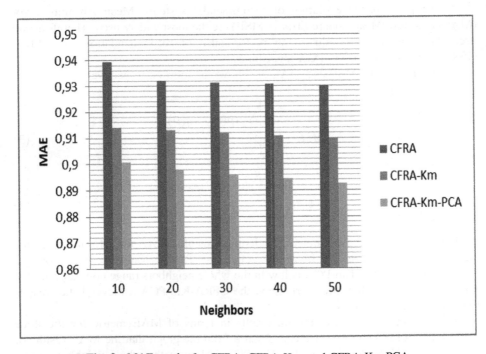

Fig. 2. MAE results for CFRA, CFRA-Km, and CFRA-Km-PCA.

compared to that for the CFRA algorithm and the CFRA-Km-PCA achieves better accuracy than both other algorithms.

From Figs. 1 and 2, we can conclude that the proposed algorithms, CFRA-Km and CFRA-Km-PCA, have better performance than the traditional algorithm CFRA in terms of RMSE and MAE. We can also conclude that the combination of PCA method with K-means clustering technique improved significantly the recommendation performance, which indicates that CFRA-Km-PCA is better algorithm for using in recommendation system for big data.

6 Conclusion and Future Work

In this paper, we have presented two kinds of improved collaborative filtering algorithms intended to enhance the prediction accuracy in the big data context. The first algorithm uses only the k-means clustering technique, while the second algorithm combines the advantages of both k-means clustering technique and PCA method. PCA was adapted to conduct dimensions reduction before formally conducting clustering tasks, which improved significantly the performance of k-means clustering-based collaborative filtering recommendation algorithm. The recommendation algorithms were evaluated in terms of RMSE and MAE metrics. The experimental results showed that the CFRA-Km-PCA achieved better results than both other algorithms, CFRA and CFRA-Km.

In the future, we will apply our algorithms to other datasets, and study the mechanism of the dimensions reduction coupled with other clustering techniques for improving recommendation precisions.

References

1. Rashid, A.M., Albert, I., Cosley, D., Lam, S.K., McNee, S.M., Konstan, J.A., Riedl, J.: Getting to know you. In: Proceedings of the 7th International Conference on Intelligent User Interfaces - IUI 2002 (2002)
2. Merve Acilar, A., Arslan, A.: A collaborative filtering method based on artificial immune network. Expert Syst. Appl. 36(4), 8324–8332 (2009)
3. Chen, L., Hsu, F., Chen, M., Hsu, Y.: Developing recommender systems with the consideration of product profitability for sellers. Inf. Sci. 178(4), 1032–1048 (2008)
4. Jalali, M., Mustapha, N., Sulaiman, M.N., Mamat, A.: WebPUM: a web-based recommendation system to predict user future movements. Expert Syst. Appl. 37(9), 6201–6212 (2010)
5. Smith, B., Linden, G.: Two decades of recommender systems at amazon.com. IEEE Internet Comput. 21(3), 12–18 (2017)
6. Koohi, H., Kiani, K.: A new method to find neighbor users that improves the performance of collaborative filtering. Expert Syst. Appl. 83, 30–39 (2017)
7. Pourkamali-Anaraki, F., Becker, S.: Preconditioned data sparsification for big data with applications to PCA and k-means. IEEE Trans. Inf. Theory 63(5), 1 (2017)
8. Gupta, P., Sharma, A., Jindal, R.: Scalable machine-learning algorithms for big data analytics: a comprehensive review. Wiley Interdisc. Rev.: Data Min. Knowl. Disc. 6(6), 194–214 (2016)

9. Bagchi, S.: Performance and quality assessment of similarity measures in collaborative filtering using mahout. Proced. Comput. Sci. **50**, 229–234 (2015)
10. Verma, J.P., Patel, B., Patel, A.: Big data analysis: recommendation system with Hadoop framework. In: 2015 IEEE International Conference on Computational Intelligence and Communication Technology (2015)
11. Gantner, Z., Rendle, S., Freudenthaler, C., Schmidt-Thieme, L.: MyMediaLite. In: Proceedings Of The Fifth Acm Conference On Recommender Systems - Recsys 2011 (2011)
12. Meng, S., Dou, W., Zhang, X., Chen, J.: KASR: a keyword-aware service recommendation method on mapreduce for big data applications. IEEE Trans. Parallel Distrib. Syst. **25**(12), 3221–3231 (2014)
13. Cheng, D., Rao, J., Guo, Y., Jiang, C., Zhou, X.: Improving performance of heterogeneous mapreduce clusters with adaptive task tuning. IEEE Trans. Parallel Distrib. Syst. **28**(3), 774–786 (2017)
14. Lee, O.J., Hong, M.S., Jung, J.J., Shin, J., Kim, P.: Adaptive collaborative filtering based on scalable clustering for big recommender systems. Acta Polytech. Hung. **13**(2), 179–194 (2016)
15. Bande, VM., Pakle, K.: CSRS: Customized service recommendation system for big data analysis using map reduce. In: 2016 International Conference on Inventive Computation Technologies (ICICT) (2016)
16. Zarzour, H., Al-Sharif, Z., Al-Ayyoub, M., Jararweh, Y.: A new collaborative filtering recommendation algorithm based on dimensionality reduction and clustering techniques. In: 9th International Conference on Information and Communication Systems (ICICS) (2018)
17. Dakhel, GM., Mahdavi, M.: A new collaborative filtering algorithm using k-means clustering and neighbors' voting. In: 11th International Conference on Hybrid Intelligent Systems, HIS (2011)
18. Herlocker, J.L., Konstan, J.A., Terveen, L.G., Riedl, J.T.: Evaluating collaborative filtering recommender systems. ACM Trans. Inf. Syst. **22**(1), 5–53 (2004)
19. Aggarwal, Charu C.: Neighborhood-based collaborative filtering. In: Charu, C. (ed.) Recommender Systems. TIRS, pp. 29–70. Springer, Cham (2016). https://doi.org/10.1007/978-3-319-29659-3_2
20. Villalba, S.D., Cunningham, P.: An evaluation of dimension reduction techniques for one-class classification. Artif. Intell. Rev. **27**(4), 273–294 (2007)
21. Adeniyi, D.A., Wei, Z., Yongquan, Y.: Automated web usage data mining and recommendation system using K-nearest neighbor (KNN) classification method. Appl. Comput. Inform. **12**(1), 90–108 (2016)
22. Hallinan, B., Striphas, T.: Recommended for you: the Netflix prize and the production of algorithmic culture. New Media Soc. **18**(1), 117–137 (2014)

Author Index

Printed in the United States
By Bookmasters